International Symposium on Earth Observation for One Belt and One Road 2016 (EOBAR)

IOP Conference Series: Earth and Environmental Science
Volume 57

Beijing, China
16 – 17 May 2016

Editors:

Xinqu Li
Jie Liu
Zhongchang Sun

ISBN: 978-1-5108-3743-0
ISSN: 1755-1307

Printed from e-media with permission by:

Curran Associates, Inc.
57 Morehouse Lane
Red Hook, NY 12571

Some format issues inherent in the e-media version may also appear in this print version.

Printed by Curran Associates, Inc. (2017)

For permission requests, please contact the Institute of Physics
at the address below.

Institute of Physics
Dirac House, Temple Back
Bristol BS1 6BE UK

Phone: 44 1 17 929 7481
Fax: 44 1 17 920 0979

techtracking@iop.org

Additional copies of this publication are available from:

Curran Associates, Inc.
57 Morehouse Lane
Red Hook, NY 12571 USA
Phone: 845-758-0400
Fax: 845-758-2633
Email: curran@proceedings.com
Web: www.proceedings.com

Table of contents

Volume 57

International Symposium on Earth Observation for One Belt and One Road (EOBAR)

16–17 May 2016, Beijing, China

Published online: 7 March 2017

Preface

Agriculture and Water

GEOSS Data Capacity Building along Belt and Road (DCB)

Climate Change

Disaster

Energy and Mineral Resources

Space Technologies for World Heritage Sites

Urban Growth and Environment

Space Observation of the Maritime Silk Road

Spatial Data Processing, Applications and Services

Earth Observation for Urban Growth and Environment

Space Atmosphere-Land Surface Interaction

Spatial Data Processing

Digital Earth for OBOR

Preface

The proceedings were written by participants of the International Symposium on Earth Observation for One Belt and One Road (EOBAR) held 16-17 May, 2016 in Beijing, China. Sponsored by the Division of Earth Sciences of the Chinese Academy of Sciences (CAS) and hosted by the Institute of Remote Sensing and Digital Earth, CAS, the conference has aimed to provide an international platform under the theme "quicker, better, broader understanding of One Belt and One Road (OBOR)" for scientists engaged in OBOR researches and earth observation applications.

The main topic of the symposium has been well represented by the 58 papers selected in the peer review. The symposium chairman was Huadong GUO (Institute of Remote Sensing and Digital Earth, Chinese Academy of Sciences). The Chairs of Scientific Committee was Guanhua XU (Ministry of Science and Technology of the People's Republic of China), Yong CHEN (China Earthquake Administration) and Valery. G. Bondur (Institute for Scientific Research of Aerospace Monitoring "AEROCOSMOS", Russian Academy of Sciences). The Organizing Committee was chaired by Bingfang WU (Institute of Remote Sensing and Digital Earth, Chinese Academy of Sciences) and Yongwei LIU (Institute of Remote Sensing and Digital Earth, Chinese Academy of Sciences).

We appreciate the efforts of all the authors who contributed to this conference, and particularly acknowledge the accepted papers for their high quality. We would like to thank the Division of Earth Sciences of CAS, Bureau of International Cooperation of CAS, Group on Earth Observations (GEO), Chinese Academy of Sciences National Natural Science Foundation, without whose generous support, the symposium could not have taken place. Special thanks also go to the staff at IOP Publishing Ltd., UK.

Xinwu LI, Jie Liu, Zhongchang Sun
Institute of Remote Sensing and Digital Earth
Chinese Academy of Sciences, China

Peer review statement

All papers published in this volume of *IOP Conference Series: Earth and Environmental Science* have been peer reviewed through processes administered by the proceedings Editors. Reviews were conducted by expert referees to the professional and scientific standards expected of a proceedings journal published by IOP Publishing.

The information extraction of Gannan citrus orchard based on the GF-1 remote sensing image

S Wang[1], **Y L Chen**

[1] School of Architectural and Surveying Engineering, Jiangxi University of Science and Technology, 341000 Ganzhou, China
1125551714@qq.com

Abstract. The production of Gannan oranges is the largest in China, which occupied an important part in the world. The extraction of citrus orchard quickly and effectively has important significance for fruit pathogen defense, fruit production and industrial planning. The traditional spectra extraction method of citrus orchard based on pixel has a lower classification accuracy, difficult to avoid the "pepper phenomenon". In the influence of noise, the phenomenon that different spectrums of objects have the same spectrum is graveness. Taking Xunwu County citrus fruit planting area of Ganzhou as the research object, aiming at the disadvantage of the lower accuracy of the traditional method based on image element classification method, a decision tree classification method based on object-oriented rule set is proposed. Firstly, multi-scale segmentation is performed on the GF-1 remote sensing image data of the study area. Subsequently the sample objects are selected for statistical analysis of spectral features and geometric features. Finally, combined with the concept of decision tree classification, a variety of empirical values of single band threshold, NDVI, band combination and object geometry characteristics are used hierarchically to execute the information extraction of the research area, and multi-scale segmentation and hierarchical decision tree classification is implemented. The classification results are verified with the confusion matrix, and the overall Kappa index is 87.91%.

1. Introduction

The orange region of Gannan in Jiangxi province has become the largest planting area in the world, which production is the third in the world, the first in China. To 2011, Ganzhou navel orange planting area of the city reached 119000 Hm^2, which annual output reached 1363900 T. The brand of Gannan navel orange has been famous at home and abroad (Wang J H, Ganzhou Citrus Research Institute, 2015). How to quickly and efficiently access the Gannan orange planting area and other related information has important significance for guiding the local fruit production and industrial planning. The development of remote sensing technology provides advanced technical means for obtaining the information of the Gannan orange planting area.

With the development of remote sensing technology, remote sensing image classification of vegetation experienced the development stage of visual interpretation, supervised classification and unsupervised classification, artificial neural network, fuzzy mathematics, expert system classification, decision tree and object-oriented interpretation, which provides some new techniques for classification of vegetation. In the classification of high resolution remote sensing images, due to the irregularity of vegetable coverage areas relative to the artificial objects, there is a certain degree of difficulty in classification, and the lower classification accuracy is always a difficult problem to solve. Yuan J G[1] has done a research on the forest vegetation classification based on NDVI (Normalized Difference Vegetation Index), which showed higher classification accuracy. Brian D. wardlow et al (2008)[2] has demonstrated that the classification of crops in the United States using hierarchical classification method based on the time series data of MODIS-NDVI achieved a high accuracy. Sulong et al [3], under the support of GIS(Geographic Information System), in the use of aerial photographs and TM images, has done the classification of the mangrove forest in Malaysia into 14 types and the accuracy respectively are 91.2% and 87.8%.

In recent years, as a large-scale comprehensive mean of earth observation, remote sensing technology reflects the characteristic of high spatial resolution, high spectral resolution and high time resolution (Li D R, 1995, 2002). Traditional image classification based on pixel spectral information is difficult to be applied to high resolution remote sensing image. Therefore, the object-oriented method based on the image began to rise, offsetting the limitations of the traditional method which uses pixels as the basic classification and processing unit, which used the objects consisting of adjacent pixels containing more semantic information as processing unit. Han S S et al [4] has shown that the object-oriented method is more accurate than the traditional maximum likelihood method based on TM and ETM. He Y H et al [5] has found in the study of land investigation based on CBERS-02 that object-oriented method has higher accuracy than other methods of classification. In the experiment of classification based on IKONOS image, Sun X X et al [6] apply the object-oriented method combined with the hierarchical classification method to improve classification accuracy, which employed the object shape feature. Recently, the object-oriented classification method is applied to the classification and extraction of vegetation. There were a lot of experts and scholars applying the object-oriented method to the classification of vegetation. Harken and Su-gumaranr [7] execute the classification of wetland using nonparametric object-oriented method and the classification accuracy was 92.13%. In China, Zhou C Y [8] and Jiang H [9] have executed the study of the classification of vegetation. Zhang X R et al[10] have done the extraction of scrub vegetation and the accuracy is up to 84.7%. Thus, the information extraction of citrus orchard based on the high resolution remote sensing image with the idea of object-oriented classification has high feasibility.

This paper, based on GF-1 remote sensing image, using the object-oriented classification theory combined with the decision tree classification model, choosing the best segmentation scale for image segmentation after pre-processing the image, doing the data statistics of spectral and object features of different objects on sample selected, has executed the study of the extraction of Gannan orange tree area using different features including the spectral characteristics and the object features, and obtained a satisfactory classification results.

2. General situation of study area

Study area (Xunwu County) is located at the junction in Jiangxi, Fujian and Guangdong provinces, the southeast of Jiangxi Province(Geographical coordinates: N24° 30′ 40″ -25° 12′ 10″ , E115° 21′ 22″ -115° 54′ 25″), with the total area of 2311.38 Km^2, many mountains and hills among them. The mountain area accounted for 75.6% of the total area. The region belongs to the south of the subtropical red soil region. The red soil is widely distributed in mountain and hilly areas, which soil fertility is better. The soil is generally acid, and the main component is iron oxide, aluminium oxide and quartz, which is suitable for kinds of orange, navel orange and other crops.

Figure 1. General situation of study area.

Such as shown in Figure 1, in hilly areas, due to the navel orange orchard and woodland, farmland and other background objects are staggered with each other and mosaic in space to constitute a complex mixture, also influenced by the shadow of the mountain, the automatic extraction is considerably difficult. The larger region in the image is the navel orange orchard planting area (abbreviation of orchard below), and the middle narrow part for buildings is the urban area(abbreviation of urban below) , and the white area for mining is the bare area (abbreviation of nudation below), and the road network is complex, no big rivers and lakes. According to the land use classification system of the resource and environment data centre of China Science and Technology

School, also according to according to the characteristics of the study area and preliminary interpretation, we can ensure the classification system of Xunwu, as shown in Table 1.

Table 1. Land use classification system.

The name	Description
Farmland	The land for planting crops, including cultivated land, new development land, consolidation and reclamation land, leisure land
Orchard	The land for planting fruit trees, widely distributed in the study area on both sides of the valley and hilly terrain
Forest	The forest land of arbors, shrubs, bamboos and other coastal mangrove, including land for herbage growth
Urban	The town for the residence of the various types of housing land and ancillary facilities land, including ordinary residential, apartments, villas and other land, and the rural homestead for life
Nudation	Towns, villages, factories, mines and internal unexploited land; and the surface soil, no vegetation covered, or the rocks and gravels on the surface. The coverage is more than or equal to 70%; surface is covered with sand, basic non-vegetated land
Road	The land of strip shape, its length greater than the width, the width changed relatively small, the smaller curvature

3. Data source and Preprocessing

3.1. Data source

This study selected the GF-1 remote sensing data covered with less than 20% cloud, and the data contains panchromatic image with resolution of 2m (sensor PMS1) and R, G, B, NIR 4 band multiband image with resolution of 8m (sensor PMS2), as shown in Table 2.

Table 2. GF-1 satellite data information.

Satellite name		Spectral range(μm)	Resolution(m)	Sensor	Note
GF-1	Panchromatic	0.45-0.90	2	PMS1	Cloud cover<20%
	Multiband	0.45-0.52 0.52-0.59 0.63-0.69 0.77-0.89	8	PMS2	

3.2. Preprocessing

Firstly, execute the radiation correction and geometric correction on the GF-1 remote sensing image data, furthermore, in order to obtain more accurate extraction all kinds of information from high resolution image, do the fusion processing to obtain more abundant image information to enhance the object feature to improve the classification accuracy.

4. Multi-scale segmentation

The segmentation technology is one of the core technologies of the high resolution image information extraction. The image is decomposed into non-overlapping regions as the basic unit of object-oriented analysis to make the analysis and understanding of the higher level possible (Zhang Y J, 2005). The image information extraction and subsequent analysis are determined directly by optimal segmentation scale. It is very important for the extraction of the target feature, the measurement and classification of the target object and the high level processing [11]. Multi-scale segmentation is, from a single pixel, to merge image objects into the high level objects by the bottom-to-up method. Through the calculation of pixel heterogeneity and homogeneity in size, the small image objects can be merged into some large homogeneous image objects after several times of merging, which can be understood as a similar pixel merged to form the image object stepwise optimization process. The result of the segmentation is determined by the segmentation parameter, which is composed of the scale parameter, the shape parameter and the compactness parameter.

Heterogeneity equation f :

$$f = w \times h_{color} + (1-w) \times h_{shape}$$
$$h_{shape} = w_{compact} \times h_{compact} + (1-w_{compact}) \times h_{smooth}$$

where w = spectral weight (0< w <1) h_{color} = spectral heterogeneity
h_{shape} = shape heterogeneity $w_{compact}$ = compactness weight (0< w <1)
h_{smooth} = smooth degree $h_{compact}$ = compactness

As shown in Equation, heterogeneity equation is determined by spectral heterogeneity and shape heterogeneity, and shape heterogeneity is composed of smooth degree and compactness.

The scale parameter can be evaluated by the reference polygon. If the segmentation results are the same as the reference polygons, the multi-scale parameters can be used to realize feature extraction and can achieve good segmentation results. If the reference polygons are divided into different parts by the segmentation results, it shows that the segmentation scale is smaller. If the reference polygon is formed into the outer covering by the minimum area of the partition, the segmentation scale is too large. The shape parameters and compactness parameters can be determined by analyzing the contour of the segmentation. If the segmentation results divided the object of different attributes into the same object, or the object of the same attributes into different features, illustrating that the shape parameter and the compactness parameter are too small. Through many experiments, the segmentation parameters are selected as follows: scale parameter is 100, shape parameter is 0.1, and compactness parameter is 0.5.

5. Feature analysis and information extraction

5.1. Object semantic feature

Length-width ratio A :

$$A = L/W$$

where L = length W = width

The brightness mean value of the object :

$$Brightness = \sum_{i}^{n} B_i$$

where B = brightness value of the pixel i = pixel number n = sum of the pixels contained in the object

Normalized Difference Vegetation Index (NDVI):

$$NDVI = \frac{NIR - Red}{NIR + Red}$$

where Red = red band mean value of the object NIR = near infrared band mean value of the object

5.2. Statistic and analysis of object feature

Through the statistical analysis of the DN of the 4 bands and NDVI values on all samples, the average brightness, 4 band reflectance and NDVI values of different vegetation types are obtained. Draw spectral curve and NDVI mean curve, as shown in Figure 2, Table 3.

Table 3. Terrain class brightness statistics.

Terrain class	Forest	Orchard	Urban	Road	Nudation	Farmland	Hill shade
Maximum value	191.5	210.3	217.8	311.3	510.4	180.3	154.5
Minimum value	158.8	182	165.8	211	274.2	165.8	109
Mean value	177.3	197.3	192.9	281.7	370.5	174.6	124.5

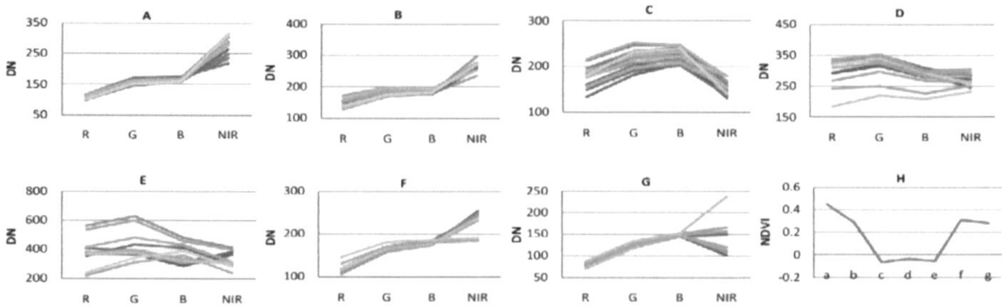

Figure 2. Spectral curve of terrain class and NDVI mean curve.

Where A. Forest B. Orchard C. Urban D. Road E. Nudation F. Farmland G. Hill shade
H. NDVI (a. Forest b. Orchard c. Urban d. Road e. Nudation f. Farmland g. Hill shade)

5.3. Hierarchical classification of decision tree model

This paper uses the decision tree method, which is not only flexible, intuitive, clear, efficient, robust [12-14], but also can integrate various supplementary information effectively into the decision tree models for classification to obtain the ideal classification result[15-16]. In-depth understanding of the environment of the study area characteristics and spectral, spatial characteristics is the key to the establishment of decision tree classification model. Therefore, the basic indispensable work of establishing the decision tree classification is the analysis of spectral characteristics of typical ground objects and other characteristic variables (Zhao Y S, 2013). According to the above features, combined with the decision tree hierarchical classification, the feature space function threshold rule set model is established to extract the feature information.

1) Distinguish between vegetation and non-vegetation

The NDVI can better distinguish between vegetation and non vegetation. Because of the influence of mountain shadow, the threshold of NDVI 0.22 is smaller than normal vegetation extraction threshold and Mean (NDVI) >0.22 satisfies the classification rules for forest, farmland, orchard and hill shade, does not meet the road, nudation and urban.

2) Distinguish between hill shadow and other vegetation

Through sample statistics analysis, the R band DN mean of hill shadow object is smaller, so select classification rules Mean (R) <87 to distinguish and meet the rules for the hill shadow, otherwise forest, farmland and orchard.

3) Distinguish between forest and farmland, orchard

By sample analysis of forest and farmland, orchard in the B band has a high discrimination, so select classification rules Mean (B) <170 to distinguish and meet the rules for forest, or farmland and orchard.

4) Distinguish between farmland and orchard

Taking it into account that generally there is a certain distance between the orange trees in the orchard to increase fertility absorption and photosynthesis and farmland has higher vegetation coverage degree. Therefore, the brightness of orchard is slightly higher than farmland. Combined with statistical feature, select classification rules Brightness<181 to distinguish, meet for farmland, do not meet for orchard.

5) Distinguish between road and other non-vegetation

Due to that road showed as strip, and the object has a smaller density relative to other segmentation, and the length-width ratio is relatively large, so select classification rules Density<1.1 and Length/width>3.1 to distinguish, satisfy the conditions for the road, not satisfied for other non-vegetation.

6) Distinguish between nudation and urban

Nudation is generally exposed surface, with significantly higher than urban brightness values, so choose classification rules Brightness<223 to distinguish, satisfy the conditions for urban, otherwise it is nudation.

The above classification process can get the decision tree classification model as shown in Figure 3.

7) According to the field survey, we found that the shadow of the hill is generally the object around the types of objects, so the shadow of the hill is classified into the objects which have the longest public side with it.

6. Classification results and accuracy evaluation

After spectral analysis of remote sensing image and main object classification and merging, the classification results are derived as shapefile vector format, combined with ArcGIS data processing software to produce the main terrain classification results figure, such as Figure 4.

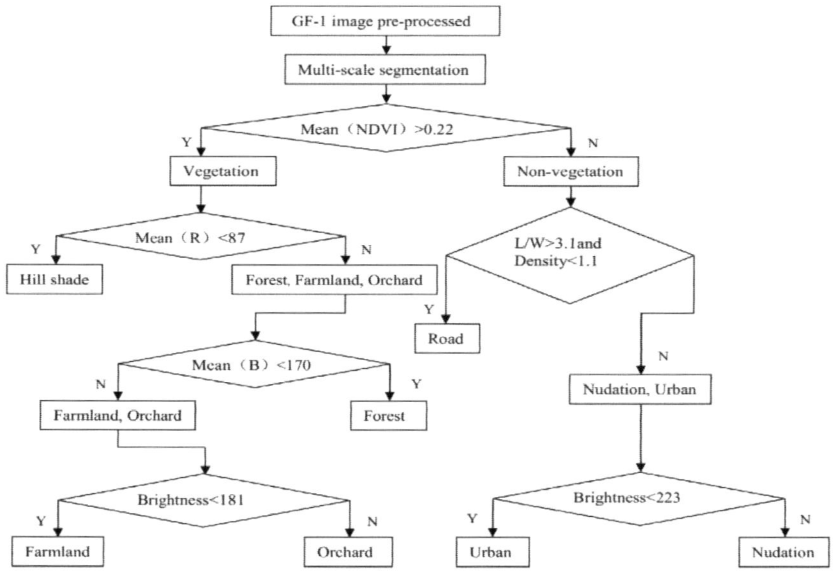

Figure 3. Classification decision tree model.

Accuracy evaluation is usually achieved by comparing the classification results with the reference data, and there are two kinds of data acquisition methods, one is the field survey, the other is to use the higher scale image or other information. This study uses the method of field survey. Through field survey we select 500 sample points, and the samples are generated by the sample points. Classification results extracted according to the experimental classification method have been compared with general classification extracted according to samples, and the quality evaluation confusion matrix is used to calculate the accuracy and the overall accuracy of classification is 91.94%, kappa coefficient is 0.8871, such as Table 4.

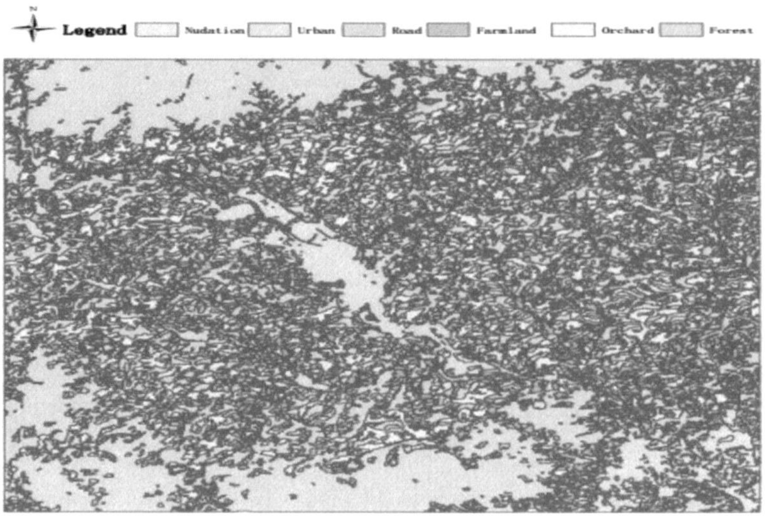

Figure 4. The result of classification.

As seen from Table 4, (1) The classification accuracy of forest is very high because of the obvious spectral feature, (2) The classification accuracy of farmland is very low due to that farmland is mixed in some forest and orchard, and its spectral feature have lower discrimination, (3) By the integration of multi-scale segmentation, object-oriented classification method and decision tree model, we can obtain higher classification accuracy of orchard.

Table 4. Classification accuracy evaluation.

Class name	Forest	Orchard	Urban	Road	Nudation	Farmland	Sum	User(%)
Forest	246	11	0	0	0	4	261	94.25
Orchard	4	86	2	0	0	1	93	92.47
Urban	0	1	240	4	20	1	266	90.23
Road	0	0	1	13	2	0	16	81.25
Nudation	0	0	1	0	53	0	54	98.15
Farmland	6	4	0	0	0	16	26	61.54
Sum	256	102	244	17	75	22	716	
Producer(%)	96.09	84.31	98.36	76.47	70.67	72.73		91.34
Overall Accuracy = 91.34% Overall Kappa = 0.879								

7. Conclusions

The orange tree planting is an important agricultural industry of Gannan region, and timely effectively obtaining the relevant information of the orange orchard has important significance to guide the local fruit production and industrial planning.

We choose the object-oriented classification method, based on GF-1 remote sensing image, through multi-scale segmentation experiment, selecting the optimal segmentation parameters, taking not only the statistical spectral characteristics, but also the shape, size and topological relationship into consideration for the information extraction, to establish the decision tree classification model, obtaining the higher information extraction accuracy. The classification results prove fully the feasibility and superiority of object-oriented classification method.

The method also has some limitations and deficiencies. ①Since the data source for imaging is selected on December 29, 2013, non citrus fruit leafy season, the orange trees were sparse causing a certain degree of difficulty of the extraction of citrus fruit. ②For irregular non artificial objects, there is a lot of regular texture information in high resolution remote sensing images, so texture information will be very helpful to extract information of vegetation. Therefore, in our subsequent study of extraction, the main work is to combine texture information with other features of citrus orchard information to obtain a higher accuracy.

8. References

[1] Yuan J G 1999. Study on Remote Sensing classification of forest vegetation. *Journal of Hebei Normal University (Natural Science),* 1999(2), pp 274-7.

[2] Wardlow B D and Egbert S L 2008. Large-area crop mapping using time-series MODIS 250m NDVI data: An assessment for the U.S. Central Great Plains. *Remote Sensing of Environment,* 112(3), pp 1096-116.

[3] Sulong I, Mohd-Lokman H, Mohd-Tarmizi K and Ismail A 2002. Mangrove Mapping Using Landsat Imagery and Aerial Photographs: Kemaman District, Terengganu, Malaysia. *Environment Development & Sustainability,* 4(2), pp 135-52.

[4] Han S S, Li H T and Gu H Y 2009. The Study on Land Use Change Detection Based on Object-oriented Analysis. *Remote Sense Information,* 2009(3), pp 23-9.

[5] He Y H, Shi L S and Zhang R H. et al. 2007. Application Research of CBERS-02 Data in Land-use survey. *China Land Science,* 21(2), pp 51-7.

[6] Sun X X, Zhang J X and Liu Z J 2006. The river and the road extraction from IKONOS panchromatic images using object-oriented classification method. *Science of Surveying and Mapping,* 31(1), pp 62-4.

[7] Harken J and Sugumaran R 2005. Classification of Iowa wetlands using an airborne hyperspectral image: a comparison of the spectral angle mapper classifier and an object-oriented approach. *Canadian Journal of Remote Sensing,* 31(2), pp 167-74.

[8] Zhou C Y, Wang P, Zhang Z Y and Qi C T 2008. Urban land use classification based on object-oriented information extraction technology. *Remote Sensing Technology and Application,* 23(1), pp 31-5.

[9] Jiang H, Zhou W B and Liu Y 2008. Research and Application of the Poyang Lake WetLand Classification Using Remote Sensing. *Remote Sensing Technology and Application,* 23(6), pp 648-52.

[10] Zhang X R, Liu L S, Zhang Y L, Zhang J P, Nie Y and Zhang H 2010. Extraction of Shrub Vegetation by Object-Oriented Classification Method Based on ENVI ZOOM in High-Altitude Area: A Case of Dingri County. *Geography and Geo-Information Science,* 26(4), pp 104-8.

[11] Zhang Y J 1996. A Classification and Comparison of Evaluation Techniques for Image Segmentation. *Journal of Image and Graphics,* 1(2), pp 151-8.

[12] Yu X Y, Zhang A D and Hou X Y 2012. Decision tree classification of orchard information extraction from TM imagery in Jiaodong Peninsula of China. *Science of Surveying and Mapping,* 37(4), pp 57-60.

[13] Xiong W C, Wei B, Sun Z P, Shen W M and You D A 2010. Australian Forest Fire Disaster Monitoring Based on CCD and IRS Data of HJ-1-B. *Remote Sensing Technology and Application,* 25(2), pp 178-82.

[14] Guo C H, Xue-Yong Q I, Gong Y L and Zhu L Y 2010. Study on the Technology and Method of Forest Fire Monitoring by Using HJ Satellite Images. *Remote Sense Information,* 2010(4), pp 85-8.

[15] Vogelmann J E, Sohl T L and Howard S M 1998. Regional Characterization of Land Cover Using Multiple Data Sources. *Photogrammetric Engineering & Remote Sensing,* 64(1), pp 45-57.

[16] Pielke R A, Lee T J, Copeland J H, Eastman J L, Ziegler C L and Finley C A 1997. Use of USGS-Provided Data to Improve Weather and Climate Simulations. *Ecological Applications,* 7(1), pp 3-21.

[17] Luo W and Kuang R Y 2014.Orchard information extraction of Dongjiang Source region with HJ satellite data. *Science of Surveying and Mapping,* 39(8).

[18] Wang Z W and Shi J Z, 2013. Assessment of vegetation by object-oriented classification and integration of decision tree classifier in Yushu. *Acta Pratacul Turae Sinica*, 22(5), pp 62-71.

[19] Mu F Y and K Han 2012. Object-oriented Remote Sensing Wetland Vegetation Classification and Information Extraction. *Journal of Anhui Agriculture Science,* 2012(12), pp 7574-6.

[20] Guo Y, Xinfang Y U, Jiang D, Wang S and Jiang X 2012. Study on Forest Classification Based on Object-Oriented Techniques. *Journal of Geo-Information Science, 14(4),* pp 514-22.

[21] Chang H, Zhan F L, Yang G D, Niu X F, Zhang X Q, Shao P and Tang T Q 2015. Research on Objects-oriented Information Extraction Technology for High Resolution Remote Sensing Image. *Bulletin of Surveying and Mapping,* 2015(1), pp 99-101.
[22] Zhang X. M, Sheng Y, Nan Z T, Zhao L, Zhou G Y and Yue G Y 2011. Vegetation classification of alpine grassland based on decision tree approach in the Wenquan area of the Qinghai-Tibet Platean. *Pratacul Tural Science,* 28(12), pp 2074-83.

Acknowledgements

This work is supported by science and technology projects from Jiangxi Provincial Office of Education Grant #GJJ10483.

Testing efficacy of monthly forecast application in agrometeorology: Winter wheat phenology dynamic

B Lalic [1], D Jankovic [2], Lj Dekic [3], J Eitzinger [4], A Firanj Sremac [1]

[1] Faculty of Agriculture, University of Novi Sad, Dositej Obradovic Sq. 8, 21000 Novi Sad, Serbia

[2] Forecasting and Warning Service of Serbia in plant protection, Temerinsnka 131, 21000 Novi Sad, Serbia

[3] Republic Hydrometeorological Service of Serbia, Kneza Viseslava 66, 11000 Belgrade, Serbia

[4] Institute of Meteorology, University of Natural Resources and Life Sciences, Gregor-Mendel-Straße 33, 1180 Vienna, Austria

branka@polj.uns.ac.rs, lalic.branislava@gmail.com

Abstract. Use of monthly weather forecast as input meteorological data for agrometeorological forecasting, crop modelling and plant protection can foster promising applications in agricultural production. Operational use of monthly or seasonal weather forecast can help farmers to optimize field operations (fertilizing, irrigation) and protection measures against plant diseases and pests by taking full advantage of monthly forecast information in predicting plant development, pest and disease risks and yield potentials few weeks in advance. It can help producers to obtain stable or higher yield with the same inputs and to minimise losses caused by weather. In Central and South-Eastern Europe ongoing climate change lead to shifts of crops phenology dynamics (i.e. in Serbia 4-8 weeks earlier in 2016 than in previous years) and brings this subject in the front of agronomy science and practice. Objective of this study is to test efficacy of monthly forecast in predicting phenology dynamics of different winter wheat varieties, using phenological model developed by Forecasting and Warning Service of Serbia in plant protection. For that purpose, historical monthly forecast for four months (March 1, 2005 - June 30, 2005) was assimilated from ECMWF MARS archive for 50 ensemble members and control run. Impact of different agroecological conditions is tested by using observed and forecasted data for two locations - Rimski Sancevi (Serbia) and Groß-Enzersdorf (Austria).

1. Introduction

Timing and length of plant growing season (GSL) is result of growing plant response on actual weather conditions. Changes in vegetation cover and its time of appearance and duration affect the seasonal changes of energy and water balance as well as aerodynamic conditions at biosphere-atmosphere interface. Notified effects can produce significant impacts on temperature and precipitation, wind share and atmospheric circulation patterns of specific region. Consequently, it additionally affects GSL producing strong feedback between atmosphere and biosphere.

Changes in GSL affect directly or indirectly all aspects of agricultural production including crop management (cropping schedules and rotation, irrigation and fertilisation schedules, timing of harmful organism appearance) and forestry. On longer time scale, permanent increase in GSL affects atmospheric-biospheric carbon budget, too. Linderholm [1] offered excellent overview of studies, covering different periods from 1930 till 2004, reporting shifting in timing and increasing length of the plant growing seasons, based on phenological, climatological and satellite data. Recent trends of winter air temperature anomalies indicate huge increase of winter temperature particularly in the Northern hemisphere mainly affecting Europe and Central Asia but with significant inclusion of USA regions (source: NOAA, Climate monitoring, with respect to a 1981-2010 base period). The most disturbing is the latest anomaly varying from 1.5-2.5 °C in Europe till 5 °C in some regions in Asia.

According to the Index of Early leafing plants used by USA National Phenology network, in the prevailing part of the USA spring 2016 comes 2-20 days earlier in comparison to a 1981-2010 base period (source: USA NPN, Gridded Data Products). In Serbia and many other European countries we are witnessing similar trends. According to recent observations made by observational network of Forecasting and Warning Service of Serbia in plant protection (PIS), in some regions of Northern Serbia (Vojvodina) are registered shifts in phenology varying from - 26 to - 73 days (Table 1). These shifts bear high risk for crop and fruit production from spring frost and early activities of harmful organism. Therefore, strong need is exerted for future more sophisticated forecasting of phenology dynamics based not on fixed phenology calendar (which become less useful under fast changing climate conditions), but on application of numerical weather prediction models in combination with phenological models. Limited number of papers dealing with short-range and monthly weather forecast (MWF) application in agrometeorological and crop forecasting underlining relevance and importance of the topic [2, 3, 4, 5].

Objective of our study is to test efficacy of MWF in predicting appearance of plant growing stages one month ahead in the most important production areas of Austria and Serbia. In this paper are presented results obtained for winter wheat at Rimski Sancevi (Serbia) and Groß-Enzersdorf (Austria) obtained using MWR for March - June 2005.

Table 1. Shift in appearance of "four tillers detectable" - growing stage of winter wheat in Serbia.

Region	Dates	Change (days)
Novi Sad	13.03.2015. 14.01.2016.	- 58
Bačka Topola	19.03.2015. 20.01.2016.	- 58
Pančevo	12.03.2015. 15.02.2016.	- 26
Ruma	22.02.2015. 31.12.2015.*	- 53
Sombor	03.03.2015. 21.12.2016.*	- 73

2. Material and Method

2.1. Study area

Two locations are selected in most important agricultural production areas in Austria (Groß-Enzersdorf - 48° 12' N 16° 33' E) and Serbia (Rimski Sancevi - 45° 3' N 19° 8' E) in which winter wheat is permanently grown as a main crop for many decades. Both locations are situated on flat terrain of southern and south western part of Pannonian lowland, although Groß-Enzersdorf weather is strongly influenced by presence of Alpine mountains on W and SW. Typical climate of study areas is continental and moderate continental with annual temperature of 11.5°C in Rimski Sancevi and 10.7°C in Groß-Enzersdorf and annual precipitation of 647 mm in Rimski Sancevi and 550 mm in Groß-Enzersdorf for 1981-2010 period.

2.2. *Meteorological data*

Observed and simulated values of maximum (T_{max}) and minimum (T_{min}) air temperatures, for selected locations, are used in this study in order to calculate growing degree-days (GDD) accumulation for winter wheat growing season.

Observed data. Historical records of selected meteorological data for Rimski Sancevi and Groß-Enzersdorf weather stations are obtained from national weather service (Hydrometeorological Service of Republic of Serbia and Central Institute for Meteorology and Geodynamics of Austria (ZAMG)).

Monthly weather forecast (MWF) data. Monthly EPS products of ECMWF (European Centre for Medium range Weather Forecast) for March 1, 2005 till June 30, 2005 in the form of 51-member ensemble (50 ensemble members and control run) are used as boundary conditions for WRF NMM model run. Downscaling was developed using WRF NMM model 3.3 version with horizontal resolution of about 20 km. From the monthly weather forecast variable list, for the purpose of this study were selected maximum and minimum air temperature.

All calculations performed using 50 ensemble members, are averaged in order to obtain ensemble averages and in further text denoted with "EA". Results obtained using observed data and control run are denoted with "OBS" and "CR", respectively.

2.3. *Phenology model*

Growing stage appearance of selected winter wheat varieties is analysed following phenology dynamic analysis concept designed and conducted by Forecasting and Warning Service of Serbia in plant protection (PIS_PHEN). This concept is based on continuous observation of plant growing stages according to BBCH scale [6]. From 13 winter wheat varieties grown on 58 locations during 2014/2015 growing season, 4 varieties were selected (Zvezda, Apac, Simonida, NS_40s) with the longest time series of observed phenological stages (21-23 during the whole season). At each location was set an automatic weather station equipped with sensors for air and soil temperature, relative air humidity, precipitation and leaf wetness measurements. Selected varieties have been grown at 15 different locations, leaving opportunity to calibrate and validated phenological model even with one-season data. Accumulated growing degree-days (GDD) were calculated using measured temperatures above 0 °C (from emergence) and 5 °C (from beginning of tillering) as a lower threshold temperature what is common practice in calculation of phenology dynamics. (see for example [7]). In further text these low temperature limits will be denoted with GDD_0 and GDD_5, respectively. Phenological analysis was based on BBCH growing stages observations from emergence (9) till ripening (80). Accumulated GDD's are identified for each variety corresponding to specific BBCH growing stage for two lower threshold temperatures. MWF and measured meteorological elements are used for calculation of day of appearance of certain BBCH stage (BBCH_D), for four selected varieties, for Rimski Sancevi and Groß-Enzersdorf locations.

2.4. *Verification statistics*

The verification methodology, based on RMSE and SPREAD calculation, was used to evaluate, ensemble based, prediction of growing stage appearance. Applying criteria of verification of ensemble based probabilistic weather forecast [8], it quantifies reliability of phenology dynamics simulations obtained using MWF. The differences between results obtained using MWF and observed meteorological data are measured using RMSE of ensemble mean of these products in the form

$$RMSE = \frac{1}{N}\left[\sum_{i=1}^{N}\left(\bar{A} - A_{OBS}\right)^2\right]^{1/2} \tag{1}$$

where $\bar{A} = 1/N \sum_{i=1}^{N} A(i)$ is ensemble mean, A_{OBS} is observed value of variable of interest, N is number of ensemble members.

The deviation of ensemble forecast from their mean is important attribute of MWF and MWF based calculation of growing stages. From that reason ensemble SPREAD is calculated for both MWF and related products using formula

$$SPREAD = \left[\frac{1}{n-1} \sum_{i=1}^{N} \left(\bar{A} - A(i) \right)^2 \right]^{1/2} \quad (2)$$

where A(n) is value of variable of interest for *i*th ensemble member.

Comparison of Equations (1) and (2) leads to conclusion that an ideal MWF will be expected to have the same size of RMSE and SPREAD since it implies that for each ensemble member, forecasted value is equal to observed ones. According to this conclusion, the simulation is performed more realistic if RMSE and SPREAD values are close to each other.

Relative deviation (RLD) of simulated values can be adequate quantifier of control run forecast data

$$RLD = \frac{\left| A_{CR} - A_{OBS} \right|}{A_{OBS}} \cdot 100\% \cdot \quad (3)$$

3. Results and Discussion

The verification statistics related to MWF and MWF related products, on longer scale, should provide information on the applicability of MWF for certain purposes. In case of plant phenology dynamics prediction, of particular importance is impact of meteorological elements prediction uncertainty (T_{min} and T_{max} in this case) on calculated values of BBCH_D.

In Tables 2 and 3 are presented verification statistics for T_{min} and T_{max} temperatures for Rimski Sancevi and Groß-Enzersdorf locations from March 1, 2005 till June 30, 2005 obtained using ENS and CR data sets. Negligible difference between RMSE and SPREAD for T_{min} at both locations indicates excellent skill of ensemble forecast in predicting low temperatures. Low performance of ensemble forecast in forecasting T_{max}, particularly in Groß-Enzersdorf is result of two warm advections which were "invisible" in the forecast for March and April. Detailed inspection of grid area from which current forecast is assimilated according to latitude and longitude of weather station indicate that model selected location is set at 218 m altitude, while in reality it is 156 m. In indicated grid area, there are significant differences in the orography resulting in a strong west to east temperature and precipitation gradient, where mountainous hilly regions dominates in the west and flat lowland region in the eastern part of the relevant grid (where the weather station Groß_Enzersdorf is located).

Table 2. RMSE and SPREAD for monthly average maximum and minimum temperatures from March 1, 2005 till June 30, 2005 obtained using 50 ensemble members (Table denotation: M-March, A-April, M-May, J-Jun).

Setting	RMSE (°C)				SPREAD (°C)			
	M	A	M	J	M	A	M	J
				T_{min}				
Groß-Enzersdorf	2.9	1.6	1.0	0.8	1.2	1.1	0.9	0.7
Rimski Sancevi	1.3	1.2	1.0	1.6	1.2	1.2	0.9	0.7
				T_{max}				
Groß-Enzersdorf	5.1	4.0	2.4	2.8	1.5	1.4	1.3	1.3
Rimski Sancevi	3.0	1.9	1.7	1.5	1.7	1.4	1.4	1.5

Table 3. RMSE and RLD for monthly average maximum and minimum temperatures from March 1, 2005 till June 30, 2005 obtained using control run (Table denotation: M-March, A-April, M-May, J-Jun).

Setting	RMSE (°C)				RLD (°C)			
	M	A	M	J	M	A	M	J
				T_{min}				
Groß-Enzersdorf	2.9	1.3	0.6	1.0	-	19.1	8.0	7.3
Rimski Sancevi	1.8	0.4	0.7	1.7	-	7.9	7.3	9.7
				T_{max}				
Groß-Enzersdorf	5.7	4.3	1.9	0.8	24.1	12.0	6.4	3.7
Rimski Sancevi	4.7	2.6	0.2	2.8	20.2	9.0	2.1	6.6

Obtained results indicate much better skill of ensemble forecast for Rimski Sancevi then for Groß-Enzersdorf, but in respect to meteorological elements, results obtained for T_{min} are much better than for T_{max}.

Efficacy of MWF in predicting growing stages is tested by calculating ensemble RMSE and SPREAD for day of appearance (BBCH_D) of BBCH stages from 20 to 99 for Rimski Sancevi and Groß-Enzersdorf using ENS and CR data sets. Low performance of MWF for both threshold temperatures for all varieties can be identified only for Groß-Enzersdorf up to BBCH phase 40 for GDD_0, i.e. BBCH phase 60 for GDD_5. This drop down in MWF skill is caused by forecast deviation from real temperatures in the month March. A month later, rate of GDD accumulation is

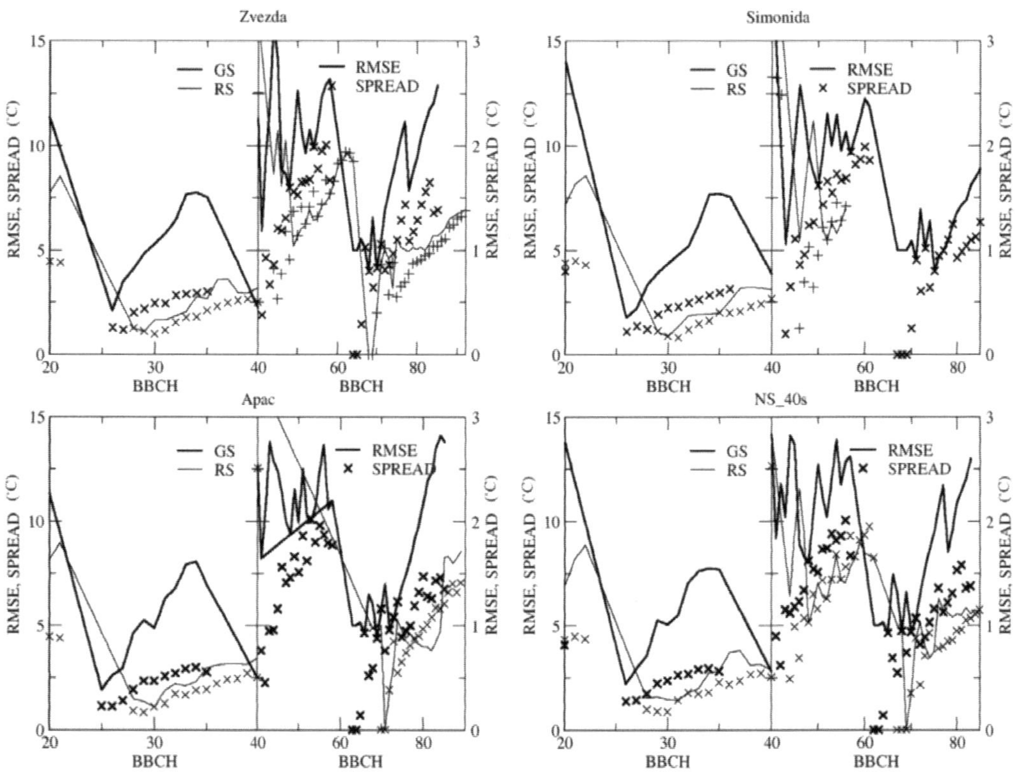

Figure 1. RMSE (line) and SPREAD (cross) of BBCH_D calculated using OBS and ENS data sets for 0 °C lower temperature threshold for Rimski Sancevi (RS) and Groß-Enzersdorf (GS) for the 4 cultivars.

high enough to reduce impact of underestimated GDD during March. Reduction of impact is slower in case of GDD_5 because contribution of period with high deviation of forecast is more pronounced than in case of GDD_0. For the rest of the vegetation period, for all varieties, difference between RMSE and SPRED is negligible, particularly in case of Rimski Sancevi location. More uniform values for RMSE and SPRED of BBCH_D calculated using GDD_5 in comparison to GDD_0 can be result of low temperature effects or photosynthetic radiation limit on crop development (i.e. stopping or delaying these) which are not taken into account. Since these limiting factors on crop development are, however, more pronounced during the winter and early spring months when accumulation of GDD_0 can already take place, variation in BBCH_D forecast quality is more exerted if it is based on GDD_0 than GDD_5.

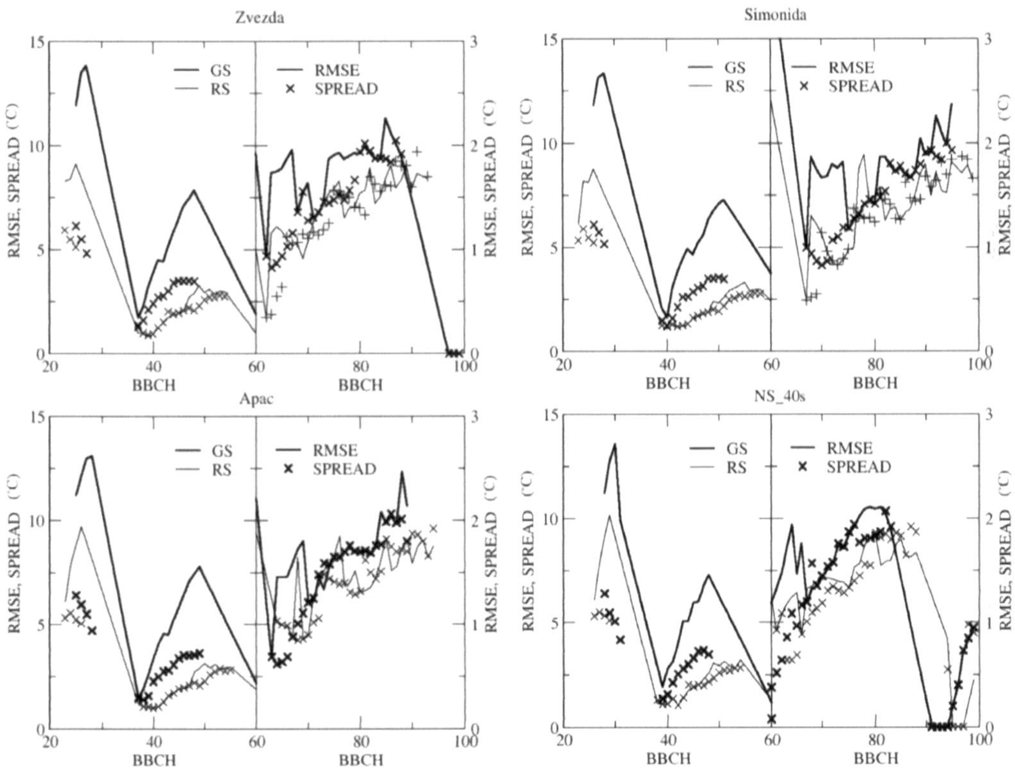

Figure 2. RMSE (line) and SPREAD (cross) of BBCH_D calculated using OBS and ENS data sets for 5 °C lower temperature threshold for Rimski Sancevi (RS) and Groß-Enzersdorf (GS) for the 4 cultivars.

Results obtained using CR data set are analysed by calculating RMSE and RLD for all BBCH stages and threshold temperatures (Figure 3 and 4). RMSE obtained for both threshold temperatures are in order of magnitude of RMSE obtained for ENS data sets for all cultivars. Relative deviation of up to 2% indicates good skill of even control run forecast in predicting growing stage appearance.

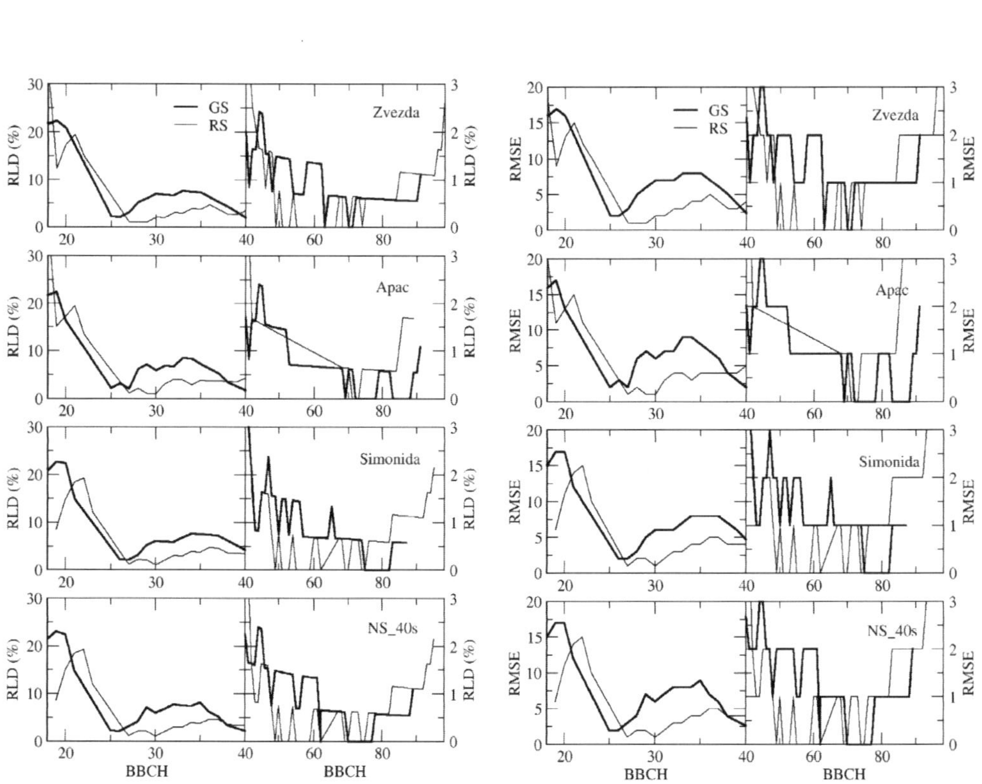

Figure 3. RLD (left) and RMSE (right) of BBCH_D calculated using OBS and CR data sets for 0 °C lower temperature threshold for Rimski Sancevi (RS) and Groß-Enzersdorf (GS) for the 4 cultivars.

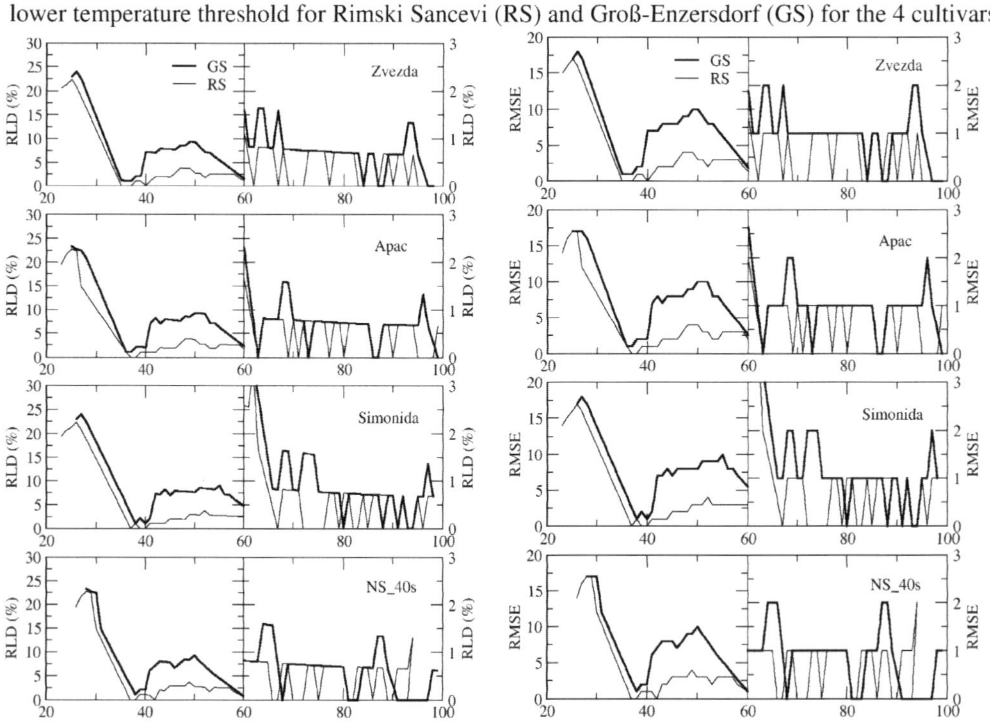

Figure 4. RLD and RMSE of BBCH_D calculated using OBS and CR data sets for 5 °C lower temperature threshold for Rimski Sancevi (RS) and Groß-Enzersdorf (GS) for the 4 cultivars.

4. Conclusions

Every day we are witnessing effects of climate variability and changing climate, expressed as prevailing weather conditions. Important step in adopting to effects of climate variability and change is to provide site specific and reliable information on the future weather conditions at various time scales for stakeholders. Any improvement in long term numerical weather prediction (monthly, seasonal) makes adaptation actions more effective. In case of agriculture, application of monthly and seasonal weather forecast is of particular importance. The best effects of this application can be achieved using long term weather forecast as an input data in well calibrated and validated forecasting tools such as physiological, phenological, agrometeorological, crop yield and pest and diseases models.

Results presented are first outcomes of a strategic study related to modelling phenology dynamics of harmful organism and host plants in order to be used in agrometeorological modelling and integrated pest management of different scales. For that purpose, a current phenological model for winter wheat was applied for two locations in representative agricultural production areas in Serbia and Austria for which MWF for March-June 2005 was available, in order to test efficacy of MWF in comparison to results obtained using observed data.

High skill of MWF in forecasting low temperatures, especially in spring, is promising from frost prediction point of view. Namely, in years with significant shift of phenological phases to earlier time, frost prediction and protection becomes one of the most challenging problems.

Lower performance of MWF in forecasting high temperatures in case of Groß-Enzersdorf is actually result of one warm advection episode which was omitted by the forecasting model. Most probably it is result of difference in altitude between model grid points and exact location of weather station. In order to clarify this problem we are preparing to make more testing, using monthly and seasonal forecast for ten consecutive years.

During the actual growing season of 2016 following a warm winter the system is operationally tested. Up to now, monthly forecast for March and April 2016 for Rimski Sancevi predicted accurately minimum and maximum temperatures and absence of late frost.

[1] Linderholm H W 2006 *Agric. Forest Meteorol.* **137** 1-14
[2] Detlefsen N 2006 *Tellus* **58A(5)** 558-564
[3] Baigorria A G Jones J W and O'Brien J J 2008 *Agric. Forest Meteorol.* **148** 1353-1361
[4] Calanca P Bolius D Weigel A P and Liniger M A 2011 *J. Agric. Sci.* **149** 15-22
[5] Lalic B Francia M Eitzinger J Podrascanin Z and Arsenic I 2016. *Meteorol. Appl.* **23** 50-56
[6] BBCH Monograph 2001 *Growth stages of mono-and dicotyledonous plants* 2nd Edition ed U Meier (Quedlinburg: Federal Biological Research Centre for Agriculture and Forestry) p 158
[7] Zhu, Y., Toth, Z., 2008. Ensemble Based Probabilistic Forecast Verification, *Proc. 19th Conf. on Predictability and Statistics* (*New Orleans)* (American Meteorological Society), pp 1-6
[8] Chmielewski F M and Kohn W 2000 *Agric. Forest Meteorol.* **102** 253-261

Acknowledgments

The research work presented was realized as a part of the project "Studying climate change and its influence on the environment: impacts, adaptation and mitigation" (43007) financed by the Ministry of Education and Science of the Republic of Serbia within the framework of integrated and interdisciplinary research for the period 2011-2016. Participation to the EOBAR conference and presentation of results is supported by "H2020 SERBIA FOR EXCELL" project. This project has received funding from the European Union's Horizon 2020 research and innovation programme under grant agreement No. 691998.
ely following the last numbered section of the paper.

IOP Conf. Series: Earth and Environmental Science 57 (2017) 012003 doi:10.1088/1755-1315/57/1/012003

A phenology-based classification for crop in Great Mekong Subregion based on MODIS data

Tingting LV[1], **Zui TAO**[1], **Xiang ZHOU**[1], **Xiaoyu SUN**[2], **Aqiang YANG**[1], **Banghui YANG** [1]

[1] Institute of Remote Sensing and Digital Earth, CAS, 100094 Beijing, China
[2] National Marine Environmental Forecasting Center, Beijing 100081, China

lvtt@radi.ac.cn

Abstract. The difference and the complementarity on agriculture decides that there is a wide-ranging cooperation on agriculture between China and GMS. So the research on GMS's agriculture has important significance. The objective of this research is to produce the phenology-based classification map for crop in GMS by investigating the spatial and temporal patterns of multiple cropping systems. Although the spatial resolution of MODIS-NDVI data is not very high, it is still useful for detecting very large-scale crop phenomenon. In this paper 16-day time-series MODIS data from 2001 to 2015 were used to exploer the imforamtion of crop phenology in GMS. After constructing a phenology-based approach, four types of cropping systems, namely single-cropped rain-fed crop, double-cropped irrigated crop, double-cropped rain-fed crop and triple-cropped crop, were classified in the region

1. Introduction

Agricultural land is one of the most important land uses in Earth especially in the Greater Mekong Subregion (GMS). The contribution of agriculture to gross domestic product (GDP) of the countries in the region has been declining over the past decade; however, the sector still plays a very important role in food security and providing employment for its population. Over the past several decades changes in land use and agricultural intensification have come at an environmental cost in GMS. Regular and timely monitoring of crop system at regional to local scale will result in efficient use of agricultural land resources, as well as provide valuable information in agricultural management to meet the needs for food while safeguarding the environment.

Remote sensing offers a feasible tool for crop classification by providing valuable information regarding the extent, status of agricultural systems at various spatial and temporal scales. A lot of researches about regional vegetation phenology by using time series vegetation index (VI) data were conducted(Agnes Beguea et al.,2014;Tingting Lv et al., 2008;Nguyen-Thanh Son et al., 2014;Toshihiro Sakamoto et al., 2006;Sangram Ganguly et al., 2010;Wen-bin WU et al., 2010;Xiangming Xiao et al., 2005;Zhuokun Pan et al., 2015). VI data acquired from high spatial resolution satellite images such as Landsat TM/ETM+, etc. do not provide sufficient information on seasonal change characteristics of cropland due to infrequent temporal coverage. So the generally used VI data for crop phenology research are available at coarse and medium spatial resolution produced by MODIS,AVHRR and SPOT-VGT data. The advantages of using coarse and medium spatial and high temporal resolution VI data are (i) increased ability to monitor the phenological change of crop plants, and (ii) the possibility of generating consistent large area crop cover maps.

To date, many researchers have applied time-series analysis techniques to the NDVI time-series to obtain the information of seasonal vegetation changes characterized. In this paper the main objective of this study is to develop a phenology-based mapping approach to explore the characteristics of cropping systems in GMS, using the MODIS NDVI time-series from 2001 to 2005 and 2011 to 2015.

2. Data and Method

2.1. Study Area

The GMS , comprising five countries of Cambodia, the People's republic of China (PRC, specifically Yunnan Province and Guangxi Zhuang Autonomous Region), the Lao People's Democratic Republic (Lao PDR), Myanmar, Thailand, and VietNam, is located within the limits of 92.19 E to 109.47 E and 5.6 S to 28.54 S in the Southeast Asian. The study area is a natural economic area bound together by the Mekong River, with an area of 2.57 million square kilometers. The region has a tropical climate and two distinct seasons. The dry season is from December to April, while the wet season from May to November provides approximately 80% of the total annual precipitation. Major food crops in the study area include rice, maize, vegetables, and cassava.

In this paper the deltas and low-lying floodplains of the Red, Mekong, Chao Phraya and Irrawaddy Rivers and Khorat Plateau are chosen as the study area, which are the main agricultural regions in GMS. According to the statistic by FAO, only the deltas and low-lying floodplains produced half of region's rice production although they constitute only 10% of the total land area.

2.2. Data

The satellite sensor and ancillary data used in this research were MODIS NDVI time-series. The NDVI data used in this paper are the Version 5 MODIS/Terra 250m resolution 16-day Normalized Difference Vegetation Index (MOD13Q1 V005) spanning from 2001 to 2005 and 2011 to 2015. They were acquired from Land Processes Distributed Active Archive Center (LP DAAC) at no cost. Statistic data on province-level were used in this paper to validate the result of the method.

2.3. Method

There are four main steps involved in classification scheme:1)Reconstructing of time-series MODIS NDVI data by applying Savitzky-Golay filter to remove deviations from the vegetation cycle due to residual cloud and atmospheric effects and so on;2) Masking non-crop area by calculating for each pixel how many times the NDVI values within each year were less than 0.4 or greater than 0.75 and then setting a threshold to separate the crop and non-crop area;3)Producing the cropping intensity map by checking how many peaks and heading date of NDVI profile on a pixel-by-pixel basis, which then be used to separate the crop area into single-cropped area, double-cropped area and triple cropped area and then separate the double- and triple-cropped area into rainy and dry growing season;4)Validation analysis.

2.3.1. Time series NDVI reconstructing. Extraction of the phenology of the crop requires high quality time series data. Even though the constraint view angle - maximum value composite method has been applied on the 16-day composite NDVI time series, the data remain significant residual effects and noises levels. In order to reduce the impact of the noises and improve the accuracy of the extraction result it is important to reconstruct high quality NDVI data. The reconstruct processing in this paper consists of the following steps:

- detecting of useless pixels (according to the Pixel Reliability flags in MDOIS VI products) and low reliable values (identified if they exceeded a change threshold in time series NDVI);
- replacing bad data and low reliable points with boxcar averages of the specified width;
- applying Savitzky-Golay(S-G) smoothing filtering to smooth NDVI time-series and making a comparison between the filtered NDVI time-series and the NDVI time-series acquired from step;
- replacing the points in the original NDVI time-series that are below a user-defined threshold with the value of the filtered NDVI time-series;
- next iteration is needed and again the NDVI time-series is searched for possible bad data until no new points are being found.

2.3.2. Masking no-crop area. By analyzing the NDVI time-series of different land covers in the study area we found that: 1) NDVI values of water,soil and rock are in general fluctuated between -0.1 and 0.4; 2)Forest generally displays a long period of high NDVI in a year compared to other land cover types; 3) Cropland generally displays a distinctly higher annual standard deviation compared to forest due to high greenness during the growing season and extremely low greenness following harvest. According to these characteristics we firstly calculated the number of NDVI time-series whose value is less than 0.4 on pixel basis. If the number is less than 20 and the maximum value of NDVI is less than 0.5 within 23 periods(one year), that pixel is determined to non-vegetation. The forest area could be identified by calculating the number of NDVI time-series whose value is greater than 0.75 on pixel basis.If the number is greater than 16 within 23 periods(one year), that pixel is determined to forest. Then all other pixels were categorized into crop class.

2.3.3. Extracting cropping intensity. NDVI time-series can provide a seasonal trajectory of vegetation phenology. For example, single cropping system exhibits a unimodal NDVI periodicity while double cropping system exhibits a bimodal NDVI periodicity. So by detecting the numbers and the date of the peaks of the smoothed NDVI profile on a pixel-by-pixel basis, cropping intensity and the date when maximum NDVI occurred can be acquired. The peak finding algorithm used to identify this parameters has been described in details in the paper(Lv Tingting, Liu Chuang,2010).

3. Result and Analysis

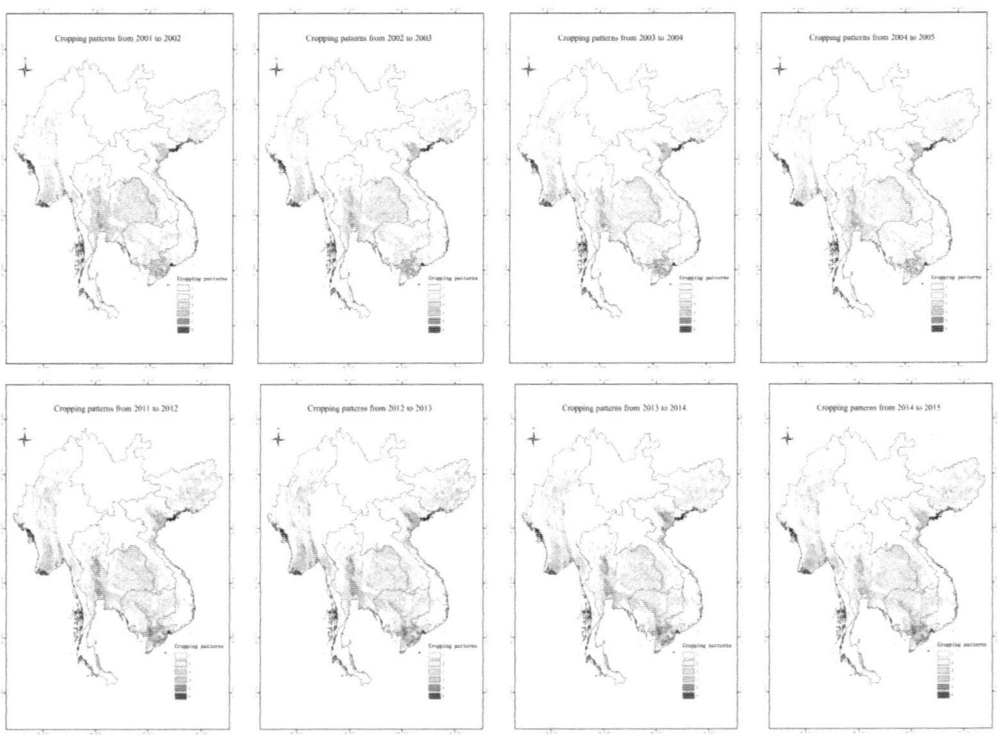

Figure 1. Cropping intensity map in study area

Figure 1 is the cropping intensity map in the deltas and low-lying floodplains of the Red, Mekong, Chao Phraya and and Khorat Plateau from 2001 to 2005 and from 2011 to 2015.The legend '1,2,3,4,5,6,7' in Figure 2 means the number of the peaks of the NDVI profile estimated per two years. From it we can see that the spatio-temporal evolution of single, double and triple cropping system in different periods. The single cropping system mainly distributed in Khorat Plateau, upper part of the floodplains of the Irrawaddy Rivers and marginal upland of the Chao Phraya basin. Double cropping system was concentrated in deltas of the Red, Mekong, Chao Phraya River and triple cropping system was particularly allocated in the central part of the delta of Mekong river.

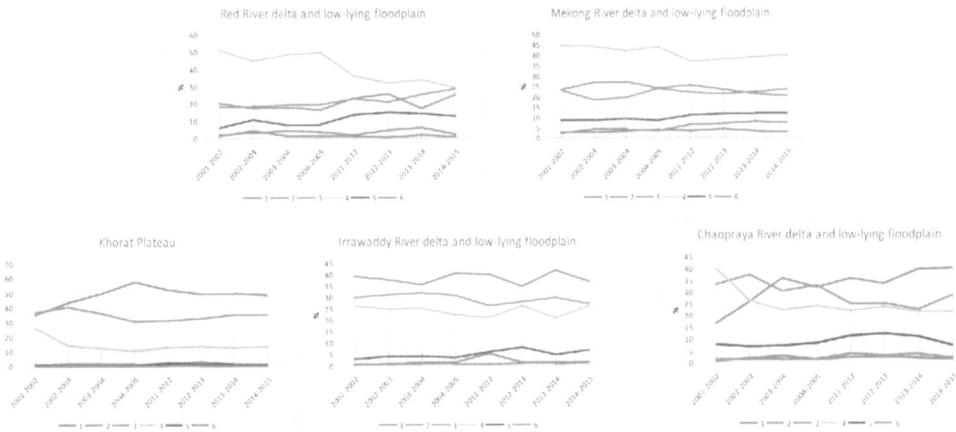

Figure 2. Statistic chart of cropping intensity in deltas,low-lying floodplains and Plateau of GMS

Figure 2 is the statistic charts of cropping intensity in the deltas and low-lying floodplains of the Red, Mekong, Chao Phraya and Irrawaddy Rivers and Khorat Plateau from 2001 to 2005 and from 2011 to 2015. The results indicated that the double cropping system are the domain cropping pattern in red and Mekong rivers delta and low-lying floodplain with higher percentage compared with other cropping pattern and the percentage decreased from 2011 to 2015. In Khorat plateau and Irrawaddy river basin single cropping is a typical feature of rain fed subsistence agricultural practice which counts for about 40% of the total crop land in these regions. In Chao Phraya and Mekong river basin the cropping systems is a little complex.The percentage of double cropping pattern decreased from 2001 to 2015 while single cropping pattern increased gradually.

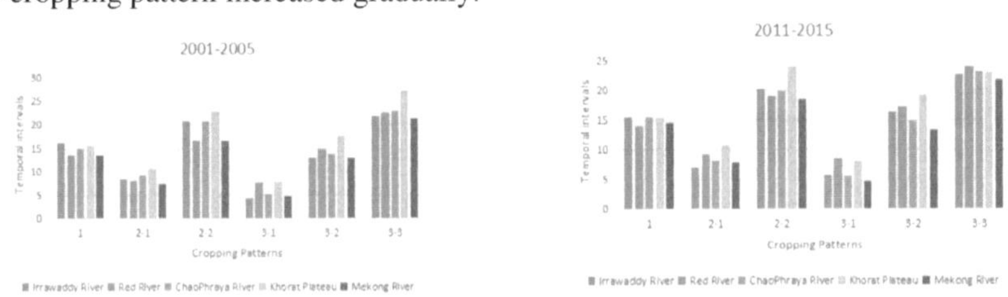

Figure3. Statistic chart of date when maximum NDVI occurred in the study area

Figure 3. is the statistic chart about time when peaks appear in NDVI curves. The horizontal (category) axis, also known as the x axis, of a chart displays the cropping patterns. The legend '1' means single cropping system.The legend '2-1'means that two growing season were estimated while '-1'means the first growing season while '-2'means the second growing season. The y axis displays the temporal intervals of the 16-day NDVI data.We can see that in the study area the peaks on the single cropping line mostly appeared between July and September. The peaks for double cropping system in the rainy season mostly appeared from October to December while peaks for double cropping system in the dry season mostly appeared from from April to June.This phenomena can be attributed to that irrigated could have different planting time in the dry season while rain-fed crop's planting time is mainly determined by the time of onset of rainfall. The peaks for triple cropping system mostly concentrated in three periods: from March to May, July to October and from November to February the next year. The onset of the date of maximum NDVI of crop may vary inter-annually which corresponds to crop variety and agricultural activities scheduling.

Figure 4. Phenology-based classification of crop in study area

Based on analysis above a phenology-based approach was constructed to classify four types of cropping systems in the study area: single-cropped rain-fed crop, double-cropped irrigated crop, double-cropped rain-fed crop and triple-cropped crop. If a single cropping pattern had a peak between July to November, it was classified as single-cropped rain-fed rice . If one of peaks for double cropping system occurred between January to June, it was classified as double-cropped irrigated crop and the other double cropping system deemed as double-cropped rain-fed crop. It must be mentioned that triple cropping system general sparsely distributed in the study area which characterized by smaller crop field sizes. So the result of this cropping system derived from coarse-spatial-resolution MODIS data has higher uncertainty. Considering this the sub-class for triple cropping system did not detect.

4. Validation

Due to lack of ground reference data we used crop statistic data from 2001 to 2010 acquired on national level to validate the result. Figure 5 is a comparison of cropping intensity from MODIS-NDVI and statistical data on provincial-level. Because the boundary of study area does not match the administrative boundary of province in GMS, only statistic data of provinces which are total included in the rectangle area were used in this paper.From figure 6 we could seen that in most of provinces the estimated intensity by MODIS data demonstrates a good association with statistical data at county-level.

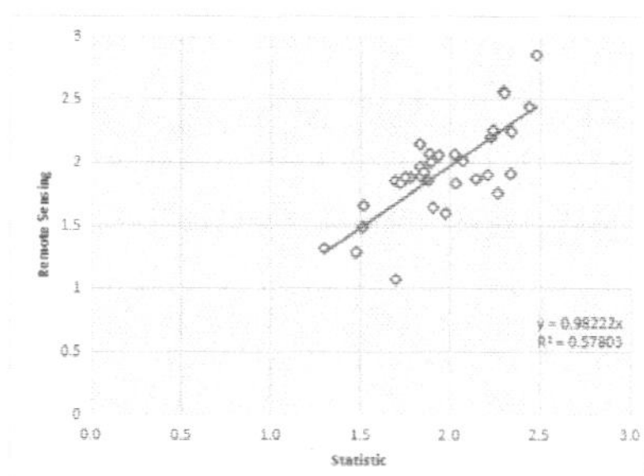

Figure 5. Comparisons of cropping intensity from MODIS-NDVI and statistic data on provincial-level

Acknowledgments

This study is financed by the National Natural Science Foundation of China (NO.41201224) and the State High-Tech Development Plan of China (2013AA12A303).

References

[1] Begue, A; Vintrou, E; Saad, A; Hiernaux, P 2014 Differences between cropland and rangeland MODIS phenology (start-of-season) in Malib *International Journal of Applied Earth Observation and Geoinformation* **31** pp. 167-170.

[2] Lv Tingting , Liu Chuang 2010 Study on extraction of crop information using time-series

MODIS data in the Chao Phraya Basin of Thailand *Advances in Space Research* **45** pp.775-784.

[3] Nguyenthanh Son , Chifarn Chen, Chengru Chen , Huynhngoc Duc and Lyyu Chang 2014 A Phenology-Based Classification of Time-Series MODIS Data for Rice Crop Monitoring in Mekong Delta, Vietnam *Remote Sensing* **6** pp. 135-156.

[4] Sakamoto T, Phung CV, Kotera A, Khang ND, Yokozawa M 2009 Analysis of rapid expansion of inland aquaculture and triple rice-cropping areas in a coastal area of the Vietnamese Mekong Delta using MODIS time-series imagery *Landscape and Urban Planning* **92** pp. 34-46.

[5] Ganguly, S; Friedl, MA; Tan, B; Zhang, XY; Verma, M 2010. Land surface phenology from MODIS: Characterization of the Collection 5 global land cover dynamics product *Remote Sensing of Environment* **114** pp.1805-16.

[6] Sakamoto T, Nguyen NV, Ohno H, Ishitsuka M, Yokozawa M 2006 Spatio-temporal distribution of rice phenology and cropping systems in the Mekong Delta with special reference to the seasonal water flow of the Mekong and Bassac rivers. *Remote Sensing of Environment* **100** pp.1-16.

[7] Wenbin W, Peng Y, Huajun T, 2010 Characterizing spatial patterns of phenology in cropland of China based on remotely sensed data *Agricultural Sciences in China* **9** pp. 101-112.

[8] Xiao X, Boles S, Liu JY, Zhuang DF, Frolking and S, Li CS 2005 Mapping paddy rice agriculture in southern China using multi-temporal MODIS images *Remote Sensing of Environment* **95** pp. 480–492.

[9] Zhuokun P ,Jingfeng H, Qingbo Z 2015 Mapping crop phenology using NDVI time-series derived from HJ-1A/B data *International Journal of Applied Earth Observation and Geoinformation* **34** , pp. 188-197.

Glacier parameter extraction using Landsat 8 images in the eastern Karakorum

H Wang [1], R Yang [2], X LI [3] and S CAO [4]

[1] School of Land Science and Technology, China University of Geosciences(Beijing), Haidian District, Beijing 100083, China
[2] Beijing Jiao tong University School of Civil Engineering, Haidian District, Beijing 100044, China
[3] Institute of Remote Sensing and Digital Earth, Chinese Academy of Sciences, Beijing 100094, China
[4] Shandong University of Science and Technology, College of Geomics, Qingdao 266590, China
Email:14121165@bjtu.edu.cn

Abstract. Changes in glacial melt and river runoff directly impact upon water resources and utilization. Glacier area data provide the basis of analysing glacier change. In this study, Landsat images from the end of the 2014 melt season were analysed using remote sensing and geographic information systems (GIS) technology. We compared the band ratio method and Normalized Difference Snow Index (NDSI) method for extracting glacial parameters. The results show that the band ratio method is better for glacier boundary extraction. However, for debris covered glaciers, textural analysis is needed for area extraction and needed to revise boundary extraction results through the visual interpretation of the remote sensing image Accurate area extraction is particularly important for glaciers , as these are major contributors to runoff.

1. Introduction

Investigating glacier change primarily relies on identifying accurate glacial boundaries from remote sensing images. Accurate area extraction is particularly important for glaciers of > 2 km^2, as these are major contributors to runoff. At present, there are numerous automatic and semi-automatic glacier identification method, with the most commonly used automatic methods including statistical thresholding, unsupervised and supervised classification, principal component analysis, band ratios, and the Normalized Difference Snow Index (NDSI) method. With the rapid development of remote sensing technology and greater understanding of glacier change, automatic boundary extraction methods have become possible.

Sidjak [1] used a combination of bands in a supervised classification to extract glacier areas and found that only the PC2-4 and TM4/TM5 combinations, and the NDSI were able to deal with shadowed areas and moraine-covered glaciers. Paul [2] compared supervised classification and unsupervised classification using the NDSI method, and the TM3/TM5, and TM4/TM5 bands, and found that only the TM4/TM5 method effectively identified shadowed glacier areas. Shangguan [3] compared different classification methods and found that: (1) TM3/TM5, TM4/TM5, and TM3/TM5 gave the highest accuracy when identifying thin snow cover; (2) supervised and unsupervised classification can both identify of icy water, but fail to identify shadowed areas; and (3) the identification accuracy of NDSI is low for shadowed glacier areas. Zhiguo [4] found that when extracting the outline of densely shadowed and moraine covered glaciers, TM3/TM5 is more accurate than NDSI and TM4/TM5, but it still unable to consistently extract moraine covered glaciers. In contrast, Song [5] argued that the NDSI index is better than ratio methods for marine glacial extraction; however, the differences in these conclusions may reflect differences between continental and oceanic glaciers. In summary, past studies have shown that both NDSI and band ratio methods are suitable for bare ice extraction, and that appropriate method selection for a given glacier of region depends on the spatial resolution and image quality.

Despite these past studies, method suitability varies with glacier type and region; therefore, understanding which method to choose remains a research focus. In this study, we compared the accuracy of the NDSI and band ratio methods using Landsat 8 Operational Land Imager (OLI) data on two glaciers (the Sizagaer glacier, Tibet Autonomous Region, and the Kowloon glacier, Xinjiang Uygur Autonomous Region) from different regions and with different sizes and morphological characteristics.

2. Materials and methods

2.1 Study area

The Sizagaer glacier is located in the Ali area of the Tibet Autonomous Region, China (N34°17'20.74", E79°37'49.43"). The glacier has a maximum elevation of 6515.6 m, a mean elevation of 6102.2 m, and a glacier terminus elevation of 5569.9 m. The glacier is 79.58 km in length and covers an area of 23.8 km^2. The ice is west facing, has a mean slope of 14.8%, and is not covered by moraine.

Kowloon glacier is located in the Hetian area of the Xinjiang Uygur Autonomous Region, China (N34° 34'17.54", E78° 26' 21.09"). The glacier has a maximum elevation of 6520.9 m, a mean elevation of 5901 m, and a terminus elevation of 5284.2 m. The glacier is 6.78 km in length and covers an area of 8.9 km^2. The ice faces to the north, has a mean slope of 17.3%, and is not covered by moraine.

2.2 Data types and analysis

Glacier extraction was performed using a Landsat 8 OLI image, a Landsat Thematic Mapper (TM) image, and high-resolution Google Earth images. The Landsat 8 image was acquired on 20 August 2014 along Worldwide Reference System (WRS) path 146 and WRS row 36. The image is mostly free of clouds (0.95% cloud cover) and contains little seasonal snow cover. High-resolution Google Earth images were used for 3D glacier visualization, which helped with manual glacier boundary delineation. Second glacier catalogue data were used to determine detailed feature information at the time of image acquisition. The Landsat 8 OLI land imager includes nine bands (Table 1), which have been adjusted from those in previous sensors (i.e., TM and ETM+) to avoid atmospheric absorption features; for example, a relatively large adjustment of Band 5 to exclude 0.825 m of moisture absorption features.

We completed radiometric calibration and atmospheric correction of the remote sensing image, and used the Landsat TM3/TM5 and TM4/TM5 band-ratios to automatically delineate the target glaciers. The NDSI was calculated as:

$$(B3-B6)/(B3+B6) \quad (1)$$

Landsat 8 band ratios were calculated as:

$$\text{Band ratio 1} = B4/B6 \quad (2)$$

$$\text{Band ratios 2} = B5/B6 \quad (3)$$

where B3, B4, B6, refers to the values of the 3rd, 4th, and 6th bands of the Landsat 8 image, respectively. Data analysis was performed using the ARCGIS and ENVI software.

Table 1. Spectral ranges of remote sensing sensors

OLI		ETM+		TM	
band	Spectral range/μm	band	Spectral range/μm	band	Spectral range/μm
B1 Coastal	0.433–0.453				
B2 Blue	0.450–0.515	B1 Blue	0.450–0.515	B1 Blue	0.45—0.52
B3 Green	0.525–0.600	B2 Green	0.525–0.605	B2 Green	0.52—0.60
B4 Red	0.630–0.680	B3 Red	0.630–0.690	B3 Red	0.63—0.69
B5 NIR	0.845–0.885	B4 NIR	0.775–0.900	B4 NIR	0.76— 0.90
B6 SWIR 1	1.560–1.660	B5 SWIR 1	1.550–1.750	B5 SWIR 1	1.55—1.75
B7 SWIR 2	2.100–2.300	B7 SWIR 2	2.090–2.350	B7 SWIR 2	2.08 —2.35
B8 Pan	0.500–0.680	B8 Pan	0.520–0.900		
B9 Cirrus	1.360–1.390				

3. Results and discussion

The results showed that visual interpretation of all three data types could effectively distinguish between cloud, ice, and snow (Figure 1). By comparing clouds and glacier spectral curves, we found that both glaciers and clouds were highly reflective in visible wavelengths; however, in 1.6 um near infrared band reflectance snow produced low reflectivity while clouds produced high reflectivity. These results confirm the band ratio operations can be used distinguish clouds from glaciers. In addition, the band ratio method was better than the NDSI. We also compared the spectral curves of water and glaciers and found significant differences among the red band, the near infrared wave band, and the infrared wavelengths. Based on the ratio operation of the red band, near-infrared bands, and the infrared wavelengths, spectral differences between water and glaciers were further magnified.

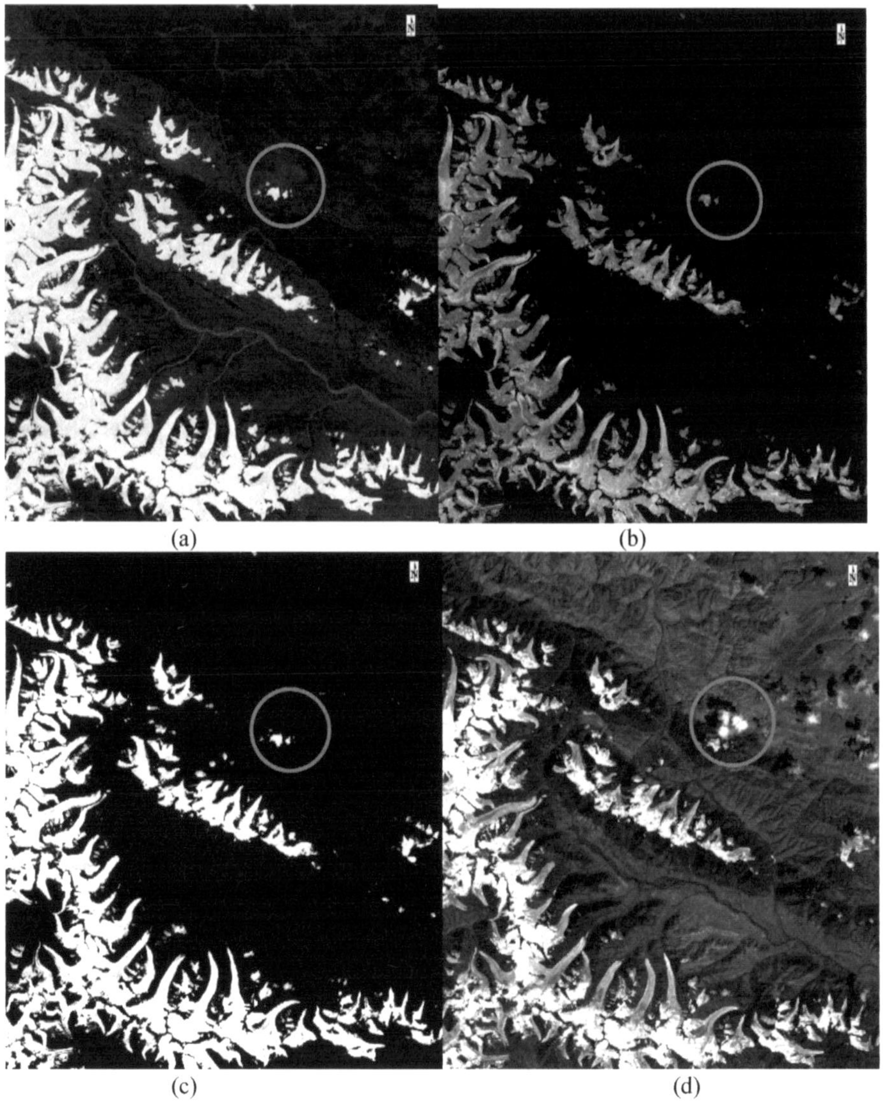

Figure 1. Remote sensing images obtained using different band operations: (a) Normalized Difference Snow Index (NDSI); (b) B4/B6 (c) B4/B6; (d) Landsat Image. Red circle encloses clouds.

3.1 Glacier area extraction

Combined with visual interpretation of remote sensing images, that data show that the NDSI threshold value is ~0.70–0.95. Without the influence of compound glaciers, the contrast between high reflectivity visible light and near infrared bands, and low reflectivity infrared wavelengths was highlighted. To reduce the influence of glacial lakes, the band ratio method was used refine the glacier threshold to 7. After automated delineation, we visually confirmed the results and manually adjusted for shadow, seasonal snow, proglacial lakes, and debris cover. Using this method, we were able to extract the boundaries of both the Sizagaer glacier (Figure 2; Table 2) and Kowloon glacier (Figure 3; Table 2).

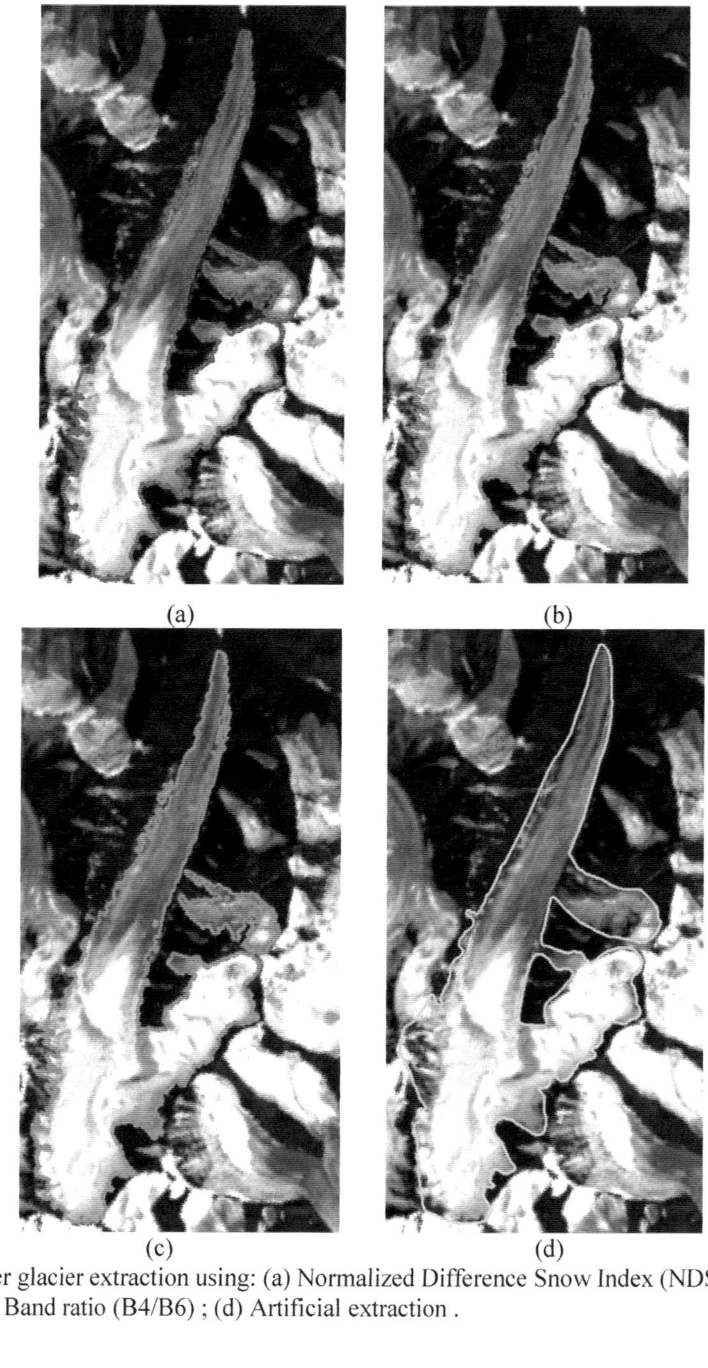

Figure 2. Sizagaer glacier extraction using: (a) Normalized Difference Snow Index (NDSI); (b) Band ratio (B5/B6); (c) Band ratio (B4/B6) ; (d) Artificial extraction .

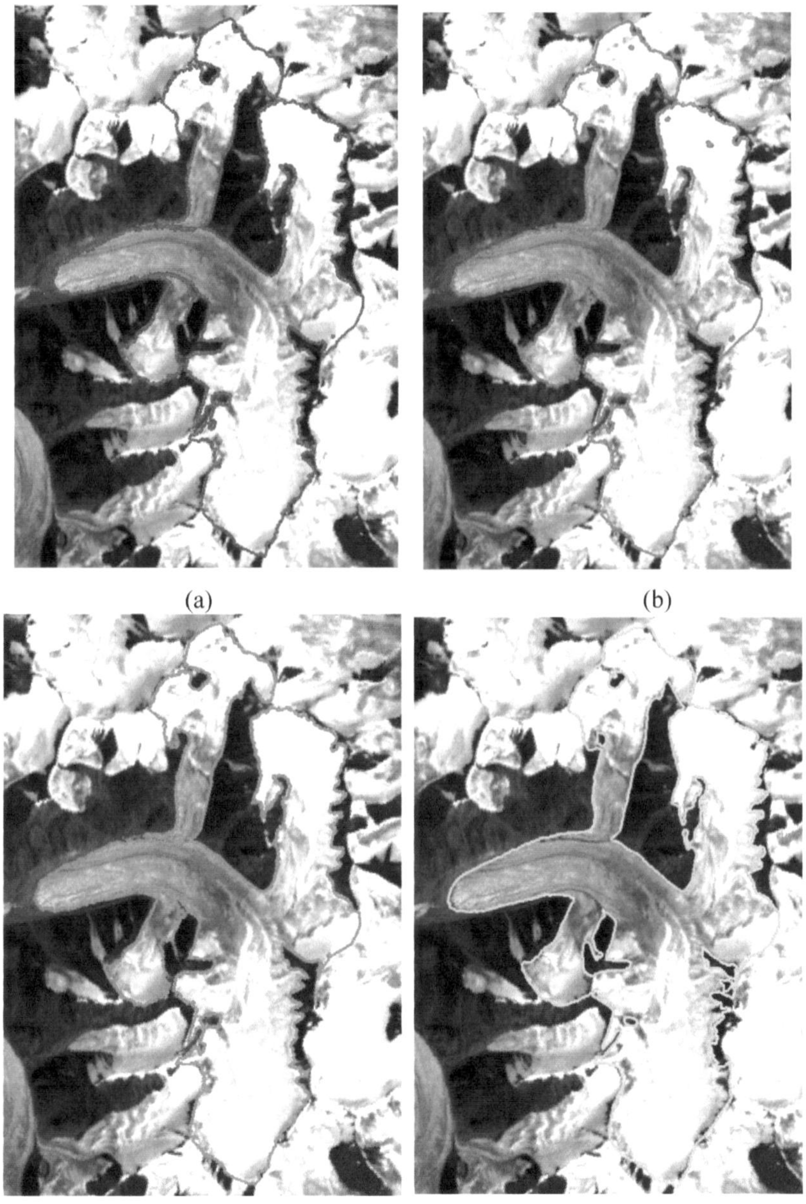

(a) (b) (c) (d)

Figure 3. Kowloon glacier extraction using: (a) Normalized Difference Snow Index (NDSI); (b) Band ratio (B5/B6); (c) Band ratio (B4/B6); (d) Artificial extraction.

Table 2. Glacier area extraction using different methods[a]

	NDSI	B5/B6	B4/B6	Artificial extraction
Sizagaer glacier /m²	7813252.79	7667881.39	7614462.60	8916504.43

Kowloon glacier / m^2	22095625.89	21864908.14	21840155.06	23807296.93

In addition, high resolution images from Earth Google are used to manually delineate the glacier boundary. And the glacier area extracted by these four methods was calculated in ARCGIS.

3.2 Discrimination of other objects

The test area contained abundant proglacial lakes; therefore, the results also allowed us to consider the automated classification of proglacial lakes and other surface features (Figure 4). The results show that the B5/B6 band ratio method is best at accurately distinguishing glaciers and pro-glacial lakes; however, snow remains indistinguishable from lakes.

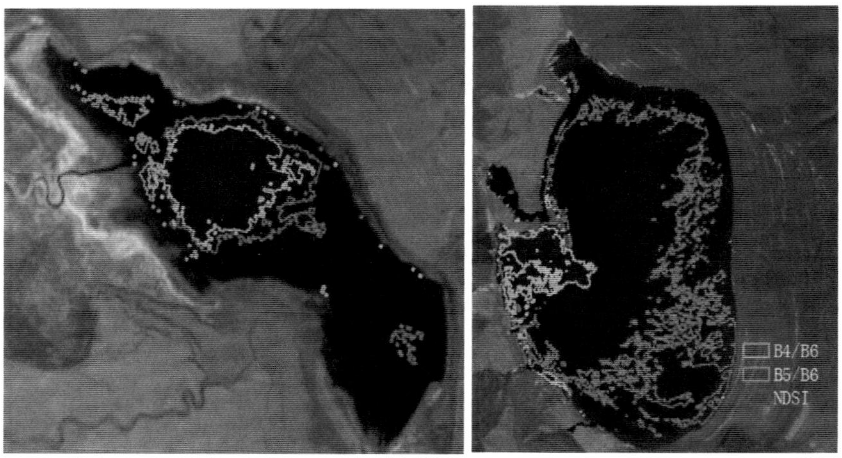

Figure 4. Delineation of lakes using different wave band operation. Blue denotes the B4/B6 band ratio, green denotes the B5/B6 band ratio, and red denotes the Normalized Difference Snow Index (NDSI).

4. Conclusions

In this study, we used four different methods to extract glacier boundaries. As the choice of threshold for each method affects the final result, we first followed the objective logic method to analyse the data and select the optimal threshold for each band operation. The results showed that the NDSI method extracts the smoothest glacial boundaries and provides the highest resolution images of small-scale glacier features; however, the method fails to distinguish glaciers and lakes. Using the band ratio method, extraction using the B5/B6 ratio was superior to that of the B4/B5 ratio; however, it also failed to completely eliminate the influence of lakes.

This study uses Landsat 8 to extract the characteristics of the glacier. Due to a single glacier types in the study area, for the glacier that have seasonal snow, turbid/frozen/multihued proglacial lakes and debris cover, how to improve the band operation for different types of glacier boundary extraction remains to be further research and analysis.

Acknowledgements

The authors would like to acknowledge the 13th Five Year Plan Project of Chinese academy of sciences: National Macro Decision-Making Space Information Support (Grant No. Y6SG0400CX and Y6SG040030), and the Key Laboratory of Tibetan Environment Changes and Land Surface Processes. CAS for the support of this research. We also thank Geospatial Data Cloud for providing Landsat data at no cost.

References

[1] Sidjak, R.,Wheate R. Glacier mapping of the Illecillewaet icefield, British Columbia, Canada, using Landsat TM and digital elevation data[J]. International Journal of Remote Sensing, 1999, **20(2)** 273-284.

[2] Paul F 2003 The new Swiss glacier inventory 2000 / Application of remote sensing and GIS. *Int. J. Remote Sens.* **14(17)** 3209–20

[3] Shangguan D, Liu S, Ding Y, Li J, Zhang Y, Ding L, Wang X, Xie C and Li G 2007 Glacier changes in the west Kunlun Shan from 1970 to 2001 derived from Landsat TM/ETM+ and Chinese glacier inventory data *Ann. Glaciol.* **46** 204–8

[4] Zhang G L, Pan B T, Wang J, et al. Research on the Glacier Change in the Gongga Mountain Based on Remote-Sensing and GPS from 1966 to 2008[J]. Journal of Glaciology & Geocryology, 2010, **32(3)** 454-460.

[5] Song B, Yuan-Qing H E, Pang H X, et al. Identifying Automatically the Debris-covered Glaciers in China's Monsoonal Temperate-Glacier Regions Based on Remote Sensing and GIS[J]. Journal of Glaciology & Geocryology, 2007.

[6] Auden J B 1937 The Snout of the Gangotri Glacier, Tehri Garhwal *Rec. Geol. Surv. India* **72** 135–40

[7] Ahlmann H W 1935 Contribution to the physics of glaciers. *Geog. J.* **86(2)** 97–113

[8] Benn D I and Evans D J A 1998 *Glaciers and Glaciation* (New York: Arnold)

[9] Benn D I and Warren C R 2000 Rapid growth of a supraglacial lake, Ngozumpa Glacier, Khumbu Himalaya, Nepal. *Debris-Covered Glaciers (Proceedings of a workshop held at Seattle, Washington)* 177–85

[10] Bhambri R and Bolch T 2009 Glacier mapping: a review with special reference to the Indian Himalayas *Prog. Phys. Geog.* **33**, 672–704

[11] Bhambri R, Bolch T and Chaujar R K 2011 Mapping of debris-covered glaciers inthe Garhwal Himalayas using ASTER DEMs and thermal data *Int. J. Remote Sens.* **32(23)** 8095–119

[12] Bolch T, Menounos B and Wheate R 2010 Landsat-based glacier inventory of western Canada, 1985–2005 *Remote Sens. Environ.* **114** 127–37

[13] Hewitt K 1998b Glaciers receive a surge of attention in the Karakoram Himalaya *EOS Transactions American Geophysical Union* **79(8)** 104–6

[14] Hewitt K 2004 Geomorphic hazards in mountain environments *Mountain Geomorphology* ed P N Owens and O Slaymaker (London: Arnold), pp 187–218

[15] Hewitt K 2005 The Karakoram anomaly? Glacier expansion and the ‘‘elevation effect,’’ Karakoram Himalaya *Mountain Research and Development* **25(4)** 332–40

[16] Kääb A, Paul F and Maisch M 2002 The new remote-sensing-derived Swiss glacier inventory: II. First results *Annals of Glaciology* **34** 362-6(5)

[17] Liu S, Shangguan D, Ding Y, Han H, Xie C, Zhang Y, Li J, Wang J and Li G 2006 Glacier changes during the past century in the Gangrigabu mountains, southeast Qinghai-Xizang (Tibetan) Plateau, China *Ann. Glaciol.* **43** 187–93

[18] Yao T (ed) 2007 *Map of the Glaciers and Lakes on the Tibetan Plateau and Adjoining Regions* (Xi’an: Xi’an Cartographic Publishing House)

[19] Zhang X 1984 Recent variations of some glaciers in the Karakoram Mountains *The International Karakoram Project 1* ed K J Miller (Cambridge: Cambridge University Press) pp 39–50

Analysis of dynamic change of Hongjiannao Lake based on MNDWI

Hui Yue, Ying Liu, Haoren Wang and Wei Zhang

College of Geomatics, Xi'an University of Science and Technology, Xi'an 710054, Shaanxi, China

E-mail: 13720559861@163.com

Abstract. In order to understand recent changes in inter-annual trends in water area of Hongjiannao Lake and to explore influence factors affecting the changes, the Landsat data during the year from 1988 to 2014 was processed by Modified Normalized Difference Water Index (MNDWI) and the water body boundary of Hongjiannao Lake was extracted. Then the water area was calculated in ARCGIS, and also the variation was analysed. The results show that the water area of Hongjiannao Lake is overall decreasing from 1988 to 2014. The possible causes of its shrinking by preliminary analysis are mainly continuous underground mining, climate drought, upstream reservoirs and so on.

1.Introduction

Lake area and water fluctuation in China had a shrinking trend over the past decades, the decrease of the lake area not only results in wetland shrinking in the lake district, but also damage the stability of the river's lake aquatic ecosystems [1]. Lake water in arid and semi-arid area is a vital recourse and plays an important role in ecological balance. Some studies pointed out that some lakes have shrunk or vanished in North China and reaches of the Yangtze River have lowered to be driven by climate variability and agricultural irrigation [2]. Therefore, it is necessary to extract water bodies from non-water features and evaluate the spatio-temporal distributions and changes of land surface water [3-4]. Remote sensing is an effective approach in monitoring water resources because of providing large-scale, long-term, repeated and real-time information [5-7]. McFeeters [8] noted that high resolution Quick Bird imagery combined with Normalized Difference Water Index (NDWI) can extract swimming pools for mosquito abatement. Xu [9] replace the NIR band with the middle infrared (MIR) band, and then establish the Modified NDWI (MNDWI). The open water features can be enhanced using this index, and the built-up land noise, as well as vegetation and soil noise, can be suppressed and even removed efficiently. This paper used MDNWI to extract Hongjiannao Lake area from the year 1988 to 2014 based on Landsat and HJ-CCD data.

2.Study area and data

2.1 Study area

Hongjiannao Lake is the largest desert freshwater lake in semi-arid and arid region of China, which played an important role in adjusting the local climate, water resource balance and biological diversity. It located at the border of Shaanxi Province and Inner Mongolia Autonomous Region, centered at (110°52′30″E, 39°06′N). The position in the south-west of Shendong mining area, which was one of China's biggest coal-producing regions (Fig.1). The lake-water area was 61.93 km^2 in 1978. However, some studies found that the area of Hongjiannao Lake decreased rapidly in the past decades, influenced the ecological and economic functions of Hongjiannao Lake, many possible reasons exist for the rapid decrease in lake-water area. These reasons include that climate change, human activities and especially the underground mining activities. Therefore, it is necessary to obtain the accuracy area and monitoring the dynamic change.

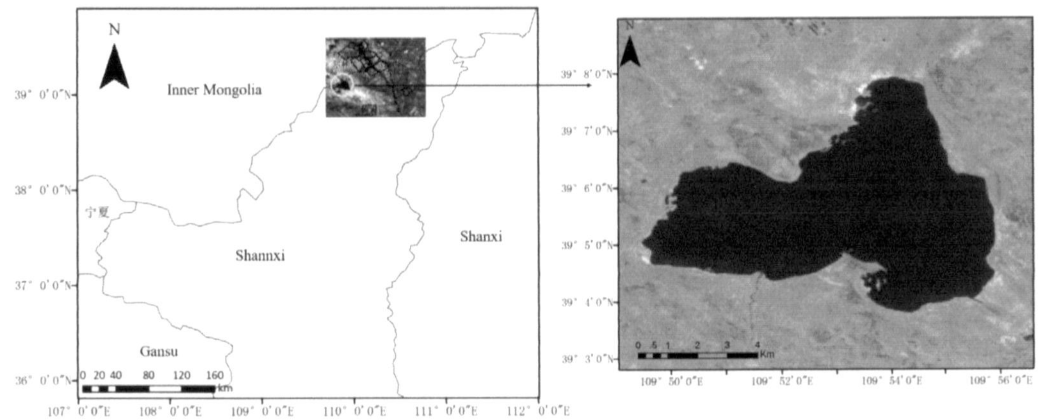

Figure 1. Location of Hongjiannao Lake.

2.2 Data

Selected 1988-2014 Landsat images which containing Green and MIR bands that used to extract water information to reduce the impact of coarse resolution. The cumulative 26 years' total of 23 scene satellite remote sensing data which listed on table 1, including 21 scene data from Landsat satellites and 2 scene HJ-CCD satellite data will be used to supplement. Data acquisition time are mainly concentrated in cloudless summer, but as a result of Landsat and HJ satellite revisit cycle is 16 days and 2 days respectively, so it is hard to get the same remote sensing images. The Landsat data download mainly from International Scientific Data Service Platform, the University of Maryland Global Land Cover Database Exchange Sites, the US Geological Survey and Chinese Academy of Sciences Institute of Earth Observation Data Sharing Plan; HJ-CCD data download in China Resources Satellite Application Centre.

Table 1. Remote Sensing Data using in paper

Sensor	Time	Sensor	Time	Sensor	Time
TM	1988.9.24	TM	1998.7.2	TM	2006.9.10
TM	1989.9.11	ETM+	1999.10.17	TM	2007.8.12
TM	1990.8.29	ETM+	2000.7.31	TM	2009.6.30
TM	1992.7.17	ETM+	2001.5.31	HJ1A-CCD2	2010.8.5
TM	1993.5.17	ETM+	2002.8.6	HJ1B-CCD2	2011.7.30
TM	1994.8.24	ETM+	2003.5.21	OLI	2013.9.13
TM	1995.6.8	TM	2004.7.2	OLI	2014.7.30
TM	1996.6.10	TM	2005.7.5		

3.Method

The Normalized Difference Water Index (NDWI) from Mcfeeters, which uses MIR (TM5) instead of NIR (TM 4) to construct the MNDWI. The MNDWI has been tested in the ocean, lake and river areas with the background of built-up lands and vegetated lands, and with both clean and polluted water bodies using Landsat TM/ETM+ imagery. Land categories of Hongjiannao lake region can be roughly divided into water, vegetation, bare land and buildings, etc., use the NDWI index model is utilized to extract water body, bare land and buildings in the background will be wrong fall into water body, reach satisfactory result of extraction. The MNDWI can depress the built-up land information effectively while high lighting water information, and accurately extract the water body information from the study areas. This reveals that the MNDWI can significantly enhance the water information, especially in the area mainly with built-up and bare land as background. In the end, transform to extract the waters diagram for vector data, using the ArcMap field calculator waters polygon area calculation to get the lake area. The formula of NDWI and MNDWI showed as below:

$$NDWI = (Green - NIR)/ (Green + NIR) \quad (1)$$

$$MNDWI = (Green - MIR)/ (Green + MIR) \quad (2)$$

Where Green, NIR and MIR are the surface reflectance values of the green, NIR and MIR bands, respectively.

4.Results

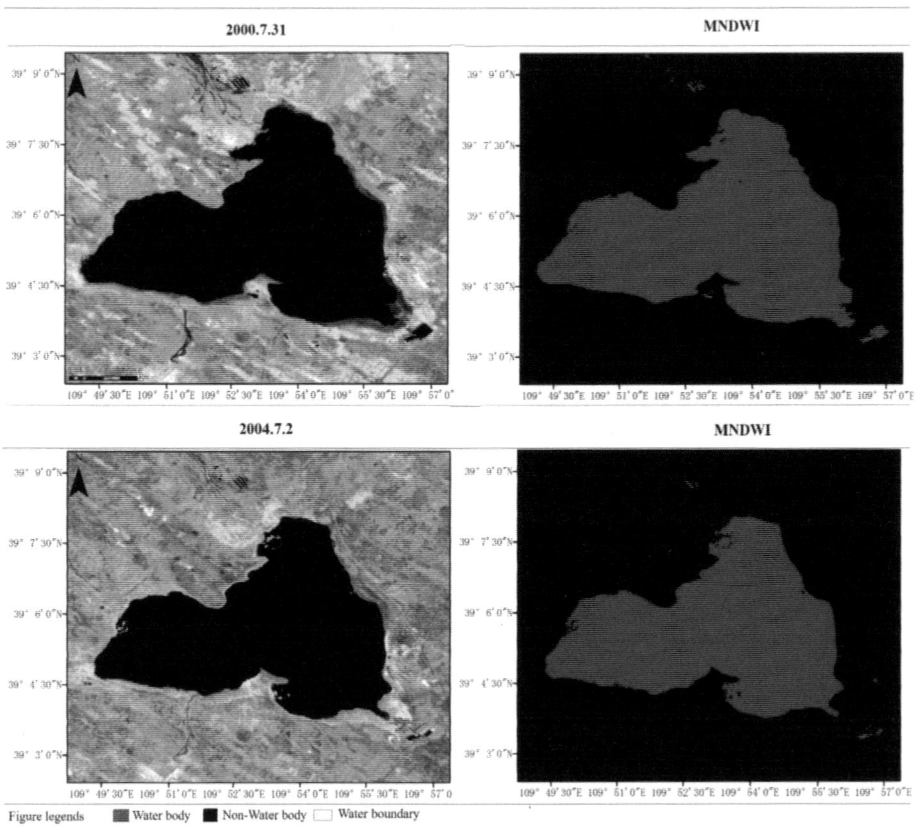

Figure 2. Map of water body extraction using MNDWI

MNDWI can extracted Hongjiannao lake boundary accurately, when extracting water body around the Hongjiannao lake (e.g., above the Hongjiannao lake area and bottom right of the lake), water boundary was showed clearly in Figure 2. Using the Hongjiannao measured lakes area of 2000 to verify the accuracy of water extraction method, MNDWI error was 2.07%. MNDWI was utilized to extract the Hongjiannao lake area in each year which showed in Figure 3, the lake area was generally a linear decrease trend (R^2=0.86), but its change trend can be subdivided into two stages: from the year of 1988 to 1999, the area of the Hongjiannao lake area showed a trend of fluctuations, the whole change is not obvious, but the partial variation is sharp; since 1999, the Hongjiannao lake area was a linear decreasing trend. In the year of 1988 and 1999, the lake area was 49.92 Km^2 and 47.92 Km^2 respectively, lake area decreased by 4.02%. In the year of 2009 and 2014, the lake area was reduced to 37.71 Km^2 and 31.69 Km^2 respectively, compared with the year of 1988 lakes area decreased by

24.45% and 24.45% respectively. Overall, Hongjiannao lake area is shrinking in the past 26 years, the average reducing speed was 0.87 Km^2/a.

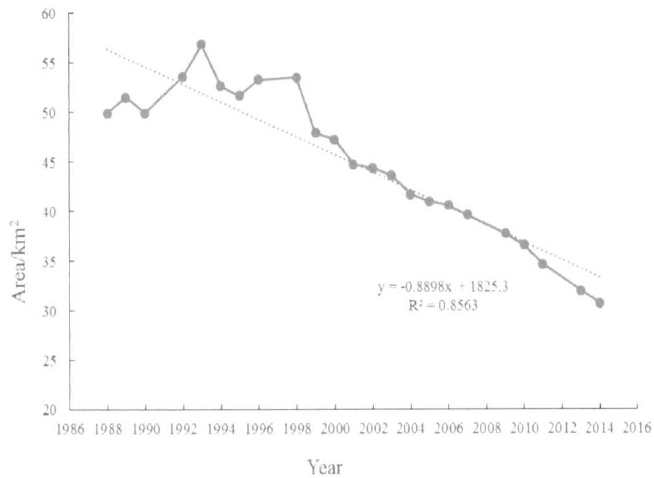

Figure 3. The area change of Hongjiannao Lake from 1988 to 2014

5.Discussion

As for the causes of the Hongjiannao lake atrophy, there are three main categories drought climate, upstream river closure and irrigation water consumption. In particularly, Hongjiannao Lake was 20-30 km away from Shendong core mining area. With the strength of the mining and mining method of mechanized, and the increase of mining depth, the shallow groundwater water bottom continuity is damaged which lead to changes in the local groundwater flow field, it would also affect the level and area of Hongjiannao lake. Meteorological factors such as rainfall and evaporation may not the main factors leading to the reduction of Lake area, irrigation, underground mining and reservoir construction may be the main reasons, all of these reasons need to be further investigated in the future research.

6.References

[1] Bai J, Chen X, Li J L 2011 *Environ Monit Assess.***178** 247-256.
[2] Ma R H, Duan H T, Hu C M, Feng X Z, Li A N, Ju W M, Jiang J H, Yang G S 2010 *Geophys. Res. Lett.* 2010 **37** L24106.
[3] Wang M, Shi W, Tang J 2011 *Remote Sens. Environ.* **115** 841-854.
[4] Klein I, Dietz A J, Gessner U 2014 *Int. J. Appl. Earth. OBS.* **26** 335–349.
[5] Sun F, Sun W, Chen J, Gong P 2012 *Int. J. Remote Sens.***33** 6854-6875.
[6] Ma B D, Wu L X, Zhang X X, Li X C, Liu Y, Wang S L 2014 *Int. J. Appl. Earth. OBS.* **33**109-118.
[7] McFeeters S K 2013 *Remote Sens.***5** 3544-3561.
[8] McFeeters S K 1996 *Int. J. Remote Sens.* **17** 1425-1432.
[9] Xu. H Q 2006 *Int. J. Remote Sens.* **27** 3025-3033.

Acknowledgements
This work was jointly supported by The National Natural Science Foundation of China (Grant No. 41401496), China Postdoctoral Science Foundation Funded Project (Grant No. 2016M592815) and Student's Platform for Innovation and Entrepreneurship Training Program (201610704120).

Ice flood velocity calculating approach based on single view metrology

X Wu[1]and L Xu[2]

1.China Railway SIYUAN Survey & Design Group Co. Ltd, Hubei Wuhan, 430063, China;

2. Yellow River Institute of Hydraulic Research, Yellow River Conservancy Commission, Henan Zhengzhou, 450003, China

Abstract: Yellow River is the river in which the ice flood occurs most frequently in China, hence, the Ice flood forecasting has great significance for the river flood prevention work. In various ice flood forecast models, the flow velocity is one of the most important parameters. In spite of the great significance of the flow velocity, its acquisition heavily relies on manual observation or deriving from empirical formula. In recent years, with the high development of video surveillance technology and wireless transmission network, the Yellow River Conservancy Commission set up the ice situation monitoring system, in which live videos can be transmitted to the monitoring center through 3G mobile networks. In this paper, an approach to get the ice velocity based on single view metrology and motion tracking technique using monitoring videos as input data is proposed. First of all, River way can be approximated as a plane. On this condition, we analyze the geometry relevance between the object side and the image side. Besides, we present the principle to measure length in object side from image. Secondly, we use LK optical flow which support pyramid data to track the ice in motion. Combining the result of camera calibration and single view metrology, we propose a flow to calculate the real velocity of ice flood. At last we realize a prototype system by programming and use it to test the reliability and rationality of the whole solution.

1 INTRODUCTION

Yellow River, which is the second longest river in China and the fourth longest in the world, is 3,395 km long. Originating in the Bayankala Mountains in Qinghai Province in eastern Tibet, it flows through nine provinces of China and empties into the Bohai Sea. The Yellow River basin has an east-west distance of 1900 km, and north-south distance of 1100 km. Total basin area is 752443 km².

Yellow River is also the river where the ice flood occurs most frequently in China, thus, the ice flood forecasting has great significance for the river flood prevention work. In various ice flood forecast models, the flow velocity is one of the most important parameters. But its acquisition heavily relies on manual observation or deriving from empirical formula. In recent years, with the high development of video surveillance technology and wireless transmission network, the Yellow River Conservancy Commission set up the ice situation monitoring system, in which live videos can be transmitted to the monitoring center through 3G mobile networks.

In this paper, an approach to get the ice velocity based on single view metrology and motion tracking technique using monitoring videos as input data is proposed. First of all, River way can be approximated as a plane. On this condition, we analyzed the geometry relevance between the object side and the image side and presented the principle to measure length in object side from image. Secondly, we use LK optical flow which support pyramid data to track the ice in motion. Combining the result of camera calibration and single view metrology, we propose a flow to calculate the real velocity of ice flood. At last we realized a prototype system by programming and used it to test the reliability and rationality of the whole solution.

2 METHODS

2.1 Principle of single view measurement and motion detection

According to the perspective projection model, one point $M_i(X_i, Y_i, Z_i, 1)$ in space can be projected to a two-dimensional image plane through a projection matrix P. We call the projected point $m_i(u_i, v_i, 1)$. The geometric relationship between the two points is:

$$\lambda m_i = PM_i = (p1 \quad p2 \quad p3 \quad p4)M_i \qquad \text{Formula 2-1}$$

λ is a non-zero proportionality factor.

When $M_i(X_i, Y_i, Z_i, 1)$ becomes a point on the plane W and set the plane W to $Z_w = 0$, the Formula 2-1 becomes:

$$\lambda \begin{pmatrix} u_i \\ v_i \\ 1 \end{pmatrix} = P \begin{pmatrix} X_i \\ Y_i \\ 0 \\ 1 \end{pmatrix} = (p_1 \quad p_2 \quad p_3) \begin{pmatrix} X_i \\ Y_i \\ 1 \end{pmatrix} \qquad \text{Formula 2-2}$$

$H = (p_1 \quad p_2 \quad p_3)$ is called homography matrix.

When the homography matrix H is known, any point on the image can be projected to the object plane. Thus we can calculate the distance and the angle between two points in the object plane. This is the basic principle of the single view measurement[1].

In the context of this application, motion detection in the video is undoubtedly an important issue on which target tracking, target classification, target behavior analysis and many other applications depend. Consequently, an accurate, robust, fast and effective target detection method has been the goal pursued by the computer vision researchers. Although, a variety of detecting algorithm for different applications have been proposed, due to the complexity and variety of applications, it is difficult to have a universal algorithm. At present, background subtraction method, inter frame difference method and optical flow method[2] are relatively mature methods that are commonly used.

The instantaneous motion projected by spatial moving object on the image plane is called the optical flow. Visual psychology identify that human beings perceive the movement and structural information of the moving object by observing the optical characteristics on the object surface. Similarly, when a camera is used to photograph a target with relative motion, the motion of the brightness characteristic is called optical flow. In other words, when the moving object is projected to the image plane, the optical flow is formed.

Optical flow intuitive performance target motion states and processes, in turn, it can be determined by the optical flow motion targets. In computer vision, optical flow has vital research value. Based on it, segmentation, identification, tracking, and many other tasks can be done[3].

Lucas-Kanade algorithm[4] (also known as the LK algorithm) is used to calculate dense optical flow at the beginning. However, since the algorithm can be easily applied to a set of points from input image, it subsequently became a crucial method for seeking sparse optical flow. In contrast, LK algorithm only requires partial information around the points of interest surrounding a small window, thus it can be applied to sparse content.

There is a fatal flaw in basic LK optical flow method. That is the presence of incoherent motion. We need a large window to capture its movement, but this will be in conflict with the assumptions in the LK algorithm. We can use the image pyramid, however, to deal with this contradiction. The function of the approach is to track on a larger spatial scale initially. After that, it can be used at the next level as staring point and we will repeat the process until it reaches the bottom of the pyramid. With this strategy, we can reduce the possibility of movement which does not satisfy the motion assumptions and achieve faster and longer motion tracking as well.

2.2 Solution

In general, when choosing detection algorithm, the background of the moving object and the motion characteristics of the target should be taken into consideration. Depending on the operating mode of the camera, the moving object can be divided into background static background and dynamic background. Motion characteristics mainly refer to the color, size, shape, quantity, stiffness and other characteristics of the moving object itself.

The background subtraction method is expedient for the application of fixed or invariant background. Since the moving object background in this paper is the changeable river water, the method is not suitable. The main advantages of inter-frame difference method are: simple algorithm, program of low complexity and suitable for real-time monitoring. However, feature point set obtained by this method is not the most appropriate set of tracked feature points. As the optical flow method does not need to predict any motion information in advance, it is applicable for the application in this article.

Considering the motion tracking, the size, shape and quantity of the target to be tracked can be changed at any time. Under the impact of water, ice tends to be gathered and dispersed frequently.

Tracking method based on active contour cannot be applied to this situation in which the shape of target changes frequently. Besides, method based on the regional and deformable template tracking requires the use of a template or model in advance. That is extremely difficult for real-time application.

Considering that what we concern is the overall speed of the target in the scene and we do not treat ice blocks as individuals, tracking method based on feature points is the most appropriate.

We propose an ice flood velocity calculation process shown in Figure 2-1:

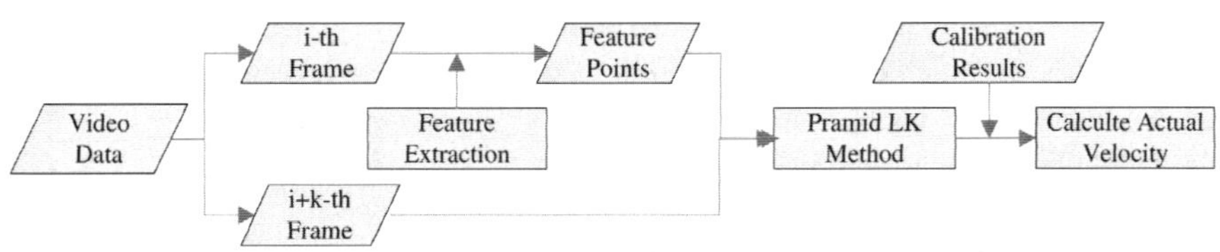

Figure 2-1 Velocity calculation flow

The specific calculation steps are:

(1) Get the video frame rate(FPS, unit: frames / sec);

(2) Get the i-th frame image and the i+k-th frame images from video, k can be understood as a video object sampling interval;

(3) Feature extraction from $f_i(x, y)$, get the feature point set, recorded as P_i. We used the feature points which are easy to be tracked by the strong corner point proposed by Shi[5] as feature points;

(4) Set P_i from $f_i(x, y)$ as the input feature points to be tracked. Use LK optical flow supporting pyramid to predict the position of P_i in $f_{i+k}(x, y)$, recorded as D_i;

(5) Calculate the actual distance between a pair of feature points from P_i and D_i by using the single view measurement principle. Record as S_{ij};

(6) Calculate the average displacement of the points in the P_i. Record the point set which have been successfully tracked in P_i as Q_i. Assume Q_i's size is n, then the average displacement of a frame can be calculate by Formula 2-3;

$$S_i = \frac{\sum_{j \in Q_i} s_{ij}}{n} \qquad \text{Formula 2-3}$$

(7) Calculate the time difference between $f_i(x, y)$ and $f_{i+k}(x, y)$ by Formula 2-4;

$$t_i = \frac{60 \times k}{FPS} \qquad \text{Formula 2-4}$$

(8) Calculate the ice flood velocity V_i by Formula 2-5;

$$V_i = \frac{S_i}{t_i} \qquad \text{Formula 2-5}$$

(9) Calculate the average speed of the flow in a period of time by Formula 2-6;

$$V = \frac{\sum_{i=1}^{m} V_i}{m} \qquad \text{Formula 2-6}$$

(10) Display the tracking results. Mark the racking results on the next frame by arrows with direction.

3 EXPERIMENT AND ANALYSIS

3.1 Experimental background

In order to verify the feasibility and reliability of the our method, we conduct experiment at ten observation stations(Labawan, Toudaoguai, Fuan-brige, Caojiawan, Dengkou, Fujiagedu, Huajiangyinzi, Sanhuhekou, Bayangaole, Sanchakou)in the Yellow River basin in February 2013. Laptop, sign board, Interphone, diving equipment, safety rope and other equipment are used in the experiment.

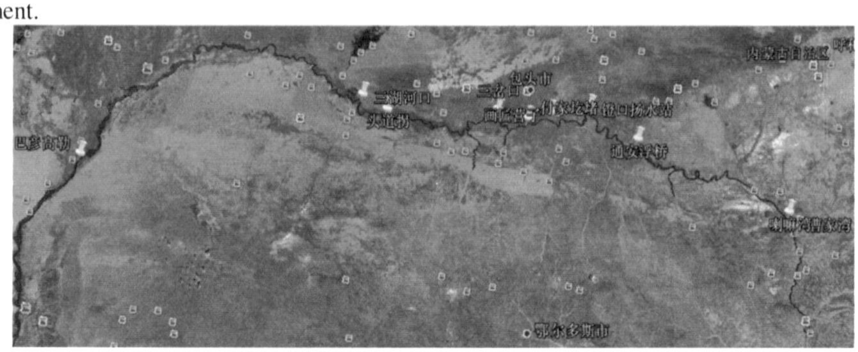

Figure 3-1 Experimental location map

3.2 Design of sign board

In the process of ice velocity calculation, we need to obtain the longitudinal information on the river plane using the single view measurement method. The key to the realization of the single view measurement is to acquire the homography martix between the reference plane and the image plane. The calculation of the matrix is based on point correspondence or line correspondence.

No known geometrical information can be utilized in the river channel, thus the camera calibration can be carried out by using the self-made sign board. In the designing process of the sign board, the following points must be taken into account:

(1) Data source is from the existing ice monitoring system. Due to the restriction of 3G transmission cost, the original video data resolution is very low (352*480), thus sign board should not be too small;

(2) To be placed on the river ice, the sign board must be waterproof. Taking into account the possible wind conditions in the field work environment, the sign plate should also be tough;

(3) Taking into account the site may be obscured, reflective, the feature points on the flag board

must be redundant.

In this paper, the sign board we design as shown in Figure 3-2. It is made by waterproof materials, light, tough and easy to take along. Sign board has four 1 meter wide and 0.5 meters long rectangle, a total of 12 easily identified critical corners, the shortest distance between them is 0.5 meters.

Figure 3-2 The sign board

3.3 Estimation of accuracy

In order to estimate the accuracy of the measurement, the following experiments are designed. The image was captured from the Sanhuhekou observation station, as shown in Figure 3-3. Use the sign board to calculate the homography matrix and calculate the length of the L1-L8 which actual length are 50cm. Obtain the results as shown in Table 3-1

Figure 3.3 Experimental field.

Table 3-1 Length measurement results

Measurement Target	L1	L2	L3	L4	L5	L6	L7	L8
Calculated Length	46.82	45.46	47.59	47.2	43.02	45.62	43.58	45.52
Relative Error	0.07	0.10	0.05	0.06	0.16	0.10	0.15	0.10

As can be seen from table 3-1, the overall relative error is higher than that of indoor experiments, reaching 9%. From figure 3-4 we can find, deviation of measurement results of L5-L8 are even greater and the relative error is significantly higher than L1-L4. If the coplanar condition is kept better, the error should be close to the average relative error of L1-L4, which is about 6%.

3.4 Experiment and analysis

Using the break-up period of ice flow video as experimental data, we realize a prototype system by programming and use it to test the reliability and rationality of the whole solution. In some video data, the ice flow is not obvious, but our program has also calculated the results.

The direction of motion of the feature points which have been tracked have been displayed on the second frame by arrows. Thus we can judge the tracking results are correct. The tracking results of Caojiawan,Labawan and Sanhuhekou as shown in Figure 3-4 to Figure 3-6.

Figure 3-4 Tracking results at Caojiawan

Figure 3-5 Tracking results at Labawan

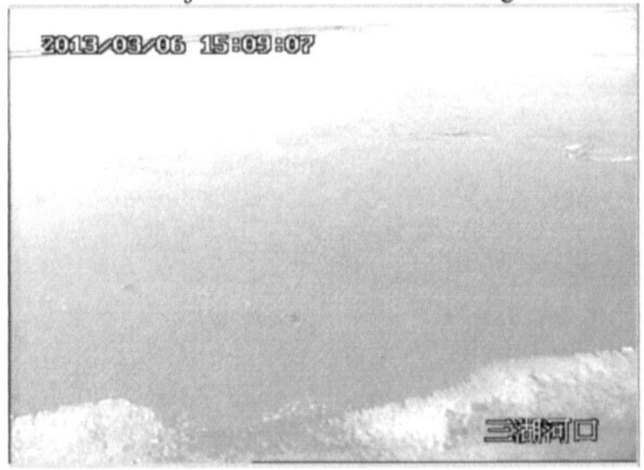

Figure 3-6 Tracking results at Sanhuhekou

In Figure 3-8, we recorded the speed of ice by setting the sampling interval to 2 seconds.

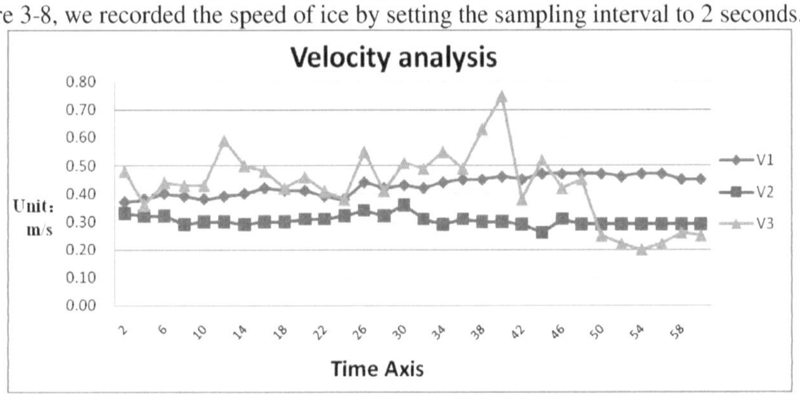

Figure 3-7 Velocity analysis

From the figure above we can find the calculated results from Caojiawan and Labawan are stable and reliable while the calculated results from Sanhuhekou has frequent volatility. Through the analysis of the video, we think the limited number of ice targets lead to this result.

4 CONCLUSION

In this paper, an approach to get the ice velocity based on single view metrology and motion tracking technique using monitoring videos as input data is proposed. Specifically, the following work has been done:

1 The characteristics of flowing ice block have been analyzed in this paper. Considering its features, we identify the LK optical flow method using strong corner feature points as tracking targets is the most suitable method.

2 This paper presents a field camera calibration method based on sign board. Relying on the calibration results, we can calculate the actual velocity of ice block's displacement in the video.

3 We realize a prototype system by programming and use it to test the reliability and rationality of the whole solution.

IOP Conf. Series: Earth and Environmental Science **57** (2017) 012006 doi:10.1088/1755-1315/57/1/012006

REFERENCES

[1] Criminisi A 2001 *Single-view metrology* (London: Springer)p 69-105

[2] Elgammal A, Duraiswami R, Harwood D, et al. Background and foreground modeling using nonparametric kernel density estimation for visual surveillance. *Proc IEEE.* 90(7):1151-1163

[3] Simoncelli E P, Adelson E H, Heeger D J 1991 *Computer Society Conf. on Computer Vision & Pattern Recognition (Hawaii)* (Probability distributions of optical flow) p 310-315

[4] Lucas B D，Kanade T．An 1981 *Proceedings of the 7th int. joint conf.on Artificial intelligence (Seattle)* (iterative image registration technique with an application to stereo vision) p 443-450

[5] Shi J，Tomasi C．Good 1994 *Computer Society Conf. on Computer Vision and Pattern Recognition (Vancouver)* (features to track) p 593-600

Long-term change of protective forest during the past four decades at Dongshan Island, southeastern China

Caiyun Zhang[1, 2,3*], **Jing Yan**[1,3], **Shaoling Shang**[1,3]

[1] State Key Laboratory of Marine Environmental Science, Xiamen University, Xiamen, China, 361102
[2] Key Laboratory of the Ministry of Education for Coastal and Wetland Ecosystems, Xiamen University, Xiamen, China, 361102
[3] Key Laboratory of Underwater Acoustic Communication and Marine Information Technology Ministry of Education, Xiamen University, Xiamen, China, 361102

*Corresponding e-mail: cyzhang@xmu.edu.cn

Abstract The Dongshan Island is a typical protective forest area of the southeastern coastal region of China. In this work, we extract the variation of the Dongshan Island's protective forest area since the 1970s from the remote sensing data of the LandSAT and the Chinese HJ 1A/B satellites, and we examined the possible reasons for the variation. The results showed that the maximum likelihood classification of the extracted remote sensing data of the Dongshan Island coastal protective forest had a total accuracy of 92.6%. In the last 40 years, the area of the Dongshan Island coastal protective forest experienced a waving variation of decrease (1973-1984), increase (1984-1999) and decrease (1999-2008), in which human activity was the predominant impact factor for this fluctuation pattern. Since the early 1960s to the 1980s, due to the Cultural Revolution and the later deforestation and land reclamation activities, the protective forest of Dongshan Island was severely damaged. The total area declined sharply from 2134.4 ha in 1964 to 1515.74 ha in 1984, and the northern area of the island had the most significant decrease. After 1988, the Fujian Province began to aggressively establish the coastal protective forest system, and the protective forest acreage increased significantly to 3370.22 ha in 1999, an increase of nearly 1.2 times based on the figure of 1984. In the past five years, the area of protective forest declined slightly again, mainly because of the natural aging of the protective forest ecosystem, tourism development and the booming aquaculture industry.

1. Introduction

The coastal protective forest is an essential safeguard of the coastal environment that promotes the sustainable economic development of the coastal region, due to its important effects in sand-fixing, soil and water conservation, typhoon resistance, amelioration of the living environment, maintaining biodiversity, microclimate improvement, etc. (Beese & Bryant, 1999; An et al., 2006; Erefur et al., 2008; Borsie et al., 2011). Under the stress of natural forcing and human activity, how were the protective forests changing globally? Acquiring the answer of this question would certainly help people predict the potential changes in the future and seek strategies for sustainable development. According to the analysis data provided by FAO 2010(FAO,2010), the total extent of forests which are designated for protection of soil and water was estimated to be 330 million hectares, increased by 59 million hectares between 1990 and 2010 and equivalent to eight percent of the global forest area. Asia has the highest proportion of forests with a protective function (26%), followed by Europe(9%). The recent increase is primarily due to large scale planting in China for protective purposes.

China is undergoing very rapid economic development, whereas it has a long coast line (~18,340 km for mainland and ~11,159km for island). Although the Chinese government decided in 1988 to construct the coastal protective forest system covering the entire coast line (Li, 2004), this system faces enormous challenges where forest destruction is not effectively controlled under the booming economic development. The infrastructure and industrial land use has rapidly increased since 2000, and forest land is in fact constantly encroached or illegally occupied by real estate development, tourism and resort construction, farming, quarrying, sand and earth excavation, etc. Consequently, the protective forest continuously declined in acreage and deteriorated in quality. Details are yet clear without a valid way to do the survey.

Here in this study, we took a small island, Dongshan, located in the southeastern China, an area subject to typhoon and storm surge hazard, as a case for assessment of protective forest change (figure 1). So far few relevant reports are available to make quantitative analysis of the areage change of protective forest in this region. One example is that the calculation showed the total Fujian coastal protective forest area was about 336,000ha in 2006 which was based on 10 MODIS images solely (Zhang et al., 2007). We used both in situ and remote sensing data, covering the time span of the 1970s-2000s. Our objective is to demonstrate how protective forests change in a typical coastal area of China during the past four decades.

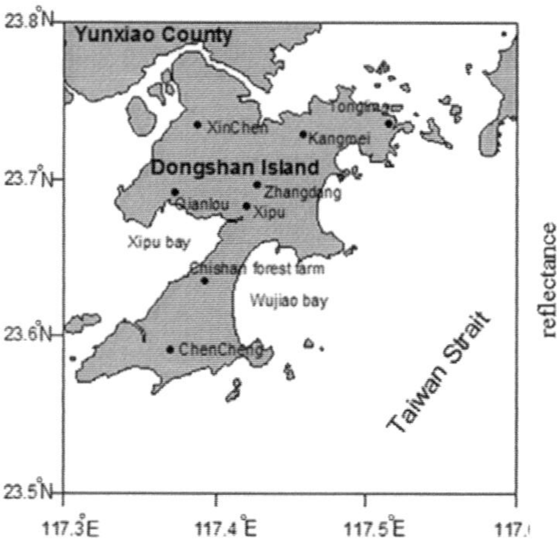

Figure 1. The map of Dongshan Island at Fujian province

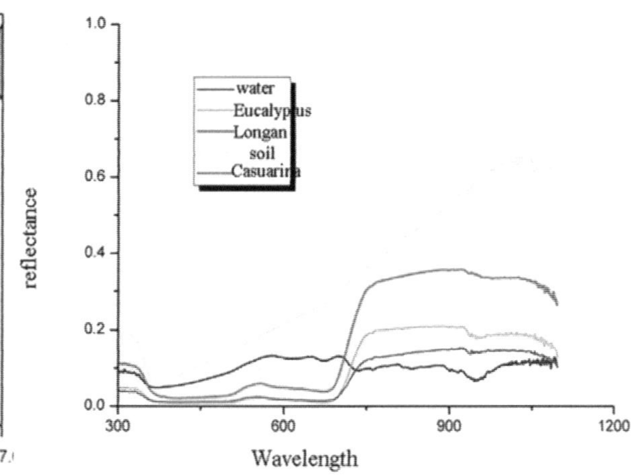

Figure 2. Spectral curve of the major tree species and other land feature in the Dongshan Island

2. Detection of Protective Forest

2.1. In situ measurements of reflectance spectrum

We recorded the spectra of the major land features of the Dongshan Island by using a GER 1500 portable spectroradiometer (Spectra Vista Corporation) covering the UV,Visibale and NIR wavelengths from 350nm to 1050nm. All reflectance were measured outdoors under natural light without cloud during 8am to 3pm. the selected sample regions are required to be homogeneous. The spectroradiometer located at the top of the target and had the angel of 45° with vertical slope. Same features were measured eight times and averaged as representative of feature spectrum. For the trees such as *Casuarina, Eucalyptus* and *Dimocarpus*, because we could not record the spectrum directly on the canopy, we picked a certain amount of the twigs and laid them on the ground with leaves facing up and with no gap, ensured their surroundings were open and unblocked, and then recorded the spectrum of these quasi-leaves. The measurements at each sample point were completed in 5–10 minutes.

It can be seen from the measured spectra (figure 2) that vegetations such as *Casuarina, Eucalyptus, Dimocarpus* have distinct spectral curves and are clearly discernable from water or bare soil. Between different vegetation, the reflectance of *Casuarina* is similar to *Eucalyptus* and *Dimocarpus* in the visible band but significantly lower in the near-infrared band.

2.2 Classification of targets

Satellite data from the U.S. Land Remote Sensing Satellite (LANDSAT) and the Chinese small satellites for environment and disaster monitoring and forecasting (HJ-1A/B) were used to assess the distribution of protective forests. Details of the data are shown in Table 1.

ENVI 4.5 software (ITT Visual Information Solutions Corporation) was used for the data processing. The atmospheric correction was realized through the built-in FLASSH atmospheric correction module in ENVI. According to the local latitude and the season of acquired images, Tropical Regional Model was selected from 6 provided atmospheric models. The aerosol type is for rural area. The geometric registration between images used the TM images of September 25, 2006 as the benchmark image. After the geometric precision correction and the polynomial correction using ground control points (GCP), the corrected benchmark image was used for adjusting other images to control the overall error within less than one pixel. For the LANDSAT MSS data of 79 m resolution, during registration we used the nearest neighbor method to reset their spatial resolution to 30 m.

The in situ measured spectrum data (figure 2) demonstrated that different targets showed different colors on satellite RGB composites. It is thus straightforward to identify seven targets basically based on differences in color, including: (1) protective forest (Casuarina, mixed species forest, etc.), (2) orchards and farmland with high vegetation cover, (3) bare soil and farmland with low vegetation cover, (4) land for urban use, (5) dams, (6) salterns and aquaculture farms, (7) tidal zone and water. Meanwhile, the normalized difference vegetation index (NDVI) was also added as one judgment for classification. The NDVI is calculated as NDVI=(NIR-VIS)/(NIR+VIS), where NIR is the near infrared band and VIS is the visible band. Finally, a classic supervised classification method-the maximum likelihood classification method (Richards,1999) was used, with the seven target feature and NDVI as input, to classify different targets on each image. Results are shown in figure 3.

Table 1. Description of satellite data used for classification

Sensor	Date	Resolution
LANDSAT MSS	October 30, 1973; October 30, 1984	79 m
LANDSAT TM	September 30, 1999; September 25, 2006	30 m
HJ-1B	November 14, 2008	30m

Note that visual examination is necessary since mistakes may occur during classification as difference land features might give identical spectrum and the same land feature might generate divergent spectrum. In addition, although the protective forest in Dongshan is dominated by Casuarina, there might be other species such as Dimocarpus mixing in the forest.

3. Validation

We take the November 14, 2008 image extracted from the environmental satellites as an example to demonstrate the accuracy assessment of our classification. The test image of the assessment mainly used the February 28, 2008 QuikBird image of 2.44 m resolution, and the area of comparison is mainly in the northeast Dongshan Island. According to the confusion matrix, the classification is overall quite satisfactory with a total accuracy of about 92.6% and a Kappa coefficient of 0.90. The classification of water bodies had the best accuracy of 100% and that of the protective forest was about 88.3%. Because the classified remote sensing images had a low spatial resolution of only 30 m, there might have been some mixed pixel that negatively affected the accuracy in information extraction. Nevertheless, we further compared the protect forest area derived from remote sensing image classification with the statistics from the Zhangzhou Forestry Bureau (figure 4). It can be seen that in 1999, 2006 and 2008 when both data are available, the data extracted from remote sensing agree well with the statistics. We thus conclude that our classification is reliable and can be used for analyzing the variation of the protective forest in the Dongshan Island since the 1970s.

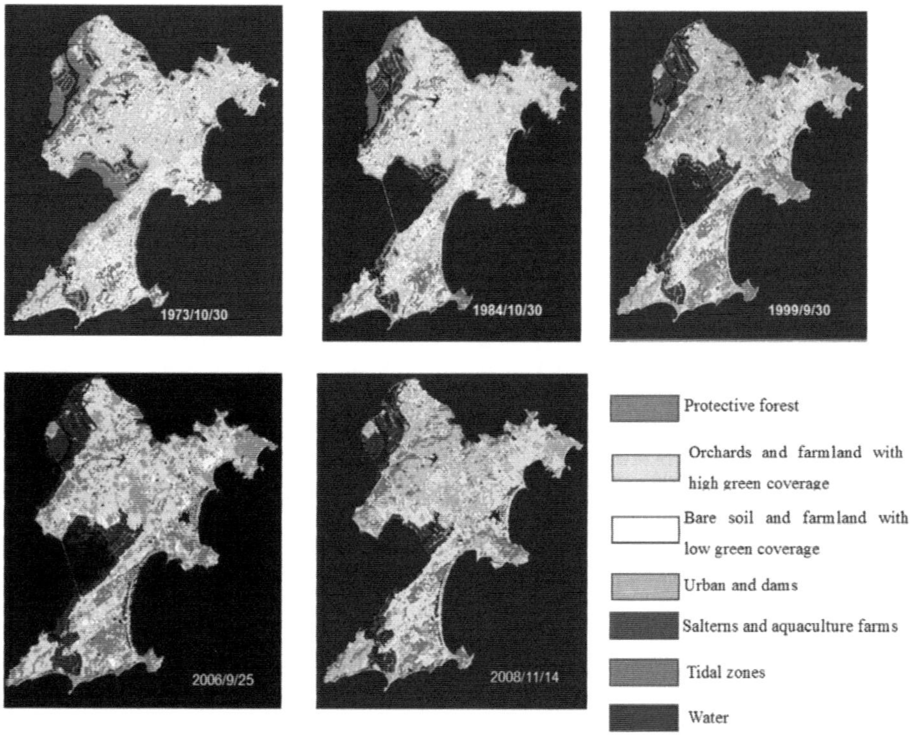

Figure 3. Extracted protection forest information from remote sensing image at different sampling time orchards

4. Temporal and spatial variation of the protective forest in the Dongshan Island

It can be seen from figures 3 and 4 that the protective forest area decreased from 2134.40 ha in the early 1960s down to 1871.22 ha in the early 1970s, which is a 12.33% decrease. In 1984 the protective forest area dropped further to 1515.74 ha, which is only about 71% of the acreage in the early 1960s. However, the protective forest area increased substantially afterwards to a maximum of 3370.22 ha in 1999, which is 2.22 times of the acreage in 1984. Since 1999, the protective forest area declined again. The acreage in 2008 is 550.49 ha less compared with that of 1999, which is a decrease of about 16%. The average annual change rate of area of protective forest during 1984-2008 is about ~3.6%, which less than the annual change rate (~4.9%) of protective forest plantations in China during 1990-2000. However, it is comparable with that of Russion Federation(~3.2%) and India (~3.7%) during 1990-2000 (FAO, 2005).

We arbitrarily divide the Dongshan Island by 23°40'N into the northern part and the southern part (figure 5). It can be seen that the northern part had more protection forest than the southern part, but their areas showed different variation trends. From 1973 to 1984, the northern part had declining protection forest area, whereas in the southern part the protection forest area increased slightly. From 1984 to 1999, the protection forest area increased substantially in both parts. From 1999 to date, both the southern and northern parts showed reduced protection forest area, but the magnitude of reduction was much greater in the southern part (about 23%) and less in the northern part (8.9%).

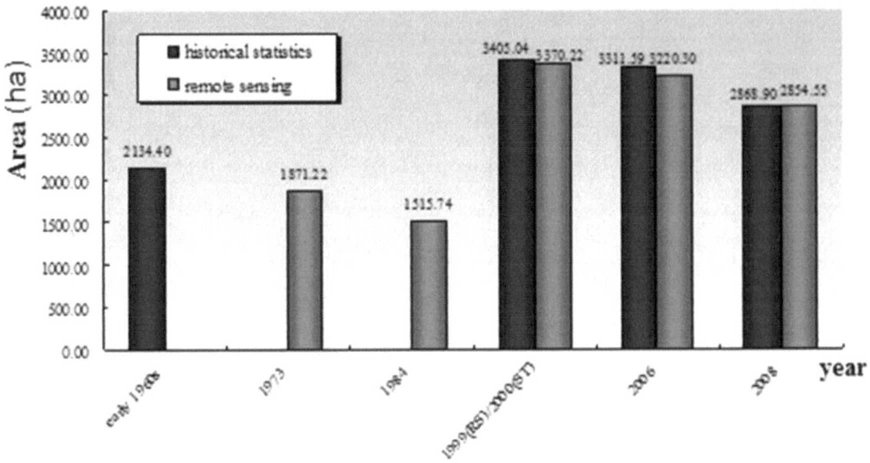

Figure 4. Comparison of protection forest area extracted from remote sensing data (orange bars) with historical statistics (blue bars) of the Dongshan Island

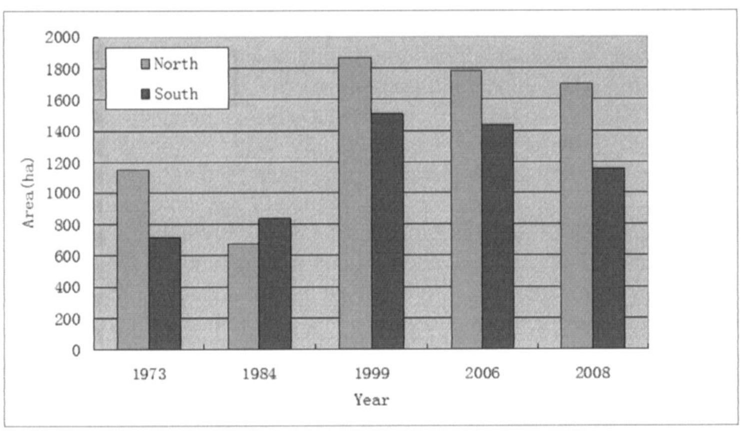

Figure 5. Comparison of the protective forest area in the northern and southern parts of Dongshan Island

Figure 6 shows the variation of protective forest land usage in 1973 and 2008, in which the green region represents additional forest planting and the red region represents forest land converted to other use. In the 35 years, for all the forest planted in 1973, 37 ha (nearly 1.98%) was converted into urban land, 537 ha (nearly 28.71%) changed into farmland for the cultivation of longan, lychee or other cash crops, and 54 ha (2.89%) changed into salterns or aquaculture farms. At the same time, an additional 983 ha protective forest has been planted in other areas, and the total protective forest area reaches 2854 ha in 2008, which increased by 53% compared with that of 1973.

From 1973 to 1984, the protective forest of the Dongshan Island had a sharp decrease of about 355.48ha, in which about 27% converted into farmland and other vegetations, 20% became bare soil, and 8% converted into urban land. During this period, the Chinese "reform and opening-up" policy had just begun and there was massive urbanization. Especially, in the northern part of the Dongshan Island, over the 10 years, the urban land use increased quickly from 1600 ha in 1973 to 3800 ha in 1984. During the construction boom, the local forests were converted to farmland and the expansions of quarries and aquaculture farms were disorderly, which resulted in the significant damage to the protective forest and exacerbated soil erosion. Survey has shown that more than 70% of the soil erosion was due to irresponsible quarrying.

Figure 6. Variation of protection forests from 1973 to 2008. Green regions indicate additional planting. Red regions indicated forest area changed for other land use

In order to improve the coastal environment and promote the economic development in coastal areas, since 1988 the Fujian province initiated the construction of the coastal protective forest system to build a comprehensive multi-species defense network across all coastal counties. Through these efforts, the basic afforestation of barren hills was completed by 1992. From 1988 to 1990, 977.4 hectares of coastal protective forest were built in the Dongshan Island that met standard requirements. Various breakthroughs ensued since then, such as the trunk updating and new forest management projects in 1993 and the new construction projects at sandy and windy coastal areas in 1996. In 2001, the construction of the Fujian coastal protective forest system became financed by national bonds. From 2001 to 2005, the treasury financed 127.13 million Yuan, which effectively supported the construction projects (Lin, 2006). The result of the ten years' efforts was remarkable. Based on the remote sensing classification results, the extracted protective forest area of Dongshan Island in 1999 is 2.22 times than that of 1984, in which 18.14% came from bare soil and 13.3% came from urban land use change. Clearly, the construction of protective forest system greatly improved the ecological environment of the island.

The Dongshan Island protective forest declined steadily since 1999 and especially notably in the latest five years. The main reasons can be rationalized as follows: (1) Aging and natural degradation of the protective forest ecosystem. Casuarina is the major species in the Dongshan Island protective forest system, which was introduced early and is currently entering the phase of aging and natural death. Meanwhile, due to inefficient management, planting multiple consecutive generations of Casuarina has degraded soil capacity. Combining with the impact of perennial sand attacks and other infrastructure construction, the Casuarina at many sections are now discontinuous, leaving gaps in the forest backbone. Therefore, new seedlings need to be constantly planted and other species should be introduced to update and strengthen the forest backbone. (2) Booming aquaculture. In recent years, due to the development of aquaculture, a number of abalone, shrimp and eel farms were built in the Dongshan county along the forest backbone, which impaired the original protection vegetation system. The Chencheng town in the southern Dongshan Island is the most evident example, where the abalone farming started since 1991 and has rapidly developed into nearly 100 farms throughout the town due to favorable economic return. Each abalone farm took approximately 0.26 ha, and most of the farms were required by the aquacultural needs to be located in the protective forest backbone, which resulted in the massive destruction of the protective forest in the Chencheng town (Wei, 2011). According to the remote sensing classification results, in the southern Dongshan Island the aquaculture farms and salterns were 1900 ha in 2008, which was three times of that in 1999. Thus, the total areage of protective forest in the southern Dongshan Island decreased more repidly than the northern part (figure 5). (3) In addition, during the rapid economic development and growing tourism in recent years, land use for construction of tourist resorts expanded greatly in a lot of towns. The inadequate management has also virtually eroded the coastal protective forest and resulted in the reduced forest area.

5. Conclusions

Using the multi-spectral data from the LandSAT TM and HJ-1A/B satellites, in this paper we extracted the information of the coastal protective forest in the Dongshan Island of the Fujian province by supervised maximum likelihood classification. The overall accuracy of the extracted data was 92.6% and the forest area extracted from remote sensing data agreed well with the historical statistics. Therefore, the multi-temporal satellite images can be used to quickly and efficiently monitor the changes in the Dongshan Island and quantitatively determine the temporal and spatial distribution of the coastal protective forest, which provides scientific data to support the effective assessment and planning of forest planting.

On this basis, we chose the data in 1973, 1984, 1999, 2006 and 2008 and analyzed the changes of the coastal protective forest in the Dongshan Island over the 35 years. We found that the area of the coastal protective forest in the Dongshan Island experienced a decrease in 1973–1984, then increased in 1984–1999 and declined again in 1999–2008. Human activity has been the major reason of the variations in the protective forest. From 1973 to 1984, due to farmland cultivation and disorderly quarrying, the protective forest of the Dongshan Island lost about 355.48ha, and the reduction was most notable in the northern part of the island. From 1984 to 1999, as the Fujian provincial government implemented the coastal protective forest construction project, the extensive artificial afforestation increased the protective forest area in the Dongshan Island by 1.22 times over 10 years. From 1999 to 2008, the protective forest acreage declined again due to the natural aging of Casuarina, booming aquaculture, tourism development, etc. Therefore, it is necessary to strengthen the forest belt by adjusting the tree species to enhance the biological diversity of the forest, exercise administrative power to manage and coordinate the development of forest and other land use, promote the construction of the coastal protective forest system and reduce coastal forest degradation, so as to eventually realize the long-term ecological, economic and social benefits.

Acknowledgments

This study was supported by the National Basic Research Program of China (973 program Grant No. 2009CB426306) and the project of Xiamen Southern Oceanographic Center (Grant No. 13GFW001NF15 and 15PZB009NF05). We also express our gratitude to Dr.Yimin Lin and Yanmin Luo for their help on field investigation.

References

[1] An S Q, Zhu F M, Zhang J H, Chen X L, Liu M S and Huang C 2006 Alternative methods for sustainably managing coastal forests as silvo-pastoral systems *Ecol. Eng.* **26**(3) 195-205

[2] Beese W J and Bryant A A 1999 Effect of alternative silvicultural systems on vegetation and bird communities in coastal montane forests of British Columbia *Canada Forest Ecol. Manage.* **115** (2-3) 231-242

[3] Borsje B W, Wesenbeeck B K V, Dekker F, Paalvast P, Bouma T J, Katwijk M M V and Vries M B D 2011 How ecological engineering can serve in coastal protection *Ecol. Eng.* **37** 113-122

[4] Erefur C, Bergsten U and Chantal M 2008 Establishment of direct seeded seedlings of Norway spruce and Scots pine: Effects of stand conditions, orientation and distance with respect to shelter tree, and fertilisation *Forest Ecol. Manage.* **255**(3-4) 1186-1195

[5] FAO (Food and Agriculture Organization of the United Nations) 2005 *Global Forest Resources Assessment 2005 main report* Chapter 6 pp 95-106

[6] FAO (Food and Agriculture Organization of the United Nations) 2010 *Global Forest Resources Assessment 2010 main report* Chapter 6 pp 109-117

[7] Li W 2004 Degradation and restoration of forest ecosystems in China *Forest Ecol. Manage* **201** 31-41

[8] Lin L Y 2006 The research of construction and protection on the protective forest of coast in Fujian province *Forestry Prospect Des. (Fujian)* (2) pp 170-174.

[9] Richards J A 1999 *Remote Sensing Digital Image Analysis* (Berlin: Springer-Verlag) p 240

[10] Wei J W 2011 On the Status Quo and Countermeasures in Coastal ShelterForest Construction in Dongshan County *Sichuan Forestry Explor. Des.* (1) pp 75-76.
[11] Zhang C, Zhu Xl, Chen H, Pan W, Gao J and Huang C 2007 A Study on Monitoring of Littoral Shelter Forest by Remote Sensing in Fujian *Chin. J. Agrometeorology* **28**(4) 446-449

Environmental suitability evaluation for human settlements in Bosten Lake Basin

A Maimaiti [1,2], L M Wang [1], J Zhang [1] and Z L Song [1]

[1] Institute of Remote Sensing and Digital Earth, CAS,100094, Beijing, China

[2] Ministry of Education Key Laboratory of Oasis Ecology, College of Resources and Environmental Sciences, Xinjiang University ,830046, Urumqi, China

E- mail: lmwang@ceode.ac.cn

Abstract. Environmental Suitability of Human Settlement in arid and semi-arid area is defined as an efficient eco-geographical landscape that has dense vegetation or stable water and comfortable climate condition which is applied to arid region's human living environment. In this work, the ESHS in the Bosten Lake Basin were evaluated based on remote sensing images, socioeconomic data, grid data and annual average climate temperature data. The results revealed that the index of ESHS in the Bosten Lake Basin ranges from 12.8 to 86.73. The proportion of high suitability area, middle suitable area, low suitable area and not suitable area accounts for 9.7%, 9.6%, 41.4%, and 39.3% respectively of the total study area. Our results indicated that natural environment of Bosten Lake is rather fragile, the mountainous area in northwest and desert area in southeast is basically do not fit for human settlement. Suitable regions for human settlement were mainly distributed around Bosten Lake in the form of batches, while others are scattered.

1. Introduction

Human living environment refers geographical space that closely related to human survival activities. With the development of social productivity, the environmental quality of living space became major concern of human beings. As a basic space for human existence and activities, the human settlement environment has attracted much research interest to seek the best living environmental condition by applying different approaches to evaluate the human environmental suitability [1]. The geographical environment is the major factor that determines where human live and develop [2].

People in Northwest China are suffering from serious environmental and ecological challenges because of distinct natural environment and historical factors [3]. Bosten Lake Basin locates in the intersecting and transitional zone between Tianshan Mountains and Taklimakan Desert, which is a sensitive place of ecological and environmental change [4].

In recent years, number of ecological and environmental problems including reduction in downstream water supplies, falling groundwater levels, land desertification and salinization, as well as vegetation loss, have been caused in Bosten Lake Basin as a result of global climatic change and uncontrolled human activity [5]. In order to understand coordinated development of the ecosystem and socio-economy in the Bosten Lake Basin, the stability and suitable scale of this area was analyzed in this study based on periods of remote sensing images and a social statistic data.

2. Materials and methods

2.1 Study Areas

Bosten Lake Basin (41°56′ N – 44° 40′ N; 86°14′ E – 90°56′ E) is located on the southern slope of the Tianshan Mountains and lies in the southeastern part of Yanqi Basin (figure.1). The Yanqi Basin borders the Tianshan Mountains and the Kuruktag Mountains in the north and west, in the south linked with Taklimakan desert. Bosten Lake is one of the largest inland freshwater lakes in China with a surface lake area of 1000 km^2, maximum water depth of 16.2 m and an average depth of 8 m. The study area covers five counties, as well as Korla City.

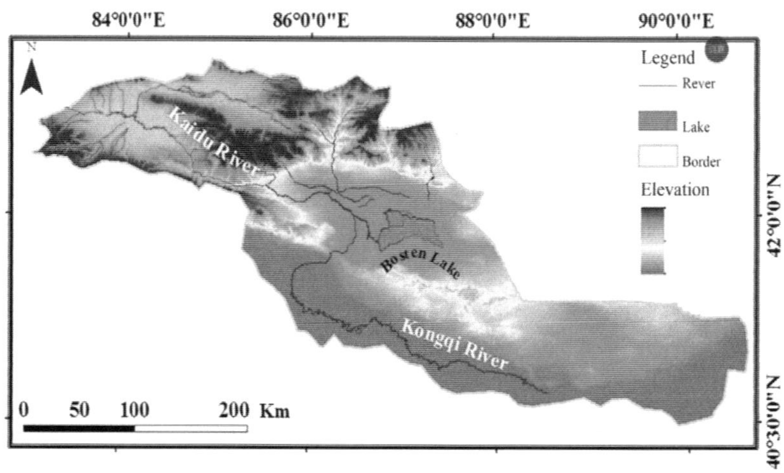

Figure 1. Location of the Bosten Lake Basin.

2.2 Data Sources

The DEM in this study were derived from the institute of Geographical Sciences and Natural Resources, Chinese Academy of Sciences. The Monthly Normalized Difference Vegetation Index (NDVI) data derived from Moderate Resolution Imaging Spectrometer (MODIS). The grid data of average annual temperature and precipitation were generated from station-based discrete point data using an Inverse Distance Weighted (IDW) spatial interpolation method. The All datasets were re-projected to Universal Transfer Mercator Projection System (UTM) Zone 45N with World Geodetic System (WGS-1984) datum.

2.3 Study Methods

Biotic and abiotic factors such as terrain, water, climate and vegetation have different spatial combinations and the spatial combinations control or modulate the suitability of human settlement and the associated socioeconomic development [5]. In this work, the DRLS model, NDVI model, THI model and WRI model were taken as main factor to decide human settlement environment index HEI model [6]. HEI model is shown below:

$$HEI = \alpha \times SRDLS + \theta \times SNDVI + \beta \times STHI + \chi \times SWRI \quad (1)$$

In this equation, *HEI* is the Human Settlements Environment Index, which value is between 0 ~ 100; where *SRDLS*, *SNDVI*, *STHI*, *SWRI* stand for a standardized relief degree (α), standardized vegetation index (θ), standardized temperature index (β) and standardized humidity index (χ), α, θ, β, χ represent weight of four indexes, respectively [7]. The weight of each indexes were calculated by using the equation in the below, and calculated index weights were given in Table 1.

$$Cov_{ij} = \frac{\sum_{k=1}^{N} (Z_{ik} - \mu_i)(Z_{jk} - \mu_j)}{N - 1} \quad (2)$$

Table 1. Correlation matrix.

Layer	Population	RDLS	NDVI	WRI	THI
Population	1.000	-0. 45	0. 69	0.644	0. 403
RDLS		1.000	0.688	0.389	-0.333
NDVI			1.000	0.729	-0.736
WRI				1.000	-0.942
THI					1.0000

3. Results and discussion

3.1. Spatial Distribution of HEI in Bosten Lake Basin

Result shows the spatial distribution of HEI in Bosten Lake Basin at 1 km × 1 km grid size, ranges from 12.8 to 86.73 (figure 2a - 2b). It is noticeable that the environmental situation from the northwest to the middle part of the Bosten Lake Basin, from the mountain to plateau, from desert to the river valley, the natural environment suitability is increasing respectively.

3.2. Natural environment suitability for human settlement in Bosten Lake Basin

Through the different natural assessment model analysis and the weight assignment, to get the HEI settlement map (figure 2a - 2b) and its corresponding values. According to the HEI and its corresponding index values of RDLS, NDVI, THI, WRI distribution, we set the criterion for natural environment suitability evaluation and classify the whole territory into four types including non-suitable, low suitable, middle suitable and high suitable area. Results showed that human living environments at different scale of natural suitability were distributed different parts of study area.

(1) High suitable area, where the HEI are between 68.36 and 86.73, is mainly distributed around Bosten lakeside and in the Korla city. The high suitable area is about 7257km^2, which equals to the 9.7% of all research area.

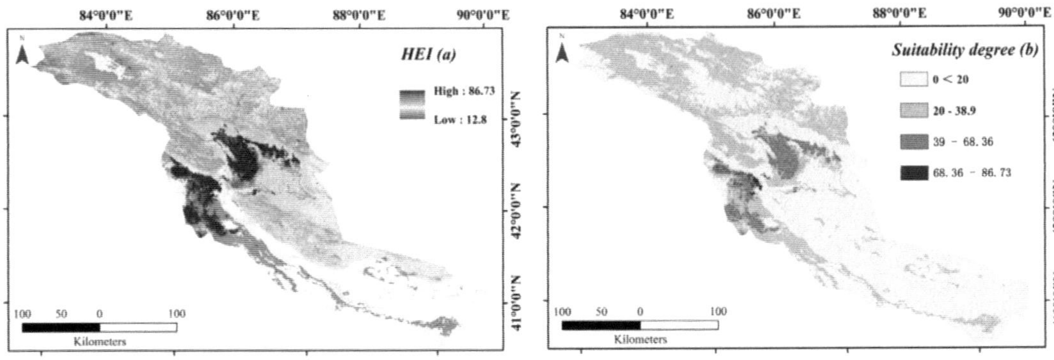

Figure 2. Human settlement suitability and descriptions in the Bosten lake basin.

(2) Middle suitable area, where the HEI is between 39 to 68.36, is mainly distributed in Konque River basin with a settlement of the main population. Because of the natural elements such as terrain, weather condition, water resource and land use are very well combined, so it is suitable for human settlement. This middle suitable area is about 7168 km^2 and account for 9.6% of the total area.

Table 2. Evaluation result of suitability for human settlements in the Bosten Lake Basin.

Comfor table level	Mean suitab ility index	Suitability characteristics	Land area (km^2)	Percen tage (%)
High suitable area (I)	87	Best suitable, not restricted by natural factors, most suitable for human living	7257	9.7
Middle suitable area (II)	60	Generally suitable, slightly influenced by natural factor	7168	9.6
Low suitable area (III)	30	Marginally suitable, influenced by natural factors to some extent, not ideal for human settlement	30960	41.4
Non-suit able area (IV)	5	Extremely unsuitable, largely limited by natural factors with very limited conditions for humans to survive and conduct productions	29483	39.3

(3) Low suitable area, where the HEI is between 20 to 38.9 and the Human Settlement environment is restricted by land use condition and water resource condition, the terrain is relatively flat. However, because of reduced vegetation cover, increased sand area, land use types and higher evaporation, which cause a negative effect on the human settlement environment. Low suitable places are mainly distributed in southeast desert places and not very suitable for human settlement. This low suitable area is about30960 km^{2}and account for 41.4% of the total area.

(4) Non-suitable area, mainly distributed in the Northwest Mountain. The region is located in the high hills and inter-mountain basins, which is about 29483 km^{2}and accounts for 39.3% of the total area, where the HEI is less than 20. In this area, Human settlements environment is restricted by terrain and climate, which is not suitable for human settlement.

4. Conclusions

In this paper, using GIS software, on the basis of simulation test and empirical model, the spatial patterns of the human settlement condition of the Bosten Lake basin are investigated in 1km×1km grid dimension. The conclusions can be drawn as below.

(1) The suitability of the human settlements environment in the Bosten Lake Basin is increasing gradually from the northwest and southeast to the middle part of the Bosten Lake Basin. The high suitable region is 7257 km^2, and accounts for 9.7% of the total basin area. The middle suitable region is 7168 km^2, and accounts for 7.6% of the total basin area. The less suitable region is 30960 km^2, and accounts for 41.4% of the total basin area. The non-suitable region is 29483 km^2, and accounts for 39.3%.

(2) Using GIS software raster calculations, statistical analysis, and repeated experiments and revision tools, some important impact factors of the suitability for the human settlements environment in the Bosten lake Basin have been used to analyze their spatial pattern. These methods can accurately take into account of regional differences in the basin and provide a reasonable quantitative visualization. The ideology contributes to the suitable distribution and flows of population. Therefore, it is a key method for the basin to promote the coordinated development of population, resources and eco-tope.

Acknowledgements

This research was supported by Director Youth Fund Project of Institute of Remote Sensing and Digital Earth CAS, (Y6XS6000CX) and Western light Project of CAS (2015-XBQN-A-07）.

References

[1] Cheng Z, Ce L, Qi T, Hui L, Qin W and Fui H 2001 Environmental change in Bosten Lake and its relation with the oasis reclamation in Yanqi Basin: *Geographical Research*. **20** 14-23.

[2] Choguill C L 1996 Toward sustainability of human settlements: *Habitat International*. **20** 5-8.

[3] Dong X B, Yang W K, Sergio U, Yan M C and Zhang X S 2012 The impact of human activities on natural capital and ecosystem services of natural pastures in north Xinjiang, China: *Ecological Modelling*. **225** 28-39.

[4] Feng Z M, Yang Y Z, Zhang D and Tang Y 2009 Natural environment suitability for human settlements in china based on GIS: *Journal of Geographical Sciences*. **19** 437-446.

[5] Huang X Z, Chen F H, Fan Y X and Yang M L 2009 Dry late-glacial and early holocene climate in arid central Asia indicated by lithological and palynological evidence from Bosten Lake, China: *Quaternary International*. **194** 19-27.

[6] Wei L, Yue C, Hai X, and Yong G 2003 A study on water environmental protection and sustainable utilization of Bosten Lake, Xinjiang: *Geographical Research*. **22** 185-191.

[7] Wei W, Shi P J, Zhou J J, Wang X F and Wang X P 2013 Environmental suitability evaluation for human settlements in an arid inland river basin: A case study of the Shiyang River Basin: *Journal of Geographical Sciences*. **23** 331-343.

Comparative analysis of NPP changes in global tropical forests from 2001 to 2013

S Yin [1,2], X Li [1] and W Wu [1]
[1] Institute of Remote Sensing and Digital Earth, Chinese Academy of Sciences, No.9 Dengzhuang South Road, Haidian District, Beijing 100094, China
[2] University of Chinese Academy of Sciences, No.19A Yuquan Road, Shijingshan District, Beijing 100049, China
Email: yinsy@radi.ac.cn

Abstract. Net primary production (NPP) is the difference between total photosynthesis (gross primary production, GPP) and total plant respiration in an ecosystem. NPP is a key component of the terrestrial carbon cycle and is important in global climate research. Tropical forests, distributed mainly in Central Africa, Central and South America, and Southeast Asia, are among the most important ecosystems on earth. They are very important to analyses of the global carbon budget and to the projection of future climatic changes. In this study, we analyzed and compared the temporal and spatial changes of NPP within the three dominant areas of tropical forest from 2001 to 2013 by using data from the Moderate Resolution Imaging Spectroradiometer (MODIS). We found that Central and South America has the highest annual mean NPP, statistically, while the average NPP shows an increasing trend both in Central and South America and Central Africa but a decreasing trend in Southeast Asia.

1. Introduction

Net primary productivity (NPP) is the rate at which all plants in an ecosystem produce net useful chemical energy. It is equal to the difference between the rate at which the plants in an ecosystem produce useful chemical energy (gross primary production, GPP) and the rate at which they use some of that energy during respiration. NPP is a fundamental ecological variable, not only because it measures energy input to the biosphere and terrestrial carbon dioxide assimilation, but also because of its significance in indicating the environmental health of an ecosystem and the status of a wide range of ecological processes.

Tropical forests are disproportionately important in the global carbon budget, representing an estimated 59% of the global carbon pool in forests [1], 20% of global terrestrial carbon stocks, and 30% of global terrestrial NPP, despite accounting for less than 7% of global dry-land surface [2]. In addition, tropical forest regions have experienced significant environmental changes in past decades, such as extreme weather events, changes in surface radiation budgets due to cloud cover changes [3, 4], and changes in global atmospheric CO_2 concentration [2, 5]. In response to these environmental challenges, tropical ecosystems may have experienced changes in their carbon budgets. To date, scientific understanding of the responses of tropical ecosystems to environmental changes is still incomplete [5], and so an improved understanding of current variations of the tropical forest NPP is critically needed.

Although the spatial variation of tropical carbon processes has been studied by many scientists, and tropical forest NPP has been assessed using multiple approaches (from plot-scale extrapolations to satellite- and model-derived estimates, among others), all large-scale NPP estimates are modeled (either explicitly or implicitly). While field-based estimates may be considered more robust in that they are based on actual small-scale measurements of plant growth, the limits of spatial and temporal coverage mean that NPP estimates cannot be adequately validated or calibrated in tropical forests [6].

Geoinformatics, including remote sensing, is an approach with many practical applications including analyses across time and space [7]. Remote sensing has been used to monitor GPP and NPP dynamics at regional and global scales [8, 9]. The Moderate Resolution Imaging Spectroradiometer (MODIS) satellite instrument produces one of the most reliable data sources at a global scale. This dataset provides operational, near-real-time calculation of global GPP at a 250 m spatial resolution with daily coverage [9]. The resulting data products have been validated as being able to capture spatial and temporal GPP and NPP patterns across various biomes and climate regimes, and the data are consistent with ground flux tower-based GPP and field-observed NPP estimations [10, 11]. The availability of MODIS GPP and NPP data provides a unique opportunity for examining spatial patterns within the global NPP/GPP ratio [12].

In order to better understand the spatial and temporal dynamics of NPP in global tropical forests in the context of climate change, we used MODIS data to analyze changes in NPP within the three main distribution areas of tropical forests, Central Africa, Central and South America, and Southeast Asia, from 2001 to 2013.

2. Data and methods

2.1 Study area

In order to analyze the distribution and variation of tropical forest NPP, we focused on forests between the Tropics of Cancer and Capricorn, as shown in Figure 1. Considering that forest area may differ from year to year, we used MODIS land-cover data for each year from 2001 to 2013 to determine the specific boundaries of the study area.

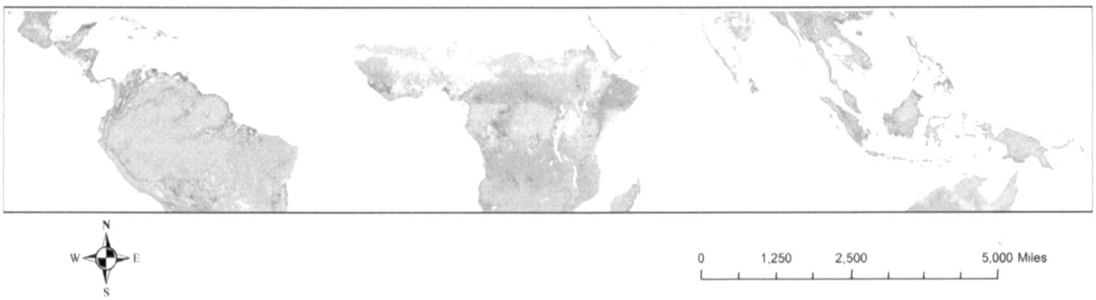

Figure 1. Study area: tropical forest regions extracted from MODIS land-cover data are shown in green.

2.2 Data

In this study, we used the MODIS land cover product (MCD12Q1), NPP product (MOD17A3), and vegetation continuous fields product (MOD44B). Each dataset covers the entire study area and is available from 2001 to 2013.

The MCD12Q1 product identifies a suite of land-cover types by mapping global land cover using spectral and temporal information derived from satellite imagery. It is produced using a supervised classification algorithm that is calibrated using a database of high-quality land cover training sites. This product supplies global maps of land cover at annual time steps and 500 m spatial resolution, consisting of five land cover data of different classification systems including the International Geosphere Biosphere Programme (IGBP) scheme [13, 14, 15, 16], which we used in this study to delineate the study region for global tropical forests.

The MOD17A3 product provides continuous estimates of GPP/NPP across Earth's entire vegetated land surface, supplying global maps of NPP at annual time steps and 1 km spatial resolution. A simple light-use efficiency model (MOD17) is at the core of the GPP component of the algorithm, as described in [17, 18]. Since the MODIS instrument was placed in orbit in 1999, there have been multiple updates of its land products [19]. In this study, we used version 55 of the NPP product, which corrects previous problems with cloud-contaminated MODIS LAI-FPAR inputs to the MOD17 algorithm. Therefore its accuracy has been assessed and uncertainties in the product are well-established via independent measurements made in a systematic and statistically robust way that represents global conditions.

The MOD44B Vegetation Continuous Fields (VCF) product is a sub-pixel-level representation of surface vegetation cover estimates globally. This product provides a continuous, quantitative portrayal of land surface cover with improved spatial detail at annual time steps and 250 m spatial resolution. The VCF product has three components of ground cover: percent tree cover, percent non-tree vegetation and percent bare [20, 21, 22, 23, 24]. The three components are each stored in separate layers so they can be used independently to look at a particular type of ground cover or collectively to look at the entire surface [25]. In this study, we used the percent tree cover layer of the version 51 VCF product, which has improved training data and data mining software which resulted in much greater accuracy in the final product without human intervention.

In order to obtain the spatial and temporal variation of global tropical forest NPP, we followed four steps. Firstly, the three MODIS datasets for the study area from 2001 to 2013 were downloaded and mosaicked; secondly, the annual study areas were extracted from the land cover type dataset; thirdly, NPP and VCF data were clipped according the study area; finally, spatial and temporal variations of NPP and VCF were analyzed and compared for the three major tropical forest regions.

3. Results and discussion

3.1 Spatial and temporal variation of NPP within Central and South America

This study area includes the enormous catchments of the Amazon and Orinoco rivers and extends into Central America [26]. The average NPP of this region shows an overall downward trend (Figure 2) with a change-trend slope of around -2.0 $g{\cdot}C{\cdot}m^{-2}{\cdot}a^{-1}$, meaning that the total NPP in this region collectively declined (on average) about that amount annually from 2001 to 2013. The forest area in this region also shows an overall decreasing trend (Figure 2) with a change-trend slope of around -0.11 million $ha{\cdot}a^{-1}$. We suggest that the reduction in forest area has partly influenced the reduction in collective NPP.

Tree cover can indicate the growth status of forests. While the land-cover type can only show whether one pixel is forest or not, tree cover reflects what percentage of the pixel is covered by canopy. The western region of the study area has experienced the most dramatic decrease in NPP from 2001-2013, in contrast to the stable or increasing trend in the eastern region. However, changes in tree cover over the same period (Figure 3) seem to show no clear geographic correlation. For example, along southeast edge of the region, tree cover has decreased while NPP has increased. In addition, the areas with the most severe decline in NPP show stable or rising tree cover.

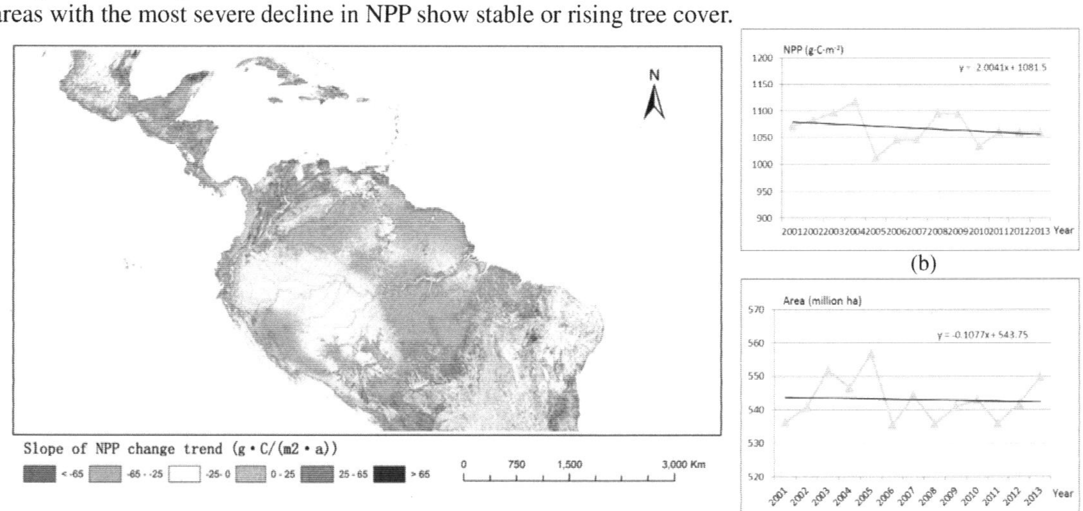

Figure 2. Distribution map (a) showing the slope of NPP change trend in the Central and South America study region. Figures on the right show change curves of NPP (b) and forest area (c) for the same region.

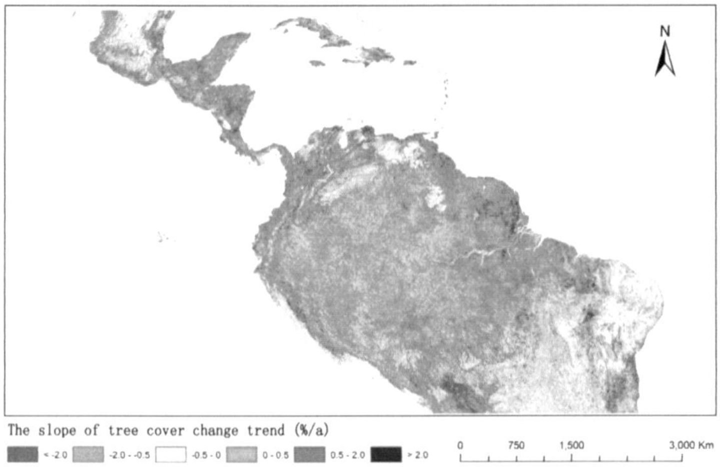

Figure 3. Distribution map showing the slope of tree-cover change trend in the Central and South America study region.

3.2 Spatial and temporal variation of NPP within Central Africa

This study area is located in the central two-thirds of Africa, including the basins of the Congo, Niger, and Zambezi rivers [26]. The average NPP of this study region shows an upward trend (Figure 4) with a change-trend slope of around 9.98 $g{\cdot}C{\cdot}m^{-2}{\cdot}a^{-1}$, meaning that the total NPP in this region has increased, on average, 9.98 $g{\cdot}C{\cdot}m^{-2}$ annually from 2001 to 2013. The total forest area in this study region is declining (Figure 4) with a change-trend slope of around -0.41 million $ha{\cdot}a^{-1}$. Most parts of the study area have experienced an increase in NPP from 2001 to 2013 (Figure 4), in contrast to the reduction in tree cover seen over much of the area (Figure 5). In the west-central area, there is an especially strong decline in tree cover with a relatively slight decline in NPP. However, in the central area, the NPP shows an upward tendency despite a decline in the tree cover.

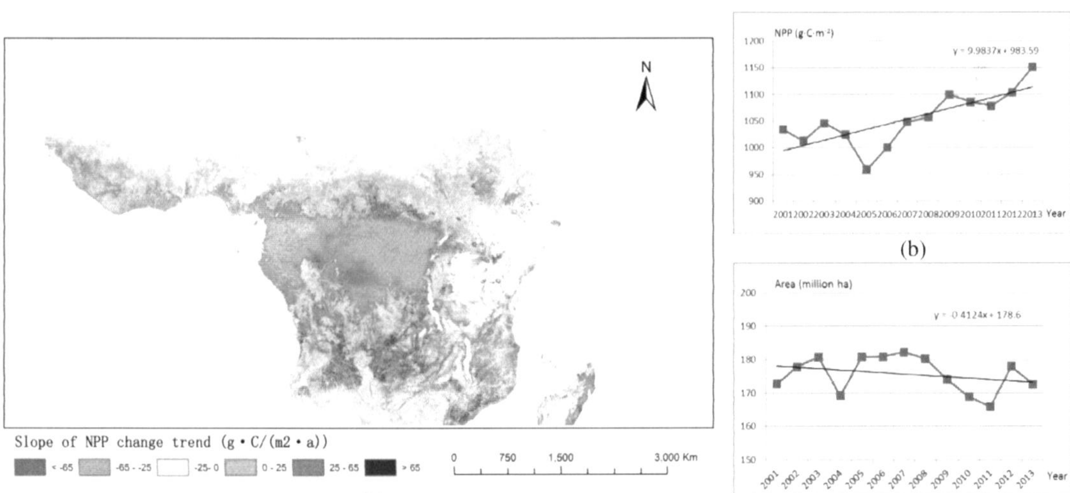

Figure 4. Distribution map (a) showing the slope of NPP change trend in the Central Africa study region. Figures on the right show change curves of NPP (b) and forest area (c) in the same region.

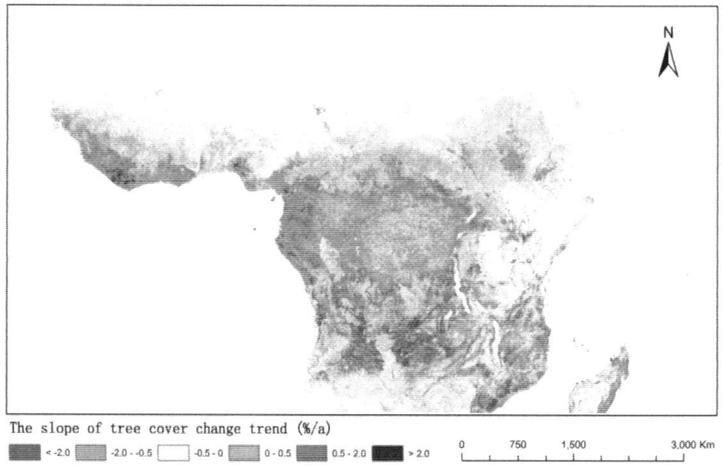

Figure 5. Distribution map showing the slope of tree-cover change trend in the Central Africa study region.

3.3 Spatial and temporal variation of NPP within Southeast Asia

This study area covers portions of India and Southeast Asia, including Myanmar (Burma), Thailand, and Malaysia along with islands including those in the Malay Archipelago, particularly Sumatra, Borneo, New Guinea, and those of the Philippines. These areas all have significant rainforest regions, as do parts of northern Australia and islands of the West Indies [26]. The average NPP of this study region shows a downward trend (Figure 6) around -14.75 g·C·m^{-2}·a^{-1}, meaning that the total NPP in this region declined, on average, about 14.75 g·C·m^{-2} annually from 2001 to 2013. The total forest area in this region shows a clear upward trend.

Many areas, including the Malay Peninsula, Sumatra, Borneo, Celebes, and northwestern New Guinea, show significant decline in NPP from 2001 to 2013. With the exception of Sumatra, most of these areas also show an increase in tree cover during the same period (Figure 7). We suggest that deforestation plays a minor role in NPP change in this area, and instead that forest degradation may have more to do with the declining NPP.

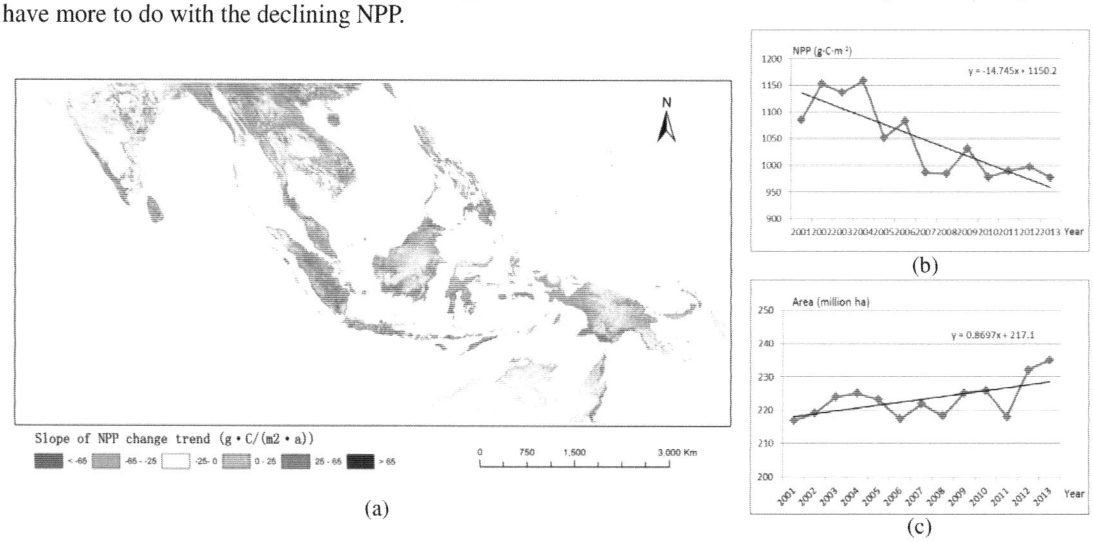

Figure 6. Distribution map (a) showing the slope of NPP change trend in the Southeast Asia study region. Figures on the right show change curves of NPP (b) and forest area (c) in the same region.

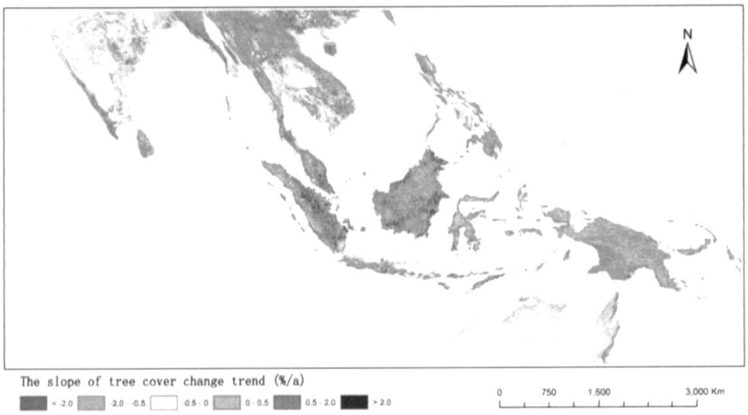

Figure 7. Distribution map showing the slope of tree-cover change trend in the Southeast Asia study region.

3.4 Comparing NPP changes between the three major tropical forest regions

The Central and South America study region has the largest physical area (543.00 million ha), followed by Southeast Asia (223.19 million ha) and the Congo (175.71 million ha). The total forest area in the three regions did not significantly change over the study period. The Southeast Asia study region shows a dramatic decline in NPP, along with having the lowest average NPP (about 1046.99 g·C·m^{-2}). The Central and South America study region shows a relatively stable NPP, and has the largest average NPP (about 1067.45 g·C·m^{-2}). The Central Africa study region is unique in showing an upward NPP tendency, with an average NPP of about 1053.48 g·C·m^{-2}. As for percent tree cover, the Central and South America study region has the highest coverage, followed by the Central Africa study region and the Southeast Asia study region. It is apparent that there is no obvious relationship between the inter-annual variation of NPP and tree cover in these study regions, which suggests that NPP is influenced by many other biochemical and meteorological factors.

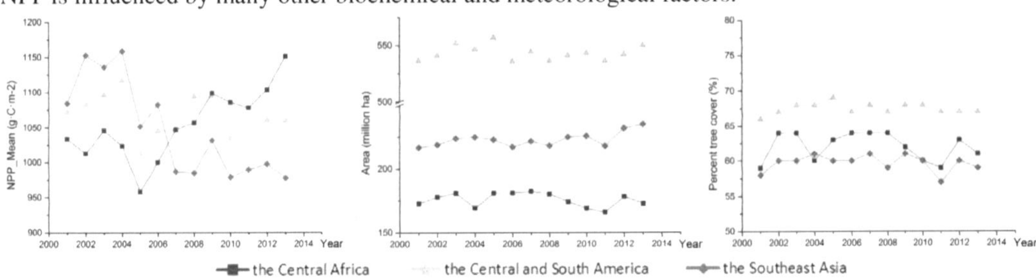

(a) (b) (c)

Figure 8. Change curves for NPP (a), forest area (b), and tree cover (c) within the three study regions.

Comparing the changes in NPP and tree cover (Figures 9 and 10), we observed that in all three study regions, areas with severe fluctuations in NPP have relatively small variations in tree cover. Under most conditions, the most dramatic reduction in tree cover appears along the edges of tropical forest areas, while central parts of these areas show only a slight change. In comparison, changes in NPP tend to be less fragmented. In general, the different slopes change gently from one level to another across a wide extent of the study areas. We suggest that the differences in distribution between NPP and tree cover may reflect the connection of tree cover to human activities and of NPP to other biogeographic and biochemical conditions.

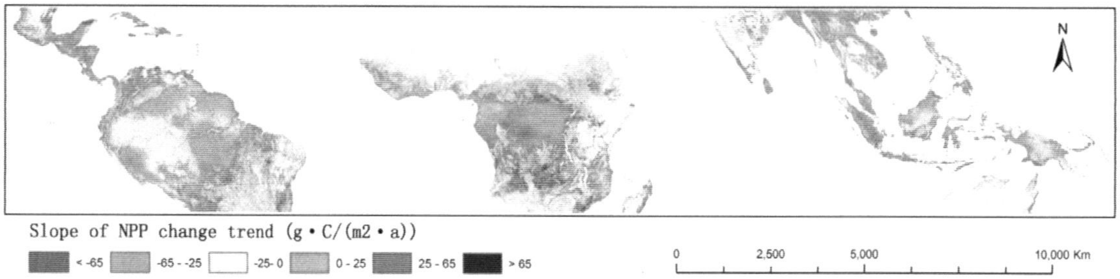

Figure 9. Distribution map of NPP slope-change trend for the three study regions.

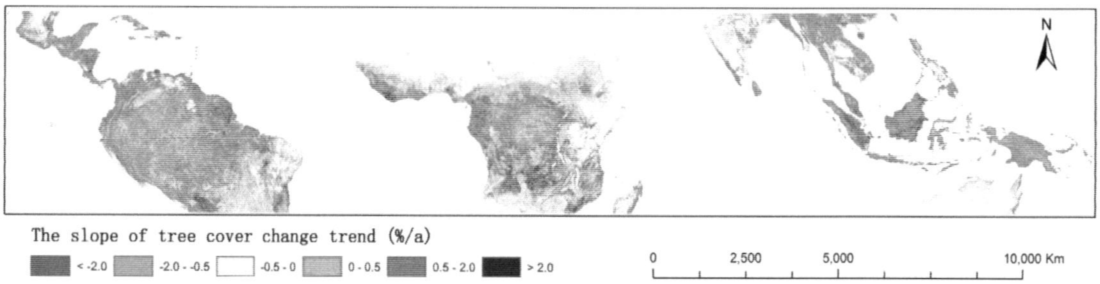

Figure 10. Distribution map of tree cover slope-change trend for the three study regions.

4. Conclusion

From 2001-2013, the Central and South America study region had the highest forest area and tree cover, with a relatively steady NPP. The Central Africa study region had the lowest forest area and second-highest tree cover, and was the only region with an upward NPP trend. The Southeast Asia study region had the second highest forest area and the lowest tree cover, with the largest decrease in NPP. Areas with severe fluctuations in NPP generally had relatively small variations in tree cover, while sections with severe reduction in tree cover tended to also show a decline in NPP. Moreover, compared to changes in tree cover, the change of NPP is more continuous, which we suggest is due to the influence of human activities on tree cover while changes in NPP are mainly connected to other biogeographic and biochemical conditions. However, the mechanisms behind these dynamics require further exploration.

Acknowledgements

The MODIS data products were retrieved from the online Data Pool, courtesy of the NASA Land Processes Distributed Active Archive Center (LP DAAC), USGS/Earth Resources Observation and Science (EROS) Center, Sioux Falls, South Dakota, https://lpdaac.usgs.gov/data_access/data_pool.
We would like to thank Editage (www.editage.com) for English language editing.

References

[1] Dixon R K, Brown S, Houghton R A, Solomon A M, Trexler M C and Wisniewski J 1994 Carbon pools and flux of global forest ecosystems *Science* **263** 185–190

[2] Prentice I C, Farquhar G D, Fasham M J R, Goulden M L, Heimann M, Jaranillo V J, Kheshgi H S, Le Quere C, Scholes R J and Wallace D W R 2001 The carbon cycle and atmospheric carbon dioxide *Climate Change 2001: The Scientific Basis. Contribution of Working Group I to the Third Assessment Report of the Intergovernmental Panel on Climate Change* ed J T Houghton, Y Ding et al. (Cambridge: Cambridge University Press) 183–237

[3] Cess R D and Udelhofen P M 2003 Climate change during 1985–1999: cloud interactions determined from satellite measurements *Geophys. Res. Lett.* **30** 1019

[4] Wielicki B A, Wong T, Allan R P, Slingo A, Kiehl J T, Soden B J, Gordon C T, Miller A J, Yang S K and Randall D A et al. 2002 Evidence for large decadal variability in the tropical mean radiative energy budget *Science* **295** 841–44

[5] Sarmiento J L and Gruber N 2002 Sinks for anthropogenic carbon *Phys. Today* 55 30–36

[6] Marvin D, Asner G P, Knapp D E, Anderson C B, Martin R E, Sinca F and Tupayachi R 2014 Amazonian landscapes and the bias in field studies of forest structure and biomass *Proc. Natl. Acad. Sci. U.S.A.* **111** E5224–32 doi:10.1073/pnas.1412999111

[7] Lu J X, Song W L, Qu W, Fu J E and Pang Z G 2015 The cross time and space features in remote sensing applications *[C] Remote Sensing of the Environment: 19th National Symposium on Remote Sensing of China* International Society for Optics and Photonics 96690J-96690J-8

[8] Nemani R R, Keeling C D, Hashimoto H, Jolly W M, Piper S C, Tucker C J, Myneni R B and Running S W 2003 Climate-driven increases in global terrestrial net primary production from 1982 to 1999 *Science* **300** 1560–63

[9] Zhao M, Running S W and Nemani R R 2006 Sensitivity of Moderate Resolution Imaging Spectroradiometer (MODIS) terrestrial primary production to the accuracy of meteorological reanalysis *Journal Geophys. Res.* **111** DOI: 10.1029/2004JG000004

[10] Zhao M, Heinsch F A, Nemani R R and Running S W 2005 Improvements of the MODIS terrestrial gross and net primary production global dataset *Rem. Sens. Environ.* **95** 164–76

[11] Heinsch F A, Zhao M, Running S W, Kimball J S, Nemani R R, Davis K J, Bolstad P V, Cook B D, Desai A R and Ricciuto D M et al. 2006 Evaluation of remote sensing based terrestrial productivity from MODIS using regional tower eddy flux network observations *IEEE Trans. Geosci. Rem. Sens.* **44** 1908–25

[12] Zhang Y, Xu M, Chen H and Adams J 2009 Global pattern of NPP to GPP ratio derived from MODIS data: effects of ecosystem type, geographical location and climate *Glob. Ecology Biogeogr.* **18** 280–90

[13] Belward A S, Estes J E and Kline K D 1999 The IGBP-DIS global 1-km land-cover data set DISCover: A project overview *Photogramm. Eng. Rem. Sens.* **65** 1013–20

[14] Scepan J 1999 Thematic validation of high-resolution global land-cover data sets *Photogramm. Eng. Rem. Sens.* **65** 1051–60

[15] Friedl M A, McIver D K, Hodges J C F, Zhang X Y, Muchoney D, Strahler A H, Woodcock C E, Gopal S, Schneider A, Cooper A, Baccini A, Gao F and Schaaf C 2002 Global land cover mapping from MODIS: algorithms and early results *Rem. Sens. Environ.* **83** 287–302

[16] Friedl M A, Sulla-Menashe D, Tan B, Schneider A, Ramankutty N, Sibley A and Huang X 2010 MODIS Collection 5 global land cover: Algorithm refinements and characterization of new datasets *Rem. Sens. Environ.***114** 168-182

[17] Running S W, Nemani R R, Heinsch F A, Zhao M, Reeves M and Hashimoto H 2004 A continuous satellite-derived measure of global terrestrial production *BioScience* **54** 547−60

[18] Heinsch F A, Reeves M and Bowker C F 2003 User's Guide, GPP and NPP (MOD 17A2/A3) Products, NASA MODIS land algorithm http://www.forestry.umt.edu/ntsg/

[19] Turner D P, Ritts W D, Cohen W B, Gower S T, Running S W, Zhao M, Costa M H, Kirschbaum A A, Ham J M and Saleska S R 2006 Evaluation of MODIS NPP and GPP products across multiple biomes *Rem. Sens. Environ.* **102** 282-92

[20] Carroll M, Townshend J, Hansen M, DiMiceli C, Sohlberg R and Wurster K 2010 MODIS vegetative cover conversion and vegetation continuous fields *Land Remote Sensing and Global Environmental Change* ed B Ramachandran and C O Justice (New York: Springer) pp 725-745

[21] Hansen M C, DeFries R S, Townshend J R G and Sohlberg R 2000 Global land cover classification at 1 km spatial resolution using a classification tree approach *Int. Journal Rem. Sens.* **21** 1331-64

[22] Hansen M C, Sohlberg R, Dimiceli C, Carroll M, DeFries R S and Townshend J R G 2002 Towards an operational MODIS continuous field of percent tree cover algorithm: examples using AVHRR and MODIS data *Rem. Sens. Environ.* **83** 303-319

[23] Hansen M C, DeFries R S, Townshend J R G, Carroll M, Dimiceli C, and Sohlberg R A 2003 Global percent tree cover at a spatial resolution of 500 meters: First results of the MODIS vegetation continuous fields algorithm *Earth Interactions* **7** 1-15

[24] Hansen M C, Townshend J R, DeFries R S, and Carroll M 2005 Estimation of tree cover using MODIS data at global continental and regional/local scales *Int. Journal Rem. Sens.* **26** 4359-80

[25] Townshend J, Hansen M, Carroll M, DiMiceli C, Sohlberg R, and Huang C 2011 User Guide for the MODIS Vegetation Continuous Fields product Collection 5 version 1 University of Maryland Maryland

[26] Rainforest, 2016 *Britannica Online for Kids* Retrieved 23 April 2016 from http://kids.britannica.com/comptons/article-203313/rainforest

Simulation of actual terrain ocean tide in the bays with SPH

Wang Tao

Guangzhou institute of Geographic, Guangzhou, China

E-mail: wangtao8869@163.com

Abstract. The real-time simulation of large-scale fluid scenes is of great value in both research and application. Water when relatively still has a well-defined surface; however, water changes its shape as it moves. In the case of ocean waves, features on the water's surface move, but the water itself does not travel. The simple surface topology can become arbitrarily complex when the water becomes turbulent. Splashing, foaming, and breaking waves are complex processes best modeled by particle systems and volumetric techniques, but these techniques are inefficient in nonturbulent situations. Ocean tide in different bays can experience rotary tidal currents or other situation. We choose the physical-based SPH(smoothed particle hydrodynamics) fluid simulation method. SPH method belongs to particle method which is has no grid. The advantages of SPH are as follows: simulating liquid convection by particles directly to eliminate numerical fluctuation at free interface; grids unnecessary avoid grid distortion and reconstruction; simulating the fluid problem of significant transformation, especially in dealing the problems such as maximum distortion, the interface of motion material, the deformation boundary and free surface flow. Application of rapid neighboring particle search method, set the number of Department of Physics, as well as the presentation and rendering of fluid material, and finally use the Lagrangian method SPH system initialization and calculate the fluid density, pressure, internal forces and external forces, define the time integration and collision handling. With the analysis of physical-based ocean tide simulation, we can create the animation of the environment, and predict damage of ocean tide.

1. Introduction

In the geographical information system, there are no real 3d display system that can simulate ocean tide in the real terrain scene.

There are two kind of methods for ocean simulation. One is water surface based on the statics by using height simulation, another is physics based fluid simulation[1, 2, 3, 4].

In grid-based numerical methods, mesh generation for the problem domain is a prerequisite for the numerical simulations. For the Eulerian grid methods like FDM, constructing a regular grid for irregular or complex geometry has never been an easy task, and usually requires additional complex mathematical transformation that can be even more expensive than solving the problem itself[5-9].

Grid-based methods have been favored the most to solve the mathematical equations for fluid flow, but often they lack the ability to create interactive fluid simulations together with detailed fluid surfaces[1].

The key idea of the meshfree methods is to provide accurate and stable numerical solutions for integral equations or PDEs with all kinds of possible boundary conditions with a set of arbitrarily distributed nodes without using any mesh that provides the connectivity of these nodes or particles[10-15].

SPH is a meshfree particle method, the problem is discretized with particles without a fixed connectivity. Treatment of large deformation is relatively much easier, Discretization of complex geometry for the MPMs is relatively simpler as only an initial discretization is required, refinement of the particles is expected much easier to perform than the mesh refinement.

Smoothed particle hydrodynamics (SPH) is a meshfree, Lagrangian particle method of modeling fluid flows. SPH was first invented to solve astrophysical problems in three-dimensional open space, in particle polytrope. 1983, Reeves first introduce the particle to the field of computer graphics. Muller use it to simulate the movement of fluid in 2002.Today, the SPH method is being used in many areas, such as the simulations of binary stars and stellar collisions, collapse[2].

In this article, we try to simulate the actual terrain ocean tide in the bays with SPH.

2. Algorithm

In summary, the governing equations for dynamic fluid flows can be written as a set of partial differential equations in Lagrangian description. The set of partial differential equations is the well-known Navier-Stokes equations, which state the conservation of mass, momentum and energy. (expressed in equation (2.1)and (2.2))

$$\rho(\frac{\partial}{\partial t}+\mathrm{u}\bullet\nabla)u=-\nabla p+\mu\nabla\bullet(\nabla u)+f \tag{2.1}$$

$$\nabla\bullet u=0 \tag{2.2}$$

where ρ stands for the density of the fluid, u is velocity, μ is viscosity, f is external force.

Lagrangian also use Navier-Stokes equation to simulate the movement of fluid, it is expressed in equation (2.3):

$$\rho\frac{du}{dt}=-\nabla p+\mu\nabla^2 u+f \tag{2.3}$$

Left side of the equation is internal and external forces, forces are expressed in euqation (2.4) as:

$$F=f^{\mathrm{internal}}+f^{\mathbf{external}} \tag{2.4}$$

In the SPH method, the state of a system is represented by a set of particles, which possess individual material properties and move according to the governing conservation equations. SPH particles have some properties, such as: location accelaration carry material properties, and are allowed to move in light of the internal interactions and external forces.

Values are interpolated from a discrete set of points.

（1）Density

Density is expressed as equation (2.5)：

$$\rho_i=\rho(r_i)=\sum_j \rho_j\frac{m_j}{\rho_j}W(r_i-r_j,h)=\sum_j m_j W(r_i-r_j,h) \tag{2.5}$$

where W is kernel function， m is the mass， h is radius, j is the number of particles i in the radius of kernel. r_i， r_j are the position of particle i and j.

(2)Internal force

Internal force arise from the particle interaction .such as: pressure and viscosity.
pressure is expressed as equation (2.6)：

$$f_i^{\text{pressure}} = -\sum_{j \neq i} \frac{p_i + p_j}{2} \frac{m_j}{\rho_j} \nabla W(r_i - r_j, h) \tag{2.6}$$

viscosity is expressed as equation (2.7) :

$$f_i^{\text{viscosity}} = -\mu \sum_{j \neq i} u_j \frac{m_j}{\rho_j} \nabla^2 W(r_i - r_j, h) \tag{2.7}$$

(3)External force

External force include gravity and buoyancy. It is expressed in equation (2.8).

$$f^{\text{external}} = \sum_{\text{n}} f^{\text{n}} \tag{2.8}$$

where gravity is expressed as equation (2.9):

$$f_i^{\text{gravity}} = \rho_i g \tag{2.9}$$

buoyancy is expressed as equation （2.10）:

$$f_i^{\text{buoyancy}} = b(\rho_i - \rho_0) g \tag{2.10}$$

where g is acceleration due to gravity, b is coefficient of buoyancy.

(4)Time integration

In order to simulate the fluid, it is necessary to set the time step. Leap-Frog is used for time integration. The advantage of the Leap-Frog algorithm is its low memory storage required in the computation and the efficiency for one force evaluation per step[1].

In the SPH method, since the smoothing function has a compact support domain, only a finite number of particles are within the support domain of dimension kh of the concerned particle, and used in the particle approximations. These particles are generally referred to as nearest neighboring particles for the concerned particle. The process of finding the near nearest particles is commonly referred to as nearest neighboring particle searching. Unlike a grid-based numerical method, where the position of neighbor grid-cells are well defined once the grids are given, the nearest neighboring particle in the SPH for a given particle can vary with time[1, 2].

Tree search algorithm works well for problems with variable smoothing lengths. It involves creating ordered trees according the particle positions. Once the tree structure is created, it can be used efficiently to find the nearest neighboring particles.

3. Algorithm

Here we use the open source code SPH Fluid Simulator for CPU and GPU to simulate the ocean tide. The workflow of the SPH is shown as follow figure 1:

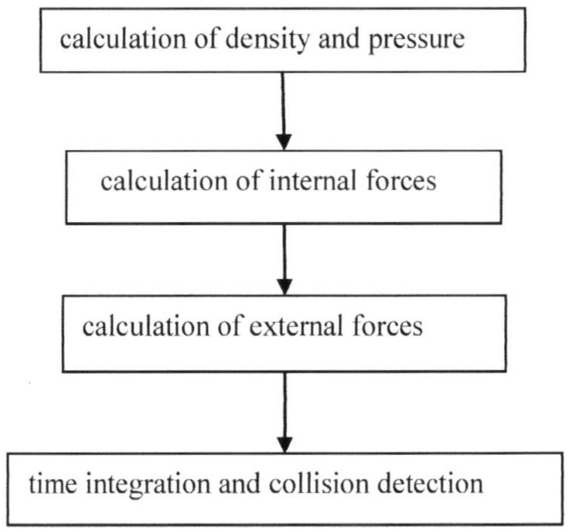

Figure1. Workflow of SPH

Simulation results are shown in the figure 2 .

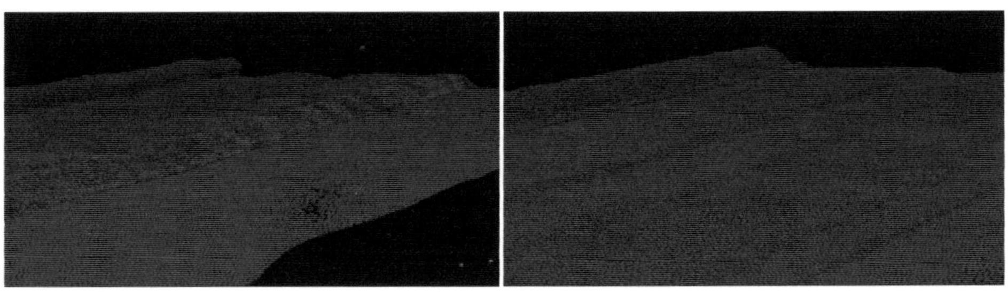

Figure 2. Simulation of ocean tide with SPH

4. Conclusion

In this article, we use SPH algorithm to simulate ocean tide in the bays. It shows a good result. But there are a lot of work to do in the future, how to use the real terrain scene to simulate the ocean tide in the bay. How to rewrite the source code to accelerate. How to render the environment.

References

[1] Li H Z 2013 *lagrangian fluid motion simulation using smoothed particle hydrodynamics* Master Thesis (Hefei: Anhui university)
[2] Liu G R and Liu M B 2003 *Smoothed Particle Hydronamics-a Meshfree Method* (Singapore: World scientific publishing company)
[3] Zheng X 2009 *an investigation of improved sph and its application for free surface flow* Doctor Thesis (Harbin: Harbin engineering university)
[4] Rick P 2012 *Computer Animation Algorithms & Techniques* (San Francisco: Morgan Kaufamann)

[5] Robert B 2008 *Fluid Simulation for Computer Graphics* (Wellesley: A K Peters)
[6] David L 2003 *Level of Detail for 3d Graphics* (San Francisco: Morgan Kaufmann)
[7] Zhang Q 2013 *Modeling and Rendering Natural Scene based on Fluid Simulation* Master Thesis (Shanghai: East china normal university)
[8] Lin H 2011 *opengl visualization based on the smoothed particle hydrodynamics method* Master Thesis (Hefei: Anhui university)
[9] Zhang S T and He S J 2011 Ocean tide flow field with actual terrain based on cloud *computing environment computer engineering and design.* 32:2432~2435
[10] Markus G 2007 *Point-based Graphic* (San Francisco: Morgan Kaufmann)
[11] Zheng X 2005 *research on algorithms of smoothed particle hydrodynamics* Master Thesis (Harbin: Harbin Engineering University)
[12] Jia L 2008 *simulation of natural phenomena based on particle system* Master Thesis (Wuhan: Wuhan university of technology)
[13] He J 2010 *study on the real-time fluid simulation with complex boudaries* Master Thesis (Hangzhou: Zhejiang university)
[14] Dai J Q 2012 *Simulate typical natural phenomenon based on fluid model* Master Thesis (Shanghai: East china normal university)
[15] Li C H 2011 *Time modeling and rendering of ocean environment* Master Thesis (Shanghai: East china normal university)

IOP Conf. Series: Earth and Environmental Science 57 (2017) 012011 doi:10.1088/1755-1315/57/1/012011

Exploring multi-scale forest above ground biomass estimation with optical remote sensing imageries

U Koju [1], J Zhang[1], H Gilani [2]

[a] Institute of Remote Sensing and Digital Earth, CAS, 100094 Beijing, China

[b] International Centre for Integrated Mountain Development, Kathmandu, 44600, Nepal

E-mail: jhzhang@ceode.ac.cn , upamakoju@radi.ac.cn

Abstract. Forest shares 80% of total exchange of carbon between the atmosphere and the terrestrial ecosystem. Due to this monitoring of forest above ground biomass (as carbon can be calculated as 0.47 part of total biomass) has become very important. Forest above ground biomass as being the major portion of total forest biomass should be given a very careful consideration in its estimation. It is hoped to be useful in addressing the ongoing problems of deforestation and degradation and to gain carbon mitigation benefits through mechanisms like Reducing Emissions from Deforestation and Forest Degradation (REDD+). Many methods of above ground biomass estimation are in used ranging from use of optical remote sensing imageries of very high to very low resolution to SAR data and LIDAR. This paper describes a multi-scale approach for assessing forest above ground biomass, and ultimately carbon stocks, using very high imageries, open source medium resolution and medium resolution satellite datasets with a very limited number of field plots. We found this method is one of the most promising method for forest above ground biomass estimation with higher accuracy and low cost budget. Pilot study was conducted in Chitwan district of Nepal on the estimation of biomass using this technique. The GeoEye-1 (0.5m), Landsat (30m) and Google Earth (GE) images were used remote sensing imageries. Object-based image analysis (OBIA) classification technique was done on Geo-eye imagery for the tree crown delineation at the watershed level. After then, crown projection area (CPA) vs. biomass model was developed and validated at the watershed level. Open source GE imageries were used to calculate the CPA and biomass from virtual plots at district level. Using data mining technique, different parameters from Landsat imageries along with the virtual sample biomass were used for upscaling biomass estimation at district level. We found, this approach can considerably reduce field data requirements for estimation of biomass and carbon in comparison with inventory methods based on enumeration of all trees in a plot. The proposed methodology is very cost effective and can be replicated with limited resources and time.

1. Introduction

Since long there has been a viable debate on GHG (Green House Gas) emission especially the carbon as the prime cause behind the perceptible global warming (Angelsen, 2008; Herold et al., 2011). Forest account for about 80% of the exchange of carbon between terrestrial ecosystems and the atmosphere(Uddin et al., 2015). It is estimated that deforestation and degradation of forest ecosystems contributes to one-fifth of the total quantum of anthropogenic greenhouse gas emissions on the globe(Lorenz & Lal, 2010). It is also reported that carbon sequestration by afforestation and avoiding deforestation could be effective in countering global warming at a low cost than the massive energy conservation and innovation adjustments we will need to minimize the use of fuel energy guzzling world economies(Smil, 2010). Due to this monitoring of forest above ground biomass (as carbon can be calculated as 0.47 part of total biomass) has become very important. Forest above ground biomass is the major portion of total forest biomass, hence a very careful consideration should be taken in its estimation. It is hoped to be useful in addressing the ongoing problems of deforestation and degradation and to gain carbon mitigation benefits through mechanisms like Reducing Emissions from Deforestation and Forest Degradation (Kim et al., 2010; Lu, 2006; YousifAli & Hammad, 2011).

Nepal is a rich country in forest resources occupying 40% (5.83 million hectares) of its total area by forest (DOF 2012). (FAO, 2010) estimated the level of forest living biomass in Nepal in 2010 at 484 million tons (t) (359 million tons above ground biomass [AGB] and 126 million tons below ground biomass [BGB]), but these values are aggregated at the national level, and not detailed enough to use for planning purposes. Also, the proposed introduction of a mechanism for REDD+ (Reducing Emissions from Deforestation and Forest Degradation) and associated requirements for monitoring, reporting, and verification (MRV) requires detailed information on forest biomass and changes in carbon stocks.

Remote sensing techniques, using different sensors and methods, offers as a better tool for estimating Forest Above ground Biomass. Remote sensing data has the advantage of being capable of obtaining spatial distribution of forest biomass at reasonable cost with acceptable (DeFries et al., 2007). Many efforts have been made to estimate forest biomass and carbon stock using different platforms (air-borne and space-borne) and sensors (optical, radar, and Lidar) (Gibbs, Brown, Niles, & Foley, 2007). Also, several methods have been proposed for estimating forest biomass using remote sensing techniques that make use of a combination of regression models, vegetation indices, and canopy reflectance models (Baccini et al., 2012; Bastin et al., 2014; Cutler et al, 2012; Frazier et al, 2014; Hussin et al., 2014; Kim et al., 2010; Ploton et al., 2012).

Medium and coarse spatial resolution imageries has always been a potential AGB estimator at National and regional scale, but in the sites with complex biophysical environment, it possess problems like mixed pixel and data saturation (Goetz et al., 2009). On the other hand, the High spatial resolution data provide more accurate results than medium resolution but they are expensive and have less area coverage per tiles and demands high processing, thus they are not appropriate for use in developing countries. (Lu, 2006) has suggested that combining remotely sensed data derived at different scales (coarse to fine resolution) could improve the accuracy of biomass estimation at national and global scales.

Multi scale biomass estimation with multiple resolution optical imageries can hence be one of the best option for forest above ground biomass estimation for the developing and REDD+ participant countries. This paper describes a methodology of using multi-resolution imageries i.e. the very high and medium resolution satellite datasets together with a very limited number of field plots. A crown projection area (CPA) vs. biomass model was developed and validated at the watershed level. Open source google imagery was used to derive CPA and Biomass of virtual plots for scaling up at district level. Using data mining technique, different parameters from Landsat imageries along with the virtual sample biomass were used for upscaling biomass estimation at district level. This approach could considerably reduce the requirement for field level data for estimation of biomass and carbon in comparison with inventory methods based on enumeration of all trees in an area. The proposed methodology is very cost effective and can be replicated with limited resources and time. It is relatively straightforward to obtain meaningful estimates of biomass in their forest areas.

The specific objectives of the study were

- to develop and validate a CPA (delineated and extracted from VHRI satellite imageries) vs. Biomass (based on the field data) model
- to design virtual plots using open-source Google Earth imagery

- to develop a multi-regression model through the data mining technique for scaling up based on the virtual plot data and parameters extracted from remotely-sensed datasets
- carry out biomass mapping for the entire district

2. Materials and Method

2.1. Study Area

The overall study area is Chitwan District of Nepal and within this the Kayerkhola watershed was selected for detailed development of the model (Figure 1.). The whole district has area of 2,218 km^2 with forest area of 1308 km^2 whereas the watershed occupies area of 80 km^2 and with forest area of 59 km^2. The study area was selected on the basis of accessibility, data availability, variation in terrain, and ongoing implementation of a REDD+ pilot project

The study area has mixed forests with the dominant species *Shorea robusta* (sal) found on most areas. Other major species are *Schima wallichii* (chilaune) and a few associated species

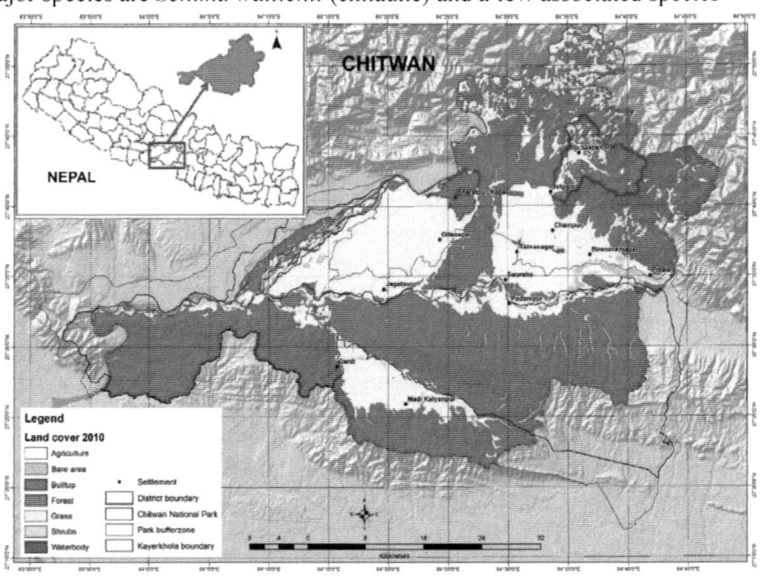

Figure 1. Study area

2.2. Data and Software

At the watershed level cloud and haze free GeoEye-1 images captured on 15 December 2012 were used. Landsat-8 Operational Land Imager (OLI) images from 9 and 18 November (winter), on row/path 142/041 and 141/41were used for scaling up at district level. The GE very high resolution Digital globe imageries from 2013 were used as the virtual plots for upscaling. A digital elevation model (DEM) was extracted from topographic sheets with a horizontal resolution of 30 m with extracted products like slope, aspect, and hill shade, to understand the topography of the area. An ordinary global positioning system (GPS) receiver was used for location identification in the field study, tree height was measured with a TruPulse 360B and DBH was measured with a measuring tape.

eCognition developer was used for the delineation of the tree crown from the very high resolution GeoEye imageries. The pre-and post-processing of the Geo-Eye and Landsat imageries were done with the use of Erdas Imagine and ENVI software. General spatial analysis processing were done with the use of Arcmap. Other statistical analysis were done using the R.

2.3. Research Method

The overall methodology of this work is summarized in the form of a flow chart in Figure 2. The method followed for has been described briefly in the sub sections below.

The research method followed in this work can be broadly divided into 3 distinct parts i.e. Field work, Geospatial analysis (Remote Sensing and GIS) and statistical modeling.

Field work was performed to obtain the in-situ data of DBH and other measurements. The geospatial operations were carried out to delineate the tree crown from the VHRI and GE virtual plots and to obtain the forest biomass at the watershed and the district level. The statistical analysis was done to develop the relation between the CPA delineated in the VHRI imageries with biomass from the field plots and the biomass generated from the filed plots with that to the Landsat imagery reflectance variables.

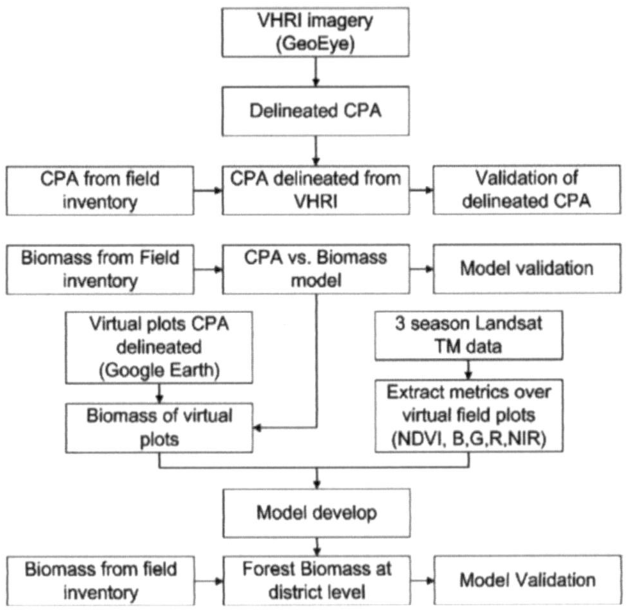

Figure 2. Methodological flow chart

2.3.1. Field data collection. A total of 58 stratified randomly selected inventory sample plots of 500m2 each were collected within the watershed area for the field based biomass estimation. Out of it 38 plots were used to develop the regression model; and a further 20 plots with the same area were used for validation. The position, height, and DBH of every tree with DBH > 5 cm in each plot were recorded.

2.3.2. Pre-processing of remotely sensed data. Individual band-wise GeoEye-1 images were ortho-rectified using rational polynomial coefficient (RPC) files along a horizontal 20 m topographic DEM by applying a cubic convolution method in zone 44 of the Universal Transverse Mercator (UTM) coordinate system, with datum and spheroid from the World Geodetic System (WGS) 84. The spectral information at lower resolution (2 m) was pan-sharpened with the high spatial resolution information (0.5 m) from the panchromatic image. The two GeoEye-1 images were independently ortho-rectified and fused with their respective multi and panchromatic spectral bands, but positional differences were observed when they were overlaid. To overcome this, their rectification was done taking 26 points were taken as ground control points (GCPs) in both datasets. This resulted in an overall root mean square error (RMSE) of 1.2 m for the panchromatic image and 1.5 m for the multispectral image. Prior to segmentation, low pass median filters was applied to avoid over-segmentation and smooth the appearance (Platt and Schoennagel 2009). In this study, convolution 3 × 3 low-pass filters were used to reduce local variation, remove noise, enhance tree features, and improve the quality of the analyzed satellite images.

The Landsat TM images were downloaded and a layer stack prepared to make multi-spectral images for visualization. The Normalized Difference Vegetation Index (NDVI) was extracted using the NIR and RED bands

2.3.3. Delineation of CPA. In eConition developer, Object based image analysis (OBIA) with a region growing technique was used for CPA delineation at the watershed level. In this method, tree tops are identified as maxima and the shadows between trees as minima. The segments are 'grown' from these maxima and the valleys act as boundaries. The first step in region growing was to create minimum size homogeneous objects through 'chessboard segmentation'; the brightest pixels were then identified as seed pixels (tree tops). Regions were 'grown' from the seed pixels up to the local minima, resulting in homogeneous objects based on predefined homogeneity criteria (Hussin et al., 2014; Ke & Quackenbush, 2008; Kim et al., 2010; Shih & Cheng, 2005). Validation of the delineation was done using manually delineated tree crowns (visual interpretation of the images in a 1 ha grid).

2.3.4. Modeling field based Biomass with delineated CPA. Using the allometric equation developed by (Sharma, 1990) the DBH (in centimetres) of each tree recorded in the field was converted into Biomass (tons). CPA was derived from the GeoEye-1 image of 2012 for the same plot areas using the region growing technique. After that the regression relation between the CPA and biomass was modelled.

2.3.5. Development of GE virtual Plots and biomass estimation at district level. Thirty-five virtual plots of 1 ha (10,000 m2) randomly distributed on a Google Earth were generated for scaling up across the entire district. The number of crowns and the crown area were observed in each plot by visual analysis and digitizing. Their biomass value were then calculated using the regression equation, derived in the watershed before thus taking the diversity and different sizes of tree crowns into account.
The forest biomass value for the district level was derived using the CART, and MARS data mining techniques which used the biomass values for the virtual plots calculated from the CPA using the regression equation, together with additional parameters extracted for each of the three seasons from the virtual plots in the Landsat images. The parameters used were blue, green, red, and near infrared reflectance, elevation and NDVI values.

3. Result and Discussion

3.1. CPA delineation and development of CPA vs Biomass regression model for watershed.
The CPA delineated through OBIA was compared with the CPA extracted manually within a 100 x 100 m grid for accuracy assessment. The results is shown in Figure 3; there was an 83% match (coefficient of determination R2 = 0.83) between the values.

After then, the CPA extracted from the satellite image of 2009 was regressed with the Biomass derived from the DBH measured on the ground in the field plots. Out of the 58 field sample plots collected 38 samples were used for deriving the relation. The two values showed a linear relationship with the coefficient of determination R2 = 0.76 (Figure 4.).

The remaining 20 field samples were then used for the validation of the developed model. The extrapolated map had an accuracy of 85% based on observed and predicted Biomass values.

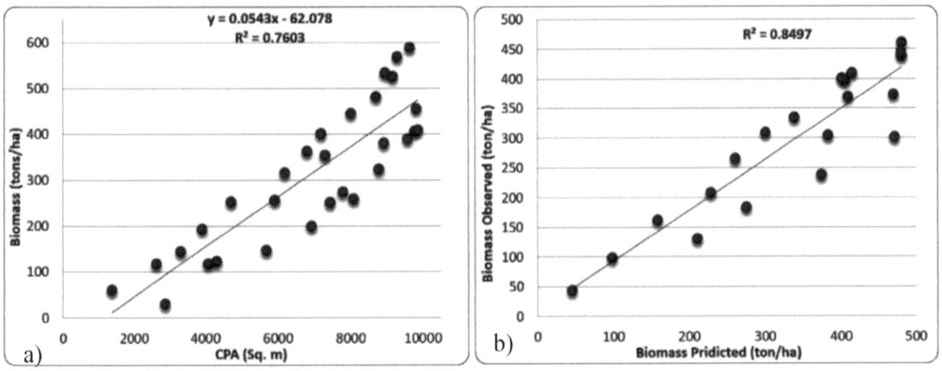

Figure 3. a) Relation between CPA and Biomass, b) Validation of model a)

3.1.1. Forest Biomass estimation at district level. The Classification and Regression Trees (CART) mapping model was used to map forest biomass at the district level. CART is a non-parametric decision tree learning technique that produces either classification or regression of trees, depending on whether the dependent variable is categorical or numeric, respectively.

At first a total of 18 parameters were fed into the CART system to see out which parameters will define the forest biomass in a best way. Out of these 18 parameters 15 parameters were that from the Landsat imageries the bands. Red, Green, Blue, Near Infra-Red and NDVI of the summer, winter and rainy season respectively. The other three parameters taken were the Dem, slope and aspect.

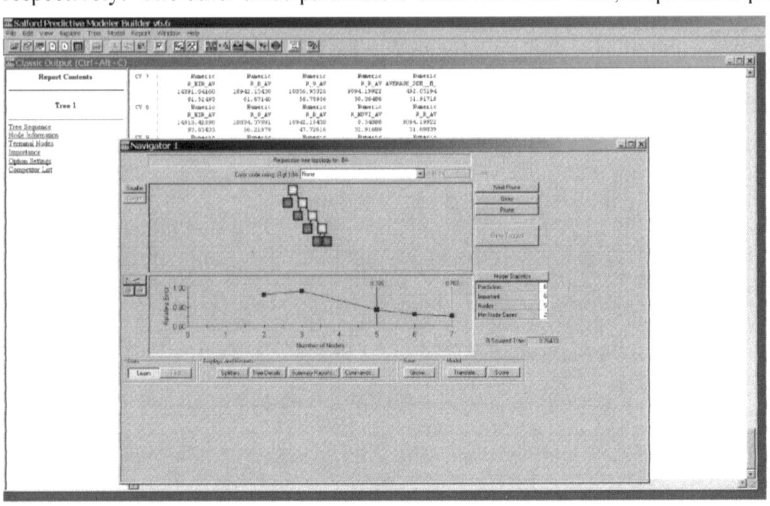

Figure 4. Finding and selecting the best defining parameters

CART find out the 5 parameters as the finest one which define the forest biomass in the best fit way (see Figure 6.) with coefficient of determination o.76. These five parameters are Rainy season NIR, rainy season Blue band, DEM, Rainy season Green band and Rainy NDVI in the respective order based on their higher influence (Figure 7.).

After that all these five major parameters along its biomass value for the plots were fed into MARS (Multivariate Adaptive Regression Splines) to generate the Multivariate model that will help to interpolate Biomass at district level.

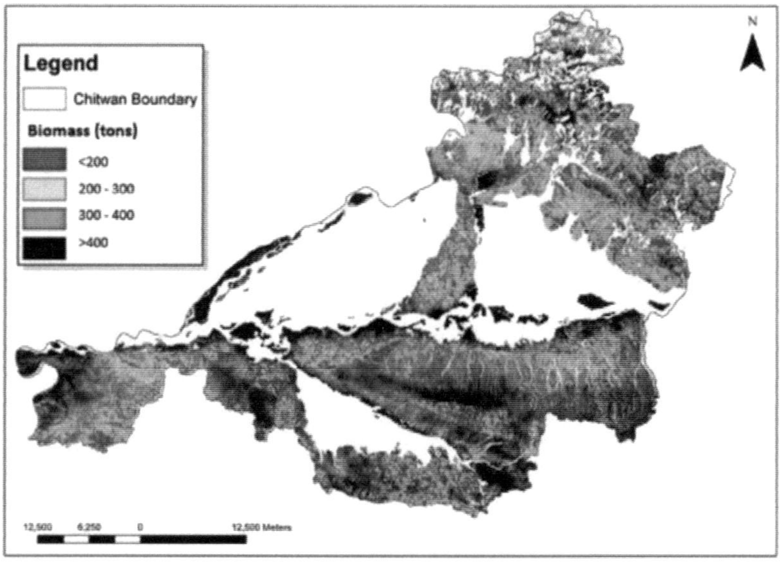

Figure 5. Forest Biomass map for chitwan district

4. Conclusion

The methodology for Above Ground Biomass mapping at both watershed and district level is convenient and easy to replicate. The use of the google earth virtual plots is very much cost effective and efficient. This approach could considerably reduce the requirements for field data for estimation of biomass and carbon in comparison with inventory methods based on enumeration of all trees in an area. Many researchers (Hussin et al., 2014; Karna et al., 2015) have explored different forms of allometric equations and found that the allometric equation from (Sharma & Pukkala, 1990) is one of the preferable one for the field biomass estimation in Nepal.

The next step in this research will be to expand the methodology to entire country and also to the HKH level with the use of other lower resolution imageries as MODIS. Instead of using commercial very high resolution satellite imagery, we are proposing to use less expensive imagery such as CARTOSAT-2 which will minimize the cost of estimation without compromising the quality. The proposed methodology is very cost effective and can be replicated with limited resources and time. The virtual plotting techniques will be tested to confirm that the method can be used by general foresters to obtain meaningful estimates of crown cover and biomass in their forest areas for reporting purposes.

References

Angelsen, A. (2008). Moving Ahead with REDD Issues , Options and Implications.

Baccini, A., Goetz, S. J., Walker, W. S., Laporte, N. T., Sun, M., & Sulla-Menashe, D. (2012). Estimated carbon dioxide emissions from tropical deforestation improved by carbon-density maps. *Nat Clim Change.*

Bastin, J. F., Barbier, N., Couteron, P., Adams, B., Shapiro, A., Bogaert, J., & De Canni??re, C. (2014). Aboveground biomass mapping of African forest mosaics using canopy texture analysis: Toward a regional approach. *Ecological Applications*, *24*(8), 1984–2001. http://doi.org/10.1890/13-1574.1

Cutler, M. E. J., Boyd, D. S., Foody, G. M., & Vetrivel, A. (2012). Estimating tropical forest biomass with a combination of SAR image texture and Landsat TM data: An assessment of predictions between regions. *ISPRS Journal of Photogrammetry and Remote Sensing*, *70*(April 2016), 66–77.

http://doi.org/10.1016/j.isprsjprs.2012.03.011

DeFries, R., Achard, F., Brown, S., Herold, M., Murdiyarso, D., Schlamadinger, B., & de Souza Jr., C. (2007). Earth observations for estimating greenhouse gas emissions from deforestation in developing countries. *Environmental Science & Policy*, *10*(4), 385–394. http://doi.org/http://dx.doi.org/10.1016/j.envsci.2007.01.010

FAO. (2010). *G LOBAL F OREST R ESOURCES C OUNTRY R EPORT.*

Frazier, R. J., Coops, N. C., Wulder, M. A., & Kennedy, R. (2014). Characterization of aboveground biomass in an unmanaged boreal forest using Landsat temporal segmentation metrics. *ISPRS Journal of Photogrammetry and Remote Sensing*, *92*(JUNE), 137–146. http://doi.org/10.1016/j.isprsjprs.2014.03.003

Gibbs, H. K., Brown, S., Niles, J. O., & Foley, J. a. (2007). Monitoring and estimating tropical forest carbon stocks: making REDD a reality. *Environmental Research Letters*, *2*(4), 045023. http://doi.org/10.1088/1748-9326/2/4/045023

Goetz, S. J., Baccini, A., Laporte, N. T., Johns, T., Walker, W., Kellndorfer, J., … Sun, M. (2009). Mapping and monitoring carbon stocks with satellite observations: a comparison of methods. *Carbon Balance and Management*, *4*(1), 2. http://doi.org/10.1186/1750-0680-4-2

Herold, M., Román-Cuesta, R. M., Hirata, Y., Laake, P. Van, Asner, G., & Heymell, V. (2011). A review of methods to measure and monitor historical forest degradation. *Unasylva*, *238*, 16–24.

Hussin, Y. A., Gilani, H., Van Leeuwen, L., Murthy, M. S. R., Shah, R., Baral, S., … Qamer, F. M. (2014). Evaluation of object-based image analysis techniques on very high-resolution satellite image for biomass estimation in a watershed of hilly forest of Nepal. *Applied Geomatics*, *6*(1), 59–68. http://doi.org/10.1007/s12518-014-0126-z

Ke, Y., & Quackenbush, L. J. (2008). Comparison of individual tree crown detection and delineation methods. *ASPRS 2008 Annual Conference Proceedings*, 11. Retrieved from http://www.asprs.org/publications/proceedings/portland08/0038.pdf

Kim, S.-R., Kwak, D.-A., Lee, W.-K., oLee, W.-K., Son, Y., Bae, S.-W., … Yoo, S. (2010). Estimation of carbon storage based on individual tree detection in Pinus densiflora stands using a fusion of aerial photography and LiDAR data. *Science China. Life Sciences*, *53*(7), 885–97. http://doi.org/10.1007/s11427-010-4017-1

Lorenz, K., & Lal, R. (2010). *Carbon sequestration in forest ecosystems*. Springer.

Lu, D. (2006). The potential and challenge of remote sensing ‐ based biomass estimation. *International Journal of Remote Sensing*, *27*(7), 1297–1328. http://doi.org/10.1080/01431160500486732

Ploton, P., Pélissier, R., Proisy, C., Flavenot, T., Barbier, N., Rai, S. N., & Couteron, P. (2012). Assessing aboveground tropical forest biomass using Google Earth canopy images. *Ecological Applications : A Publication of the Ecological Society of America*, *22*(3), 993–1003. http://doi.org/10.1890/11-1606.1

Sharma, E. R. (1990). *Volume equations and biomass prediction of forest trees in Nepal*. Forest Survey and Statistics Division.

Shih, F. Y., & Cheng, S. (2005). Automatic seeded region growing for color image segmentation. *Image and Vision Computing*, *23*(10), 877–886. http://doi.org/http://dx.doi.org/10.1016/j.imavis.2005.05.015

Smil, V. (2010). *Energy myths and realities: bringing science to the energy policy debate*. Government Institutes.

Uddin, K., Gilani, H., Murthy, M. S. R., Kotru, R., & Qamer, F. M. (2015). Forest Condition Monitoring Using Very-High-Resolution Satellite Imagery in a Remote Mountain Watershed in Nepal. *Mountain Research and Development*, *35*(3), 264–277. http://doi.org/10.1659/MRD-JOURNAL-D-14-00074.1

YousifAli, H., & Hammad, G. (2011). Mapping carbon stocks in community forests of Nepal using high spatial resolution satellite images. (Earth observation and climate change.). *ICIMOD, Sustainable Mountain Development*, *60*, 22–24.

Dynamic drought risk assessment using crop model and remote sensing techniques

H Sun [1,2,*], **Z Su**[1,2], **J Lv** [1,2], **L Li** [3], **Y Wang** [1,2]

[1] Research Centre on Flood and Drought Disaster Reduction of the Ministry of Water Resources of China, Beijing 100038, China
[2] China Institute of Water Resources and Hydropower Research, Beijing 100038, China
[3] College of Life Science and Technology, Beijing Normal University, Beijing, 100875, China
[*]Corresponding author: E-mail: sunhq@iwhr.com (H. Sun)

Abstract. Drought risk assessment is of great significance to reduce the loss of agricultural drought and ensure food security. The normally drought risk assessment method is to evaluate its exposure to the hazard and the vulnerability to extended periods of water shortage for a specific region, which is a static evaluation method. The Dynamic Drought Risk Assessment (DDRA) is to estimate the drought risk according to the crop growth and water stress conditions in real time. In this study, a DDRA method using crop model and remote sensing techniques was proposed. The crop model we employed is DeNitrification and DeComposition (DNDC) model. The drought risk was quantified by the yield losses predicted by the crop model in a scenario-based method. The crop model was re-calibrated to improve the performance by the Leaf Area Index (LAI) retrieved from MODerate Resolution Imaging Spectroradiometer (MODIS) data. And the in-situ station-based crop model was extended to assess the regional drought risk by integrating crop planted mapping. The crop planted area was extracted with extended CPPI method from MODIS data. This study was implemented and validated on maize crop in Liaoning province, China.

1. Introduction

Drought is a common occurrence natural disaster, which directly and primarily affects agriculture production. The Drought Risk Assessment (DRA) of agricultural has important significance for ensuring food security. Generally, DRA of a given region is to evaluate its exposure to the hazard and the vulnerability to extended periods of water shortage. This kind of analysis is to estimate the static risk of the region according to the characteristics of locality, normally it is an annual risk.

However, due to the slow occurrence/slow-onset characteristic of droughts (Wilhite, 1991), it is difficult to detect the emergence of droughts, monitor the situation, assess the impact and evaluate the risk. Specifically, the impacts on agriculture will be very different if the drought occurs during crop growth periods (Kulshreshtha, 1989). Therefore, it is necessary to estimate the drought risk taking into account of the crop growth process and water stress conditions real-time.

There are a large number of studies on DRA (Wilhite, 2007; Zhou, 2010), but most of these studies are concerned with *static* DRA (Gong, 2010; Wilhite, 2000; Wu, 2004). Kumar (Kumar, 2007) indicated that crop yield is a comprehensive result of drought stress, so it can be used to quantify drought risk. Wu (Wu, 2004) suggested that the yield losses under drought stress can be used to measure the drought severities. These studies strongly support our method of using the crop model to assess the drought risk which described in this paper. Some similar ideas, such as Jia and Wang (Jia, 2012) using EPIC model to evaluate the maize drought risk of China, Yu and Li (Yu, 2014) using DNDC model to analyse the impact of drought. However, these studies mainly focus on the crop simulation process, but not the drought risk assessment.

In this paper, we propose to use crop model and remote sensing techniques for assessing drought risk dynamically. The main contributions of this paper focus on: 1) a Dynamic Drought Risk Assessment (DDRA) method based on crop model was put forward; 2) the remote sensing technique was used to help to promote the performance of the DNDC model and mapping the crop planted area.

2. Materials and Methodology

2.1 Framework of DDRA

The DDRA method we proposed mainly includes the crop modelling and remote sensing techniques. The crop model dynamically simulates the crop growth process and predicts the crop yield with a scenario based method. The remote sensing techniques will play two roles in the framework, which on the one hand is to retrieve the Leaf Area Index (LAI) to calibrate the crop model, and on the other hand is to extract the crop planting area. And the dynamic drought risk is quantitatively assessed with the predicted yield losses.

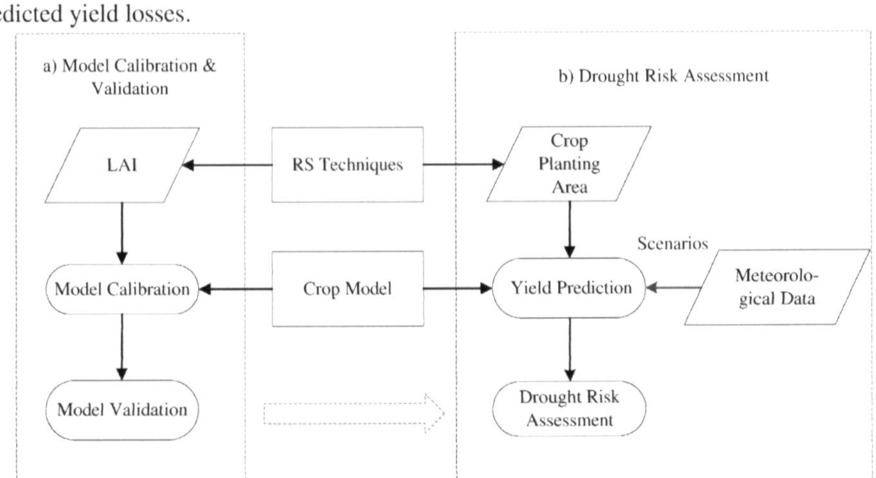

Figure 1. DDRA framework using crop model and remote sensing techniques.

The DDRA includes two procedures: a) model calibration & validation, and b) drought risk assessment. First of all, the model parameters were re-calibrated with remote sensing techniques, in which the model was revised partially according to the LAIs retrieved from remotely sensed images. After the development and validation of the model, it was implemented to predict the final crop yields dynamically. As the future climate conditions are unknown, we used a scenario-based method to make the assessment dynamically. It is significant to know how many area planted to assess the regional drought risk, which the annual crop planting area was extracted with a remote sensing techniques.

By this hybrid method combined of crop model and remote sensing techniques, the performance of the simulation was significantly improved, the in situ assessment was successfully extended spatially, and the drought risk can be assessed dynamically by the daily rolling simulation process.

2.2 Study Area and Data Description

The study area is Liaoning Province, located in northeast of China. It has 14 prefecture cities, 19 districts and 81 counties with agriculture planting. The annual precipitation ranges from 600 to 1100 mm, and the average annual temperature ranges from 7 to 11 °C. It is a drought prone area, especially the severe droughts hit the northwest of the region more frequently. In recent years, more than 70% of the total farmland was maize of rain fed cultivated, which is out of the effect of irrigation during the analysis process of drought issue.

The data we utilized in the study mainly include meteorological data and remotely sensed images. The meteorological data consists of 820 rainfall stations and 27 meteorological stations with observations of temperature, precipitation, etc. from 1995-2009. The Thiessen polygon method was employed to partition the meteorological stations into separate zones, such that the rainfall stations in the identical zone had the corresponding temperature. The MODIS products of MOD15A2 (LAI) to and MOD13A1 (vegetation index) were used to calibrate the model and extract crop planted area, respectively.

2.3 Crop Model and Model Calibration

2.3.1 DNDC Model: The crop model we adopted is DNDC (DeNitrification-DeComposition) model (Li, 1992). It was widely applied in crop simulation and yield prediction (Han, 2014; Nyckowiak, 2013). The performance of the model has been verified for field and regional scales all over the world (Zhang, 2002). The model input consists of meteorological data, soil properties, land use and cultivation management, etc. The meteorological observations are the driven factors for the model running. Most of the parameters are available by measuring and surveying, while some need to be calibrated by the historical data to promote the performance of the model.

2.3.2 Model Calibration: The primary principle of model calibration is to make the simulated values as closed as possible to the observed ones. It is better to use multiple variables to perform the model calibration. In the study, we calibrated the model by variables of yield and Leaf Area Index (LAI), successively. First of all, the historical yields data were used to tune the parameters in a coarse process. And then, the remotely sensed retrieved LAI data were employed to make a fine calibration. As the yield-based calibration could not provide the 'monitoring' of the crop growth procedure, and a small number of samples (one record each year) are available for calibration. Thus, the LAI retrieved from remotely sensed images were used to re-calibrate the model. Many studies indicated that MODIS-LAI product is generally underestimated (Fensholt, 2004; Tian, 2002). The relationship between the MODIS-LAI and the ground truth vary with the crop and the season. It is proved that the MODIS-LAI is about 33%~53% under the synchronous ground measured LAI value (Yang, 2010). Here, we modify the MODIS-LAI by multiply a constant k, and given k equal to 1.5.

2.4 Crop Planted Mapping from Remote Sensing Images

An alternative remote sensing techniques for crop planted mapping could not only provide the result fast and cost saving, but also demonstrate the spatial distribution of crop planted. Based on the crop planted map, the site-based model could be extended to the regional one. A number of previous studies have attempted to map cropping planted area using remote sensing techniques in China (Li, 2003; Xiao, 2006). By now, the primary technique is the phenology-based remote sensing technique. However, the main challenge is to identify crop phenological cycles from time series of satellite-derived vegetation indices, such as the Enhanced Vegetation Index (EVI) or Normalized Difference Vegetation Index (NDVI) (Ganguly, 2010; Zhang, 2003). Here, we adopted a novel method to map the maize planted area, which was proposed by Pan and Li (Pan, 2012) and Li and Friedl (Li, 2014). Basically, it set up a Crop Proportion Phenology Index (CPPI) to quantitative analyse the relationship between the MODIS vegetation index (VI) time series and winter wheat crop area, and a time series mapping crop cycles in China was raised based on MODIS-EVI data.

Here, we implemented the maize phenology analysis from time series of MODIS-EVI data, and identified the planted area by CPPI specifically for maize. A series of 8-day MODIS product from 2000 to 2010 were used to mapping the crop planted area, and the Landsat TM images were sampled as ground-truth.

2.5 Drought Risk Assessment

2.5.1 Scenarios Analysis: Scenarios analysis ensure the accomplishment of the yield prediction. Each time to predict crop yield is a full simulation from sowing to harvesting. Based on the current time point, in the past time, the meteorological observations for the model input are available, however, in the future, the weather conditions are unknown. The scenarios are the assumption for future unknown weather conditions from now on to the harvest time. In this study, a series of scenarios were required to represent different drought conditions. Here, we specified 8 scenarios: 1) no effective rainfall in 5, 10, 20, 30 and 50 days, and without water stress afterwards, noted as S5, S10, S20, S30 an S50, respectively; 2) 3 typical years represent the wet, medium, and dry years, noted as SW, SM, and SD, respectively; 3) without water stress from now on to the harvest time, noted as S0.

2.5.2 Yield losses: The drought risk was quantitatively estimated by simulated yield losses. It is regarded that the simulated yield losses is entirely caused by drought disaster. Just because in the simulation process, the crop model integrated all the impact factors of drought intrinsically, and the ultimate yield losses completely come from water stresses. However, the main challenge is determine the normal yield. The normal yield we defined here is the potential maximum yield that the crop grow without any stress (such as water, temperature, fertilizer, etc.) during the whole growing season. Thus, the normal yield is completed determined by the crop physiological characteristics and the soil properties. Therefore, we simulated the normal yields by running the model under the condition of no stress preliminarily. Then the yield losses was directly obtained as the portion that the predicted yield less than the normal yield.

2.5.3 Drought Risk Classification: The drought risk classification carries out the grades of the drought risk severities according to the quantitative yield losses. Here, we categorized the drought risk by the yield losses rate instead of the amount. The drought risk is divided into 4 levels, which are I, II, III and IV with the yield losses rate of less than 10%, range from 10% to 25%, range from 25% to 35% and more than 35%, respectively.

3. Results

3.1 Model Validation

We used the parameters calibrated by data from 1995 to 2004 to simulate the yields for each county from 2005-2009. The result shows that the model estimated yield corresponded well with the reported yield from statistical yearbook (Figure 2). And then the provincial level yield can be obtained straightforward from the county level yield by multiplying the crop planted area retrieved from remotely sensed data. The provincial level yields estimation is evaluated with RMSE of 237.7 kg/ha, and R^2 of 0.77.

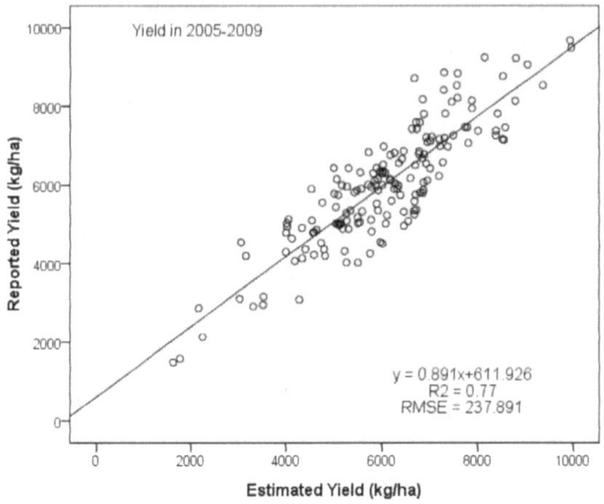

Figure 2. The comparison of the simulated yields and the historical reported yields in 2005-2009

Figure 3. Maize planted mapping of Liaoning Province in 2006

3.2 Maize Planted Mapping

A series of crop planted area in 2005-2009 was mapped using the remote sensing technique. Due to the stability of the planting structure, the annual crop sowing area would not have a great change. Therefore, here, we only present the maize planted map in 2006 (Figure 3). It can be seen that the major maize planted regions are in the north of the province, including Shenyang, Tieling, Fuxin and Jinzhou, etc., which is in agreement with the reported results from statistical yearbook (Table 2). Statistically, the planted area of each prefecture was retrieved from the map, and compared with the statistical yearbook data (Table 2). The Mean Absolute Percentage Error (MAPE) is 21.0%, 16.3%, 14.2%, 14.3% and 13.8% from 2005 to 2009, respectively.

Table 2. The maize planted area from remotely sensed images (column "RS") and the records from statistical yearbook (column "SY") for each prefectures in Liaoning Province. (unit: 10^3 ha)

Prefecture	2005		2006		2007		2008		2009	
	SY	RS	SY	RS	SY	RS	SY	RS	SY	RS
Shenyang	337.0	346.8	305.7	337.5	304.8	326.4	339.9	353.3	344.8	360.1
Dalian	180.6	191.1	185.6	197.5	182.9	189.1	188.3	195.8	189.3	203.4
Anshan	152.4	161.7	160.6	172.3	160.1	165.4	164.4	170.2	165.2	166.4
Fushun	59.1	43.7	69.2	51.2	68.2	41.7	69.7	40.5	69.8	42.2
Benxi	35.0	6.5	35.3	24.6	35.1	28.1	35.1	27.5	35	26.7
Dandong	95.2	64.2	99.4	78.5	98	84.2	99.3	60.5	101	77.4
Jinzhou	265.3	307.5	281.8	290.6	271.8	297.3	295.8	315.8	294.7	322.2
Yingkou	41.9	47.5	46	48.2	46.6	47.5	47.2	48.9	47.4	48.5
Fuxin	200.5	212.7	215.9	231.1	193.7	206.2	217.4	230.9	223.4	245.5
Liaoyang	78.6	68.9	79.1	66.3	78.7	68.9	80.6	59.8	85.3	79.6
Panjin	13.5	20.9	13	18.8	13.6	18.7	14.2	16.8	14.5	19.9
Tieling	302.1	341.6	303.2	356.8	276	331.3	323	356.4	346.5	354.4
Chaoyang	191.0	216.7	216.5	240.4	215.3	202.7	236	239.5	253.7	270.8
Huludao	124.6	145.2	126.1	151.2	123	145.2	135.2	143.6	147.8	160.3

3.3 Case study in Liaoning Province

Based on the crop model simulation, scenarios analysis and crop planted mapping techniques described above, the DDRA can be implemented day by day. Here, only the cases of June 1, 2009 (Figure 4) and July 1, 2009 (Figure 5) in Liaoning Province, China were demonstrated, respectively.

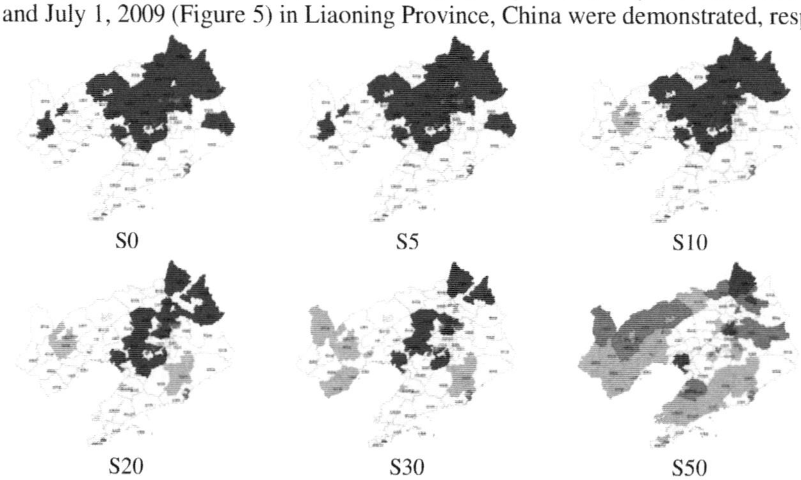

Sw Sm Sd

Figure 4. The DDRA for each scenario in Liaoning Province, China on June 1, 2009.

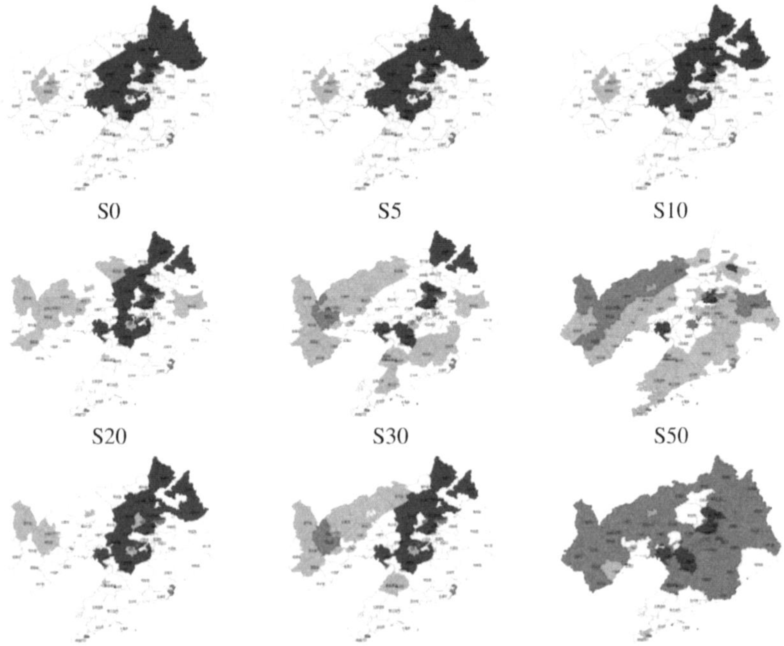

Sw Sm Sd

Figure 5. The DDRA for each scenario in Liaoning Province, China on July 1, 2009

The results show that the drought risk grades increased if no effective rainfall days prolonged. And no effective rainfall within 10 days (S0, S5 and S10) will not lead to a substantial increase in yield losses, but more than 30 days no effective rainfall (S30) will results sharp yield losses. This is because the soil moisture could meet the crop growth in a limited time, which basically in the scope of one month. According to the comparative analysis of Figure 4 and Figure 5, the drought risk had increased from June 1 to July 1. This is proved that the amount of rainfall in Liaoning is seriously low, which only 18% of average annual value. Due to the previous drought stress, even the scenario of wet year (Sw) and without water stress hereafter (S0) were of significant risk yet. Spatially, the drought risk in the northwest region is generally higher than in other region, which is coincident with the reality.

The case study shows the complete process of the method and proved its reasonability. This method can be implemented in a real-time to dynamically assess the drought risk. The result is very helpful for drought mitigation.

4. Conclusion

This study proposed a *dynamic* drought risk assessment method using crop model and remote sensing techniques. We employed the DNDC model to simulate the crop growth process and predict the yield. After calibrating the crop model by series of historical yield records, we then re-calibrated the model by LAI retrieved from MODIS data during crop growing. A scenario-based method was used to solve the problem of unknown future climate condition. Then the drought risk was quantified by the yield losses. And the future drought risk was evaluated under the assumed scenarios. However, the in situ drought risk assessment could be extended to regional analysis by integrating the crop planted mapping with a remote sensing method.

The case study for maize in Liaoning Province, China shows that the DDRA method was validated and applicable. This method is of great significance for drought mitigation in practical application. The manager could make an appropriate drought relief measurement according to DDRA results, including the current water stress of the crop and the future crop yields, etc.

However, there are a lot of sophisticated analysis in thse method. In the future, we will continuously improve the model performance, integrate the weather forecasting result in the scenarios, promote the crop mapping accuracy, and so on.

5. References

[1] Fensholt, R., Sandholt, I., Rasmussen, M.S., 2004. Evaluation of MODIS LAI, fAPAR and the relation between fAPAR and NDVI in a semi-arid environment using in situ measurements. *Remote Sensing of Environment*, **91**, pp 490-507.

[2] Ganguly, S., Friedl, M.A., Tan, B., et al., 2010. Land surface phenology from MODIS: Characterization of the Collection 5 global land cover dynamics product. *Remote Sensing of Environment*, **114**, pp 1805-1816.

[3] Gong, Z., Sun, R., Li, A., 2010. Risk prediction of agricultural drought in China, 2010 7th International Conference on Fuzzy Systems and Knowledge Discovery, FSKD 2010, August 10, 2010 - August 12, 2010. IEEE Computer Society, Yantai, Shandong, China, pp 927-930.

[4] Han, J., Jia, Z.K., Wu, W., et al., 2014. Modeling impacts of film mulching on rainfed crop yield in Northern China with DNDC. *Field Crops Research*, **155**, pp 202-212.

[5] Jia, H., Wang, J., Cao, C., et al., 2012. Maize drought disaster risk assessment of China based on EPIC model. *International Journal of Digital Earth*, **5**, pp 488-515.

[6] Kulshreshtha, S.N., Klein, K.K., 1989. Agricultural drought impact evaluation model: A systems approach. *Agricultural Systems*, **30**, pp 81-96.

[7] Kumar, V., Panu, U., 2007. Predictive assessment of severity of agricultural droughts based on agro-climatic factors. *Jawra Journal of the American Water Resources Association*, **33**, pp 1255-1264.

[8] Li, C., Frolking, S., Frolking, T.A., 1992. A model of nitrous oxide evolution from soil driven by rainfall events: 1. Model structure and sensitivity. *Journal of Geophysical Research: Atmospheres*, **97**, pp 9759-9776.

[9] Li, L., Friedl, M., Xin, Q., et al., 2014. Mapping Crop Cycles in China Using MODIS-EVI Time Series. *Remote Sensing*, **6**, pp 2473-2493.

[10] Li, X., Wang, X., 2003. Changes in agricultural land use in China: 1981-2000. *Asian Geographer*, **22**, pp 27-42.

[11] Nyckowiak, J., Lesny, J., Juszczak, R., et al., 2013. Application of LandscapeDNDC Model for Simulating Soil Water Content and Crop Yield Quantity. *Rocznik Ochrona Srodowiska*, **15**, pp 1937-1951.

[12] Pan, Y., Li, L., Zhang, J., et al., 2012. Winter wheat area estimation from MODIS-EVI time series data using the Crop Proportion Phenology Index. *Remote Sensing of Environment*, **119**, pp 232-242.

[13] Tian, Y., Woodcock, C.E., Wang, Y., et al., 2002. Multiscale analysis and validation of the MODIS LAI product: I. Uncertainty assessment. *Remote Sensing of Environment*, **83**, pp 414-430.

[14] Wilhite, D.A., 1991. Drought planning: A process for state government. *Water Resources Bulletin*, **27**, pp 19-38.

[15] Wilhite, D.A., 2000. Drought planning and risk assessment: Status and future directions. *Annals of Arid Zone*, **39**, pp 211-230.

[16] Wilhite, D.A., 2007. Preparedness and coping strategies for agricultural drought risk management: Recent progress and trends. *Springer Berlin Heidelberg*, pp 21-38.

[17] Wu, H., Wilhite, D.A., 2004. An operational agricultural drought risk assessment model for Nebraska, USA. *Natural Hazards*, **33**, pp 1-21.

[18] Xiao, X., Boles, S., Frolking, S., et al., 2006. Mapping paddy rice agriculture in South and Southeast Asia using multi-temporal MODIS images. *Remote Sensing of Environment*, **100**, pp 95-113.

[19] Yang, F., Sun, J., Zhang, B., et al., 2010. Assessment of MODIS LAI product accuracy based on the PROSAIL model,TM and field measurements. *Transactions of the Chinese Society of Agricultural Engineering*, **26**, pp 192-197(196).
[20] Yu, C., Li, C., Xin, Q., et al., 2014. Dynamic assessmen t of the impact of drought on agricultural yield and scale-dependent return periods over large geographic regions. *Environmental Modelling & Software*, **62**, pp 454-464.
[21] Zhang, X., Friedl, M.A., Schaaf, C.B., et al., 2003. Monitoring vegetation phenology using MODIS. *Remote Sensing of Environment*, **84**, pp 471-475.
[22] Zhang, Y., Li, C.S., Zhou, X.J., et al., 2002. A simulation model linking crop growth and soil biogeochemistry for sustainable agriculture. *Ecological Modelling*, **151**, pp 75-108.
[23] Zhou, K., Liu, C.N., 2010. Models of Agriculture Risk Management with Agriculture Ontology, in: Yi, X., Mi, L. (Eds.), Materials and Manufacturing Technology, Pts 1 and 2, vol. 129-131, pp 743-747.

6. Acknowledgements

The authors gratefully acknowledge funding support for this research provided by the National Natural Science Foundation Project of China (No. 51209220), Special Fund for Water Resources Research in the Public Interest (201401036), and Fund of China Institute of Water Resources and Hydropower Research (JZ0145B592016). We especially thank and cherish the memory of Dr. C. Li (1941-2015), the founder of the DNDC model, for his contribution in the work.

IOP Conf. Series: Earth and Environmental Science **57** (2017) 012013 doi:10.1088/1755-1315/57/1/012013

Infrastructure stability surveillance with high resolution InSAR

Timo Balz[1,2], Ralf Düring[3]

[1]State Key Laboratory of Information Engineering in Surveying, Mapping and Remote Sensing, Wuhan University, Wuhan 430076, China

[2]Collaborative Innovation Center for Geospatial Technology, Wuhan 430076, China

[3] Airbus DS Intelligence, Beijing 101312, China

balz@whu.edu.cn

Abstract. The construction of new infrastructure in largely unknown and difficult environments, as it is necessary for the construction of the New Silk Road, can lead to a decreased stability along the construction site, leading to an increase in landslide risk and deformation caused by surface motion. This generally requires a thorough pre-analysis and consecutive surveillance of the deformation patterns to ensure the stability and safety of the infrastructure projects. Interferometric SAR (InSAR) and the derived techniques of multi-baseline InSAR are very powerful tools for a large area observation of surface deformation patterns. With InSAR and deriver techniques, the topographic height and the surface motion can be estimated for large areas, making it an ideal tool for supporting the planning, construction, and safety surveillance of new infrastructure elements in remote areas.

1. Introduction

The construction of new infrastructure can lead to a decreased stability along the construction site. The extent and the safety risk of the decreased stability depend on the geological and geomorphological situation, but it can lead to an increase in landslide risk and deformation caused by surface motion. Interferometric SAR and the derived techniques, like differential InSAR and multi-baseline InSAR, are very powerful tools for a large area observation of such deformation patterns.

With TerraSAR-X, COSMO SkyMed, and Kompsat-5, high-resolution SAR satellites are available that are capable of providing detailed information that can be used for mapping. Thanks to their interferometric capability, precise digital elevation models (DEM) can be created that can e.g. be used during the planning and construction phase of new infrastructure projects. Even more so, with differential interferometry and derived techniques, surface motions can be estimated in a very high precision [1]. These techniques therefore allow for high-precision deformation measurements along e.g. roads and railways. This offers the possibility to detect deformations not only along a road or railway, but also deformation patterns nearby the corresponding infrastructure project that could cause problems in the near future. This is true for the risk to the traffic infrastructure itself, as well as for the surrounding infrastructure like urban settlements or energy infrastructure.

However, for the observation of long linear structures, like roads and railways, the current limitations of data acquisitions of spaceborne systems makes this task very difficult. Roads and railways do not follow the orbit directions of the SAR systems. Consequently, a large number of data acquisitions to track and trace these structures are required. However, as the acquisition capability of existing systems is limited, conflicts in the acquisition plan can (and will) occur, which typically leads to interruptions in the acquisition process, which may cause time gaps in the acquisition of adequate time series. This makes the InSAR processing difficult or, in case of large deformation with high amplitudes or temporal decorrelation, even impossible.

Next Generation X-Band SAR systems, like the follow-on mission of TerraSAR-X, will be able to use 1200 MHz spectrum range, as approved recently at the World Radio Conference in November 2015. This will constitute a game changer in terms of imaging capacity, image resolution and image quality. Specifically, larger areas can be mapped more frequently as much larger swath width can be achieved for stripmap imaging modes. Additionally, very high image resolution, which can achieve 0.25 m x 0.25 m with a very high radiometric performance, will significantly increase the density and accuracy of measurements. Thus, precise deformation measurements along long linear structures will be possible in an unprecedented way.

In the following section, we will describe the status of high-resolution SAR data today. In section 3, we demonstrate persistent scatterer interferometry (PS-InSAR) with very high resolution data. Then, in section 4, we will discuss and give and outlook to future SAR systems. Finally, conclusions are drawn.

2. Very high-resolution SAR data

With the introduction of the new SAR imaging modes for TerraSAR-X in 2014 [2], a new very high resolution SAR mode, the staring spotlight mode, was made available allowing for an increased resolution in azimuth direction up to around 25 cm. This is achieved by increasing the synthetic aperture length with an increased illumination time.

In the opposite way, Sentinel-1, achieves wide coverage using TOPS [3]. TOPS is basically the opposite of the spotlight mode. When in spotlight mode the synthetic aperture is enlarged by changing from forward squinting to backward squinting along azimuth, in TOPS the synthetic aperture is shortened by shifting from backward squinting to forward squinting during acquisition. This allows an improvement to the normal ScanSAR mode by avoiding scalloping effects.

As we can see in Figure 1, mid-resolution SAR systems, like Sentiel-1, are useful for regional surveillance, but details in urban areas are hardly recognizable. With 3m spatial resolution stripmap data, as shown in Figure 2 with an example from Kompsat-5, details in urban areas are clearly recognizable on a building block level. However, when we move to very high-resolution staring spotlight data, as shown in Figure 3, details on single buildings, are clearly identifiable, making it ideal for the surveillance of dense urban areas. Staring spotlight does, due to the nature of spotlight mode data acquisition, not allow continuous coverage in azimuth and is therefore only suitable for hot-spot surveillance, but not for large or regional scale surface motion estimation.

Using SAR for interferometric applications, like height measurement or surface motion estimation e.g. with PS-InSAR, the application can benefit from a higher resolution. Although in principle the achievable precision in height and motion measurement depends on the wavelength and not on the spatial resolution, an improved spatial resolution can be used for a better characterization of objects, DEM generation with finer resolution, or e.g. noise reduction by averaging over larger areas. So, although in theory the precision of interferometric measurements does not depend on the spatial resolution, a very high resolution can in practice tremendously improve the product quality, as we will demonstrate below in section 3.

Figure 1. Subset of the Sentinel-1 IW image of Wuhan acquired on 2016-01-24 (Copernicus Sentinel data 2016)

Figure 2. Subset of the Kompsat-5 stripmap image of Wuhan acquired on 2016-01-03 (© KARI, 2016)

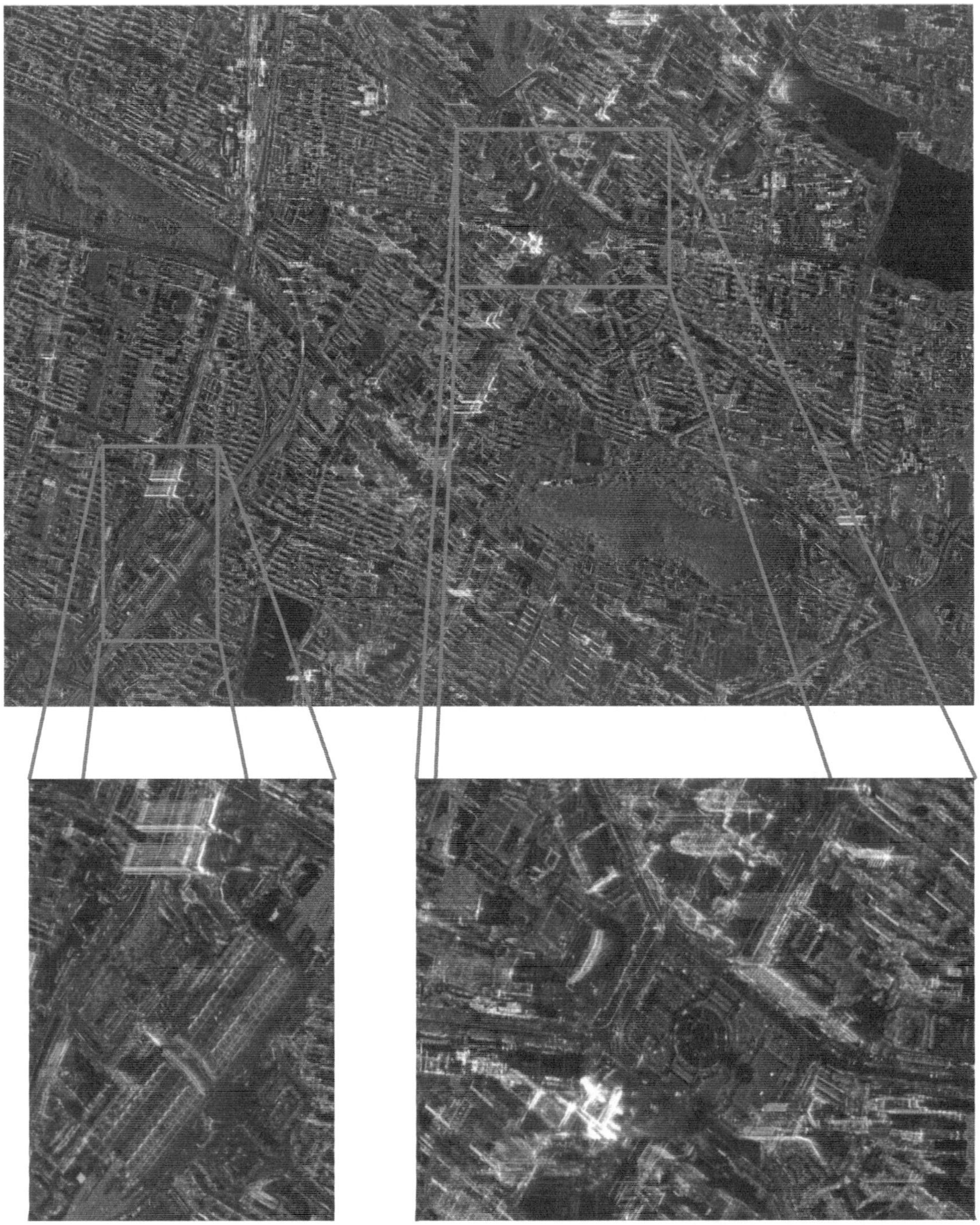

Figure 3. TerraSAR-X staring spotlight mode image of Wuhan acquired on 2016-01-12 (© DLR 2016)

3. Very high-resolution PS-InSAR

Persistent scatterer interferometery (PS-InSAR) uses point-scatterers that are interferometrically stable over time to estimate long-time deformation at these stable scatterers that are called persistent scatterers (PS) [4]. PS-InSAR can be used to estimated surface motions for urban areas and infrastructure. Below in Figure 5, we can see the estimated surface motions of the Beijing International Airport derived from a stack of 30 TerraSAR-X stripmap images with approximately 3 m spatial resolution shown in Figure 4. This demonstrates the high resolution and high precision deformation estimation with TerraSAR-X stripmap data.

Figure 4. Mean amplitude of 30 TerraSAR-X stripmap images

Figure 5. Mean line-of-sight velocity over Beijing International Airport estimated using SARProZ [5]

An even higher spatial resolution can lead to a higher PS point density and a better estimation of the atmospheric phase screen. As shown below in Figure 6, a very high PS point density can be achieved with the TerraSAR-X staring spotlight data. In the example below, we see a PS density of around 25.000 PS / km^2. With such a high PS density, several independent deformation measurements on a single building are possible, allowing for the detection of motion differences in single buildings; such motion differences are especially dangerous and it is therefore important to have a high PS point density, to be able to detect such dangerous motion patterns.

Figure 6. Mean line-of-sight velocity over Wuhan estimated using WhuPsi

4. Discussion and outlook to future systems

As demonstrated above in section 3, a high PS point density is beneficial for urban subsidence estimation and is crucial for infrastructure stability analysis. However, in spotlight mode, and even more so in staring spotlight mode, the coverage is limited. As we can see above in Figure 3 and Figure 6, only a small subset of Wuchang can be acquired in staring spotlight mode and this is just a fraction of the urban area of Wuhan. Staring spotlight is therefore valuable for the surveillance of small areas or single objects of interest, but not suitable for city-wide or regional motion estimation. However, in stripmap, ScanSAR, or TOPS mode, the spatial resolution is reduced and the achievable PS point density is significantly lower.

Therefore, we would like to achieve a high spatial resolution with a better coverage. This is even more true, if we move from city-wide surveillance to regional applications as it will be necessary for the New Silk Road. However, there are several restrictions in sensor design limiting the available resolution and swath width. One is the limited bandwidth that limits the possible spatial resolution in range direction. Before the World Radio Conference in November 2015, the bandwidth for X-band SAR systems was limited to 300 MHz, limiting the range resolution of X-band SAR systems to around 80 cm. Now, since November 2015, up to 1200 MHz spectrum range are now possible, which will allow an increased resolution in range for new SAR systems, like the follow-on mission of TerraSAR-X.

This will allow stripmap mode data acquisition in approximately 0.25 x 0.25 m resolution. With such a system, a wide area coverage with very high resolution and good radiometric quality would be possible. This is ideal for very high resolution interferometry, especially regarding the stability analysis of buildings and infrastructure.

5. Conclusions

With the launch of TerraSAR-X, high-resolution SAR data become widely available. This lead to a paradigm shift in SAR and interferometric SAR processing. Nowadays, persistent scatterer interferometry with high-resolution SAR data can be used to detect several PS points on a single building and estimate motion differences on a single building.

However, the highest resolution is achieved in spotlight mode, hindering the acquisition for large areas. We would prefer to achieve this resolution in stripmap mode, so that large areas can be acquired with high resolution. This will be possible in future SAR systems and such a high resolution sensor would provide the perfect data basis for high resolution and high precision surface motion estimation for ensuring the stability and safety of infrastructure projects, e.g. along the New Silk Road.

References

[1] Ferretti A, Savio G, Barzaghi R, Borghi A, Musazzi S, Novali F, Prati C, Rocca F 2007 Submillimeter Accuracy of InSAR Time Series: Experimental Validation *IEEE Transactions on Geoscience and Remote Sensing* 45(5) pp. 1142–1153.

[2] Mittermayer J, Wollstadt S, Prats-Iraolo P, Schreiber R 2014 The TerraSAR-X staring spotlight mode concept *IEEE Transactions on Geoscience and Remote Sensing* 52 pp. 3695-3706.

[3] Yague-Martinez N, Prats-Iraola P, Rodriguez Gonzalez F, Brcic R, Shau R, Geudtner D, Bamler R 2016 Interferometric Processing of Sentinel-1 TOPS Data *IEEE Transactions on Geoscience and Remote Sensing* 54(4) pp. 2220–2234.

[4] Ferretti A, Prati C, Rocca F 2001 Permanent scatterers in SAR interferometry *IEEE Transactions on Geoscience and Remote Sensing* 39(1) pp. 8–20.

[5] Perissin D, Wang Z, Wang T 2011 The SARPROZ InSAR Tool for Urban Subsidence/manmade Structure Stability Monitoring in China *Proc. of ISRSE 2011*, Sidney, Australia.

Mapping mountain torrent hazards in the Hexi Corridor using an evidential reasoning approach

Youhua Ran [1,2], **Jinpeng Liu** [3], **Feng Tian** [4], **Dekai Wang** [3]

[1] Cold and Arid Regions Environmental and Engineering Research Institute, Chinese Academy of Sciences, 730000 Lanzhou, China
[2] University of Chinese Academy of Sciences, 100049 Beijing, China
[3] Geologic Hazards Prevention Institute, Gansu Academy of Sciences, 730000, Lanzhou, China
[4] College of Resources and Environment, Gansu Agricultural University, 730070, Lanzhou, China

Corresponding author's e-mail address: ranyh@lzb.ac.cn

Abstract. The Hexi Corridor is an important part of the Silk Road Economic Belt and a crucial channel for westward development in China. Many important national engineering projects pass through the corridor, such as highways, railways, and the West-to-East Gas Pipeline. The frequent torrent disasters greatly impact the security of infrastructure and human safety. In this study, an evidential reasoning approach based on Dempster-Shafer theory is proposed for mapping mountain torrent hazards in the Hexi Corridor. A torrent hazard map for the Hexi Corridor was generated by integrating the driving factors of mountain torrent disasters including precipitation, terrain, flow concentration processes, and the vegetation fraction. The results show that the capability of the proposed method is satisfactory. The torrent hazard map shows that there is high potential torrent hazard in the central and southeastern Hexi Corridor. The results are useful for engineering planning support and resource protection in the Hexi Corridor. Further efforts are discussed for improving torrent hazard mapping and prediction.

1. Introduction

Water-related natural hazards such as landslides and torrent processes pose a threat to human life, property and infrastructure (Totschnig and Fuchsm 2003; Balica et al., 2009). The hazard risk may increase due to continued socio-economic development (population increase, economic wealth, human activities in hazard-prone areas and urban development) and climate change (increase of heavy precipitation, changes in natural water cycles) (Turowski et al., 2009; Keiler et al., 2010; Mazzorana et al., 2012; Ronco et al., 2014). Disaster risk assessment is an important prerequisite to support the development of strategic adaptation and prevention measures to minimize torrent impacts (Ronco et al., 2014). Thus, the concept of risk represents the possibility of the occurrence of hazards (hazard) and their potential consequences (vulnerability) (Carter, 1992; Alexander, 2000; Kienholz et al., 2004).

The Hexi Corridor is a very important part of the Silk Road Economic Belt and a crucial channel for westward development in China. It is an engineering corridor and thus many important national engineering projects pass through the corridor, such as highways, railways, and the West-to-East Gas Pipeline. The Hexi Corridor is a densely populated area in the northwest of China and is also a cultural transmission path (Chen et al., 2003; Li et al., 2010a). The Hexi Corridor is in a transition zone between the alpine cold and extreme arid regions northwest of China. The oasis is dependent on a series of inland river sources from precipitation and glacial run-off in the Qilian Mountains and groundwater. Mountain torrents frequently occur in this area. Approximately 96 water-related disasters occurred in the last thirty years of the last century in Wuwei city in the eastern Hexi corridor, causing hundreds of deaths and approximately one billion dollars in economic losses (Luo et al., 2005). The precipitation in mountain areas has clearly increased since the 1990s. The increasing extreme precipitation, intensity and frequency of extreme precipitation may exacerbate the risk of flooding and landslides (Li et al., 2010b). Therefore, mapping the risk of mountain torrents in the Hexi Corridor is a realistic need to support engineering and landscape planning and contribute to cultural heritage protection and disaster prevention.

Several methodologies have been developed to assess torrent risk. The choice of methodology largely depends on the analysis objectives (Ronco et al., 2014; Lepuschitz, 2015). As a preliminary study, the objective of this study is to provide a hazard map of torrent hazard occurrence, i.e., the physical environmental relative risk map. The related environmental factors including slope, vegetation fraction, river channel buffers, meteorological factors such as rainstorm days, maximum runoff discharge, flood disaster history, and expert knowledge were integrated using Dempster-Shafer theory within a Geographic Information System (GIS) framework.

2. Study area

The Hexi Corridor is located in northwestern China (Figure 1). It refers to the area west of the Yellow River in Gansu province. The Hexi Corridor is surrounded by the Qilian Mountains to the south, by the Mazun, Heli, and Longshou Mountains and the Badain Jaran and Tengger Deserts to the north, by the Wushao Mountains to the east, and by the Kumtagh Desert to the west (Zhang et al., 2016). It includes Jiuquan, Zhangye, Wu Wei, Jinchang, and Jiayuguan cities and approximately 19 counties such as Dunhuang, Subei, Sunan, Minle, Gulang, under the jurisdiction of Jiuquan, Zhangye, and Wu Wei cities, respectively. The total population is approximately 4,920 thousand.

The Hexi Corridor is in the convergence zone of monsoon and continental climates. It is characterized by its distinct cold and arid landscapes (glacier, alpine meadow, forest, crops and desert) dominated by climate and terrain. The altitude of the Hexi Corridor decreases from south to north. The altitude of most of the mountains in the south is greater than 4000 metres and the altitude of the alluvial and diluvial plain is between 1300-2500 metres. Along an altitude gradient, the precipitation decreases and the air temperature increases from alpine to northern plain. The annual mean precipitation in the Qilian Mountains is approximately 250 mm and mainly occurs in summer. The oasis in the plain area mainly relies on water resources from the Qilian Mountains.

The hydrogeological characteristics of the Hexi Corridor vary from the south to the north and can be divided into discrete geomorphologic units, including the piedmont alluvial plain, alluvial plain, and desert. The southern region of the basin has extensive faulting with underlying bedrock, where an aquifer comprises highly permeable cobble and gravel deposits 300–500 metres thick. From the northern edge of this diluvial fan, the aquifer becomes confined or semi-confined, 100–200 metres thick; it comprises interbedded cobble, gravel, fine sand, and clay. Farther north, the groundwater table becomes shallow. Additional details on the hydrogeological characteristics of the area can be found in Fan (1981) and Chen (1997).

The geological tectonic of the Hexi Corridor is a complex fault basin. The basement is cross cut into a block faulting body by a series of North-West-West, North-West and nearly East-West faults. The neotectonic movement in the southern Hexi Corridor, i.e., the Qilian Mountains northern piedmont, is very active. The mountains in this area rose rapidly and the piedmont depression basin continuously subsided. This formed the corridor plain with thick quaternary sediments. Quaternary sedimentary layers are approximately 700-1200 metres thick (Li and Xing, 1988). For stratigraphic lithology, bedrock dominates the southern high mountain and northern low-mid mountain regions. The lithology is mainly composed of sedimentary rock and pyrogenic rock. The former includes argillaceous sandstone, mudstone, clastic rock, quartz sandstone, and shale and the latter includes granite and diorite. Rock with developed fracture is exposed at steep positions. The bottom and lower slope areas are generally covered with loess. Alluvial loess and gravel soil are generally distributed in the intermountainous basin and on the surface of the accumulational plain. Alluvial Gobi and gravel are widely distributed in the central area of the Corridor (Li and Xing, 1988; Liu et al., 2013).

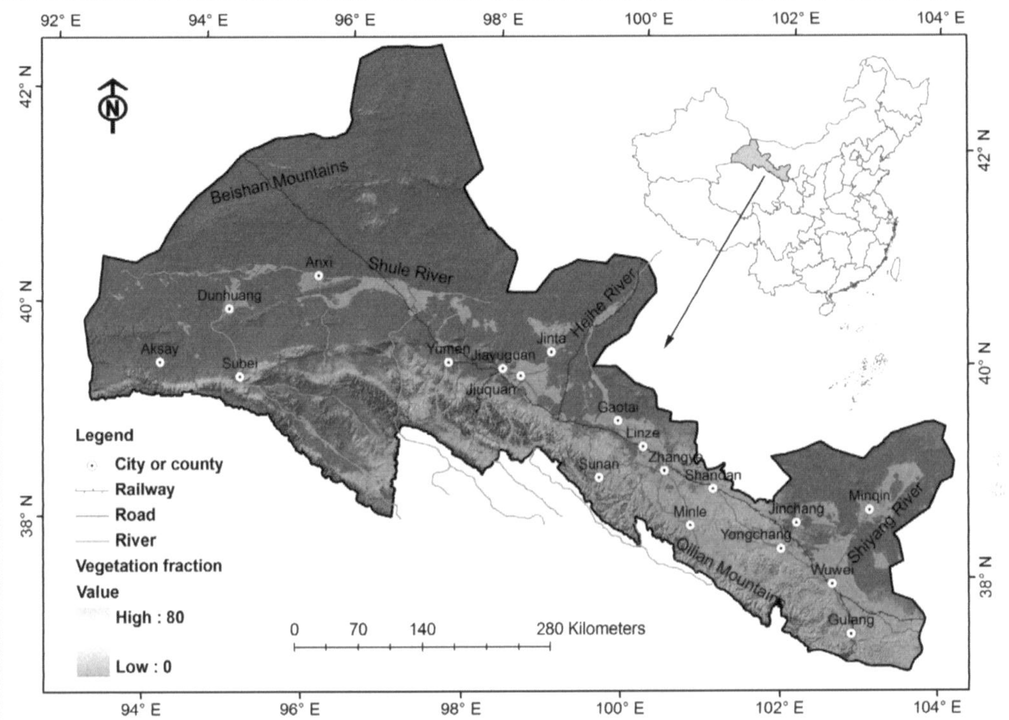

Figure 1. Location of the Hexi Corridor.

3. Materials and methodology

Previous studies show that the annual precipitation and slope are the dominant factors for torrent hazard in the Hexi Corridor (Luo et al., 2005; Zhang et al., 2007; Wang et al., 2013; Liu et al., 2013). Thus, slope, precipitation, vegetation fraction, flow accumulation, and the standard deviation of elevation were combined using an evidence reasoning method based on Dempster-Shafer theory.

3.1. Materials

The Advanced Spaceborne Thermal Emission and Reflection Radiometer (ASTER) Global Digital Elevation Model (GDEM) version 2 data with a spatial resolution of approximately 30 metres, developed jointly by the United States National Aeronautics and Space Administration (NASA) and Japan's Ministry of Economy, Trade, and Industry (METI), were used in this study. The ASTER GDEM2 is considered the highest resolution digital elevation model (DEM) among the freely accessible global DEMs (Arefi and Reinartz, 2011). It is available online through the Data Pool at the NASA Land Processes Distributed Active Archive Center (http://gdex.cr.usgs.gov/gdex/). The ASTER GDEM product is generated from automatic processing of 1.5 million ASTER Level-1A stereo pairs by applying stereo correlation methodology and covers land surfaces between 83°N and 83°S. The absolute vertical accuracy of ASTER GDEM is 20 metres at a 95% confidence level. Compared with the ASTER GDEM, the ASTER GDEM2 has significantly improved spatial coverage, horizontal resolution, horizontal and vertical accuracy (7-14 m), and water masking and includes an additional 260,000 ASTER stereo-pairs to supplement the voids and artefacts of ASTER GDEM (Tachikawa et al., 2011). The factors influencing torrent hazard, including the slope and the standard deviation (SD) of the elevation, were derived from ASTER GDEM using ArcGIS software. The accumulated flow was derived from ASTER GDEM, and precipitation was used as a weight factor using the ArcGIS hydrology module.

The mean 5 to 9 month total precipitation data from 1980 to 2011, developed by the Data Assimilation and Modeling Center for Tibetan Multi-spheres, Institute of Tibetan Plateau Research, and the Chinese Academy of Sciences (ITPCAS) (Yang et al., 2010; Chen et al., 2011) were used in this study. It has a spatial resolution of 0.1° and a temporal resolution of 3 hours. The precipitation field was produced by combining the three precipitation data sets including 740 in situ observations collected from China Meteorological Administration (CMA) operational meteorological stations, Tropical Rainfall Measuring Mission (TRMM) satellite precipitation analysis data (3B42) for 1998-2008 (Huffman et al., 2007), and the Asian Precipitation-Highly Resolved Observational Data Integration Toward Evaluation of Water Resources (APHRODITE) data (Yatagai et al., 2009); GLDAS precipitation data for 1979-2010 were used to replace TRMM 3B42 data that were not available beyond 40°N.

The vegetation fraction with a 1 km spatial resolution used in this study was estimated using a dimidiate pixel model based on the MODIS (Moderate Resolution Imaging Spectrometer) 16-day composite Normalized Difference Vegetation Index (NDVI) product (MOD13A2).

Flood disaster history data were used to validate the result.

3.2. Dempster-Shafer evidence theory

Dempster-Shafer evidence theory is an extension of Bayesian probability theory and allows for the quantification and management of data uncertainty (Gordon and Shortliffe 1985; Lee et al. 1987). The basic assumptions of Dempster-Shafer theory are that gaps exist in the body of knowledge and that belief in a hypothesis is not necessarily the complement of belief in its negation. The basic concept of evidence theory is the frame of discernment, denoted by θ. The selection of θ depends on the known knowledge, the level of understanding and what we want to know. The computing elements of Dempster-Shafer theory are the power set 2^{θ}. The degree of belief in the evidence from a source (for example, torrent hazard) in support of torrent hazard is referred to as the mass (m) committed to that. The amount of mass is often referred to as an evidence measure. A mass can be expressed as a mass function that maps each element of the power set into a real number from 0 to 1, with a larger value indicating a higher level of "belief". A mass function meets the following conditions:

$$\begin{cases} m(\varphi) = 0 \\ \sum_{A \subseteq 2^{\theta}} m(A) = 1 \end{cases} \tag{1}$$

where Φ is the null set or an empty set and m(A) is a basic probability assignment (BPA) which represents the support for every subset, A.

3.3. Generating evidence

The related environmental factors include slope, vegetation fraction, accumulated flow, and the standard deviation of the elevation as evidence of torrent hazard. This evidence, or the degree of support to torrent hazard from each data source, includes two parts:

$$M_i = W_i \times S_i \tag{2}$$

where W is the weight of variable i assigned by local expert using analytic hierarchy process (AHP) as shown in Table 1, and S is the standardization for variable i using the minimum and maximum values, except for the accumulated flow, as follows:

$$S_r = \frac{V_i - V_{\min}}{V_{\max} - V_{\min}} \tag{3}$$

The flow accumulation function of the ArcGIS hydrology module was used to calculate the accumulated flow based on ASTER GDEM data. The standardization of total precipitation over 5 to 9 months (mm) as an accumulated weight was used to calculate the accumulated flow. Then, the mean accumulated flow within a 3 km × 3 km area was normalized using a symmetric membership function shape.

Table 1. Factors influencing torrent hazard and their weights generated using AHP

Variable (i)	Commitment	Weight (W)
Slope	Belief	0.85
Vegetation fraction (%)	Disbelief	0.5
Flow accumulation	Belief	0.9
Precipitation (5-9 month)	Belief	0.9
SD of elevation	Belief	0.7

3.4. Combining the evidence

In general, the basic belief, or evidence, is represented by the mass function. The mass function generates a basic level of belief for each evidence source for the torrent hazard and then a system of combination is used to combine these basic beliefs to generate a total degree of belief for the torrent hazard. The Dempster-Shafer theory uses an orthogonal sum ($\oplus$) to compute a total belief degree using this equation:

$$m_1 \oplus m_2(Z) = \frac{\sum_{X \cap Y = Z} m_1(X) m_2(Y)}{1 - k} \tag{4}$$

where Z is a torrent hazard label and X, Y is an input data label. The sum extends over two labels whose intersection is $X \cap Y=Z$. The set of intersections represent common labels of evidence. The $m_1 \oplus m_2(Z)$ are used to determine the combined mass and are then assigned to the torrent hazard label Z.

$$k = \sum_{X \cap Y = \varphi} m_1(X) m_2(Y) \tag{5}$$

where k is a normalization constant that corrects for any mass that was committed to the empty set (Φ) and indicates the extent of conflict between the two sources considered (Shafer 1976). k=0 indicates complete compatibility and k=1 indicates complete contradiction. A value between zero and one indicates partial compatibility.

The equations above provide a method to combine two basic probability assignment functions and can be used with any number of basic probability assignment functions through repeating the application of equations (4) and (5). The implementation of combined evidence was conducted using the belief module of IDRISI 14.01 software, a GIS and remote-sensing image-processing tool developed by Clark Labs (Worcester, MA), USA.

Last, three indicators of the degree of belief, degree of plausibility, and belief interval are used to qualify the risk of torrent hazard. The degree of belief is the total degree of belief of the torrent hazard. The degree of plausibility is the evidence failing to reject the torrent hazard. The belief defines the lower boundary of the support for torrent hazard whereas the plausibility function defines an upper boundary. The range of belief and plausibility is referred to as a belief interval or an interval of uncertainty.

4. Results and discussion

4.1. Torrent hazard map and uncertainty

The degree of belief for the torrent hazard derived from five variables using the evidential reasoning method shows the distribution of torrent hazard in the Hexi Corridor (figure 2). High belief means a high risk of torrent. The high degree of belief for torrent hazard with a small belief interval shows the small uncertainty of the decision-making results (belief interval is not shown). The result is in agreement with disaster history data. The uncertainty of the results is mainly due to the limited impact factors and subjective judgement in the expert scoring process as well as the accuracy, scale-mismatch, and simple linear standardization process of the evidence.

In previous study, the precipitation factor was generally combined directly with other factors. This may be less reasonable because the spatial locations of annual or seasonal total precipitation factors and torrent occurrences do not completely match. Precipitation not only directly acts on torrents but also acts through flow accumulation. The length of flow accumulation is dependent on the spatial scale. The spatial consistency of the relative high risk area with past disaster points shows that the flow accumulation factor was successfully combined in the evidential reasoning process.

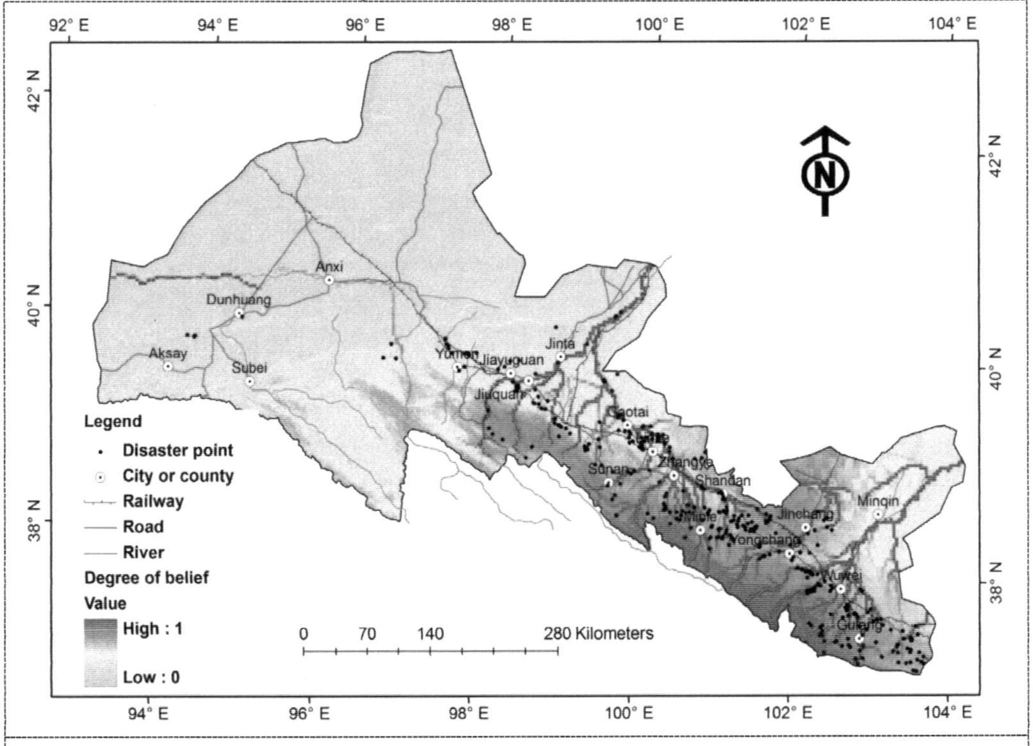

Figure 2. The degree of belief for torrent hazard in the Hexi Corridor.

4.2. Characteristics of torrent hazard in the Hexi Corridor

As shown in Figure 2, the distribution of torrent hazard in the Hexi Corridor is consistent with the landform characteristics. The high-risk areas are mainly concentrated in the Qilian Mountains, especially the upstream Shiyanghe and Heihe Rivers in the central eastern Qilian Mountains. Most torrent related disasters are distributed along the piedmont transition zone where the area is vulnerable to flood because it is close to a densely populated area. In general, the torrent hazard in the Hexi Corridor is dominated by medium- and low-risk levels, which are very sensitive to local factors and human activity. Thus, the characteristics of torrent hazard in the Hexi Corridor require higher resolution mapping in the future.

For higher resolution and higher accuracy prediction for random- and local-scale torrent hazard in the future, efforts in three aspects should be strengthened. The first is high-resolution data. Terrain and vegetation maps are available from airborne remote sensing (LiDAR: Light Detection and Ranging, and optical image) or satellite remote sensing images at sub-meter resolution (such as the high spatial resolution satellite GF-2 in China). Higher resolution precipitation products are the scarcest for prediction of torrent hazard and should be focused on in the future through cooperation with the meteorological field. The second is determining the impact thresholds of multi-factors and the non-linear relationships of the effects. The last is the study of the mechanisms of torrent hazard with a systems perspective.

5. Summary

In this study, we proposed an evidential reasoning approach based on Dempster-Shafer theory to map torrent hazard and risk. The capability of this method was demonstrated by the application to mapping mountain torrent hazard in the Hexi Corridor. A torrent hazard risk distribution map for the Hexi Corridor was generated by integrating the driving factors of mountain torrent hazard including precipitation, terrain, flow concentration processes, and the vegetation fraction. Although limited factors were integrated and they have some uncertainty, this study is a useful exploration for mapping torrent hazard based on GIS and remote sensing data. The map shows that the pattern of torrent hazard risk is consistent with past disaster locations in the Hexi Corridor. It is useful for supporting engineering planning and resource protection based on the level of probable torrent events in different locations.

However, torrents and the closely related disasters of landslide and debris flow are not only closely related to lithology and geological structure but also to trigger factors such as earthquakes, rainstorms and human activities with great uncertainty and randomness. The uncertainty and randomness are the largest obstacles to disaster prediction, especially for medium- and low-risk torrent hazard in the Hexi Corridor. For higher resolution and higher accuracy prediction and mapping of torrent hazards in the future, higher resolution remote sensing data could play an important role in furthering our understanding of the mechanisms of torrent hazard.

Acknowledgements

This paper was supported by a cooperative project between the Lanzhou Branch of the Chinese Academy of Sciences and the Geologic Hazards Prevention Institute of the Gansu Academy of Sciences: The risk assessment of torrent hazard in the Hexi Corridor.

References

[1] Arefi H and Reinartz P 2011 Accuracy enhancement of ASTER global digital elevation models using ICESat data. *Remote Sensing*, 3(7), pp 1323-1343.

[2] Balica S F, Douben N, and Wright N G 2009 Flood vulnerability indices at varying spatial scales. *Water Science and Technology*, 60, pp 2571–2580.

[3] Chen B L, Dang G M, Cui W, Zhang Y S, Dong C and Liu J S 2003 Advances in study of crustal stability in hexi corridor, northwestern China. *Journal of Geomechanics*, 9(1), pp 14-20.

[4] Chen M X 1997 The water resources related with Quaternary basin system in arid area of Northwest China. *Quaternary Science Reviews*, 16, pp 97-104.

[5] Chen Y Y, Yang K, He J, Qin J, Shi J C, Du J Y and He Q 2011 Improving land surface temperature modeling for dry land of China. *Journal of Geophysical Research*, 116, D20104.

[6] Fan X 1981 Transformation of ground-water and surface water and rational utilization of water resources in Hexi Corridor region. *Hydrogeology and Engineering Geology*, 4, pp 1-6 (in Chinese with English abstract).

[7] Gordon J and Shortliffe E H 1985 A method for managing evidential reasoning in a hierarchical hypothesis space. *Artificial Intelligence*, 26(3), pp 323-357.

[8] Huffman G J, Adler R F, Bolvin D T, Gu G J, Nelkin E J, Bowman K P, Stocker E and Wolff D B 2007 The TRMM multi - satellite precipitation analysis (TMPA): Quasi-global, multi-year, combined-sensor precipitation estimates at fine scales. *Journal of Hydrometeorology*, 8(1), pp 38–55.

[9] Jia W X, Zhang Y S and Li Z X 2014 Spatial and Temporal Change of Precipitation Extremes in Qilian Mountains and Hexi Corridor in Recent Fifty Years. Scientia Geographica Sinica, 34(8), pp 1002-1009. (In Chinese, English abstract)

[10] Keiler M, Knight J and Harrison S 2010 Climate change and geomorphological hazards in the eastern European Alps. *Philosophical Transactions of the Royal Society of London A: Mathematical, Physical and Engineering Sciences*, 368(1919), pp 2461-2479.

[11] Lepuschitz E 2015 Geographic information systems in mountain risk and disaster management.

Applied Geography, 63, pp 212-219.

[12] Li S C, Shui T and Wang H 2010a Report on prehistoric archaeological survey in hexi corridor. *Acta Archaeologica Sinica*, 2, pp 229-264.

[13] Li L P, Li Y Y, Qian L and Zhang H L 2010b The changing tendency of extreme precipitation in the east of Hexi Corridor in 1961-2005. *Journal of Glaciology and Geocryology*, 32(3), pp 497-502. (In Chinese, English abstract)

[14] Li Y L and Xing C Q 1988 Research on the fundamental characteristics of the geological structures of the Hexi Corridor and the active faults of the northern and eastern flank of the Yumushan Mountain. *Northwestern Seismological Journal*, 10(2), pp 35-47.

[15] Liu J P, Gao S M and Wang D K 2013 Research on Distribution Feature of Debris Flow Gullies and Their Susceptibility Evaluation in Hexi Area of Gansu Province. *Journal of Glaciology and Geocryology*, 35(6), pp 1610-1618.

[16] Lee T, Richards J A and Swain P H 1987 Probabilistic and evidential approaches for multisource data analysis. *IEEE Transactions on Geoscience and Remote Sensing*, (3), pp 283-293.

[17] Luo X L, Li YY, Cheng H and Li L P 2005 Analysis on causes of geological disasters in Hexi Corridor and the counter measures. *Journal of Catastrophology*, 20(3), pp 103-106.

[18] Mazzorana B, Comiti F, Scherer C and Fuchs S 2012 Developing consistent scenarios to assess flood hazards in mountain streams. *Journal of environmental management*, 94(1), pp 112-124.

[19] Ronco P, Gallina V, Torresan S, Zabeo A, Semenzinet E, Critto A and Marcomini A 2014 The KULTURisk Regional Risk Assessment methodology for water-related natural hazards–Part 1: Physical–environmental assessment. *Hydrology and Earth System Sciences*, 18(12), pp 5399-5414.

[20] Shafer G 1976 *A mathematical theory of evidence*. Princeton, NJ: Princeton University Press, P 297.

[21] Tachikawa T, Kaku M, Iwasaki, A. Gesch D, Oimoen M, Zhang Z, Danielson J, Krieger T, Curtis B, Haase J, Abrams M, Crippen R and Carabaja C 2011 *ASTER global digital elevation model version 2-summary of validation results*. NASA.

[22] Totschnig R and Fuchs S 2013 Mountain torrents: quantifying vulnerability and assessing uncertainties. *Engineering Geology*, 155, pp 31-44.

[23] Turowski J M, Yager E M, Badoux A, Rickenmann D and Molnar P 2009 The impact of exceptional events on erosion, bedload transport and channel stability in a step - pool channel. *Earth Surface Processes and Landforms*, 34(12), pp 1661-1673.

[24] Wang R Z, Li L P and Li W L 2013 Risk Division of Flood Disaster in Wuwei Based on GIS. *Journal of Arid Meteorology*, 31(3), pp 602-608.

[25] Yang K, He J, Tang W J, Qin J and Cheng CCK 2010 On downward shortwave and longwave radiations over high altitude regions: Observation and modeling in the Tibetan Plateau. *Agricultural and Forest Meteorology*, 150, pp 38-46.

[26] Yatagai A, Arakawa O, Kamiguchi K, Kawamoto H, Nodzu M and Hamada A 2009 A 44 - year daily gridded precipitation dataset for Asia based on a dense network of rain gauges. *SOLA*, 5, pp 137–140.

[27] Zhang Z, Dong Z, Li J, Qian G and Jiang C 2016 Implications of surface properties for dust emission from gravel deserts (gobis) in the Hexi Corridor. *Geoderma*, 268, pp 69-77.

[28] Zhang S C, Chu J H and Zhang T Z 2007 Calculation of Critical Rainfall Amounts and Risk Area Partition of Flood Disaster in Qilian Mountain Area. *Journal of Soil and Water Conservation*, 21(5), pp 196-200.

Volunteered Geographic Information for Disaster Management with Application to Earthquake Disaster Databank & Sharing Platform

H Chen[1,2], W C Zhang[1,4], C Deng[3], N Nie[3] and L Yi[3]

[1] Key Laboratory of Digital Earth Science, Institute of Remote Sensing and Digital Earth, CAS, 100094 Beijing, China
[2] University of Chinese Academy of Sciences, 100049 Beijing, China
[3] State Key Laboratory of Pollution Control & Resources Reuse, School of the Environment, Nanjing University, 210093 Nanjing, China

E-mail: zhangwc@radi.ac.cn.

Abstract. All phases of disaster management require up-to-date and accurate information. Different in-situ and remote sensor systems help to monitor dynamic properties such as air quality, water level or inundated areas. The rapid emergence of web-based services has facilitated the collection, dissemination, and cartographic representation of spatial information from the public, giving rise to the idea of using Volunteered Geographic Information (VGI) to aid disaster management. In this study, with a brief review on the concept and the development of disaster management, opportunities and challenges for applying VGI in disaster management were explored. The challenges, including Data availability, Data quality, Data management and Legal issues of using VGI for disaster management, were discussed in detail with particular emphasis on the actual needs of disaster management practice in China. Three different approaches to assure VGI data quality, namely the classification and authority design of volunteers, a government-led VGI data acquisition framework for disaster management and a quality assessment system for VGI, respectively, were presented and discussed. As a case study, a prototype of VGI oriented earthquake disaster databank & sharing platform, an open WebGIS system for volunteers and other interested individuals collaboratively create and manage the earthquake disaster related information, was proposed, to provide references for improving the level of earthquake emergency response and disaster mitigation in China.

1. Introduction

Disasters have long been presented as tragic disruption to the humans' lives, properties, infrastructure, economy, capital investment and development process. Disaster management is a continuous process that aims at avoiding or reducing the impact of disasters. All phases of disaster management require large volumes of accurate, relevant, on-time geo-information that systematically create and maintain by various organizations. Disaster contingency, however, have changed both in people's sense of connection to distant places and in their ability to contribute to relief efforts. Both changes are strongly relied on ITs, which support the collection, dissemination, and cartographic representation of

[4] Corresponding author: W.C. Zhang, E-mail: zhangwc@radi.ac.cn.
To whom any correspondence should be addressed.

spatial information from the public [1]. VGI, a special case of the larger Web phenomenon known as user-generated content, is the harnessing of tools to create, assemble, and disseminate geographic data provided voluntarily by individuals, has proven very successful as a means to acquire timely and detailed geographic information at very low cost. And there is nowadays a growing consensus in recognizing the role of VGI in disaster management practices [2].

The concept of VGI was first proposed by Goodchild [3]. It is the fusion of some enabling technologies, namely Web 2.0, geo-referencing, geo-tags, GPS, graphics and broadband communication, etc. Which enables a large number of non-professional users to create, store, update and manage geographic information through an online collaborative manner. In recent years, with the advantages of rapid renewal, low-cost data acquisition, easy operation and high quality of local contribution data, VGI has gradually become the third kind of geographic information acquiring method that helps fill deficiencies of the traditional mapping technology and remote sensing. Numerous studies stressed the added-value of using VGI in earthquakes [1], forest fires [4], hurricanes [5] and floods [6].

However, the research on the VGI in terms of depth and breadth now is still at the preliminary stage of exploration in China. It is very necessary to clarify the concepts of disaster management, the important position of VGI applications for disaster management and some of the outstanding problems and challenges of current researches and applications. In this study, VGI was mainly introduced to apply in the field of earthquake disaster management. The key technologies and methods of VGI data management, fusion, and distribution were elaborate, which was a useful attempt to improve the level of information in China's earthquake disaster management, and had the certain promotion and application value.

2. VGI for disaster management

2.1. Opportunities of using VGI for disaster management

For decades, different in-situ sensors have been used in monitoring of a variety of hazards parameters, such as rainfall, water level, shaking of ground, etc. Advances in information and communication technologies, make it possible for these sensors obtaining and expressing data in near real time. But most of these sensors are highly specialized and stationary, causing that they generally can only measure one or very few parameters and the pros and cons of their monitoring results have a strong dependence on their locations, which are the result of careful planning and optimisation, and do not have flexibility.

Remote sensing is another important data source of the earth observation. With the rapid development of satellite technologies, high-resolution data products and various types of active and passive remote sensing system are gradually popularization. Numerous examples have shown that remote sensing plays an increasingly important effects in all phases of disaster management, such as monitoring of the scope floods [7] and forest fire [8].

Despite recent advantages in both in-situ and remote sensing system, there are still some phenomena that cannot be sufficiently measured, and these in-situ sensors for example water pressure gauges may be destroyed due to very severe floods. For remote sensing, some optical images may not be available in severe weather conditions when sensors are often obstructed by clouds and smoke. The satellites may not pass over the affected area for several days and the ground conditions may prevent the rapid downloading of images due to Internet connections, or computer hardware and software. These gaps may be filled by VGI. As Goodchild [3] has proposed to consider "Citizens as Voluntary Sensors". They continuously observe the environment around them and obtain information. Compared with a variety of physical sensors, human beings can get more different parameters by their own senses [2]. Moreover, they can move freely, and familiar with the area. It is possible for them to interpret their surrounding circumstances accurately, and thus complementing the framework of geographic data, providing rich detail and updating in real time through multiple terminals, using voice, text, or pictures in order to have a better understanding of the complexities of disaster management characteristics.

2.2. Challenges of using VGI for disaster management

As such VGI offers substantial advantages of involving the public in gathering information relevant for disaster management, but there are also some challenges cannot be ignored, specifically in terms of data availability, data quality, data management and legal issues of using VGI for disaster management.

2.2.1. Data Availability. VGI data has a characteristic of unknown, that it is not clear beforehand about how much, which, and from where information will be supplied. Unlike a sensor network that can be planned in advance and using a structured approach to configure pathways of data acquisition. Therefore, unless its accuracy and reliability are assured, VGI should only be used as a supplement to the authoritative geographic data sources.

2.2.2. Data Quality. Data quality problem is one of the major obstacles for using VGI widely. Since VGI are usually provided by the public, most of whom are lacking professional knowledge and trainings. However, because lacking unified standards, that the information often has a certain degree of subjectivity, or even can be biases and extreme factors. As a result, the data may be biased, repeated or wrong. As a kind of geographical information, VGI has its unique nature, that is, each information has a designated location with one or several related attributes description. The location here can be expressed as different kinds of geometry objects, such as point, line, polygon, etc. which may exist distribute unevenly, difficult to query and update and other issues [9]. For disaster information, someone may exaggerate the facts for personal interests or make panics, while some valuable information may be deleted or polluted maliciously. Therefore, taking some measures to assess and improve the quality of data is one of the most important challenges of current VGI research and application.

2.2.3. Data Management. The role of VGI in disaster management can be summarized as the acquirement, processing, express and analysis of information. In the basic model of online VGI application for disaster management, high-resolution satellites images are adopted as base maps. Users interpret features to create vectors of geometric objects with related attributes description, and release them through multiple terminals to gradually accumulate and form an open sharing geographic information database. Metadata can control and tracking data effectively and has a major impact on data quality. While it is difficult for users to ensure the accuracy and completeness of metadata even in OpenStreetMap, Google Map Maker VGI project [10]. Therefore, how to build a comprehensive framework to take full use of VGI is essential.

2.2.4. Legal Issues. Successful operation of the VGI project involves complex activities among tripartite, namely web site operators, contributors and users, respectively. Issues such as intellectual property, personal privacy, legal responsibility, caused by them began to be emphasized. Summarizing problems in present studies on legal issues, including: (1) research visual angle is unitary, which is always only considered from the perspective of web site operators, but lacking omnidirectional considerations from the perspective of the government and ordinary citizens; (2) and since VGI applications are in their infancy, construction of relevant laws and regulations are still in the exploratory stage.

Therefore, although VGI for disaster management has huge potential and demand, there are still many theoretical and technical issues to be resolved, especially when working with data quality problems of VGI. It will be necessary to adopt some methods to control and improve data quality of

VGI to enable it to be an effective supplemental information into authoritative geographic information database.

3. Mechanisms for quality assurance and a case study

3.1. Mechanisms for quality assurance

3.1.1. Classification and authority design of volunteers. Usually the main functions of a complete VGI platform for disaster management, should include system management, feature updating, information inquiry, data updating, map services, metadata services, disaster information acquisition, management and sharing, etc. And rational classification of user community will exert an important and fundamental influence on VGI applications. In this paper, the users of VGI platform for disaster management are broadly classified as general public, organization, professional personnel, decision maker and system administrator, as shown in Table 1.

The general public, namely who are lacking the specialized knowledge of geographic information science, but interested in sharing disaster information spontaneously. The organization is a type of personnel with some specialized knowledge, and focuses on specific types of disasters, such as insurance institutes, charitable organizations and medical teams, which can take charge, follow-up and feedback the related disaster information. The professional personnel are people with expertise in field of geographic and disaster, and can review, correct errors and filter the information. The design makers announce measures base on VGI, such as deployment of rescue teams and relief distribution, etc. They are commanders of disaster management. And the system administrator is particularly technical person to ensure the normal operation of system and features updates. Moreover, rational configuration of the user authority bases on the above classification of volunteers, can improve the accuracy and reliability of VGI data to some extent.

Table 1. Classification and authority design of volunteers

Function	User classification and authority				
	General public	Organization	Professional personnel	Decision maker	Administrator
System management					√
Features updating					√
Information inquiry	√	√	√	√	
Map services	√	√	√	√	
Metadata services			√		
Data acquisition	√	√			
Data updating	√	√	√	√	
Information response		√		√	

3.1.2. Government-led VGI data acquisition framework. Most of current VGI applications are based on non-governmental organizations' data collection system, such as OpenStreetMap, which are lacking of effective convergence with authoritative geographic database, leading to extract VGI data very much difficult. And according to disaster management, it is necessary to provide structured data collection standards, models and process to guide the users to provide accurate and reliable disaster related information. Therefore, a four-tier government-led VGI data Acquisition framework was illustrated in Figure 1. Of which, the guiding role of government performs in the aspects of determination of disaster theme, publishing requirements of data, providing of VGI management platform and establishing regime of liability and reward, etc. In practical terms, when facing a serious disaster, the government decision makers firstly announce requirements of VGI data and specify the data classification standards and the data model, then provide platform for VGI management. And related organizations or professional personnel can get permissions to join in rapidly to deal with a large number of information from the public. Finally, the decision makers can carry out disaster response works in time based on the valuable information. Thus a framework of disaster management involved in the public, can effectively make up for lack of important and dynamic disaster related information and provide all aspects of the guidance for improving the level of disaster mitigation.

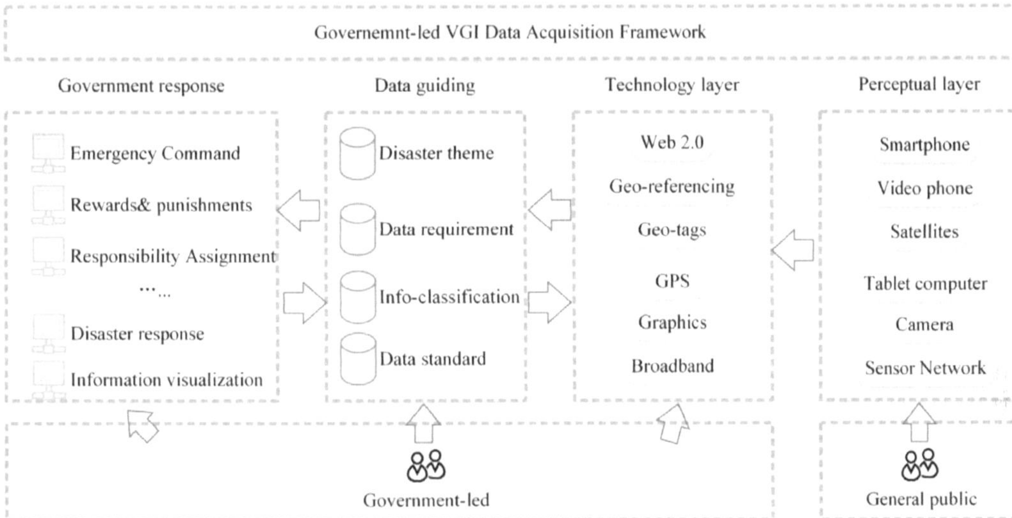

Figure 1. A government-led VGI data acquisition framework

3.1.3. Quality Assessment System for VGI. According to the actual demand of disaster management platform, we discussed the effectively quality assessment approaches from two aspects of the platform users and the data content to construct a quality assessment system. Firstly, from the perspective of the users, decision maker, organizations and professional personnel can timely update and feedback the general public's volunteered information, and to assess their abilities and reputation of contributing data, then raise or lower the privilege level of the users, namely affirming contributors who provide high quality and valuable data, while timely filtering out low-grade or junk data publishers. Secondly, from the perspective of data content, Oort [11] put forward 11 elements of spatial data quality, including lineage, positional accuracy, attribute accuracy, logical consistency, completeness, semantic accuracy, temporal quality, variation in quality, meta-quality and resolution, etc. Therefore, it is very necessary to organize a volunteered team with professional knowledge and experience to participate in activities of optimizing the accuracy and reliability of VGI data based on specific disaster management platform.

However, how to ensure data quality is still one of the research & application hotspots and difficulties for using VGI in disaster management. Three approaches we introduced here are just small part of data quality studies. With further research, more scientific and effective methods will play an important role in practice.

3.2. Case Study: A prototype of VGI oriented earthquake disaster databank & sharing platform

3.2.1. Data and methods. The platform construction requires large amount of data, including temporal-spatial information of earthquake disasters, such as locations, origin time, magnitude, depths and distribution of intensity. Basic geographic information and ecological environment data, involving multi-source satellites images, topography, land use and land cover, etc. Socio-economic data such as the type of economy, GDP, distribution of population and so on. As well as disaster information in pictures, text, videos and many other formats. These data vary in format, covering the relational data model, vector and raster data model, are very difficult to manage. In this study, we put forward an object & theme-oriented data model based on the rapid development of spatial database techniques to manage the multi-source heterogeneous data, as shown in Figure 2, in order to enhance the data expression, retrieval and simulation capabilities, which can greatly improve the efficiency of earthquake related information.

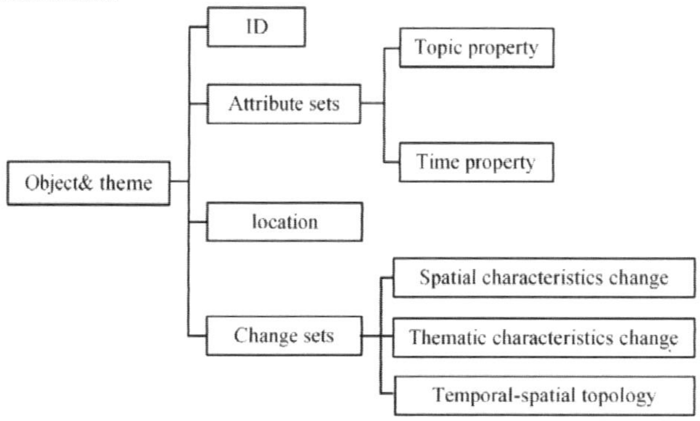

Figure 2. Object & theme-oriented data model

Seismic object & theme-oriented data model is composed of object ID, attribute sets, spatial location and change sets. Of which, object ID is a unique code used to identify the object. The attribute sets contain the object's topic and time properties, and time property represents a continuous renewal process, which characterizes the temporal variation of the object, while the topic property describes the object's natures, characteristics, which can be in multi format such as text, number, picture, video, etc. The location describes the position and shape of the object, which may be as different kinds of vector objects, such as point, line, polygon, or a row, column position of raster data. And the change sets are based on version, encapsulating the object's thematic characteristics change, spatial characteristics change and temporal-spatial topology. Of which, the spatial characteristics change mainly refer to the geometry changes and location moving, as well as the split and merge of the object, such as ground cracking, etc. While the thematic characteristics change recode changes as deltas, for example, acquisition and distribution of charity after an earthquake. The temporal-spatial topology is a reflection of the feature data quality, involving spatial topology and temporal topology and the relations of feature classes, such as causality, etc. Which can be used to ensure the quality of earthquake disaster related data and as a basis for further data mining.

The development techniques of the prototype system were mainly based on ArcGIS and its related products. We used the spatial database engine ArcSDE combining with SQL Server database as well as the object & theme-oriented data model to design an online earthquake emergency response and mitigation database. Advance techniques such as RS, communication, Internet and networking were integrated to obtain, store and manage the multi-source heterogeneous information from the public.

3.2.2. Result of the Case Study. Based on design of the data model, we constructed a comprehensive database system, as shown in Figure 3, including basic background database, earthquake disaster database, case knowledge database, disaster information database and extensional database, respectively. In practise, the basic background database, includes geographical data, socio-economic and demographic data, land use and building information, multi-source satellites images, etc. Which was used to create basic information about the disaster rapidly after a disaster happened. The earthquake disaster database was built to revolve around a broad seismic object (theme) that contained basic earthquake information, seismic network data, lifeline engineering data, etc. Which played a supplementary role in disaster assessment, decision-making. The case knowledge database mainly involved legal and regulatory information, emergency plans as well as historical earthquake experiences for current earthquake disaster management. The disaster information database is the main part of the database system, because it closely coupled with VGI from a large number of volunteers. The database revolved around a current earthquake object (theme) for dynamic data acquisition, analysis and sharing in real time, involving of basic information of the earthquake, disaster assessment information, secondary disaster information and public opinion information, etc. These data were core information for emergency dispatching and decision-making. While the extensional database was mainly for the unique circumstances of each earthquake disaster area, which can be extended in time, including data with some uncertainties, such as social relief supplies, funds of non-profits and insurance, etc. And integrating some specialized models of earthquake destruction, degree of damage, and the risk of earthquake of the factory warehouse building, etc.

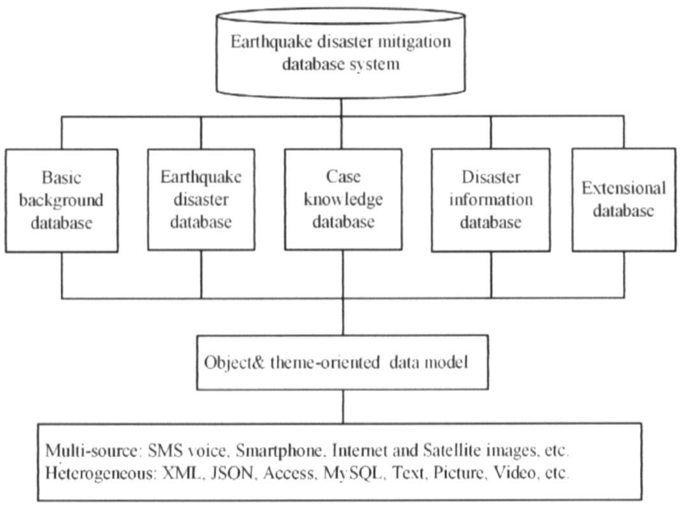

Figure 3. The main interface of VGI oriented earthquake disaster information sharing platform

While the disaster related information sharing platform is the bridge between specialization and popularization. Which made it possible for ordinary people to obtain, manage and share disaster information through multiple terminals, using voice, text, or pictures. Thus, in the prototype system, we integrated timeline list and map techniques to visualize the plenty of dynamic disaster information in order to aid decision-making, and the main interface of VGI oriented earthquake disaster information sharing platform is shown in Figure 4.

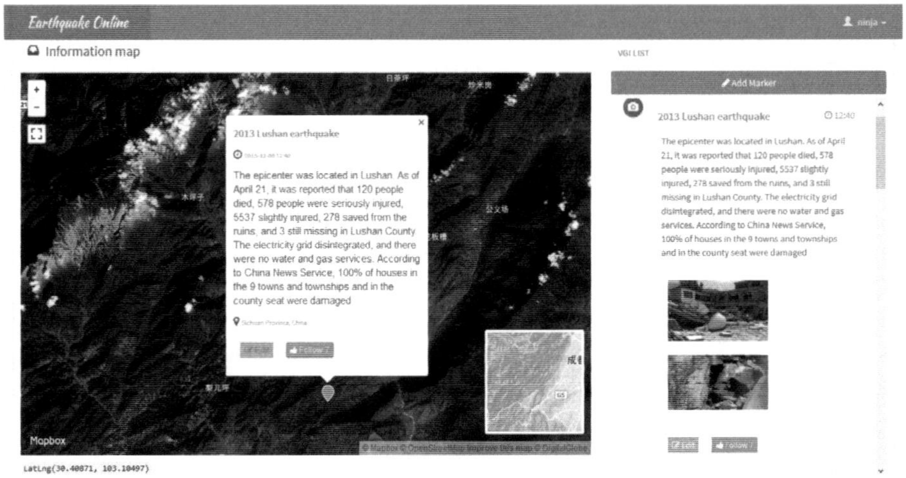

Figure 4. Earthquake disaster mitigation database system

4. Conclusion

System construction by using VGI for disaster management has huge potential and demand, and it is now in a stage of rapid development and improvement. This study systematically introduced the concepts related to disaster management, and discussed the opportunities and challenges for applying VGI in disaster management. Three different approaches to assess VGI data quality were presented and discussed in detail with particular emphasis on the actual needs of disaster management practice in China. However, they are only a tiny proportion of data quality studies. With further research, more scientific and effective methods will play an important role in practice.

Moreover, in the case study, an online integration database system for earthquake mitigation was built by using the object & theme-oriented data model. And a sharing platform mainly by using VGI from the public was developed, which acted an open WebGIS system for volunteers and other interested individuals collaboratively create and manage the earthquake disaster related information. It has shown significant advantages of information richness, timely feedback and accurate data, etc. However, as a new born concept of geographic information collection, theory research and practice of VGI are still not mature. It is necessary to carry out studies widely and thoroughly, in order to effectively integrate VGI to all phases of disaster management, and ultimately to improve the service level of disaster management.

Acknowledgements

This study is funded by CRSRI Open Research Program (CKWV2015226/KY), and the National Key Technologies R&D Program of China (No. 2016YFA0602302 and No. 2016YFB0502502).

References

[1] Zook M, Graham M, Shelton T and Gorman S 2010 Volunteered geographic information and crowdsourcing disaster relief: a case study of the Haitian earthquake *World Medical & Health Policy* **2** 7-33

[2] Poser K and Dransch D 2010 Volunteered geographic information for disaster management with application to rapid flood damage estimation *Geomatica* **64** 89-98

[3] Goodchild M F 2007 Citizens as sensors: the world of volunteered geography *GeoJournal* **69** 211-21

[4] De Longueville B, Smith R S and Luraschi G 2009 Omg, from here, i can see the flames!: a use

case of mining location based social networks to acquire spatio-temporal data on forest fires. In: *Proceedings of the 2009 international workshop on location based social networks: ACM* 73-80

[5] Hughes A L and Palen L 2009 Twitter adoption and use in mass convergence and emergency events International Journal of Emergency Management **6** 248-60

[6] Longueville B, Luraschi G, Smits P, Peedell S and Groeve T 2010 Citizens as sensors for natural hazards: A VGI integration workflow *Geomatica* **64** 41-59

[7] Tralli D M, Blom R G, Zlotnicki V, Donnellan A and Evans D L 2005 Satellite remote sensing of earthquake, volcano, flood, landslide and coastal inundation hazards *ISPRS Journal of Photogrammetry and Remote Sensing* **59** 185-98

[8] Voigt S, Kemper T, Riedlinger T, Kiefl R, Scholte K and Mehl H 2007 Satellite image analysis for disaster and crisis-management support *IEEE transactions on geoscience and remote sensing* **45** 1520-8

[9] Deren L and Xinlin Q 2010 A Brief Introduction of Data Management for Volunteered Geographic Information *Geomatics and Information Science of Wuhan University* **4** 379-83

[10] Zhigang H, Le Y and Renjie L 2015 Foreign Research Progress on Volunteered Geographic Information and Enlightenment to China *GEOMATICS WORLD* **22** 51-8

[11] van Oort P A 2006 Spatial data quality: from description to application

Flood monitoring and damage assessment in Thailand using multi-temporal HJ-1A/1B and MODIS images

S L Zhou[1,2] **and W C Zhang**[1,3]

[1] Academy Key Laboratory of Digital Earth, Institute of Remote Sensing and Digital Earth, CAS, Beijing 100094, China
[2] University of Chinese Academy of Sciences, Beijing 100049, China

E-mail: zhangwc@radi.ac.cn

Abstract. Flood is one of the most serious natural disasters in South Asia. How to monitor floods and assess damage caused is the most urgent problem for the government and disaster experts. With the advances of remote sensing, images acquired before the beginning of disaster to the very end or after the disaster from Earth-observing satellites benefit the decision making for reduction and protection of disaster dramatically. By using multi-temporal HJ-1A/1B and MODIS remote sensing data, applicability of different algorithms for flood monitoring and damage assessing was investigated in 2011 Thailand floods. Three different algorithms were adopted to monitor flood disaster events with water indices. Comparisons on the flood disaster monitoring and damage assessing by means of HJ-1A/1B and MODIS images suggested that multi-temporal HJ-1A/1B is much useful for the purpose, which demonstrated with the analysis of the thresholds estimated and problems in data post-processing. The variations of the inundated areas in the process of 2011 Thailand floods revealed were presented in this paper, and the damage caused by flooding was evaluated in three aspects, the population in the inundated region, the inundated information of different provinces and land use areas. Compared with MODIS, HJ-1A/1B images can provide more rapid and accurate flood extent and damage assessment for the disaster prevention, damage mitigation and disaster relief.

1. Introduction

Floods are the most damaging of all natural hazards, and, each year, result in major disaster response activities (such as food assistance provided by the World Food Programme). As one of the most frequent and severe natural disasters, floods cause huge human sufferings and property loss, and affect a wide range of activities related to agriculture, vegetation, human and local economics [1-3]. Climate and land cover directly affect flood and drought frequency and severity. Future changes in the Earth's water cycle will have major impacts on the growing global population, and they cannot be accurately predicted without adequate knowledge concerning present conditions and rate of change since this work requires near real time information. However, the mapped record of flooding can improve future flood risk assessment, and it can be used in tandem with hydrological modelling to facilitate weather-based flow and inundation prediction. The rapid monitoring and damage assessment of floods, as the basis of the disaster prevention, damage mitigation and disaster relief, as well as an important component of Earth's water circle, therefore, have been one of the research hotspots and a huge

[3] To whom any correspondence should be addressed.

challenge for the government and disaster experts [4].

In the last century, data analysis by the network of meteorological stations and hydrological observatories was widely used to predict and monitor floods [5]. With the development of remote sensing technique, various images from earth-observing satellites make it possible to obtain the ground information rapidly by means of sensors operating in several spectral bands such as SPOT, AVHRR, MODIS, HJ-1A/1B, RADARSAT, Envisat ASAR, which can provide the effective means to detect and assess disasters accurately [6-7]. Meanwhile, many flood monitoring algorithms by using different remote sensing data have been developed. Except common density slicing or image segmentation methods such as maximum likelihood method [8], with the spectrum information of water body, water body extraction techniques, such as spectrum-photometric method, water index algorithms, are widely used in flood related studies [9-12].

During August to December of 2011, a large-scale flood took place over the central Thailand, and this event was reported as the maximum recorded flood in recent 50 years of Thailand. During this flood, extensive areas of Thailand were inundated, about 7 industrial estates and 804 companies suffered inundation damage, which resulted in 813 dead and nearly 1.36 trillion baht loss nationwide [13-15]. By means of multi-temporal HJ-1A/1B and MODIS images acquired during this great flood, this study was aimed to map the flooded area from July 2011 to February 2012 for understanding the dynamic processes of 2011 Thailand flood, and meanwhile assessing the flood damage from multiple perspectives. The suitability of different water index methods by using different sensors were presented and compared with each other to investigate the function and limitation of water index methods on HJ-1A/1B and MODIS images in flood monitoring and assessments.

2. Study region

Located in the central part of Thailand, the Chao Phraya River basin has suffered several large-scale floods in history, and was one of the mostly inundated areas during 2011 Thailand flood. As shown in figure 1, the down-stream areas of the Chao Phraya River basin covers the provinces with huge damages in the flood, such as Bangkok and Ayutthaya, where the flooding dynamics of which can reflect the general condition of 2011 Thailand flood.

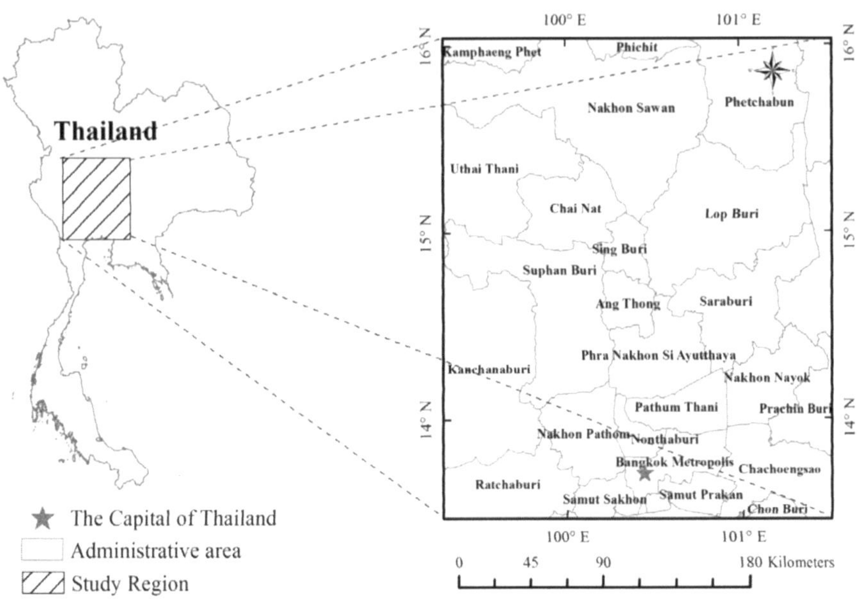

Figure 1. Study region.

3. Datasets and methods

3.1. Datasets

The HJ-1 mini-satellite constellation is a national program by the National Committee for Disaster Reduction and State Environmental Protection Administration (NDRCC/SEPA) of China, to construct a network of Earth observing satellites. HJ-1A/1B satellites include two 30m-resolution CCD optical cameras which work together to form a 700 km wide swath with a 4-day repeat cycle. Each CCD camera has three visible bands (430-520 nm, 520-600 nm, and 630-690 nm) and a near-infrared band (760-900 nm). In comparison, MODIS on board the Terra and Aqua satellites, has a viewing swath width of 2330 km, views the entire surface of the Earth every one to two days, and acquires data in 36 spectral bands between 0.405 and 14.385 μm, at three spatial resolutions -- 250m, 500m, and 1000m. Both HJ-1A/1B and MODIS data are widely used in flood monitoring. In this study, a time-series HJ-1A/1B and MODIS data set spanning from July 2011 to February 2012 were used. All the satellite images were calibrated and geometrically corrected using ENVI and MRT software.

Except the satellites images, to assess the damage of this flood, land cover map, population density data and administrative areas data are needed to provide the basis information of Thailand. A land use map of Thailand from GlobCover datasets in 2009 at the spatial resolution of 300m, administrative areas data from GADM datasets updated in 2015, and population density data in 2011 at the spatial resolution of 100 m estimated by WorldPop were chosen for the assessment.

3.2. Methods

3.2.1. Water body identification. Water body identification is the primary part of flood monitoring, mostly based on the spectral absorption characteristics of water body which differ very much from the other ground objects in visible and infrared bands [16]. In recent years, several water indices were developed by combining two or more bands for inundated area identifications, among which the normalized difference water index (NDWI) [9] is a popularized one in flooding monitoring and assessment for visible and infrared bands data applications. In this paper, inundated extent were derived from MODIS images by NDWI, the modified normalized difference water index (MNDWI), spectrum-photometric water index (SPWI) algorithms, and from HJ-1A/1B images only by NDWI methodology.

According to Mcfeeters (1996), the normalized difference water index that can enhance the water information by the difference of water spectral absorption intensity in green and near-infrared bands can be calculated as:

$$NDWI = \frac{(Green - NIR)}{(Green + NIR)} \qquad (1)$$

Where Green is the reflectance of green band and NIR is that of near-infrared band.

To improve the accuracy of water body extent identification from soil and buildings with NDWI, MNDWI was proposed by Xu (2005) based on a combination of reflectance in the green and mid-infrared wavelengths. The MNDWI can be calculated as:

$$MNDWI = \frac{(Green - MIR)}{(Green + MIR)} \qquad (2)$$

Where MIR is the reflectance of mid-infrared band.

A water spectral model was established by Ding et al. (2006) for extracting water body by using MODIS images, as regard to spectrum-photometric method. This model can be expressed as:

$$CH_1 + CH_4 > CH_2 + CH_6 \qquad (3)$$

Where CH_1, CH_2, CH_4, CH_6 is the reflectance of band 1, 2, 4, 6 in MODIS images, respectively.

According to the model, SPWI can be calculated as:

$$SPWI = CH_1 + CH_4 - CH_2 - CH_6 \qquad (4)$$

A suitable cut-off threshold was needed to extract water body by using these three water indices. The pixel values of NDWI, MNDWI, SPWI images derived from satellites images range from −1 to 1. In general, pixels with the three indices value greater than 0 represent water bodies. But due to the different water index values of the same location from various images, slight calibration of the thresholds is necessary for more accurate results [17]. Therefore, the water indices values of several sample points classified as inundated areas by the visual interpretation from HJ-1A/1B images were calculated to adjust the thresholds to distinguish water body from other objects. Tests showed that thresholds of images obtained at different time of year by the same sensor were similar. The thresholds of NDWI from HJ-1A/1B images were concentrated in 0.29, and NDWI, MNDWI, SPWI thresholds from MODIS images were concentrated in 0, 0.1, 0, respectively.

3.2.2. Data Post-processing. Compared with other water body extraction application, the monitoring of floods is more difficult because of the specific characteristics of flooded areas, which are always partly covered by water because the top of other objects may extent up to the water surface even during the flood. Thus, the value of pixels represented inundated areas is sometimes the mix of the reflectance of water and other objects, which may cause the misclassification especially by methods based on pixels. Therefore, open and close operators, the mathematical morphology methods, were used for noise removal in this paper, because open operator can remove small objects from the foreground of an image, placing them in the background, while close operator removes small holes in the foreground, changing small islands of background into foreground. Compared with results from MODIS images, the HJ-1A/1B monitoring results need more times of open and close operators performed, which reduces the degree of detection automation.

4. Results and discussions

Monitoring results showing the comparisons of monitoring data derived from MODIS and HJ-1A/1B images (table 1) are presented below.

Table 1. Inundation area derived from MODIS and HJ-1A/1B.

Method	Oct.2011		Nov.2011		Dec.2011	
	Inundated Area (km^2)	Overall accuracy (%)	Inundated Area (km^2)	Overall accuracy (%)	Inundated Area (km^2)	Overall accuracy (%)
Visual Interpretation	10879.15	/	9365.68	/	8420.66	/
MODIS-NDWI	8023.25	73.75%	10132.97	91.81%	10887.25	70.71%
MODIS-MNDWI	12639.75	83.82%	9858.00	94.74%	6868.00	81.56%
MODIS-SPWI	9433.50	86.71%	7819.25	83.49%	5925.50	70.37%
HJ-1A/1B-NDWI	9688.64	89.06%	8797.10	93.93%	7780.57	92.40%

In this paper, the visual interpretation results were regarded as "ground truth" to evaluate the accuracy of monitoring and identify the errors in the estimation. As shown in table 1, it is almost certain that HJ-1A/1B monitoring results are superior to others in this study, the mean accuracy of which is 91.80%, higher than that of MODIS results, 81.88%. Besides, the results derived from MODIS data using different methods are also distinct greatly, where MNDWI method is tested best.

The accuracy difference between HJ-1A/1B and MODIS images can be explained by the spatial resolution. Though MODIS images have more spectral bands and shorter repeat cycle, the lower spatial resolution causes the edge of flooded area and the mixed pixels harder to be defined due to the more complex mix of reflectance of various objects. And with more ground area corresponding to one pixel, the misclassification will lead to more area errors. It can also explain the variation among the results from different methods on MODIS images, which are not stable because many factors such as the thresholds estimated, the classification of mixed pixel, differs among the three methods. Therefore, the detail of monitoring process on MODIS images should be more concerned, because of the error caused by slight adjustment which cannot be ignored, though the estimation of thresholds and data post-processing for MODIS images are simpler and need fewer tests.

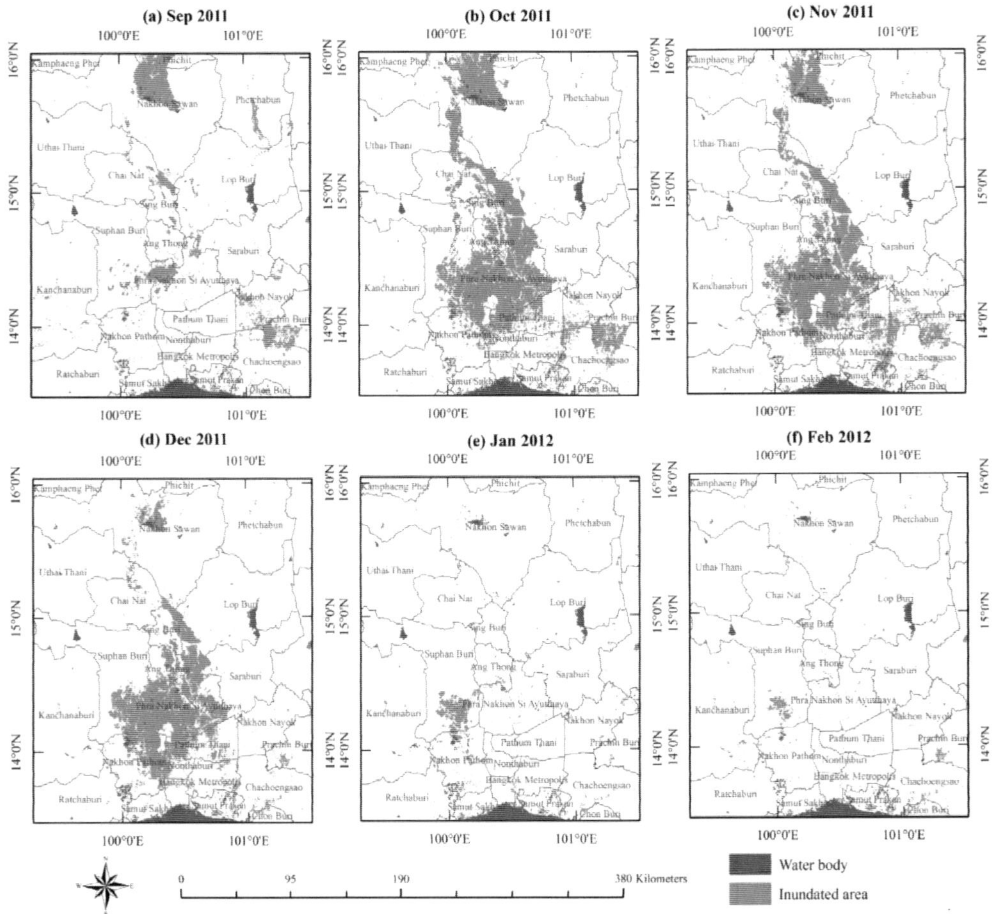

Figure 2. Maps of inundated areas derived from HJ-1A/1B images from Sep 2011 to Feb 2012. (a) Sep 2011; (b) Oct 2011; (c) Nov 2011; (d) Dec 2011; (e) Jan 2012; (f) Feb 2012.

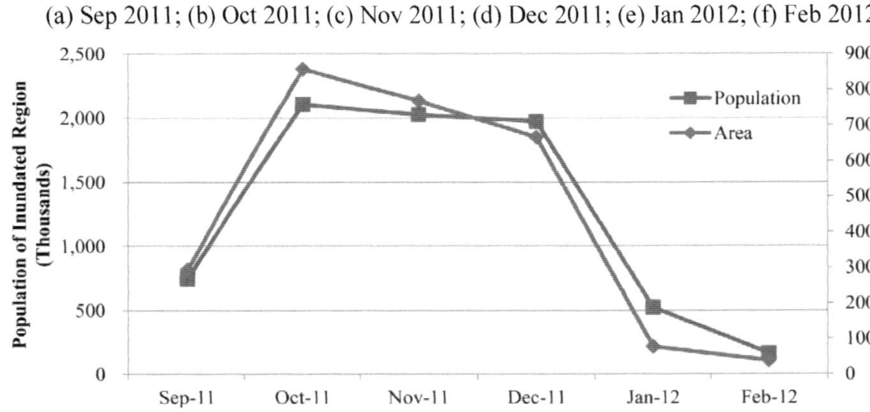

Figure 3. The Changes of inundated area and the population of inundated region from Sep 2011 to Feb 2012.

Figure 2 gives the inundated extent achieved from HJ-1A/1B images from September 2011 to February 2012, with the extraction result in July 2011 representing the normal water body extent before floods. The change of inundated area demonstrates the process of 2011 Thailand flood, which occurred in central and southern Thailand in Aug 2011 [15], quickly spread to the surrounding, especially in Phra Nakhon Si Ayutthaya Province, peaked in October and November 2011, and receded till February 2012. The inundated area and the population of inundated region are shown in figure 3, the changes of which are also consistent with the process of this flood analyzed above.

Table 2. Inundation area in different provinces.

NAME	Area（km^2）			
	Sep-11	Oct-11	Nov-11	Dec-11
Phra Nakhon Si Ayutthaya	322.73	1919.41	1750.48	1776.17
Suphan Buri	160.87	1056.37	904.92	1064.93
Lop Buri	134.49	589.42	470.12	426.73
Pathum Thani	0.00	534.88	689.78	473.61
Ang Thong	11.27	431.12	350.25	387.55
Nakhon Pathom	0.00	384.79	476.46	818.10
Sing Buri	130.24	374.03	260.85	212.82
Chai Nat	79.43	343.70	214.57	75.86
Saraburi	42.39	202.75	172.85	141.36
Nonthaburi	0.17	188.15	197.20	256.44
Nakhon Nayok	246.98	128.02	223.53	16.90
Bangkok Metropolis	20.40	41.39	260.56	155.27

Table 3. Inundation area of different land use types

Land Cover	Area（km^2）		
	Oct-11	Nov-11	Dec-11
Post-flooding or irrigated croplands	5134.25	4753.52	3861.16
Mosaic cropland/vegetation	1834.56	1540.72	1321.53
Rainfed croplands	1529.48	1341.60	1308.30
Closed to open shrubland	243.08	150.58	45.49
Closed to open herbaceous vegetation	165.31	167.87	258.50
Open broadleaved deciduous forest/woodland	132.25	135.18	111.14
Closed to open broadleaved evergreen or semi-deciduous forest	15.38	4.63	5.35
Artificial surfaces and associated areas	13.23	37.83	63.57
Closed needleleaved evergreen forest	10.89	11.83	19.88
Closed to open mixed broadleaved and needleleaved forest	9.58	22.42	41.57
Bare areas	6.18	4.55	4.09
Closed broadleaved deciduous forest	5.55	4.59	4.57
Mosaic forest or shrubland / grassland	2.28	6.60	8.21
Permanent snow and ice	0.32	0.32	0.32

Different regions and land use areas in Thailand were affected by 2011 flood to varying degrees. The statistics of flooded area of different land use types and provinces are shown in table 2 and table 3, indirectly reflecting the flood damage. As shown in table 2, a number of provinces which have significant influence among Thailand's politics, culture, economy and modern society, suffered huge damage during this flood, including Bangkok Metropolis, the capital of Thailand. Among these, Phra Nakhon Si Ayutthaya province was affected most seriously in the study region, the inundated area of which peaking at 1919.410 km^2 in October 2011, accounting for 75.37 percent of the total area. Besides, it can be found that the croplands were the land use types damaged most severely from table 3, corresponding to the report by Thai Ministry of Interior, which caused huge reduction in yield [13].

5. Conclusions

Remote sensing technology is powerful for flood monitoring and damage assessing, which is capable of providing considerable inundation information in near real time. However, the different applicability of algorithms and kinds of images in flood monitoring still limit the accuracy of monitoring results. In this paper, the comparisons of monitoring data derived from MODIS and HJ-1A/1B images in 2011 Thailand flood were presented, which show that HJ-1A/1B images are more suitable for the flood monitoring by providing higher spatial resolution information. The results also demonstrated that MNDWI index can provide more accurate results compared with NDWI and SPWI indices. Depend on the monitoring information derived from multi-temporal HJ-1A/1B images, the damage of 2011 Thailand flood was analysed, showing that the flood peaked in October and November 2011, caused huge damage in several provinces and affected croplands severely, corresponding to the news and reports by Thai Ministry of Interior. The successful application of HJ-1A/1B images in flood monitoring and assessment demonstrated the potential of this satellite data in disaster prevention, damage mitigation and disaster relief applications, however, more researches about algorithms and data post-processing are need to improve the accuracy and automation of remote sensing monitoring.

Acknowledgements

This work was financially supported by the National Key Technologies R&D Program of China (No. 2016YFA0602302 and No. 2016YFB0502502), and CRSRI Open Research Program (CKWV2015226/KY), respectively.

References

[1] Peter L, Matjaž M and Krištof O 2013 Detection of Flooded Areas using Machine Learning Techniques: Case Study of the Ljubljana Moor Floods in 2010 *Disaster Advances* **6** 4-11
[2] Jeyaseelan A 2003 Droughts & floods assessment and monitoring using remote sensing and GIS *Satellite remote sensing and GIS applications in agricultural meteorology* 291-313
[3] Ahmad R and Daniyal D 2013 Evaluating Damage Assessment of Breaches Along the Embankments of Indus River during Flood 2010 Using Remote Sensing Techniques *ISPRS-International Archives of the Photogrammetry, Remote Sensing and Spatial Information Sciences* **1** 7-11
[4] Merkuryeva G, Merkuryev Y, Sokolov B V, Potryasaev S, Zelentsov V A and Lektauers A 2014 Advanced river flood monitoring, modelling and forecasting *J Comput Sci-Neth* **10** 77-85
[5] Zhao Y and Cheng X F 2012 Review on Remote Sensing Monitoring of Flood Disaster *Sichuan Enviromment* **31** 106-109.
[6] Auynirundronkool K, Chen N, Peng C, Yang C, Gong J and Silapathong C 2012 Flood detection and mapping of the Thailand Central plain using RADARSAT and MODIS under a sensor web environment *Int J Appl Earth Obs* **14** 245-55
[7] Zhou C, Luo J, Yang C, Li B and Wang S 2000 Flood monitoring using multi-temporal AVHRR and RADARSAT imagery *Photogramm Eng Rem S* **66** 633-8
[8] Huang C, Chen Y and Wu J P 2014 Mapping spatio-temporal flood inundation dynamics at large river basin scale using time-series flow data and MODIS imagery *Int J Appl Earth Obs* **26** 350-62

[9] McFeeters S 1996 The use of the Normalized Difference Water Index (NDWI) in the delineation of open water features *Int J Remote Sens* **17** 1425-32

[10] Xu H Q 2005 A Study on Information Extraction of Water Body with the Modified Normalized Difference Water Index(MNDWI) *Journal of Remote Sensing* **9** 589-595.

[11] Ding L D, Wu H, Wang C J, Zhang Z H and Zhang Q X 2006 Study of the Water Body Extracting from MODIS Images Based on Spectrum-photometric Method *Geomatics&Spatial Information Technology* **29** 25-27.

[12] Sheng Y, Gong P and Xiao Q 2001 Quantitative dynamic flood monitoring with NOAA AVHRR *Int J Remote Sens* **22** 1709-24

[13] Komori D, Nakamura S, Kiguchi M, Nishijima A, Yamazaki D, Suzuki S, Kawasaki A, Oki K and Oki T 2012 Characteristics of the 2011 Chao Phraya River flood in central Thailand *Hydrological Research Letters* **6** 41-6

[14] Shimakage J and Yamazaki F 2012 Detection of flooded areas following the 2011 Thailand floods using ASTER images *Proc. 33rd Asian Conference on Remote Sensing (Pattaya Beach, Thailand)*

[15] Rakwatin P, Sansena T, Marjang N and Rungsipanich A 2013 Using multi-temporal remote-sensing data to estimate 2011 flood area and volume over Chao Phraya River basin, Thailand *Remote Sens Lett* **4** 243-50

[16] Guerschman J P, Warren G, Byrne G, Lymburner L, Mueller N and Van Dijk A I 2011 MODIS-based standing water detection for flood and large reservoir mapping: algorithm development and applications for the Australian continent *CSIRO: Water for a Healthy Country National Research Flagship Report (Canberra)*

[17] Ji L, Zhang L and Wylie B 2009 Analysis of dynamic thresholds for the normalized difference water index *Photogrammetric Engineering & Remote Sensing* **75** 1307-17

IOP Conf. Series: Earth and Environmental Science 57 (2017) 012017 doi:10.1088/1755-1315/57/1/012017

Landslide Identification and Information Extraction Based on Optical and Multispectral UAV Remote Sensing Imagery

Jiayuan Lin[1, 2], Meimei Wang[1], Jia Yang[1] and Qingxia Yang[3]

[1] Institute of Mountain Hazards and Environment, Chinese Academy of Sciences, 610041 Chengdu, China

[2] Plateau Atmosphere and Environment Key Laboratory of Sichuan Province, Chengdu University of Information Technology, 610225 Chengdu, China

[3] Jiuzhaigou Administration Bureau, Jiuzhaigou County, 623402 Aba Prefecture, Sichuan, China

E-mail: linjy@imde.ac.cn

Abstract. Landslide is one of the most serious natural disasters which caused enormous economic losses and casualties in the world. Fast and accurate identification of newly occurred landslide and extraction of relevant information are the premise and foundation for landslide disaster assessment and relief. As the places where landslides occur are often inaccessible for field observation because of the temporary failure in transportation and communication. Therefore, UAV remote sensing can be adopted to collect landslide information efficiently and quickly with the advantages of low cost, flexible launch and landing, safety, under-cloud-flying, and hyperspatial image resolution. Newly occurred landslides are usually accompanied with those phenomena such as vegetation burying and bedrock or bare soil exposure, which can be easily detected in optical or multispectral UAV images. By taking one typical landslide occurred in Wenchuan Earthquake stricken area in 2010 as an example, this paper demonstrates the process of integration of multispectral camera with UAV platform, NDVI generation with multispectral UAV images, three-dimensional terrain and orthophoto generation with optical UAV images, and identification and extraction of landslide information such as its location, impacted area, and earthwork volume.

1. Introduction

Landslide refers to the slope deformation phenomena caused by natural forces such as the gravity applied to slope-composed materials (Delacourt, C., et al., 2007). It is one of the most serious natural disasters, which causes enormous economic losses and casualties only second to earthquake (Derek C., 2005). The plateau and mountainous region are widely distributed in Southwest China, therefore landslides occur very frequently due to the complex geological environment and harsh weather conditions. After the earthquakes in Wenchuan, Lushan, and Ludian, the rock and soil in the near surface of the disastrous area is so loose that the risk of landslide occurrence is very high when induced by strong rainfall and surface runoff (Yin Yueping, 2009). So we need to timely carry out landslide investigation, identification and extraction of information about their locations, impacted areas and earthwork volumes, and assessment of landslide hazard in the earthquake stricken areas.

With the advantages of low cost, flexible launch and landing, safety, under-cloud-flying, hyperspatial image resolution, Unmanned Aerial Vehicles (UAVs) are very suitable for obtaining remote sensing imagery to investigate newly occurred landslides (Lin, J., 2010). Newly-occurred landslides are usually accompanied by the phenomena of lodged or covered vegetation and exposed bedrock or bare soil (Zhang, W., et al., 2010). Therefore, we can employ obvious difference between the vegetation and the surrounding environment to identify newly occurred landslides using NDVI (Normalized Difference Vegetation Index) images of the before and after landslide occurrence in the target area. The accuracy of landslide identification will be improved with the consideration of terrain conditions. The NDVI images will be generated based on the multispectral remote sensing images acquired by multispectral camera mounted on the UAV. Then information about identified newly occurred landslides such as impacted area and earthwork volume can be extracted from the optical images or the corresponding pseudo colour images.

2. UAV system and raw data

2.1. Integration of multispectral camera with UAV platform

The UAV remote sensing system employed to investigate landslides is composed of four major components: fixed-wing UAV platform (Figure 1.a), ground station control system, flight control system, and remote sensor (optical or multispectral camera, Figure 1.b). Due to the UAV's limited payload, the on-board remote sensors are restricted in their weight and size. The ADC(Agricultural Multispectral Camera) used in our study has the size of 114 mm x 77 mm x 22 mm, the weight of 200 g, and three bands (red, green, near infrared) (Tetracam Inc., 2011). The vegetation indexes including NDVI (Normalized Difference Vegetation Index), SAVI (Soil Adjusted Vegetation Index), etc. can be obtained after processing raw images using the accompanied software.

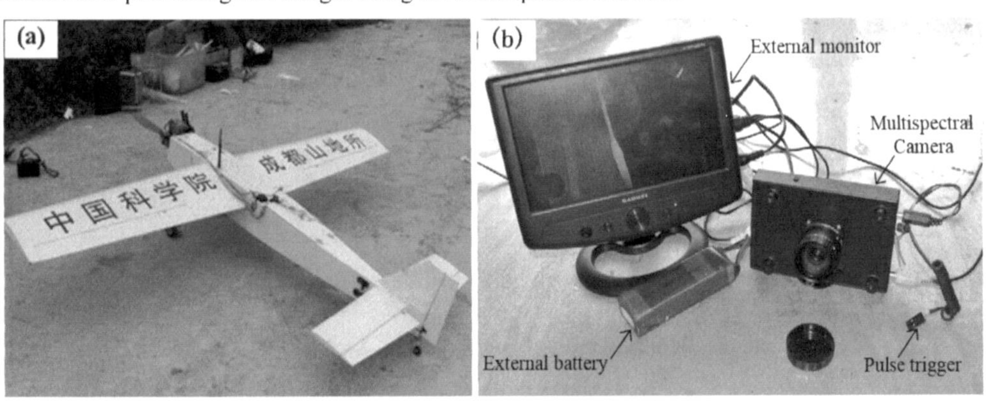

Figure 1. Fixed-wing UAV and multispectral camera.

The multispectral camera has certain differences in performance parameters and usage from the household SLR visible light camera usually used on UAV. Therefore, it is necessary to transform its hardware interfaces to be integrated with UAV remote sensing platform. In addition, multispectral camera could only be adjusted to the best working status after many field operations amounted on UAV. The specific measures and configurations (as shown in Figure 1.b) mainly include:

- Multispectral camera lens focal length and aperture adjustment. In the use of household SLR cameras on UAV, the lens focal length is usually set to infinity. However, the lens focal length of the multispectral camera should be adjusted according to the flight height of the UAV to get clear images. Its lens aperture size should be adjusted according to the environment light, which is usually completed through the external connected LED monitor after many test shots. We should choose the situation of the image looking slightly darker as the appropriate aperture setting. At the same time, the white board image should also be captured simultaneously in order to carry out radiation correction.

- Multispectral image storage format and exposure time setting. Since the multispectral camera continuously takes photographs during the UAV flight process, the time interval is relatively short, usually 3 seconds or so, and the time taken to store photographs into the memory card should also be counted in. Therefore, the storage format should be set as RAW8 (storage time of less than 1 second) and the exposure time set as 1 second in order to shorten the time of photograph taking and storage. If not set so, the unsynchronization and unmatched number between the recorded flight attitude data and photographs will possibly happen and as a result later aerotriangulation cannot be done correctly.

- Configuration of external power supply and pulse trigger device. As two batteries sold on the market can only maintain one hour continuous shooting of the multispectral camera, external power supply must be employed, which can support five-hour continuous shooting and satisfy the needs of the most UAV tasks. At the same time, the multispectral camera cannot be manually pressed to take pictures when amounted on UAV, therefore, it is necessary to configure pulse trigger device, so that pulse signals can be sent to the multispectral camera by the flight control system and complete the picture taking tasks.

2.2. NDVI generation with multispectral UAV image

The multispectral camera captures three channels grey image (Green, Red and NIR, which approximately equal to TM2, TM3 and TM4 of Landsat 7). We can use the accompanied PixelWrench2 software to obtain pseudo colour image (Figure 2.a) and vegetation index image, such as NDVI (Normalized Difference Vegetation Index) (Figure 2.b). As we know, green vegetation rich in chlorophyll has a strong reflection in the near infrared band. As Figure 2.a displays, near infrared channel replaced the red channel and vegetation in the figure are usually red colour. This is known as pseudo colour image, which is much convenient for technicians to do visual interpretation.

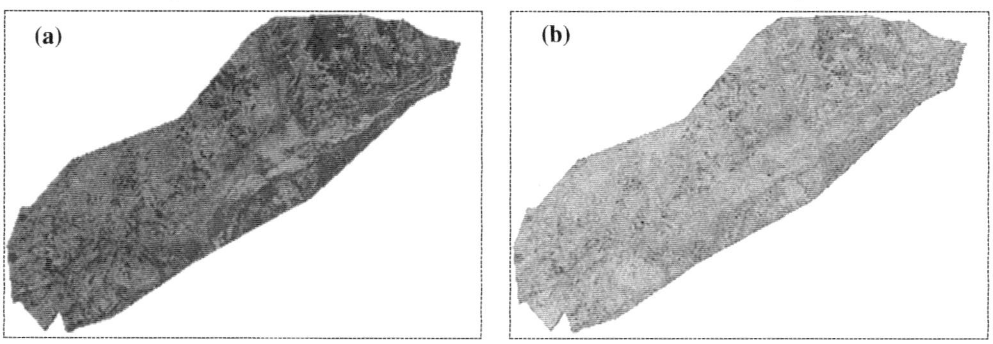

Figure 2. Multispectral pseudo colour image and corresponding NDVI image.

NDVI is most popular vegetation index. Its formula is shown below:

$$NDVI = (\rho_{nir} - \rho_r)/(\rho_{nir} + \rho_r) \quad (1)$$

Based on the above formula, the NDVI vegetation index image (Figure 2.b) can be acquired. It is a grey scale image, which is different from false colour image. The higher the NDVI index, the greater the value of the image displayed, and the denser and lusher the corresponding vegetation (Pettorelli, N., et al., 2005).

After producing the false-colour images and NDVI grey images, the panoramic orthorectified UAV image of target area can be acquired using aerial triangulation software with combination of the image spatial coordination (X, Y, Z) and attitude data (heading angle, roll angle, pitch angle). In the processing procedures the corresponding panoramic DSM (Digital Surface Model) is also produced.

2.3.2. Optical UAV image

The optical UAV images of the same area were taken in different time from those multispectral images. Compared with multispectral images, the optical images have higher spatial resolution and more textural information. Due to 60% to 80% overlap between acquired images, the good quality DEM and DOM of target area can be produced with the interior orientation parameters from digital camera calibration and the exterior orientation parameters namely the attitude data recorded in UAV flight. Here we will demonstrate the aerotriangulation in details.

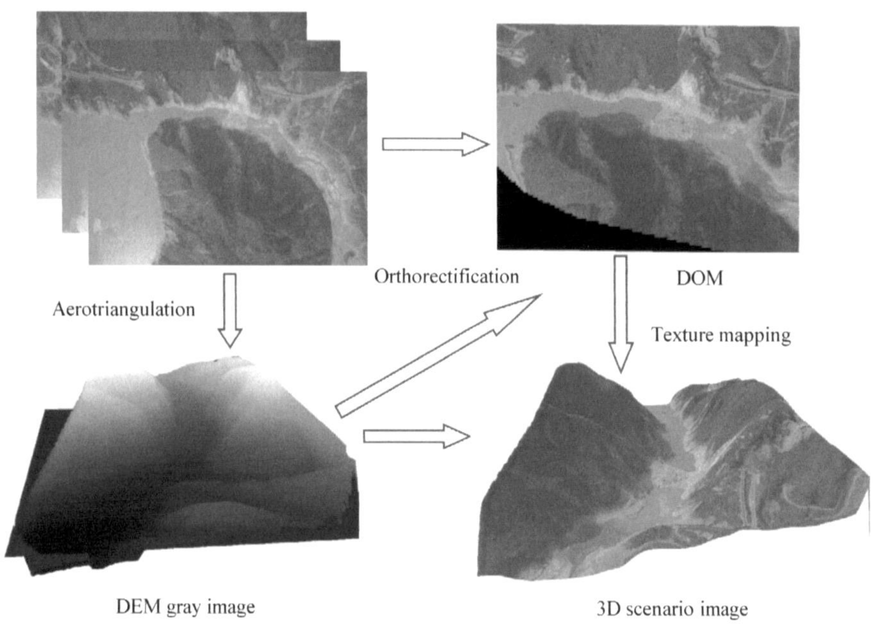

Figure 3. Generation of DOM and DEM data with optical UAV images.

- 3D terrain generation. Exterior orientation parameters recorded during UAV flight include photo reference number, latitude, longitude, relative height, roll angle, pitch angle, heading angle, and altitude above sea level. We use the modern Switzerland aerial triangulation software Pixel4UAV to generate 3D terrain data of target disaster area. Specific steps include putting UAV images and corresponding exterior and interior orientation data into the aerotriangulation software, automatically searching for matching tie points between adjacent images, and conducting aerial triangulation analytical operations to generate 3D terrain data, namely DEM (Digital Elevation Model) of target disaster area (see Figure.3).
- Image orthorectification and mosaicking. An orthophotograph is an aerial photograph geometrically corrected ("orthorectified") such that the scale is uniform: the photo has the same lack of distortion as a map (Lear, A.C., 1997). Unlike an uncorrected aerial photograph, an orthophotograph can be used to measure true distances, because it is an accurate representation of the Earth's surface, having been adjusted for topographic relief, lens distortion, and camera tilt. The generated DEM will be used to orthorectify the corresponding UAV image to generate

orthorectified image, namely DOM (Digital Orthophoto Map). As shown in Figure. 3, the DOM of the whole target area was obtained by mosaicking all DOM blocks in the UTM WGS84 coordinate system.

- Texture mapping. We can use 3D display software such as ArcScene to map the DOM of target landslide area to the DEM of the same area (see Figure. 3). Since both the DEM terrain data and DOM data are from the UAV remote sensing images, they fit each other perfectly with excellent visual effects and make the interpretation and identification of landslides more accurately and intuitively.

3. Landslide Identification and information extraction

3.1. Identification of landslide slope surface and deposit body

Since newly occurred landslides are usually accompanied by the phenomena such as falling and buried vegetation, exposed bare soil or base rock, we can utilize UAV optical images, pseudo colour images, and NDVI images generated from UAV multispectral images to identify newly occurred landslides by analysing strong differences between the landslide surface vegetation and the surrounding environment with consideration of local terrain conditions.

As shown in Figure 4, the landslide slope surface (yellow line surrounded region) and deposit body (blue line surrounded region) are sketched out on both optical image and pseudo colour image respectively. The optical image and false colour image are captured in different angle and different time. The former was captured in 2011, and the latter in 2013. Based on visual interpretation, it can be seen from the Figure 4.b that the spatial extent of the landslide slope surface has been significantly reduced after two years, and the peripheral area of the landslide slope surface and even part of the intermediate region has grown some plants.

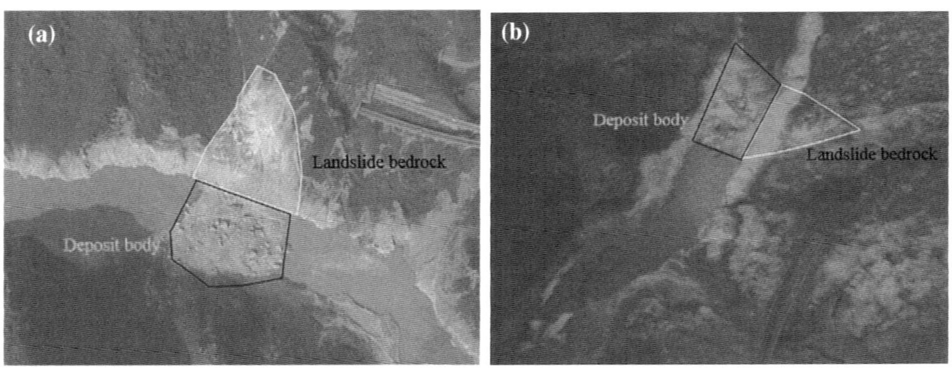

Figure 4. Identification and distinguishment of landslide bedrock surface and deposit body on optical image (captured in 2011) and pseudo color image (captured in 2013).

3.2. Distinguishment between landslide slope bedrock and deposit body

As shown in Figure 4, directly acquired UAV images are two-dimensional, so it is very easy to take landslide slope surface and deposit body as a whole if just taking bare soil or rock situation into account. To distinguish landslide slope surface from its deposit body, one indirect method is to consider the natural surroundings, where the landslide deposit body is located in the river. Another more direct method is to use the 3D scenario image (DOM mapped on DEM) of the landslide area generated by UAV images. The landslide slope surface is generally steep, and the deposit body is relatively flat. As shown in Figure 5, in the three-dimensional virtual environment, the distinguishment between landslide slope surface and deposit body are very intuitive.

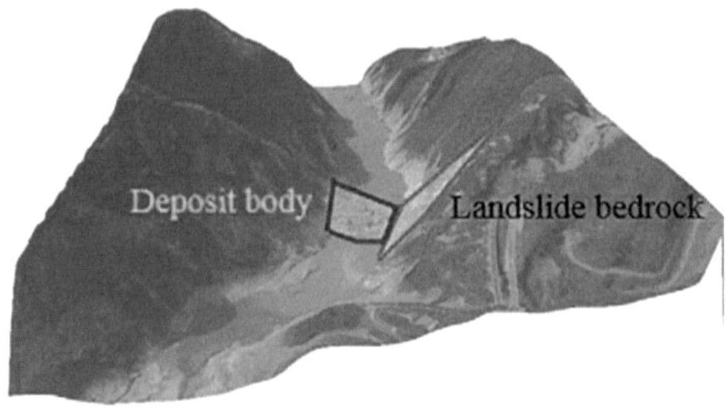

Figure 5. Distinguishment landslide bedrock surface from deposit body in 3D scenario.

3.3. Landslide impacted area measurement

As both the DOM and DEM are georeferenced, the landslide impacted area can be directly measured using the measurement tools in 3D scenario display software such as ARCScene (Kennedy, K. H., 2009). The landslide slope surface area measured from the optical image in 2011 is about 25, 000 square meters, and the deposit body area is about 31, 600 square meters. However, the landslide slope surface area measured from the pseudo colour image in 2013 is about 16, 000 square meters, and the deposit body area is about 30, 000 square meters. It can be noticed that the spatial extent of the landslide slope surface has been significantly reduced after two years, but that of the landslide deposit body has relatively slight deceased. The reason for the slope surface area change may be the vegetation growth. The peripheral area of the landslide slope surface has gradually grown some plants, but nearly none plants has grown on the deposit body. As for the deposit body area change may be the different water level in two years.

Table 1. Landslide slope surface and deposit body area in 2011 and 2013.

Year	Image type	Slope surface area	Deposit body area
		square meters	square meters
2011	RGB	25, 000	31, 600
2013	Multispectral	16, 000	30, 000
	Area Change	9,000	1,600

3.4. Landslide earthwork volume estimation

Terrain analysis can be done on the 3D terrain data, including contour generation, slope orientation, slope, and so on. In theory, the landslide earthwork volume can also be measured in 3D scenario. However, the measurement of landslide earthwork volume is different from landslide impacted area measurement. The landslide earthwork volume can be calculated by reducing the current 3D terrain from the 3D terrain that generated before the landslide occurred. It is a pity that we do not have the 3D terrain before landslide occurred. Therefore, we cannot take this high precision method to calculate the landslide earthwork volume. One alternative method, as the formula (2) shows, is to multiply the deposit body area measured from optical image with estimated deposit body height which is about 5 meters (average local river depth plus the average height of deposit body exposed above water surface). As a result, the estimated landslide earthwork volume is about 158,000 cubic meters (31,600 square meters*5 meter).

$$V_{Le} \approx A_d * (H_e + D_r) \tag{2}$$

Where VLe is the target landslide earthwork volume; Ad is the landslide deposit body area; He is the height of the deposit body exposed above the river water surface; Dr is the average local river depth.

4. Conclusion

Landslide refers to the slope deformation phenomena caused by natural forces such as the gravity applied to slope-composed materials. It is one of the serious natural disasters, which causes enormous economic losses and casualties only second to earthquake. Newly occurred landslides are usually accompanied with those phenomena including vegetation burying and bedrock or bare soil exposure. Therefore, they can be easily identified in optical or multispectral UAV images by detecting sharp vegetation change combined with local steep terrain. In our study, we take one typical landslide occurred in Wenchuan Earthquake stricken area in 2010 as an example. We demonstrate the process of integration of multispectral camera with UAV platform, NDVI generation with multispectral UAV images, orthophoto and three-dimensional terrain generation with optical UAV images, and identification and extraction of landslide information such as its the location, impacted area, and earthwork volume. The different area changes of landslide slope surface and deposit body from 2011 to 2013 and the possible reasons are discussed. We took advantage of the landslide deposit body to estimate the landslide earthwork volume without the 3D terrain acquired before the landslide occurred. Results prove the feasibility and effectiveness of applying the fixed wing UAV system to rapid identification and information extraction of newly occurred landslides under harsh natural conditions including complex terrain and deficient background materials.

Acknowledgements

This work was supported by Key Laboratory of Digital Mapping and Land Information Application of National Administration of Surveying, Mapping and Geoinformation (No. DM2016SC01), Open Research Fund Program of Plateau Atmosphere and Environment Key Laboratory of Sichuan Province (No. PAEKL-2016-C2), and the Key Technology Research and Development Program of the Science & Technology Department of Sichuan Province (No. 2012SZ0057). It was also partially supported by the Natural Science Foundation of China (No. 41271433, No. 41471281 and No.41272366).

References

[1] Delacourt C, Allemand P, Berthier E, Raucoules D, Casson B, Grandjean P, Pambrun C and Varel E 2007 *Bull. Soc. Geol. Fr.* **178** 89
[2] Derek H C 2005 *Landslides in Practice: Investigation, Analysis and Remedial/Preventative Options in Soils* (New York: John Wiley and Sons, Inc.)
[3] Kennedy K H 2009 *Introduction to 3D Data: Modeling with ArcGIS 3D Analyst and Google Earth* (New York: John Wiley and Sons, Inc.)
[4] Lear A C 1997 *IEEE Comput Graph* **17** 12
[5] Lin JY, Tao HP, Wang YC and Huang Z 2010 *18th Int. Con. on Geoinformatics* vol 2 (Beijing: IEEE) pp 1-5
[6] Pettorelli N, Vik J O, Mysterud A, Gaillard J M, Tucker C J and Stenseth N C 2006 *Trends Ecol. Evol.* **21** 11
[7] Tetracam Inc. 2011 *Agricultural Digital Camera User's Guide* (Chatsworth)
[8] Yin YP 2009 *Grate Wenchuan Earthquake: Seismogeology and Landslides Hazards* (Beijing: Geological Publishing House)
[9] Zhang WJ, Lin JY, Peng J, Lu QF 2010 *Int. J. Remote Sens.* **31** 3495

IOP Conf. Series: Earth and Environmental Science **57** (2017) 012018 doi:10.1088/1755-1315/57/1/012018

Building damage assessment using a single post-earthquake PolSAR image: a case of the 2010 Yushu earthquake

Wei Zhai [1], **Wenhao Zeng** [1, *]

[1] Monitoring Center, Gansu Earthquake Administration, 730000, Lanzhou, Gansu, China

The corresponding author's e-mail: zwxzzzdsyhq@163.com

Abstract. Earthquakes are one of the most destructive natural disasters. Efficiently and quickly acquiring building earthquake damage information can help to reduce the casualties after an earthquake. In this paper, building damage information is extracted using a single post-earthquake PolSAR image. In PolSAR images, since the undamaged oriented buildings characterized by volume scattering with weaker scattering power are very similar to collapsed buildings, the collapsed buildings are difficult to extract accurately. In this paper, the difference in the relative contribution change rate of scattering components before and after polarization orientation angle (POA) compensation is proposed to enhance the difference between collapsed buildings and oriented buildings, in order that the collapsed buildings can be extracted more accurately. The "4.14" Yushu earthquake, which occurred in Yushu County, Qinghai province of China, is used as the case study to test the proposed method, and an airborne high-resolution PolSAR image of the urban region of Yushu County is used in the experiment. The experimental results show that the accuracy of building damage information extraction can be improved by the use of the proposed method, compared with the traditional polarimetric classification.

1. Introduction

Earthquakes can give rise to great damage to life and property. In an earthquake, most of the casualties are caused by collapsed buildings, so building damage information extraction is the main task of earthquake damage information investigation. After an earthquake, the collapsed building information should be acquired in a timely manner, and can be used to guide the effective implementation of the emergency rescue, which is crucial for the reduction of casualties [1]. Radar is not affected by weather and climate, and can obtain images in bad weather and even at night [2], because of its strong penetrative power, so it has become a reliable remote sensing data source for acquiring the damage information of earthquake-stricken areas. Compared to single-polarization synthetic aperture radar (SAR), fully polarimetric SAR (PolSAR) is capable of recording the backward scattering information in four kinds of polarization mode, which contain much more information and are more helpful for understanding the characteristics of ground objects [3,4].

Most of the studies about the disaster investigation using PolSAR data were carried out mainly based on the changes of polarimetric features between pre- and post-earthquake using multitemporal PolSAR data [5-9]. However, archived pre-event PolSAR images matched with the post-event images are often difficult to obtain, and the registration of the pre- and post-event PolSAR images also costs time and manpower. Therefore, earthquake damage assessment using only a single post-earthquake PolSAR image has attracted the attention of more and more researchers. Guo et al.[10] and Li et al.[11] introduced the circular polarization correlation coefficient ρ and proposed the H-α-ρ method to extract collapsed building information using Radarsat-2 polarimetric data after the "4.14" Yushu earthquake. Zhao et al. [12] introduced the texture parameter of *homogeneity* to improve the H-α-ρ method using high-resolution airborne PolSAR data. Shen et al. [13] used the method of image retrieval based on feature template matching to extract the collapsed building information.

This study is devoted to extracting building damage information using a single post-earthquake PolSAR image. The collapsed buildings are mainly characterized by volume scattering due to the destruction of the building structure. The uncollapsed oriented buildings which are not parallel to the flight pass of the radar can cause polarization orientation angle (POA) shift, so that volume scattering becomes the dominant scattering mechanism. If the traditional polarimetric classification methods are used to extract the collapsed buildings, the oriented buildings can be easily misclassified as collapsed buildings, and the building collapse rate is likely to be overestimated, especially in built-up areas containing many oriented buildings. In view of this, the difference in the change rate of the relative contributions of double-bounce scattering and volume scattering, $CR_{Dbl\text{-}Vol}$, is introduced to help to distinguish the two types of buildings. The experiments show that the building damage assessment accuracy can be improved using the method proposed in this paper.

2. Methodology

2.1. *Analysis of building scattering components in earthquake-stricken areas*

The buildings of earthquake-stricken areas are divided into three classes in this paper, based on the characteristics of the PolSAR images: collapsed buildings, parallel buildings, and oriented buildings. The scattering power of parallel buildings is particularly strong because they form a dihedral structure with the ground, but the scattering power of oriented buildings is as weak as the collapsed buildings. Compensating the POA shift induced by the polarization basis being rotated can enhance the double-bounce scattering power of the oriented buildings. The details about the scheme of POA compensation can be found in [14]. In order to comprehend the changes in the scattering mechanism after POA compensation for the three kinds of buildings, the four scattering components generated from Yamaguchi four-component decomposition [15], before and after POA compensation, are analyzed in this section.

The relative contributions of the four components of the three kinds of building samples selected in Fig. 4, before and after POA compensation, are shown in Fig. 1. As can be seen in Fig. 1, the dominant scattering component of the parallel buildings is double-bounce scattering, and the oriented buildings and the collapsed buildings are dominated by the volume scattering component. Before and after POA compensation, the relative contributions of each scattering component for the parallel buildings basically remain the same, while the relative contribution of double-bounce scattering for the oriented buildings more than doubles, and the relative contribution of volume scattering significantly decreases. In addition, for the collapsed buildings, the relative contribution of double-bounce scattering also increases, but the increase is less than for the oriented buildings, and the decrease in the relative contribution of volume scattering is very small. In conclusion, for the oriented buildings, the scheme of POA compensation can greatly improve the relative contribution of the double-bounce scattering component, and can significantly reduce the relative contribution of the volume scattering component.

2.2. $CR_{Dbl\text{-}Vol}$

Inspired by the conclusion deduced from Fig. 1, the change rates of the relative contribution before and after POA compensation (*CR*, for short) for the double-bounce scattering component and the volume scattering component are shown in Fig. 2. As can be seen from Fig. 2, the *CR* for the double-bounce scattering component (CR_{Dbl}, for short) of the oriented buildings is more than twice as large as that of the collapsed buildings, and the amplitude of the *CR* for the volume scattering component (CR_{Vol}, for short) of oriented buildings is also larger than that of collapsed buildings. In view of this, the difference of the *CR* between the double-bounce and volume scattering components is defined as:

$$CR_{Dbl-Vol} = \frac{Dbl_{afterPAC} - Dbl_{beforePAC}}{Dbl_{beforePAC}} - \frac{Vol_{afterPAC} - Vol_{beforePAC}}{Vol_{beforePAC}} \tag{1}$$

where $Dbl_{beforePAC}$ denotes the relative contribution of the double-bounce scattering component before POA compensation;

$Dbl_{afterPAC}$ is the relative contribution of the double-bounce scattering component after POA compensation.

$Vol_{beforePAC}$ denotes the relative contribution of the volume scattering component before POA compensation;

$Vol_{afterPAC}$ is the relative contribution of the volume scattering component after POA compensation.

$CR_{Dbl-Vol}$ is also shown in Fig. 2. The difference in $CR_{Dbl-Vol}$ between collapsed buildings and oriented buildings is more distinct than the difference in CR_{Dbl} and in CR_{Vol}. The difference in $CR_{Dbl-Vol}$ of oriented buildings is much greater than that of collapsed buildings.

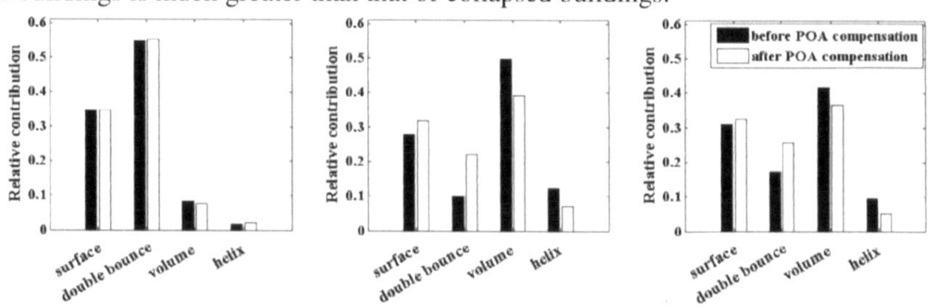

(a) Parallel buildings (b) Oriented buildings (c) Collapsed buildings

Figure 1. Relative contribution of the different scattering mechanisms of the three kinds of buildings before and after POA compensation

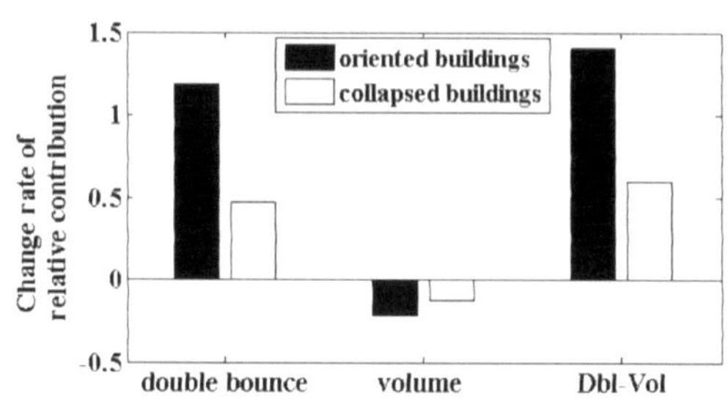

Figure 2. Change rate of the relative contribution of the different scattering mechanisms for the oriented and collapsed buildings before and after POA compensation.

As can be seen from Fig. 1, the relative contribution of the double-bounce scattering component of the oriented buildings after POA compensation cannot be increased as much as for the parallel buildings. Therefore, the classification result of the oriented and collapsed buildings generated from the Wishart supervised classification performed on the PolSAR data after POA compensation is still not very accurate. However, the difference in $CR_{Dbl-Vol}$ between the oriented buildings and the collapsed buildings can be used to distinguish the two kinds of buildings, and to correct the results of the two kinds of buildings obtained from Wishart supervised classification, so as to improve the accuracy of the collapsed building extraction. Using $CR_{Dbl-Vol}$ to separate the two kinds of buildings can be expressed as:

$$\begin{cases} if \ \ CR_{Dbl-Vol} > \varepsilon, \ x \in \text{oriented buildings} \\ if \ \ CR_{Dbl-Vol} \le \varepsilon, \ x \in \text{collapsed buildings} \end{cases} \tag{2}$$

where ε is the threshold value of $CR_{Dbl\text{-}Vol}$ set according to experiment, and x is a pixel of the PolSAR image.

3. Building earthquake damage information extraction

The process of building damage information extraction is shown in Fig. 3. Firstly, POA compensation is performed on the PolSAR data after preprocessing, and the new [*C*] matrix is obtained. Secondly, according to the case study selected in this paper, the ground objects in the earthquake-stricken area are classified as bare areas, collapsed buildings, parallel buildings, and oriented buildings, using the Wishart supervised classification [17] implemented on the PolSAR data after PolSAR compensation. Thirdly, the values of $CR_{Dbl\text{-}Vol}$ of the collapsed buildings and oriented buildings are calculated, and the threshold value ε is set to separate the two kinds of buildings according to the difference in $CR_{Dbl\text{-}Vol}$ between the collapsed buildings and the oriented buildings, as shown in Fig. 2. In this way, the new class centers for the two kinds of buildings are obtained. The clustering and iteration using the complex Wishart classifier needs to be carried out again to update the new class centers. After this, the two classes of collapsed buildings and oriented buildings generated from the Wishart supervised classification are corrected. Finally, the parallel buildings and the oriented buildings are combined as the undamaged buildings. The building collapse rate (*BCR*) is calculated at the scale of the city block, which is defined as the ratio of the collapsed buildings to the total buildings of one block. The damage levels of the buildings are divided into slight, moderate, and serious damage levels according to the threshold values of the *BCR* of each city block.

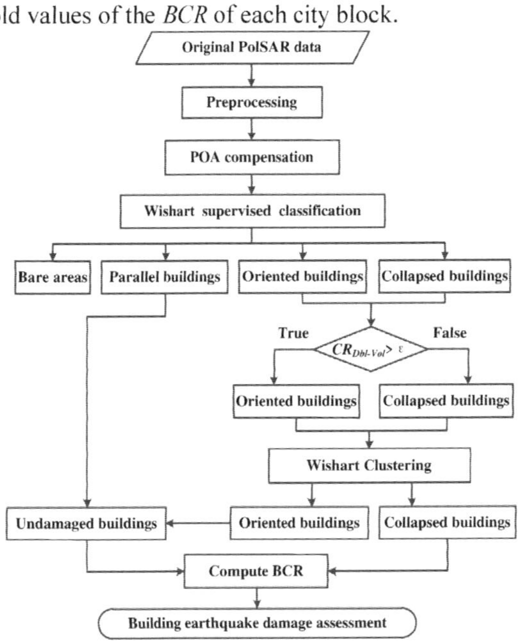

Figure 3. Experimental flow chart of building earthquake damage information extraction

4. Experiment and analysis

4.1. Experimental Data

In this paper, the selected case study is the Yushu earthquake of magnitude 7.1 that occurred in Yushu County, Qinghai province, China, on April 14, 2010. The location of the epicenter was 33.1°N and 96.6°E. There were many collapsed buildings and more than 2600 people died in the earthquake. The direct economic losses amounted to more than 22 billion CNY. The study area is the urban region of Yushu County. The vegetation is very sparse and low-level in this urban region, so it was ignored in the experiment. The buildings are mainly low-rise rural residential buildings in Yushu County.

The experimental data are the airborne high-resolution PolSAR image acquired one day after the earthquake in the P-band by the Chinese airborne SAR mapping system (SARMapper). Both the range resolution and azimuth resolution are approximately 1 m. The flight pass was in an east–west horizontal direction. The Pauli RGB image is shown in Fig. 4 with the size of 8192×4384 pixels. In the PolSAR data, the mountains surrounding the urban area were removed by masking, and only the urban region was kept. The whole urban region was divided into 72 city blocks by roads and according to the similarity of the built-up patches. The earthquake damage assessment reference information is shown in Fig. 5. The building damage levels of the 72 city blocks were divided into slight, moderate, and serious damage levels.

Figure 4. Pauli RGB color composite image of PolSAR data of Yushu County, with red (|HH-VV|), green (|HV|), and blue (|HH+VV|). The areas in red, blue, and orange boxes are the samples of the collapsed buildings, oriented building, and parallel buildings, respectively.

Figure 5. Reference map of the building earthquake damage information at the city block scale in the Yushu urban region.

4.2. Experimental Results

According to the process of building earthquake damage information extraction shown in Fig. 3, the Wishart supervised classification method was performed on the PolSAR data after POA compensation. We then calculated the $CR_{Dbl\text{-}Vol}$ for all of the samples of oriented buildings and the collapsed buildings generated from the Wishart supervised classification, and set the threshold value ε as 0.7 according to the difference shown in Fig. 2. The two kinds of buildings were reclassified based on Eq. (2), as described in Section 2.2, and clustering based on the Wishart distance was carried out. The final classification results of the oriented buildings and the collapsed buildings were obtained. The oriented buildings together with the parallel buildings were considered to be the undamaged buildings.

The results of extracting the collapsed buildings, the undamaged buildings, and the non-buildings are shown in Fig. 6, in which the red areas are the collapsed buildings, the green areas denote the undamaged buildings, and the blue areas correspond to the non-buildings. The *BCR* of each block was calculated, and the building damage levels were set as:

$$\begin{cases} \text{if } BCR \le 0.2, \text{ the block} \in \text{slight damage} \\ \text{if } 0.2 < BCR \le 0.5, \text{ the block} \in \text{moderate damage} \\ \text{if } BCR > 0.5, \text{ the block} \in \text{serious damage} \end{cases} \quad (4)$$

The map of building damage levels for the whole urban region at the block scale is shown in Fig. 7, and the accuracy evaluation of the experimental results is shown in Table 1 by reference to Fig. 5. For comparison, the results of both the proposed method and the method using Wishart supervised classification directly without POA compensation and $CR_{Dbl\text{-}Vol}$, which is termed direct Wishart supervised classification (DWSC), are listed in Table 1.

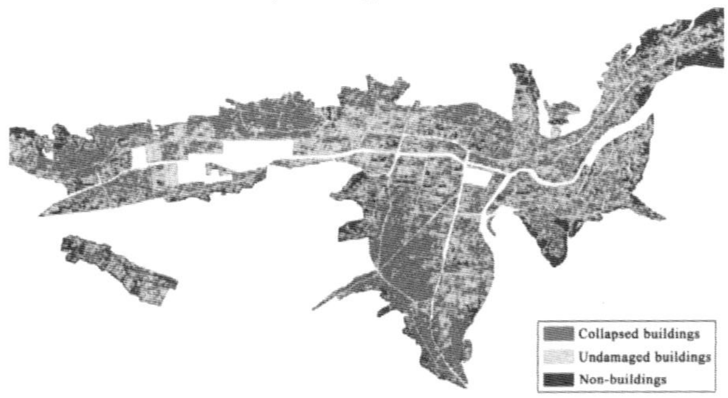

Figure 6. Experimental results distribution map of the collapsed buildings, the undamaged buildings, and the non-buildings in the earthquake-stricken urban area of Yushu County.

Figure 7. Experimental results map of the building damage levels at the block scale in the Yushu urban region.

Table 1. Accuracy evaluation of the earthquake damage information extraction of the two methods

		The proposed method			DWSC		
		(Experiment)					
		SLD	MOD	SED	SLD	MOD	SED
		(No. of blocks)					
	SLD	11	3	0	9	3	2
(Reference)	MOD	4	29	0	3	22	8
	SED	1	7	17	0	4	21
		OA: 79.2%			OA: 72.2%		

4.3. Discussion and analysis

As can be seen from Table 1, the overall accuracy of the building earthquake damage information extraction was increased by 7% using the proposed method compared to the DWSC method. There were actually 25 blocks, and 17 blocks were extracted using the proposed method, but 31 blocks were extracted by the DWSC method. Therefore, the proposed method can effectively reduce the overestimation of building earthquake damage. Most of the misclassification situations were those of the slight damage level or the serious damage level being misclassified as the moderate damage level, and the moderate damage level being misclassified as the slight damage level. In other words, the misclassification of the damage level only spanned one level, and there was only one block of serious damage misclassified as the slight damage level. In short, for the 15 misclassified blocks, the main situation of misclassification involved reduced damage levels.

Although POA compensation can enhance the double-bounce scattering power of oriented buildings, the dominant scattering mechanism of oriented buildings is still volume scattering. Therefore, there are still many oriented buildings and collapsed buildings mixed up with each other in the classification results of the Wishart supervised classification performed on the PolSAR data after POA compensation. Accordingly, the difference in $CR_{Dbl\text{-}Vol}$ between oriented buildings and collapsed buildings is used to correct the results of the Wishart supervised classification.

However, the correction can result in the oriented buildings being detected too much and can result in overestimation of the undamaged buildings. This is influenced by the selection of the $CR_{Dbl\text{-}Vol}$ threshold value ε for separating oriented buildings and collapsed buildings. In addition, due to the fact that some residual walls of the collapsed buildings form a dihedral structure with the ground, some collapsed building areas also have high double-bounce scattering power, and these areas are easily misclassified as undamaged parallel buildings. In the collapsed building areas, the double-bounce scattering power of some oriented residual walls is increased after POA compensation, and these walls may also be misclassified as parallel buildings. These are the causes of the overestimation of the undamaged buildings.

The overestimation of the undamaged buildings gives rise to a reduction in the building damage level, and the identification accuracy of the blocks with the serious damage level is reduced. However, the identification accuracy of the blocks with the slight damage level and moderate damage level is generally improved, and the overall accuracy is improved.

5. Conclusion

In this paper, building earthquake damage information has been extracted using a single post-event PolSAR image. Because the characteristics of oriented buildings are very similar to the collapsed buildings in PolSAR images, and the double-bounce scattering power of oriented buildings is weak, the scheme of POA compensation is performed on the PolSAR image to enhance the double-bounce scattering power of the oriented buildings. Although the scattering power of oriented buildings in the PolSAR image after POA compensation is increased, the dominant scattering mechanism of oriented buildings is still volume scattering. Therefore, the results of the Wishart supervised classification are not very accurate and need to be improved. Through analyzing the changes in the scattering mechanism before and after POA compensation for the three kinds of buildings, we found that the increase in the amount of double-bounce scattering component and the reduction in the amplitude of the volume scattering component of the oriented buildings are greater than for the collapsed buildings. Accordingly, the parameter of $CR_{Dbl\text{-}Vol}$ is proposed to further distinguish the oriented buildings from the collapsed buildings. Therefore, the significant difference in $CR_{Dbl\text{-}Vol}$ between collapsed buildings and oriented buildings is used to correct the results of the Wishart supervised classification for the two kinds of buildings, and the collapsed building extraction accuracy is improved. In a comparison with the DWSC method, it was found that the proposed method can improve the accuracy of building earthquake damage assessment.

Acknowledgements

This work was supported by the Earthquake Science and Technology Development Fund Program of Lanzhou Earthquake Research Institute, China Earthquake Administration (2015M02); the Object-oriented High Trusted SAR Processing System of the National 863 Subject Program; and the Airborne Multiband Polarimetric Interferometric SAR Mapping System of the National Major Surveying and Mapping Science and Technology Special Program. We would also like to thank the anonymous reviewers for their advice on improving the quality of this paper.

References

[1] Zhai W., Shen H.F., Huang C.L., Pei W.S., 2016. Building Earthquake Damage Information Extraction from a Single Post-Earthquake PolSAR Image. *Remote Sens*, 8(3), 171.

[2] Gamba P., Dell'Acqua F., Trianni G., 2007. Rapid Damage Detection in the Bam Area Using Multitemporal SAR and Exploiting Ancillary Data. *IEEE Trans Geosci Remote Sens*, 45(6), pp. 1582-1589.

[3] Zhai W., Shen H.F., Huang C.L., Pei W.S., 2016. Fusion of polarimetric and texture information for urban building extraction from fully polarimetric SAR imagery. *Remote Sens Lett*, 7(1), pp. 31–40.

[4] Zhai W, Huang C.L., 2016. Fast building damage mapping using a single post-earthquake PolSAR image: a case study of the 2010 Yushu earthquake. *Earth Planets Space*, 68(1), 1.

[5] Yamaguchi Y., 2012. Disaster Monitoring by Fully Polarimetric SAR Data Acquired With ALOS-PALSAR. *Proc IEEE*, 100(10), pp. 2851–2860.

[6] Chen S.W., Sato M., 2013. Tsunami Damage Investigation of Built-Up Areas Using Multitemporal Spaceborne Full Polarimetric SAR Images. *IEEE Trans Geosci Remote Sens*, 51(4), pp. 1985–1997.

[7] Park S.E., Yamaguchi Y., Kim D., 2013. Polarimetric SAR remote sensing of the 2011 Tohoku earthquake using ALOS/PALSAR. *Remote Sens Environ*, 132, pp. 212–220.

[8] Sato M., Chen S.W., Satake M., 2012. Polarimetric SAR Analysis of Tsunami Damage Following the March 11, 2011 East Japan Earthquake. *Proc IEEE*, 100(10), pp. 2861–2875.

[9] Watanabe M., Motohka T., Miyagi Y., Yonezawa C., Shimada M., 2012. Analysis of Urban Areas Affected by the 2011 Off the Pacific Coast of Tohoku Earthquake and Tsunami With L-Band SAR Full-Polarimetric Mode. *IEEE Geosci Remote Sens Lett*, 9(3), pp. 472–476.

[10] Guo H.D., Wang X.Y., Li X.W., Liu G., Zhang L., Yan S.Y., 2010. Yushu earthquake synergic analysis using multimodal SAR datasets. *Chinese Sci Bull*, 55(31), pp. 3499–3503.

[11] Li X.W., Guo H.D., Zhang L., Chen X., Liang L., 2012. A New Approach to Collapsed Building Extraction Using RADARSAT-2 Polarimetric SAR Imagery. *IEEE Geosci Remote Sens Lett*, 9(4), pp. 677–681.

[12] Zhao L., Yang J., Li P., Zhang L., Shi L., Lang F., 2013. Damage assessment in urban areas using post-earthquake airborne PolSAR imagery. *Int J Remote Sens*, 34(24), pp. 8952–8966.

[13] Shen J.C., Xu X., Dong H., Gui R., Song C., 2015. Collapsed Building Extraction from Single Full Polarimetric SAR Image after Earthquake. *Sci Technol Eng*, 15(14), pp. 86–91.

[14] Lee J.S., Schuler D.L., Ainsworth T.L., 2000. Polarimetric SAR data compensation for terrain azimuth slope variation. *IEEE Trans Geosci Remote Sens*, 38(5), pp. 2153–2163.

[15] Yamaguchi Y., Moriyama T., Ishido M., Yamada H., 2005. Four-component scattering model for polarimetric SAR image decomposition. *IEEE Trans Geosci Remote Sens*, 43(8), pp. 1699–1706.

[16] Lee J.S., Grunes M.R., Ainsworth T.L., Li-Jen D., Schuler D.L., Cloude S.R., 1999. Unsupervised classification using polarimetric decomposition and the complex Wishart classifier. *IEEE Trans Geosci Remote Sens*, 37(5), pp. 2249-2258.

IOP Conf. Series: Earth and Environmental Science 57 (2017) 012019 doi:10.1088/1755-1315/57/1/012019

Sentinel-1 TOPS interferometry for along-track displacement measurement

H J Jiang[1], Y Y Pei[2] and J Li[1]

[1] School of Geographic and Biologic Information, Nanjing University of Posts and Telecommunications, Nanjing 210023, China
[2] School of Civil Engineering, Anhui Jianzhu University, Hefei 230601, China

jianghouj@njupt.edu.cn

Abstract. The European Space Agency's Sentinel-1 mission, a constellation of two C-band synthetic aperture radar (SAR) satellites, utilizes terrain observation by progressive scan (TOPS) antenna beam steering as its default operation mode to achieve wide-swath coverage and short revisit time. The beam steering during the TOPS acquisition provides a means to measure azimuth motion by using the phase difference between forward and backward looking interferograms within regions of burst overlap. Hence, there are two spectral diversity techniques for along-track displacement measurement, including multi-aperture interferometry (MAI) and "burst overlap interferometry". This paper analyses the measurement accuracies of MAI and burst overlap interferometry. Due to large spectral separation in the overlap region, burst overlap interferometry is a more sensitive measurement. We present a TOPS interferometry approach for along-track displacement measurement. The phase bias caused by azimuth miscoregistration is first estimated by burst overlap interferometry over stationary regions. After correcting the coregistration error, the MAI phase and the interferometric phase difference between burst overlaps are recalculated to obtain along-track displacements. We test the approach with Sentinel-1 TOPS interferometric data over the 2015 Mw 7.8 Nepal earthquake fault. The results prove the feasibility of our approach and show the potential of joint estimation of along-track displacement with burst overlap interferometry and MAI.

1. Introduction

Sentinel-1, the new generation of European Space Agency's (ESA) C-band synthetic aperture radar (SAR) mission, provides data continuity to its previous European Remote Sensing (ERS) and ENVISAT SAR missions [1]. The Sentinel-1 mission is comprised of two identical radar satellites to form a constellation, which can significantly reduce the revisit time.

Sentinel-1's SAR instrument supports four imaging modes providing different resolution and coverage: Interferometric Wide Swath Mode (IW), Extra Wide Swath Mode (EW), StripMap (SM), and Wave (WV). Terrain observation by progressive scan (TOPS) antenna beam steering is utilized for the IW mode to provide a 250 km wide swath consisting of three sub-swaths at 5 m × 20 m resolution [2]. Due to the wide swath, the revisit cycle has been notably reduced from 35 days for ERS-1 and ERS-2 or 30/35 days for ENVISAT to 12 days for Sentinel-1, and can be reduced to 6 days when the two satellites are in orbit. The IW mode based on TOPS is set as the default operation mode for Sentinel-1.

The TOPS acquisition can overcome the problems of scalloping and azimuth-varying signal-to-ambiguity ratio of the conventional ScanSAR mode by means of steering the antenna beam in the along-track direction [3]. The antenna beam is rotated throughout the acquisition from backward to forward at a constant rotation rate, which leads to a reduction of the azimuth bandwidth, and consequently a worsening of the azimuth resolution. Since all targets are observed by the complete azimuth antenna pattern, the scalloping effect nearly eliminates, and the azimuth ambiguities and the signal-to-noise ratio become constant in azimuth. But the accuracy of along-track displacement measurement with multi-aperture interferometry (MAI) for Sentinel-1 IW mode will be much lower than that for SM mode, because its azimuth bandwidth and resolution are much lower than that of SM mode [2].

The MAI technique is an important complement to cross-track SAR interferometry (InSAR). A major limitation of InSAR for geophysical applications is that the technique is only sensitive to one dimensional motion along the satellite's line of sight (LOS), while measurements of the deformation field in three dimensions allow a better resolution of the parameters of deformation models for earthquakes, volcanic activity, and other processes. Therefore, the techniques for along-track displacement measurement should be fully exploited to improve the deformation resolving capability of Sentinel-1.

The antenna beam steering in TOPS acquisition provides another way for along-track displacement measurement. To assure that the images can be mosaiked without any gap, an overlap area is present in the azimuth direction between consecutive bursts. In the overlap area, the squint angle of the two consecutive bursts is changed from forward to backward, which results in a much larger azimuth spectral separation than the one achieved by MAI. Hence, the "burst overlap interferometry" gives a sensitive estimation for azimuth motion [4].

Both MAI and burst overlap interferometry have been used to estimate the azimuth coregistration error over stationary scenes [5]. This paper focuses on along-track displacement measurement using burst overlap interferometry and MAI with Sentinel-1 TOPS acquisitions over nonstationary scenes. A Sentinel-1 IW InSAR image pair over the 2015 Mw 7.8 Nepal earthquake fault is selected for demonstration.

2. Burst overlap interferometry and MAI

Burst overlap interferometry and MAI are spectral diversity techniques applied to the azimuth direction for along-track displacement measurement. The TOPS azimuth signal properties provide a probability for burst overlap interferometry, and require a deramping operation in interferometric processing steps. Table 1 lists the Sentinel-1 IW data parameters over the Nepal earthquake fault.

Table 1. Sentinel-1 IW data parameters over the 2015 Mw 7.8 Nepal earthquake fault

	sub-swath IW1	sub-swath IW2	sub-swath IW3
Incidence angle [deg]	34.66	39.93	44.48
Azimuth resolution [m]	22.5	22.7	22.6
Azimuth bandwidth [Hz]	327	313	314
Azimuth steering rate [deg/s]	1.59	0.98	1.40
Doppler rate [Hz/s]	-2265	-2129	-2000
Doppler centroid rate [Hz/s]	1745	1464	1539
$\Delta f_{DC}^{overlap}$ [Hz]	4813	4039	4245
Range bandwidth [MHz]	56.5	48.3	42.8
Azimuth sampling frequency [Hz]		486.49	
Azimuth spacing [m]		14	
Range sampling frequency [MHz]		64.35	
Range spacing [m]		2.3	
Wavelength [cm]		5.547	

2.1. Spectral Properties of TOPS azimuth signal

The properties of TOPS azimuth signal are a consequence of steering the antenna beam from backward to forward in azimuth direction, which produces a linear variation of the Doppler centroid frequency within each burst. The antenna beam steering introduces a Doppler centroid rate given by [3]

$$k_{rot} \approx -\frac{2v}{\lambda}\omega_r \tag{1}$$

where v is the satellite velocity, λ is the radar wavelength, and ω_r is the antenna beam steering rate.

After the raw data have been focused to SLC bursts in zero-Doppler geometry, a linear variation of Doppler centroid frequency is present in the azimuth direction. The new Doppler centroid rate in zero-Doppler geometry can be calculated as [3]

$$k_t = \frac{k_{rot} \cdot k_a}{k_a - k_{rot}} \tag{2}$$

where k_a is the Doppler rate.

Then the azimuth dependent Doppler centroid frequency can be written as

$$f_{DC} = k_t t_{az} \tag{3}$$

where t_{az} is the relative azimuth time within the burst. The Doppler variation has to be considered for interferometric processing, including spectral filtering and interpolation.

2.2. Accuracy analysis for burst overlap interferometry and MAI

Spectral diversity techniques measure along-track displacement by exploiting a differential interferogram obtained from two pairs of complex SAR images with an azimuth spectral separation. The phase of the differential interferogram is given by [6]

$$\phi_{sd} = 2\pi \cdot \Delta f \cdot x / ASF \tag{4}$$

where Δf is the spectral separation, x is the along-track displacement in units of resolution cells, and ASF is the azimuth sampling frequency.

According to equation (4), the Cramér-Rao bound in estimation of along-track displacement with MAI can be written in terms of standard deviation of resolution cells [5]

$$\sigma_{MA} = \frac{1}{2\pi\Delta f_{MA}}\sqrt{\frac{3}{N}}\frac{\sqrt{1-\gamma^2}}{\gamma}ASF \tag{5}$$

where Δf_{MA} is the spectral separation of the two sublooks, which is equal to 2/3 azimuth bandwidth when the accuracy approaches the Cramér-Rao bound, N is the number of independent averaged samples, and γ is the interferometric coherence.

Similarly, the estimation accuracy with burst overlap interferometry in terms of resolution cells is given by [5]

$$\sigma_{ovl} = \frac{1}{2\pi\Delta f_{ovl}}\frac{1}{\sqrt{N}}\frac{\sqrt{1-\gamma^2}}{\gamma}ASF \tag{6}$$

where Δf_{ovl} is the spectral separation in the burst overlap region.

As shown in table 1, Δf_{ovl} is over 20 times larger than Δf_{MA}. Usually the value of N for burst overlap interferometry is about 7% of that for MAI in Sentinel-1 IW case. Using equation (5), (6) and the parameters in table1 and taking all available pixels averaged, we find that the estimation accuracy of burst overlap interferometry is over 10 times better than that of MAI, since the accuracy improves inversely proportional to Δf, while only inversely proportional to the square root of N.

On the other hand, large spectral separation will wrap the phase ϕ_{sd}, and reduce the maximum displacement x_{max} that can be measured without ambiguity. x_{max} is given by

$$x_{\max} = \pm \frac{1}{2\Delta f} ASF \tag{7}$$

This value is equal to ±1.147 pixels for MAI and ±0.056 pixels for burst overlap interferometry with three sub-swaths IW1, IW2 and IW3 in average, using $\Delta f_{MA} = 2B_{az}/3$ (B_{az}: azimuth bandwidth) and the parameters in table 1.

2.3. Processing Strategy

The proposed approach with burst overlap interferometry and MAI for along-track displacement measurement involves the following steps:

- Coregistering the Sentinel-1 IW interferometric pair by using the precise orbit information and an external digital elevation model (DEM), such as an SRTM DEM. This geometric coregistration can achieve very good relative accuracy, both in range and azimuth dimensions. Therefore, only a constant coregistration offset remains, imposed by the limited orbit accuracy.
- Deramping the full resolution master and slave SLC images by the azimuth-variant Doppler centroid function equation (3) before cross-correlation coregistration, spectral filtering and interpolation. The consideration of the Doppler centroid frequency of the focused SLC burst is critical for interferometric processing.
- Refining the offsets obtained from geometric coregistration. The azimuth coregistration error, which will introduce an undesirable phase bias for along-track displacement measurement, is estimated by using equation (8). Regarding the range direction, cross-correlation is applied to patches distributed over the master and slave images and a linear correction of the shifts is performed.

$$x_{az} = \arg\min \left\{ \left| \arg \sum_i \gamma_i e^{j\left(\phi_{ovl,i} - 2\pi\Delta f_{ovl}\Delta x_{az}/ASF\right)} \right| \right\} \tag{8}$$

  where x_{az} is the azimuth coregistration error in units of resolution cells, ϕ_{ovl} is the interferometric phase difference between burst overlaps.
- Calculating the phase bias caused by the azimuth coregistration error for burst overlap interferometry and MAI using equation (4). The phase bias is then subtracted from the differential interferogram obtained from burst overlap interferometry and MAI.

3. Experimental results

The proposed approach is tested on a Sentinel-1 IW InSAR image pair over the 2015 Mw 7.8 Nepal earthquake fault. The image pair was acquired on 17 and 29 April 2015 in descending path 19. As shown in figure 1, The 250 km wide swath consists of three sub-swaths, covering the Himalayan Mountain and the Indo-Gangetic Plain.

The earthquake occurred on 25 April 2015 at a depth of approximately 8.2 km, with its epicentre (the red star in figure 1) approximately 80 km to the northwest of the Nepalese capital of Kathmandu [7]. The intensity contours in figure 1 portrays the distribution and severity of shaking. We can see that strong shakes are located at the top half of swath IW1 and IW2, which means ground displacements are present in this region. Therefore, the bursts over this region should be treated as nonstationary scenes.

The image pair is first coregistered using a geometrical approach. The image offsets are calculated with an SRTM DEM and the precise orbit state vectors. The precise orbit product has a 3D 1σ accuracy of 5 cm, which would turn into an azimuth error of 0.0036 pixels with an azimuth spacing of 14 m. This error is less than 1/10 of the ambiguity band of burst overlap interferometry as discussed in 2.2. Therefore the estimation of the azimuth coregistration error is not aliasing over stationary scenes when using burst overlap interferometry.

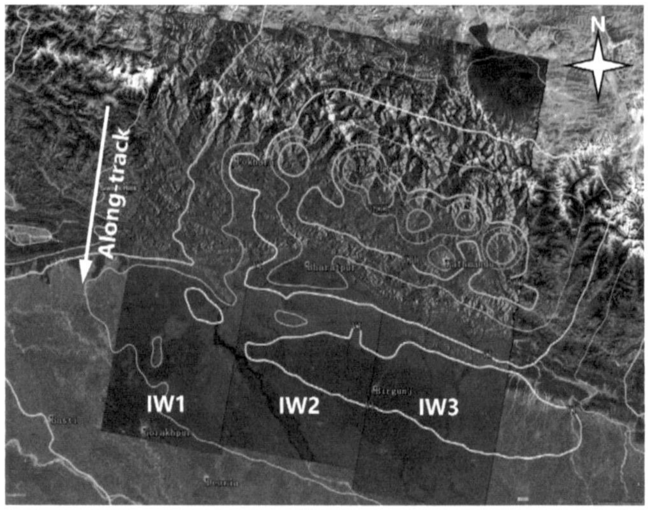

Figure 1. The Sentinel-1 IW detected SAR image over the 2015 Mw 7.8 Nepal earthquake fault. The red star marks the location of epicentre, and the coloured lines are the intensity contours of the earthquake published by USGS.

After the geometric coregistration, the full resolution master and slave SLC images are deramped to remove the phase ramp introduced by azimuth variation of Doppler centroid frequency. The image pair is then re-coregistered using cross-correlation to refine the range offsets, and the slave SLC image is resampled into the master image geometry. An interferogram is formed with the coregistered SLC images for burst overlap interferometry and LOS displacement measurement, and the phase ϕ_{ovl} is obtained from the differential interferograms within burst overlap regions. Meanwhile, the azimuth spectrums of the coregistered SLC images are filtered into two nonoverlapping bands with a spectral separation of $2B_{az}/3$, resulting in two pairs of sublook images for MAI.

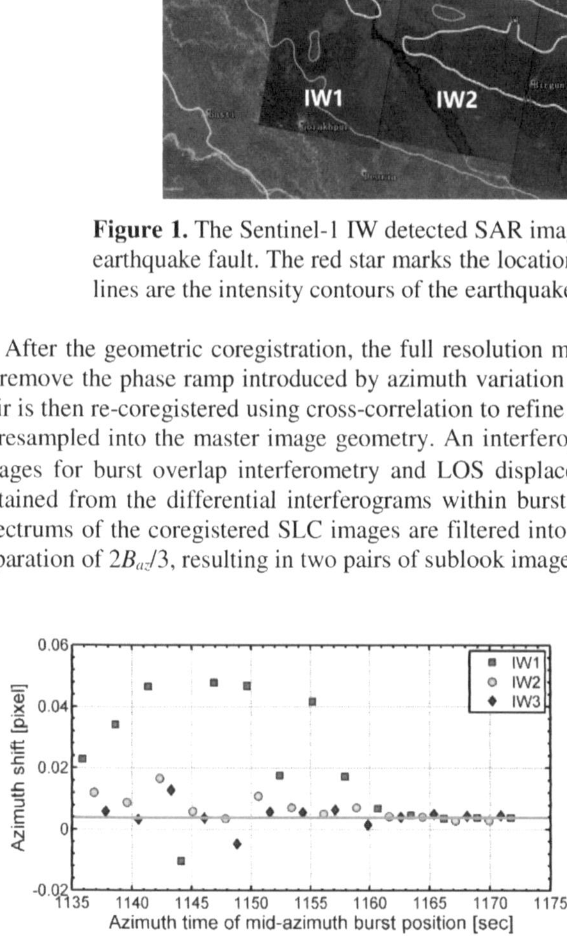

Figure 2. The azimuth shift estimated by using burst overlap interferometry. The cyan line indicates the average azimuth coregistration error of IW 1, IW2 and IW3.

The azimuth coregistration error is estimated using equation (8) over stationary regions. The results are plotted in figure 2. Each burst overlap region can give an estimate, and obviously the estimates close to the cyan line in figure 2 are related to the coregistration error, because the error should be nearly constant within each subswath after geometric coregistration. The estimated error of each subswath is 0.0038 pixels for IW1, 0.0035 pixels for IW2 and 0.0044 pixels for IW3. The average error of the three subswaths is about 5.5 cm, which is close to the accuracy of the precise orbit.

The estimates deviating from the cyan line in figure 2 are resulted from along-track displacement. After removing the azimuth coregistration error, the phase ϕ_{ovl} can represent the along-track displacement. The MAI interferogram is formed at last by a differential operation between the interferograms from the two pairs of sublook images. The phase bias due to the coregistration error is also corrected using equation (4).

Figure 3 and 4 portray the results of burst overlap interferometry and MAI respectively. It is observed that the along-track displacements mainly occur at the earthquake fault, and the displacement magnitude increases toward the centre of the fault rupture plane (the white quadrangle in figure 3 and 4). The result from burst overlap interferometry is more sensitive to along-track displacement compared with that from MAI. Note that the phase ϕ_{ovl} is wrapped over the fault rupture plane. Since ϕ_{ovl} is not continuous in azimuth direction, phase unwrapping should be carefully applied for further modelling of the deformation. Although the MAI result is less sensitive, the MAI phase is easier to unwrap, and is available over the whole image unless a strong decorrelation occurs. A combination of these two results may improve the modelling accuracy of the deformation.

Figure 3. The differential interferogram in burst overlap regions. The red star is the epicentre, and quadrangle represents the approximate surface projection of the fault rupture plane.

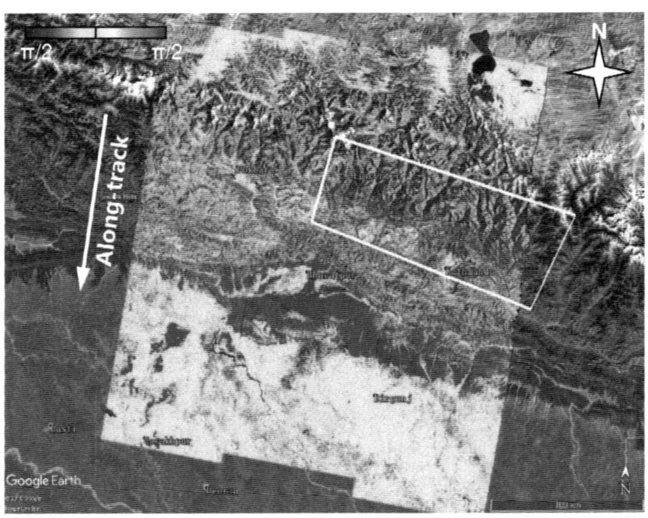

Figure 4. The MAI interferogram. The red star is the epicentre, and quadrangle represents the approximate surface projection of the fault rupture plane.

4. Conclusions

We analyze the accuracy of burst overlap interferometry and MAI for along-track displacement measurement in Sentinel-1 TOPS mode. We propose a data processing strategy to achieve very high accuracy of coregistration and to remove phase bias from the results of burst overlap interferometry and MAI. The approach is applied on Sentinel-1 TOPS interferometric data over the 2015 Nepal earthquake fault. The results prove that burst overlap interferometry is a more sensitive measurement, and a better deformation model might be built using a joint estimation with burst overlap interferometry and MAI.

Acknowledgments

This work was financially supported by the National Natural Science Foundation of China (Grant No. 41501497) and NUPTSF (Grant No. NY214197). The authors thank the European Space Agency for providing the Sentinel-1 A data.

References

[1] Salvi S, Stramondo S, Funning G J, Ferretti A, Sarti F and Mouratidis A 2012 The Sentinel-1 mission for the improvement of the scientific understanding and the operational monitoring of the seismic cycle *Remote Sensing of Environment* **120** 164-74

[2] Jung H S, Lu Z and Zhang L 2013 Feasibility of along-track displacement measurement from Sentinel-1 interferometric wide-swath mode *IEEE Transactions on Geoscience and Remote Sensing* **51** 573-8

[3] Zan F D and Guarnieri A M 2006 TOPSAR: terrain observation by progressive scans *IEEE Transactions on Geoscience and Remote Sensing* **44** 2352-60

[4] Scheiber R, Jäger M, Prats-Iraola P, Zan F D and Geudtner D 2015 Speckle tracking and interferometric processing of TerraSAR-X TOPS data for mapping nonstationary scenarios *IEEE Journal of Selected Topics in Applied Earth Observations and Remote Sensing* **8** 1709-20

[5] Prats-Iraola P, Scheiber R, Marotti L, Wollstadt S and Reigber A 2012 TOPS interferometry with TerraSAR-X *IEEE Transactions on Geoscience and Remote Sensing* **50** 3179-88

[6] Scheiber R and Moreira A, 2000 Coregistration of interferometric SAR images using spectral diversity *IEEE Transactions on Geoscience and Remote Sensing* **38** 2179-91

[7] Avouac J P, Meng L S, Wei S J, Wang T and Ampuero J P 2015 Lower edge of locked Main Himalayan Thrust unzipped by the 2015 Gorkha earthquake *Nature Geoscience* **8** 708-11

Field Hyperspectral Remote Sensing of Target Region in Xiemisitai Mountain, Xinjiang Province, China.

Q J Wang[1,2], Y M Wei[1] , Y Chen[1,2]*, X L Ma[3], H Y Zhou[4]

1. Key Laboratory of Earth Observation Hainan Province, Hainan 572029, China.

2. Key Laboratory of Digital Earth, Institute of Remote Sensing and Digital Earth, Chinese Academy of Sciences, Beijing, 100094, China.

3. China University of Geosciences (Beijing), Beijing, 100083, China.

4. Research Institute of Petroleum Exploration & Development, PetroChina, Beijing, 100083, China

Corresponding author: Y Chen, chenyu@radi.ac.cn

Abstract. A fine mineral identification model using the field Hyperspectral remote sensing was proposed to solve the problem of low mineral identification accuracy. Results show that the accuracy was improved by spectral noises removal, endmember optimization and mineral absorptions enhancement. A regional endmember library was established to improve the reliability by systematically considering of the mineral assemblage relationships. A fine mineral identification system (FMIS) was developed to help geologists to quickly identify minerals and it was applied in the Xiemisitai Mountain, Xinjiang province, China in 2014 to newly find copper mineralized points. The improved model and the FMIS system are therefore not only of great significance to improve efficiency and save cost in remote sensing mineral exploration, but also of great economic value of the local economy development in the future.

1. Introduction

Minerals, the basic unit of rocks, are stable elements or compounds under certain geological and geophysical conditions during the multi-period geological activities. Primary minerals in rocks are called rock-forming minerals, while those with changed components or structures are named as the alteration minerals. Rock-forming minerals, such as quartz, feldspar and mica, are primarily silicates, carbonates and oxides. Alteration minerals are primarily the components in hydrothermal mineralization, which are important indicators for mineral prospection. The common alteration minerals include skarnization, greisenization, sericitization and propylitization, etc. There are many techniques to determine mineral contents in laboratory, such as the slice identification, X-Ray diffraction analysis and the electron probe detection. However, large-scale field mineral identification needs a fine mineral identification model using the hyperspectral remote sensing to quickly map minerals.

Mineral inversion models, such as the linear spectral unmixing and the nonlinear spectral unmixing models (Hapke B 1981; Shkuratov Y et al. 1999) have been proposed. Classic techniques for mineral identification and rock info enhancement, such as the spectral angle mapper (Kruse F A 1993), Principle Component Analysis (Singh A and Harrison A 1985), Maximum Noises Fraction transformation (MNF) (Green A A etal. 1988) and Crosta transformation (Crosta A P and Moore J M 1989) have been established. Recently, new methods, such as spectral energy level matching(Wang Q J and Lin Q Z 2006a, Wang Q J et al. 2007), weighted spectral angle mapper (He Z H and He B B 2011; Liu K et al 2013), derivative of ratio spectroscopy (Zhao H Q et al 2013), rapid mineral quantification identification model (Li S et al 2010), improved ICA algorithm (Wu F H et al. 2013), generalized morphology-based hyperspectral image unmixing algorithm (Zhao Y et al 2015) were proposed to explore golds (Safwat G et al 2010; Ren G L et al 2013), coppers (Wang Q J and Lin Q Z 2006b; Yusuf E M et al. 2014), boron (Bernard E H and James K C 2005), geothermal deposits (Vaughan R G et al 2005; Christopher K et al 2010), martian components (Frank J AV et al 2014), oil deposits (Chen S B et al 2012) and the volcanic evolution activities (Brandmeier M et al. 2013).

2. Regional settings

As shown in figure 1., with a low hilly area in relief, the study area is located in the Xiemisitai Mountain, which is 25km southwest to the Hoboksar County, Tacheng district, Xinjiang province, China. Hebuke River passes through the county and becomes a primary water source for the Hoboksar town. With an average temperature of 3°C and precipitation of 142mm, it is a continentally arid climate, which is characterized by long winter, short summer, windiness in Spring and autumn.

As shown in figure 2., strata in this area are primarily in the Erjisite group developed in the middle Devonian with intermediate-arid intrusive rocks, volcanic tuff and agglomerate. Rocks of the strata are primarily the light gray andesite, dacite, rhyolitic brecciated crystal tuff. Composed of the meat-red potassium granite, the deeply intrusive rocks are distributed in the northern and western. However, the shallow intrusive rocks are primarily located in the southwest and southeast, which are primarily composed of the light meat-red granite porphyry, gray altered andesitic porphyrite and the purple andesite.

Structures of the study area are primarily the NE and near EW faults. They are primarily straight valleys with 3-10m width, 2-7m depth and 2-4km long.

The surface alterations are primarily epidotizations, carbonations and malachites. The malachites are primarily developed in the andesitic porphyrites in the eastern, especially obvious in the intersections of two fault groups with orientations of 20-30° and 110-120°.

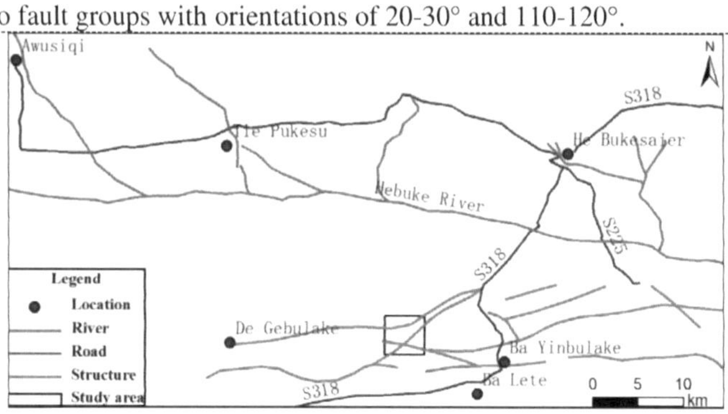

Figure 1. Location of the study area.

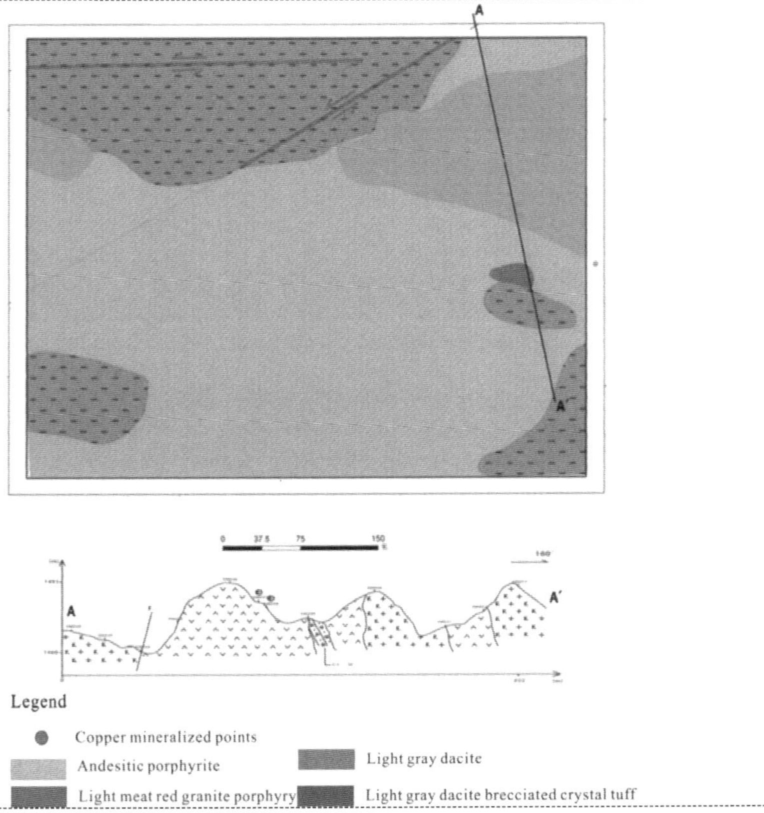

Figure 2. Geological map of the study area.

3. Method

A fine mineral identification model is established taken the measured spectra as the data resource, which were measured by the Analytical System Devices, Inc. portable field spectrometer (ASD FieldSpec®3). In which, a regional spectral library and those methods smoothing spectra and optimizing endmember selections are developed to improve the mineral identification accuracy.

3.1. Principles

In nature, especially in the process of mixed spectral analysis of rock and mineral compositions, the mixed pixels are commonly existed. The linear spectral mixture is considered as the primary mixed model (Zhao H Q et al. 2013; Wang Y J et al. 2012). It is a linear equation expressing the correlation between the mixed pixel spectra and the endmembers spectra, whose hypothesis is that a pixel spectrum is linearly composed of multiple endmembers, written by:

$$AX + \varepsilon = B$$

In which, A is a n×n array, representing n endmember spectra with n bands; X is a n×1 array, representing the endmembers weights that should be determined by the equation; ε is the residual; B is a n×1 array representing the measured spectrum with n bands in a pixel. The mathematic expression of the model can then be concluded to find an optimized solution of X to minimize the residual using the least squares method.

In order to solve the problem, spectral processing, including endmember selections and noises removal, becomes necessary. Therefore, a regional endmember library and a spectral noises removal method using the Fourier transformation become the innovation of the model.

According to above principles, a Finely Mineral Identification System (*FMIS*) V1.0 was developed using C++BuilderV6.0 software.

3.2. Precision evaluation.

As shown in table 1., according to the slice identified minerals in the laboratory, a regional endmember library was established by selecting the standard spectra from the USGS mineral library. Using the mineral contents and their standard spectra, the simulated spectra are reconstructed by the linear spectral mixture model.

Table 1. Slice identified minerals.

Samples	Spectral No.	Mineral contents (%)
4	020-024	Plagioclase,45;Quartz,30;Hornblende,20;Calcite,3;Chalcopyrite,1;limonite,1.
B-5-5	025-029	Orthoclase,70;Plagioclase,20;Quartz,5;Biotite,2;Pyroxene,2.
5	030-034	Pyroxene,70;Plagioclase,20;Potassium feldspar,10.
B-2-12	035-039	Plagioclase,60;Hornblende,15;Pyroxene,10;Biotite,5;Quartz,5;Sphene,2;Magnetite,1.
BW-1	045-049	Potassium,65;Plagioclase,15;Hornblende,10;Biotite,5;Sphene,1
B-5-17	050-054	Orthoclase,65;Plagioclase,10;Hornblende,15;Biotite,5;Pyroxene,2.
2319	055-059	Orthoclase,40;Plagioclase,35;Quartz,10,Hornblende,5,;Biotite5;Pyroxene,3;Sphene,5.
7	060-064	Quartz,25;Plagioclase,40;Potassium feldspar,20;Hornblende,15.
3	065-069	Plagioclas,60;Hornblende,20;Quartz,10;Chlorite,10.
A	070-074	Quartz,30;Potassium feldspar,45;Plagioclase,20;Hornblende,5.
2361	075-079	Plagioclase,70;Hornblende,15;Quartz,5;Biotite,3;Pyroxene,2;Magnetite,3; Ilmenite,2.
B-1-14	080-084	Plagioclase,70;Hornblende,20;Biotite,5;Quartz,3;Pyroxene,2.
2360	085-089	Plagioclase,60;Pyroxene,30;Biotite5;Hornblende,5;Magnetite,1
2356	090-094	Potassium,70;Ilmenite,10;Hornblende,5-10;Pyroxene,5;Biotite,5;Calcite,1;Hematite,1;Rutile,1.
2355	095-099	Plagioclase,60;Quartz,15;Hornblende,10;Pyroxene,10;Biotite,5
2354	101-104	Quartz,35;Potassium feldspar,50;Albite,10;Hornblende,5.
2330	105-109	Plagioclase,65;Hornblende,20;Quartz,5;Pyroxene,5;Magnetite,5;Ilmenite,3;limonite,1.
B-2-7	111-114	Plagioclase,60;Hornblende,20;Biotite,10;Quartz,5;Pyroxene,5.
B-2-42	115-119	Orthoclase,40;Plagioclase,35;Biotite,10;Hornblende,5;Pyroxene,3;Sphene,2;limonite,1.

As shown in table 2., accuracies with multi-endmember groups were calculated by comparing the FMIS inversed minerals with those simulated. From which, we can see that the accuracies reduced with the increasing endmembers. The averaged mineral identification accuracy was 90.79% and the averaged error of mineral contents inversion was 2.2%, when the endmember number is less than 11.

Table 2. Accuracies of the model under different mineral groups.

Endmembers	Accuracy(%)	Error(%)
1	100	0
2	100	0
3	100	0
4	100	0
5	99.1	0.04
6	85.45	2.57
7	83.24	3.16
8	82.3	4.72

9	79.62	5.53
10	78.2	5.84
Average	90.791	2.186

4. Data and processing

As shown in figure 3., 362 field samples distributed along 6 lines from the western to the eastern were sampled in July, 2014. Partial samples were tested by the Leica 2500 high magnification microscope. Based on the tested mineral contents, a regional endmember spectral library was established referring to the USGS standard mineral spectra. More than 1500 spectra were measured by the ASD spectrometer. Then, the spectral noises were removed and the mineral absorption parameters were determined by the FMISV1.0 software.

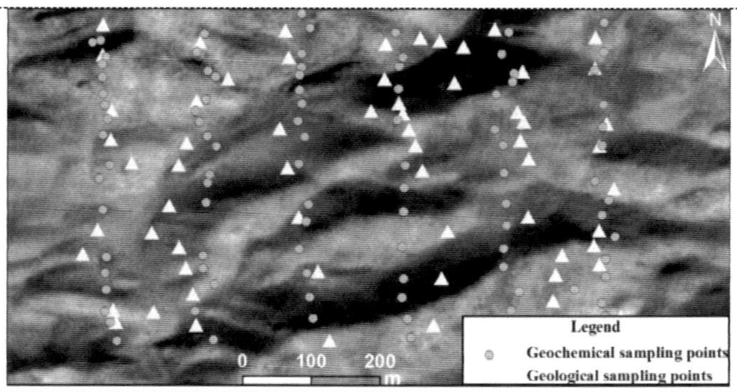

Figure 3. Location of the samples.

5. Results and discussion

As shown in figure 4., alteration minerals were mapped based on the inversed mineral contents. Results show that they were distributed in belts: from the centre to the outer, alteration minerals were potassium, silicates and propylitizations. Mineralized points were distributed near to the intersections between NW structures, propylitization belts and the silicate belts which are important indicators to delineate target areas.

As shown in table 3., the copper contents meet the industrial exploration grade. It was, therefore, delineated as a remote sensing target region, which can be further systematically evaluated by well drilling holes to determine the underground ore deposit's location, scale and grade.

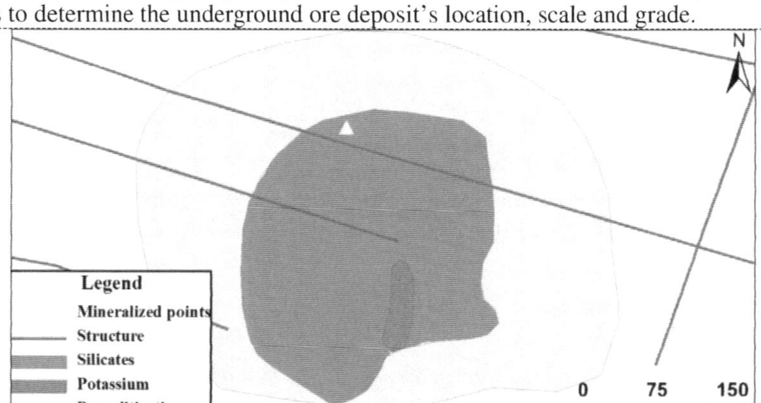

Figure 4. Alteration minerals map.

Table 3. Laboratory determined metal contents

No. 014B-181		ω (10^{-9})	ω (10^{-6})		
Analysis number	Sampless number	Au	Ag	Cu	Pb
1--8	TC-2-1-H1	4.0	0.239	87.5	17.7
1--9	TC-2-1-H2	319.3	>1	>10000	213.0
1--10	TC-2-1-H3	2.7	0.227	104.4	25.4
1--11	TC-2-2-H1	6.4	0.258	294.0	21.6
1--12	TC-2-2-H2	451.7	>1	>10000	86.8
1--13	TC-2-2-H3	63.4	>1	4186.3	56.8
1--14	TC-2-3-H1	182.8	>1	>10000	54.9
1--15	TC-2-4-H1	898.9	>1	>10000	45.1

6. Conclusions

(1) A fine mineral identification model was proposed to provide a technique for quantitative minerals inversion.

The accuracy of the mineral identification was improved by the noises removal, optimized endmember selection, and the mineral spectral absorptions enhancement. Results of mineral identification were more reliable by establishing a regional endmember library.

(2) The model was applied in the Xiemisitai Mountains, Xinjiang province, China, to newly find a copper mineralized point.

According to regional settings, the mineral prospecting regions were forecasted and validated to find a copper mineralization point. Metal contents show that the study area is prospective in exploring the copper deposits.

(3) Technique proposed in this paper can be widely used in remote sensing prospecting, remote sensing image processing and the weak lithological info (such as rocks, structures and alteration minerals, etc.) enhancement and it is, therefore, of great significance to improve the mineral exploration efficiency and save cost.

7. Acknowledgements

This research was financially supported by Hainan Provincial Department of Science and Technology (No.ZDKJ2016021), the project of Natural Science Foundation of Hainan Province (No. 20154188), Hainan Provincial Department of Science and Technology (No. ZDKJ2016021) and Sanya Institute of science and technology cooperation projects (No.sy14ys01132).

During the process of remote sensing mineral exploration model construction, helps from Wang Runsheng, professor of China Aero Geophysical Survey and Remote Sensing Center for Land and Resources; Ji Wenhua, senior engineer from the Xi'An Institute of Geology and Mineral Resources, China Geological Survey; Tian Guoliang, professor from the Institute of Remote Sensing and Digital Earth, Chinese Academy of Sciences, and Lu Yigan, the chief engineer from the Guilin Research Institute of Geology for Mineral Resources, are sincerely all appreciated.

8. References

[1] Hapke B 1981 *J. Geophys. Res.* **86** pp 3039-3054
[2] Shkuratov Y Starukhina L Hoffmann H and Arnold G 1999 *Icarus* **137** pp 235–246
[3] Kruse F A Lefkoff A B and Dietz J B 1993 *Remote Sensing of Environment* **14**(6) pp 1159-1177
[4] Singh A and Harrison A 1985 *International Journal of Remote Sensing* **6** pp 883–896
[5] Green A A Berman M Switzer P and Craig M 1988 *IEEE Transactions on Geoscience and Remote Sensing* 26(1) pp 65-74
[6] Crosta A P and Moore J M 1989 *7th Thematic Conference: Remote Sensing for Exploration*

Geology Proceedings ERIM Calgary pp 1173-1187.
[7] Wang Q J and Lin Q Z 2006a *Geology and Prospecting* **42**(3) pp 91-96
[8] Wang Q J Tian Q J and Li MX 2007 *Opto-Electronic Engineering,* **10** pp16-24
[9] He Z H and He B B 2011 *Spectroscopy and Spectral Analysis* **31**(8) pp 2200-2204
[10] Liu K Zhang L F Qin H H and Sun X J 2013 *Journal of Remote Sensing* **3** pp 609-617
[11] Zhao H Q Zhang L F Wu T X Huang C P 2013 *Spectroscopy and Spectral Analysis*, **33**(1) pp 172-176
[12] Li S Lin Q Z Liu Q J, Wang M F Wang Q J and Wei Y M 2010 *Spectroscopy and Spectral Analysis* **30**(5) pp1315-1319
[13] Wu F H Jiang X P Jiang Y 2013 *Journal of Henan Normal University (Natural Science Edition)*, **41**(5) pp 48-52
[14] Zhao Y Zhang C J and Wang D H 2015 *Journal of Heilongjiang University of Science & Technology*, **25**(3) pp 343-347.
[15] Safwat G Abduwasit G and Timothy K 2010 *Ore Geology Reviews* **38** pp 59-69.
[16] Ren G L Yang J L Yang M Li J Q Gao T Yi H Han H H and Zhao Y J 2013 *Geotectonica et Metallogenia* **37**(4) pp 765-776
[17] Wang Q J and Lin Q Z 2006b *Geography and Geo-Information Science* 22(2) pp9-12
[18] Yusuf E M Davood R and Ali H T 2014 *International Journal of Applied Earth Observation and Geoinformation* **27** pp 117-127
[19] Bernard E H and James K C 2005 *Remote Sensing of Environment* **99** pp173-186
[20] Vaughan R G Simon J H Wendy M C and James V T 2005 *Remote Sensing of Environment* **99** pp 14-58
[21] Christopher K Wendy M C and Mark F C 2010 *Remote Sensing of Environment* **114** pp 2297-2304
[22] Frank J A V Wim H B Harald M A V Tanja E Z Jelmer H P O Zulfa A O Stuart H M and Freek D V 2014 *Planetary and Space Science* **101** pp 108-117
[23] Chen S B Zhao L and Wang J N 2012 *Spectroscopy and Spectral Analysis* **32**(6) pp 1616-1619
[24] Brandmeier M Erasmi S Hansen C Höweling A, Nitzsche K Ohlendorf T Mamani M and Wörner G 2013 *Journal of South American Earth Sciences* **48** pp 296-314
[25] Wang Y J Lin Q Z Wang Q J and Li S 2012 *Spectroscopy and Spectral Analysis* **32**(8) pp 2065-2069

NPP-VIIRS DNB-based reallocating subpopulations to mercury in Urumqi city cluster, central Asia

X Zhou[1,2,4], X B Feng[2], W Dai[3], P Li[2], C Y Ju[4], Z D Bao[5], Y L Han[6]

[1] Xinjiang Institute of Ecology and Geography, Chinese Academy of Sciences, Urumqi, China, 830011
[2] State Key Laboratory of Environmental Geochemistry, Institute of Geochemistry, Chinese Academy of Sciences, Guiyang, China, 550081
[3] School of Environmental Science and Engineering, Sun Yat-Sen University, Guangzhou, China, 510006
[4] Institute of Desert Meteorology, China Meteorological Administration, Urumqi, China, 830002
[5] Department of Civil and Environmental Engineering, Lamar University, Beaumont, USA, 77705
[6] South China Sea Institute of Oceanology, Chinese Academy of Sciences, Guangzhou, China, 510310

E-mail: zhouxu@ms.xjb.ac.cn

Abstract. Accurate and update assignment of population-related environmental matters onto fine grid cells in oasis cities of arid areas remains challenging. We present the approach based on Suomi National Polar-orbiting Partnership (S-NPP) -Visible Infrared Imaging Radiometer Suite (VIIRS) Day/Night Band (DNB) to reallocate population onto a regular finer surface. The number of potential population to the mercury were reallocated onto 0.1x0.1 km reference grid in Urumqi city cluster of China's Xinjiang, central Asia. The result of Monte Carlo modelling indicated that the range of 0.5 to 2.4 million people was reliable. The study highlights that the NPP-VIIRS DNB-based multi-layered, dasymetric, spatial method enhances our abilities to remotely estimate the distribution and size of target population at the street-level scale and has the potential to transform control strategies for epidemiology, public policy and other socioeconomic fields.

1. Introduction

Mercury (Hg) is a highly potent neurotoxin that readily elevate potentially toxic levels to affect human and environmental health [1, 2]. About one third of the world's land surface is arid and semi-arid. There are not specific study on the environmental impacts of Hg in developing areas with rapid urbanization in these remote regions [3]. Urumqi city cluster is an emerging competitive key areas of central Asia. Over the past 35 years, its population grew from 1.02 to 3.11 million and its gross domestic product (GDP) increased from 0.82 to 245 billion Yuan. Coal-based industrial sector is a major source of Hg in the city cluster, especially coal-based polyvinyl chloride (PVC) manufacture [4-6]. The coal-based PVC production of the city cluster is also one of the most significant consumers of Hg in China [5]. Considerable concerns about the adverse impacts of Hg-containing wastes and fly ash related to the manufacture on environmental and human health. However, accurate and update assignment of population-related environmental matters onto grid cells within cities remains challenging.

High-resolution spatiotemporal population data can be particularly difficult to access for remote areas and low-income areas due to methodological insufficiencies [7,8]. In general, methods of censuses and surveys to measure human population can yield dramatic discrepancies when calculating important health metrics and interventions. Combining across data sources can improve population estimates in accuracy and spatiotemporal resolution. For example, the Gridded Population of the World (GPW) [9], the Global Rural Urban Mapping Project (GRUMP) [10], LandScan [11], and WorldPop [12]. Recently, mobile phone data are introduced to measure movement, but access to phone data is highly restricted [13].

Daily, serial satellite images of nighttime lights provide a direct, quantifiable indicator of human presence [14, 15]. These products mainly include Defense Meteorological Satellite Program (DMSP) – Operational Linescan System (OLS) nighttime stable light product and Suomi National Polar-orbiting Partnership (S-NPP) – Visible Infrared Imaging Radiometer Suite (VIIRS) Day/Night Band (DNB). The DMSP – OLS nighttime stable light data is limited in its application to local and within city scales due to its coarse spatial resolution (3 km pixels), overglow (the "spilling" of light from built-up areas into non-lit areas), saturation in urban areas and intra-sensor calibration problems [16]. The NPP/VIIRS–DNB launched in October 2011 overcomes saturation problem in nighttime light data. As a new-generation nighttime light data, the NPP-VIIRS data have a wider radiometric detection range and a higher spatial resolution (0.74 km pixels) compared with the DMSP-OLS data, supporting a more accurate fine-scale estimations [17].

This study proposes the NPP/VIIRS-DNB-based multi-layered, dasymetric, spatial approach to reallocate population onto 0.1 × 0.1 kilometer (km) reference grid, improving population estimates in accuracy and spatial resolution within the city cluster of the remote arid region.

2. Methods

2.1. Site description
Xinjiang Uygur Autonomous Region (XUAR) is the China's largest administrative unit and is a valuable resource hub in the central and western Asia. Urumqi, capital of the XUAR, has been an emerging area in rapid urbanization and industrialization over the most recent decade. The air quality in the Urumqi city ranked as the second poorest in China and the 61st poorest in 1,600 cities worldwide from 2010 to 2013, reported by the World Health Organization's (WHO's). Thus, the Urumqi city cluster (e.g., Urumqi and adjacent Wujiaqu and Fukang city) is one of key regions launched by the China 12th Five-Year Plan to reduce air pollution. In addition, more than 40% of soil Hg is released from coal-based chemical industries [4-6]. There is a deep concern on the particulate matter re-suspension in the windy seasons of the arid environment.

2.2. Sampling, test and data collection

2.2.1. Sampling

Sampling sites of this study were selected according to the basic type of land use from China Code for Classification of Urban Land Use and Planning Standards of Development Land (GB 50137-2011), and come from urban development land (residential, green space, municipal utilities, industrial and manufacturing, street and transportation, commercial and business facilities, administration and public services, logistics and warehouse) and non-development land. The interval of measurement be <5 km to ensure the enough high accuracy of spatial analysis. Accordingly, the average sampling grid was designed a square mesh with sampling points at approximately 5.0 km × 5.0 km with the actual sampling sites randomly selected near the target sampling sites in each grid cell. The samples were located in GPS of ArcGIS10.2 by the connection of Landsat 7 satellite images and hand-held GPSs.

Topsoil samples of the whole study area were collected from 0-20 cm depth with a steel sampler and a wooden shovel. Each sample, approximately 2 kg composite soil was collected into plastic bags prior to the laboratory processing. The soil samples were air-dried, grinded, and then sieved through <1.7 mm nylon sieve to remove stones, coarse materials, and other debris. A sub-sample of the sieved soils were grounded to < 0.15 mm by an agate grinder for total mercury (THg) analysis of the bulk soils.

2.2.2. *Testing of total mercury concentration*

THg in bulk soil were analyzed by a Lumex RA 915+ mercury analyzer (Lumex Ltd., Russia). The detection limit for THg was 0.5 $ng{\cdot}g^{-1}$. Quality control for THg analysis was addressed using certified reference material (GBW07426, Xinjiang Soil), with an average recovery of 98.0 ± 1.8% (2σ, n=5). Quality assurance and quality control of THg determinations were carried out using duplicates and certified reference materials.

2.2.3. *Data collection*

The NPP-VIIRS data were obtained from website of NOAA/NGDC (http://ngdc.noaa.gov/eog/viirs). Landsat-8 imagery were obtained from website of USGS (http://landsat.usgs.gov). Population census and other auxiliary information were available from National Bureau of Statistics of China and Bureau of Statistics of XUAR.

The NPP-VIIRS imagery is a preliminary product, containing lights from cities, towns, transportation corridors, gas flares, biomass burning and background noise, and the reflectance of light from bright surfaces. The highest DN value of 235.13 is used as a threshold to correct the outliers [16].The pixels whose DN value is larger than 235.14 in the NPP-VIIRS data are assigned as a new value. The new value is the maximal DN value within the pixel's immediate eight neighbours. The pixels with negative DN values in NPP-VIIRS data are assigned the value of 0. Data of built-up regions of the Landsat-8 Operational Land Imager (OLI) imagery are classified using the approach of Support Vector Machine.

2.3. Downscaling and data integration

The physical exposures are represented through the inventory of buildings in urban and rural areas called built-up environment. A multi-layered, dasymetric, spatial approach is used to reallocate population onto a regular finer surface (reference grid) by the following workflow:

Step1: Define population of built-up environment

The mask based on the Landsat-8 OLI imagery are used to delimit the cells representing the built-up areas. It was employed to extract the NPP-VIIRS DNB-based population data. A regression model is used to estimate population using the satellite images of nighttime lights in this study (Equation1)

$$Pop_A^{census} = f(\sum R_{VIIRS}) \quad (1)$$

where Pop_A^{census} is the statistical population of per administrative unit, $\sum R_{VIIRS}$ denotes the total radiometric of NPP-VIIRS nighttime light of per administrative unit, f is the regression coefficient.

Step2: Aggregation onto the 0.1 km × 0.1 km reference grid

There are different between the estimated population and the population census of each administrative unit due to error of the linear regression. Therefore, the percentage of the statistical population and the total population to the grid cell in administrative units is calculated using the following Equation 2 and the total population are successively corrected onto the 0.1 km × 0.1 km reference grid (Equation 3).

$$w_A = \frac{Pop_A^{census}}{\sum_{(A)} Pop(x,y)} \quad (2)$$

$$Grid\sum_{(A)} Pop(x,y)' = Grid\sum_{(A)} Pop(x,y) \times w_A \quad (3)$$

where: w_A is the percentage of the statistical population and the total population to the grid cell in per administrative unit, Pop_A^{census} the census population of per administrative unit, $\sum_{(A)} Pop(x,y)$ denotes the total population living the grid cell per administrative unit, $\sum_{(A)} Pop(x,y)'$ denotes the corrected total population living the grid cell per administrative unit.

Step3: Dasymetric of the population distribution per the element at risk

The following algorithm employs to downscaling a DPE to the grid cell for evaluating the number of potential population damage related to the elements at risk (Equation 4).

$$DPE(x,y) = Pop(x,y)' \times E(x,y) \quad (4)$$

where: $DPE(x,y)$ is population per the element at risk per cell at x,y position, $Pop(x,y)'$ denotes the corrected population living per cell, $E(x,y)$ denotes the element at risk per cell at x,y position.

After the above operation, each cell represents the number of potential exposed persons for the element at risk in a portion of built-up areas.

3. Results and Discussion

As shown the scatter diagram of linear regression analysis (Figure 1.), the DNB-POP relationship were evaluated in the district and town units. The R^2 value of the corrected NPP-VIIRS DNB and POP data was 0.8689. The spatial distribution of gridded street-level population was allocated (Figure 2. Left). The Monte Carlo modeling indicated that the range of 0.5 to 2.4 million people was reliable (Figure 2.Right). The polluted area of soil THg reached 425 square kilometer (km^2), which concentrations exceed the GradeII limits (for pH>6.5) (GB15618-1995) (EQSS). Based on the above operations, the subpopulation to Hg at risk were reallocated onto 0.1 km × 0.1 km reference grid. The hotspots of subpopulation to Hg were traced in the Urumqi inner city and industrial areas (Figure 3).

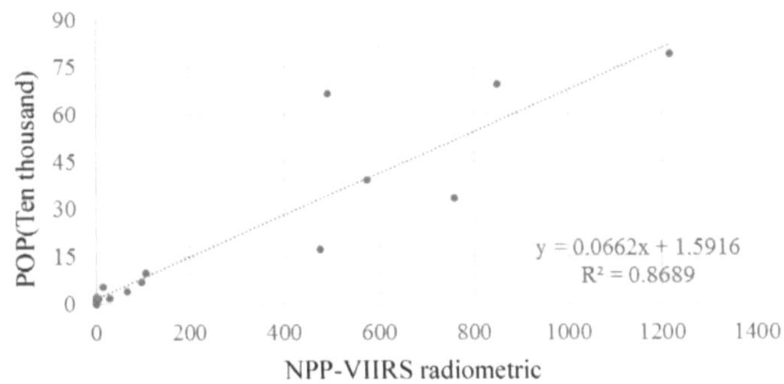

Figure1. The scatter diagram of linear regression analysis between the total nighttime light of corrected NPP-VIIRS data and the population census.

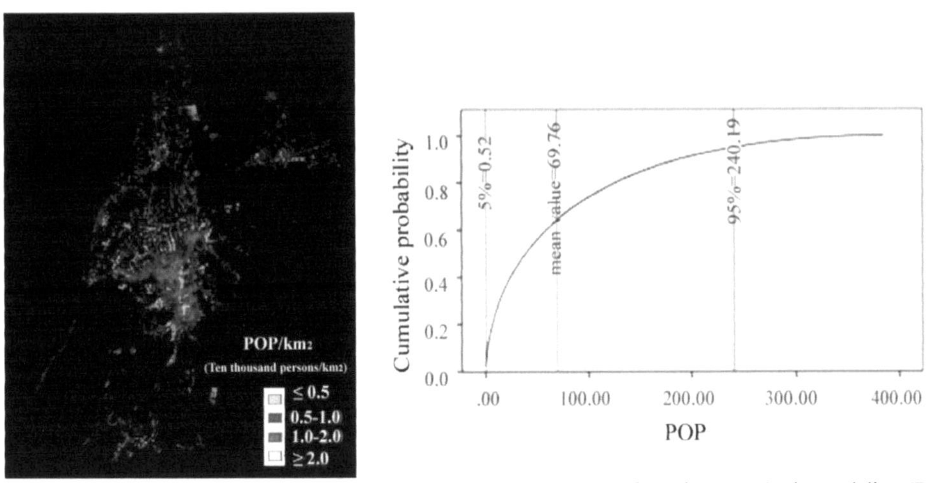

Figure2. The spatial distribution of the gridded street-level population (Left) and Monte Carlo modeling (Right).

Unlike the DMSP-OLS data, the varied DN values in NPP-VIIRS data can reflect human activities and provide a more accurate population estimations. The more reasonable estimated results within the cities mainly benefit from the higher resolution and radiometric detection range of NPP-VIIRS nighttime light data. The overestimation population of the study area may affects the accuracy of the population size due to deserts, dry riverbeds, mining fields and the original noises of NPP-VIIRS data. Some methods have be proposed for removing the noise or detecting wildfires from NPP-VIIRS data [17]. Those works would benefit to improve the quality of the new-generation nighttime light data in wider fields in the future.

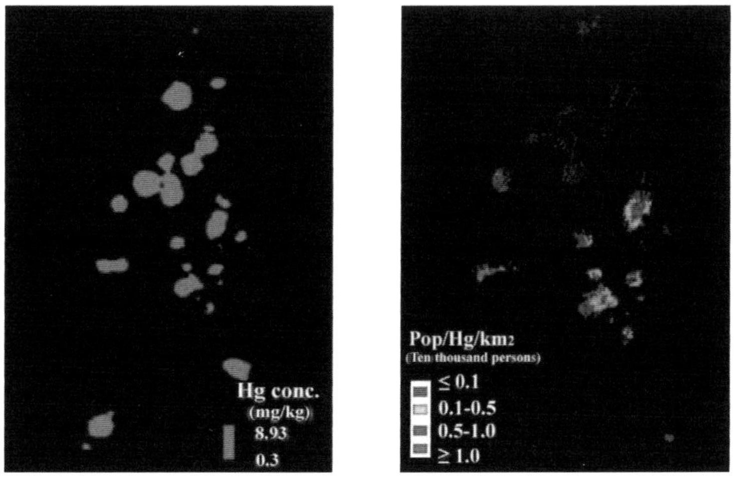

Figure3. The gridded map of Hg concentration exceed the GradeII limits (Left) and the number of the subpopulation to Hg at risk were reallocated onto 0.1 km × 0.1 km reference grid (Right).

4. Conclusions

Gaining an understanding of local subpopulation sizes and patterns helps make informed changes in the planning of epidemiology, public policy and other socioeconomic aspects. The characteristics of target susceptible subpopulations at the street-levels scale can be remotely measured with widely available NPP/VIIRS satellite imagery of anthropogenic lights, and can be applied to other health efforts in areas without detailed spatial epidemiological data. The empirical relationship between the nighttime light data and population cannot be viewed as an absolute law. The reliability of statistical data is also a major factor affecting the reallocating accuracy. There is still room for improving the data quality, and more methods could be applied to the correction process of the emerging data source in wider fields.

Acknowledgements

This project was jointly supported by the National Natural Science Foundation of China (Grant No. 21207155), the Science Foundation of the Chinese Desert Meteorology (Grant No.Sqj2015014), and the West Light Foundation of the Chinese Academy of Sciences (Grant No. RCPY 201104).

References

[1] P. Li, X. B. Feng, G. L. Qiu, L. H. Shang, Z. G. Li (2009) Mercury pollution in Asia: A review of the contaminated sites. J Hazard Mater 168:591– 601.
[2] P. Li, X. B. Feng, H. M. Chan, X. F. Zhang, B.Y. Du (2015) Human body burden and dietary

methylmercury intake: The relationship in a rice-consuming population. Environ Sci Technol 49: 9682–9689.

[3] N. Pirronel, S. Cinnirella, X. B. Feng, R. B. Finkelman, H. R. Friedli, J. Leaner, R. Mason, A. B. Mukherjee, G. B. Stracher, D. G. Streets, K.Telmer (2010) Global mercury emissions to the atmosphere from anthropogenic and natural sources. Atmos Chem Phys 10:5951–5964.

[4] X. Zhou, X. B. Feng, P. Li, S. Y. Mu, G. L. Qiu (2016) Prior carrying out the internet of things-based supervision of mercury-contained wastes from the chlor-alkali plants in Xinjiang, National Development and Reform Commission.

[5] X. Zhou, X. B. Feng, P. Li, G. L. Qiu. Mercury challenge: Coal-based polyvinyl chloride industrial chain alarm in Tien Shan (Central Asia), The 12th International Conference on Mercury as a Global Pollutant (ICMGP), 2015.6.14-2015.6.19, Jeju. Session18: UNEP Global Mercury Partnership session.

[6] X. Zhou, X. B. Feng, P. Li, W. Dai, Y. L. Han, C.Y. Ju, C. Zhang. Spatial patterns and environmental impacts of total and methyl mercury in the oasis of the Urumqi catchments, Central Asia, The first international workshop on urbanization in watersheds: Ecological and environmental responses (UWEER), 2014.10.12-2014.10.17, Xiamen. Session: Modeling and prediction of changes in urbanized and urbanizing watersheds, and strategies and policies for watershed management

[7] A. J. Tatem, Z. Huang, C. Narib, U. Kumar (2014) Integrating rapid risk mapping and mobile phone call record data for strategic malaria elimination planning. Malaria J 13(1):1-16.

[8] N. Bharti, A. J. Tatem, M. J. Ferrari, R. F. Grais, A. Djibo, B. T. Grenfell (2011) Explaining seasonal fluctuations of measles in Niger using nighttime lights imagery. Science, 334:1424–1427.

[9] U. Deichmann, D. Balk, G. Yetman. Tranforming population data for interdisciplinary usages: from census to grid. Documentation for GPW Version 2, 2001.

[10] D. L. Balk, U. Deichmann, G. Yetman, F. Pozzi, S. I. Hay (2006) Determining global population distribution: methods, applications and data. Adv Parasit 62:119–156.

[11] J. Dobson, E.Bright, P. R Coleman, R. Durfee, B. Worley (2000) LandScan: a global population database for estimating populations at risk. Photogramm Eng Rem S 66:849–857.

[12] C. Linard, M. Gilbert, R.W. Snow, A. M. Noor, A. J. Tatem (2012) Population distribution, settlement patterns and accessibility across Africa in 2010. PLoS ONE 7: 1–8.

[13] P. Deville, C. Linard, S. Martin, M. Gilbert, F. R. Stevens, A. E. Gaughan, V. D. Blondel, A. J. Tatem (2014) Dynamic population mapping using mobile phone data. PNAS 111(45): 15888–15893.

[14] C. D. Elvidge, K. E. Baugh, E. A. Kihn, H.W. Kroehl, E.R. Davis (1997) Mapping city lights with nighttime data from the DMSP Operational Linescan System. Photogramm Eng Rem S 63:727–734.

[15] P. Sutton, D. Roberts, C. D. Elvidge, K. E. Baugh (2001) Census from heaven: an estimate of the global population using nighttime satellite imagery. Int J Remote Sens 22:3061–3076.

[16] K. F. Shi, B. L.Yu, Y. X. Huang, Y. J. Hu, B.Yin, Z. Q. Chen, L. J. Chen, J. P. Wu (2014) Evaluating the ability of NPP-VIIRS nighttime light data to estimate the gross domestic product and the electric power consumption of China at multiple scales: A comparison with DMSP-OLS data. Remote Sens 6(2):1705–1724.

[17] C. D. Elvidge, M. Zhizhin, F-C. Hsu, K. E. Baugh (2013) VIIRS nightfire: Satellite pyrometry at night. Remote Sens 5(9):4423–4449.

Building the 3D Geological Model of Wall Rock of Salt Caverns Based on Integration Method of Multi-source data

WANG Yongzhi, WANG hui, LIAO Lixia, LI Dongsen

Faculty of Architectural and Surveying Engineering, Jiangxi University of Science and Technology, Ganzhou, Jiangxi 341000, China

Email:

Abstract. In order to analyse the geological characteristics of salt rock and stability of salt caverns, rough three-dimensional (3D) models of salt rock stratum and the 3D models of salt caverns on study areas are built by 3D GIS spatial modeling technique. During implementing, multi-source data, such as basic geographic data, DEM, geological plane map, geological section map, engineering geological data, and sonar data are used. In this study, the 3D spatial analyzing and calculation methods, such as 3D GIS intersection detection method in three-dimensional space, Boolean operations between three-dimensional space entities, three-dimensional space grid discretization, are used to build 3D models on wall rock of salt caverns. Our methods can provide effective calculation models for numerical simulation and analysis of the creep characteristics of wall rock in salt caverns.

Keywords. GIS, integration method of multi-source data, wall rock of salt caverns, 3D geological models

1. Introduction

There are many advantages of salt rock, such as the behaviors of very low permeability, excellent creep behavior, and damage recovery, and so on(Consenza, 1999; Chunhe Yang, 1999). Moreover, the property that salt rock can be dissolved in water makes the excavation of salt caverns which are underground spaces formed by rock salts mining more easily and economically. Due to the advantages of salt caverns, many countries try to use them to build gas storages, oil storages, hazardous waste underground repositories, compressed air energy storages, and so on. (Sehalge, 1993; Thomas, 2000; Xu Xinqiao, 2006). Compared with the thick salt formed by marine deposit in other countries, salt rock in China is layered structure formed by lacustrine deposit, with the characteristics of small thickness salt bed, multiple inter-beds mudstone and shallow buried deep, and so on(YANG Chun-he, 2005, 2009; YANG Qiang, 2011). As a result, it is really important to research the stability of salt rock geologic lithology and wall rock of salt caverns when energy underground storages are built with the salt caverns under this condition. In the mean time, finding the way to construct the 3D models of wall rock of salt caverns that reflect the real geometry is one of the prime issues of the study that mentioned above.

Conventional instruments cannot meet the needs of measurement, because the volume of salt caverns is generally large and deep, special measuring instruments and devices are applied in the detection (Tian Zhonglan, 2009). At present, sonar detection measurement is the common method to detect the shape of salt caverns (James, 2001; Liu Guanling, 2007; Wei Donghou, 2007;). By the visualization analysis of salt caverns surface models, abnormal changes of the caverns can be found, and then the caverns processing parameters can be adjusted in time for controlling the shape of salt caverns, determining the caverns volume and guaranteeing for the quality of caverns. At present, 3D models of wall rock of salt caverns are constructed by numerical simulation software, such as finite element and finite difference. These pieces of software which are low inefficiency 3D modeling function, and actual geological environment is simplified largely. Also, the structure of the surface layers of rock salt is not fully considered and sonar data is underutilized sufficiently. And they only divide geological environment by the rules grid in the ideal state. Therefore, with the constraint of points, lines, faces and bodies, the actual geological environment is very complex.

In this paper, the rough 3D models of salt rock strata and salt caverns in the study areas are constructed based on 3D spatial modeling technique. During 3D models implementing, multi-source data, such as basic geographic data, DEM, geological plane, geological section, engineering geological data, and sonar data are used. Then, 3D geological models of wall rock of salt caverns are established by 3D spatial analysis method. The research achievements will significantly promote the economic and social development, and can provide a scientific basis and technical support for the construction of salt caverns, especially for constructing and using salt caverns safely and rationally.

2. Organization and Management for Multi-source Data of 3D Geological Modeling of Wall Rock of Salt Caverns

There are many characteristics of data onto 3D geological modeling of wall rock of salt caverns, such as data source pluralism, data formats diversity, connection complexity, mass, and so on. For these characteristics, Geodatabase data model is used for the relative management of spatial information and attribute information on fundamental spatial data, drilling data, sonar interpretation data, etc. In this way, the management and integration of multi-source data is realized and the needs of 3D geological modeling construction of wall rocks of salt caverns and their analytical application are guaranteed efficiently. As shown in figure 1.

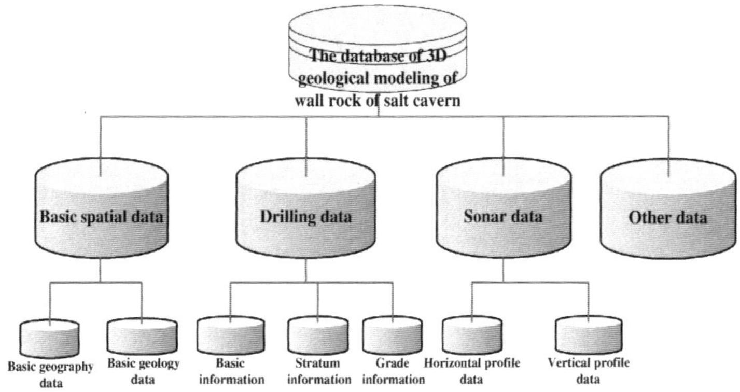

Figure 1. The organization structures graph of the multi-source data onto three-dimensional geological modeling on wall rock of salt caverns.

3. Construction of 3D Models of Salt Rock Strata Based on Multi-source Data

Currently, drilling data is used as primary data source in the construction of stratum 3D models. However, the model which is relatively accurate, complete, and with a good visualization is difficult to be constructed by the original drilling data, due to the drilling data distribution is extremely uneven, and very sparse in many modeling research areas. In this paper, a method of 3D modeling on salt rock strata based on multi-source integration is suggested. This method can make up for the deficiency in 3D modeling based on drilling data. Firstly, multi-source including foundation geographic data with regional boundary and DEM, basic geological data with geological plane and geological section, and drilling data are integrated. Secondly, a series of processing operations are implemented, including fitting the surface of salt strata, building the tin model of initial surface and suturing the multi-DEMs, through the methods of spatial discrete, spatial interpolation and Delaunay triangulation. At last, generalized tri-prism (GTP) is used as an intermediate unit to build intermediate GTP models on study areas (Wu Li-xin, 2003; CHE De-fu, 2006). Then, 3D models(B-Rep) of salt rock strata is constructed successfully after the combination of intermediate cells. Three-dimensional geological modeling procedure of this paper's method is shown in figure 2.

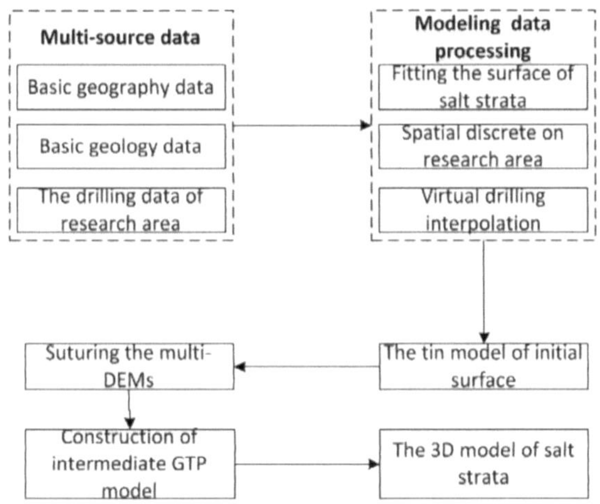

Figure 2. Three-dimensional geological modeling flow chart of salty based on drills.

The established initial surfaces TIN models play an important role in the process of constructed 3D salt strata modeling above, and the surface morphology of rock salt can be expressed with different levels of resolution based on initial surface TIN models. Using the standard Delaunay triangulation construction algorithm (D-TIN), a D-TIN model is generated based on initial surface TIN models, and it has the ability of expressing the topological associations with different strata in the researching area. The initial surfaces TIN models can be taken as a "template" which can determine the topological relation of the stratum on the modeling domain and calculate all the information about the stratum, along with the extension direction of drilling from top to bottom. Therefore, determined and consistent topological relation of all levels strata can be ensured. In addition, the topological relation of all strata units is established based on initial surface TIN models, then the continuous and global surface on the modeling domain is formed. Thus, the thought of multi-DEMs is adopted for lithology of different salt strata. A triangle among the triangulation of top stratum interface is regarded as the starting point. Subsequently, all the triangles of each TIN surface are treated cyclically. After all triangles have been treated, the triangulation of overlying and underlying geological interface in the same stratum is sutured, as shown in figure 3.

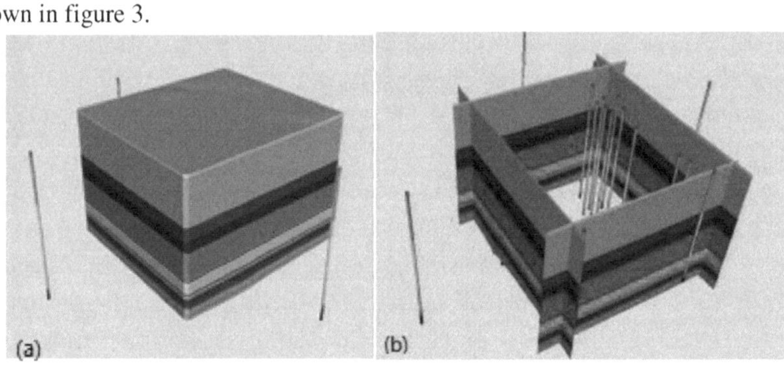

Figure 3. (a) Three-dimensional geological model instance of some research region, (b)Three-dimensional fence diagrams.

4. The Dynamic Update of Three-dimensional Geological Models of Wall Rock of Salt caverns Based on 3D GIS

3D geological modeling of salt caverns is a dynamic process in which more multi-source is added and modeling methods are refined constantly. In this study, 3D models of salt caverns surface are established by using enriching drilling data, geological section and sonar data. Then, 3D geological models of wall rock of salt caverns are updated and optimized dynamically by means of intersection detection in three dimensional space, Boolean operation among the three-dimensional entities, three-dimensional grid discretion and so on.

Based on the data structure of HalfEdge (WANG Yong-zhi, 2012), B-Rep surface model is used for restructuring topological relations of plan and vertical sonar salt caverns data. Thus, three-dimensional surface models of salt caverns are obtained, as shown in figure 4.

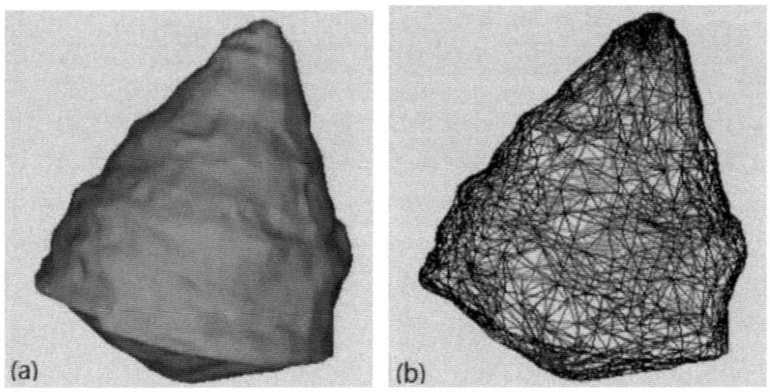

Figure 4. (a)Solid display of three-dimensional surface model of a cavern, (b)Wire-frame display of three-dimensional surface model of a cavern.

The 3D geological model of wall rock of salt caverns is the rest that the 3D model of salt rock stratum minus the 3D surface model of salt caverns. Therefore, it is needed to do intersection detection and Boolean operation for the stratum 3D model of salt rock and the 3D surface model of salt caverns in the process of building 3D surface model of salt caverns. To address the problem of bad real time capability of intersection detection and computing in complex geography scene, a fast intersection detection algorithm which was named SweepPlane (Wang Yongzhi, 2011) for three-dimensional objects was proposed. In the first step of the algorithm flow, event-points and points list were constructed according to the distribution of spatial objects in the scene. Secondly, a series of virtual space scanning planes were established using spatial sweeping strategy. The state of space objects was divided into death state, active state and dormant state, during the scanning plane is moving. And only the space objects of activated state were used for the intersection test. Thus, unnecessary intersection calculation can be omitted and the efficiency of Boolean operation between spatial entities was improved. Then, Boolean operation module is executed with the condition that the three-dimensional model of salt rock strata is inputted as the target body, and the 3D surface model of salt caverns is inputted as a tool body. Also, a Boolean subtraction operation should be selected. After the module of Boolean operations working, the three-dimensional surface models of wall rock of salt caverns are obtained.

In the three-dimensional analysis of underground space, Geo-science calculation models should be loaded into 3D space models to allow the numerical calculation and simulation. In addition, the 3D spatial surface models must be subdivided to obtain 3D discrete elements that support geo-science analysis and calculation. Due to their many excellent characteristics, tetrahedral elements are used widely in geo-computation, such as the finite element method and finite volume method. In order to address the needs for creep simulating and analyzing wall rock of salt caverns, the advancing front method is expanded. A tetrahedral mesh discretization method (WANG Yongzhi, 2012) that considers the external boundary and interior feature constraints (such as point, line, polygons and hole) in wall rock of salt caverns is employed to obtain high quality meshes for Geo-science analysis of wall rock of salt caverns. A 3D model of a rock salt cavity after subdivided by tetrahedral grid is shown in figure 5. figure5 (a) is solid display of the model. figure5 (b) shows the result after the model imported into Flac3D.

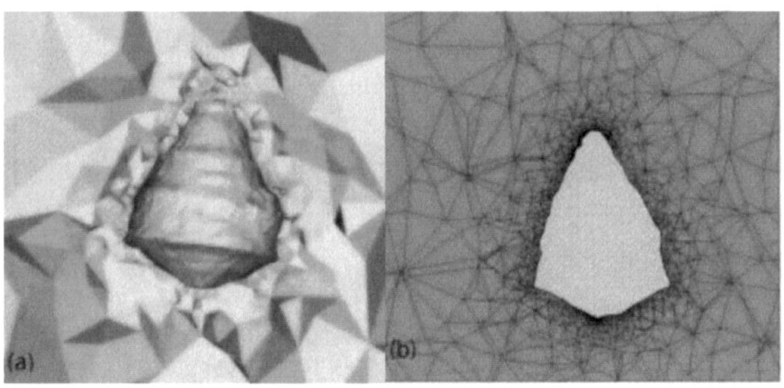

Figure 5. (a)Solid display of rock salt cavity tetrahedral element of 3D model, (b)Wire-frame display of rock salt cavity tetrahedral element of 3D model.

Using the 3D models of wall rock of salt caverns in this paper, the complex geological constraints are taken into consideration, geological form is expressed more clearly, and internal and external boundaries are fitted well. Moreover, the topological relation between each grid cell in geological body of rock salt cavity is realized, a good geometric structure is provided for irregular data field, and a reliable computational grid is provided for the numerical simulation and analysis of the wall rock of salt caverns.

5. Conclusion

In order to address the needs for analyzing geological characteristics of salt rock and stability of salt caverns, the integration methods of 3D geological modeling of wall rock in salt caverns are explored. The multi-source data, such as basic geographic data, DEM, geological plan, geological section, engineering geological data, and sonar data, are used to build the 3D models. In this study, a rough 3D framework model of salt rock geological bodies in salt caverns was built by using 3D GIS spatial modeling theory and method. Using the methods of 3D spatial analyzing, 3D models of salt rock bodies are executed, and then effective calculation model is provided for numerical simulation and analysis of creep deformation of wall rock in salt caverns. These methods can provide scientific basis and technical support for the construction of salt caverns, especially for using salt caverns safely and rationally.

6. Acknowledgements

Thanks the supports of NSFC (No. 41361078), the Open Research Fund of Key Laboratory of Digital Earth Science, Institute of Remote Sensing and Digital Earth, Chinese Academy of Sciences (No. 2013LDE001), Youth Science Foundation of JiangXi Provincial Department of Science and Technology (No. 20142BAB217027) .

References

[1] CHE De-fu, CHEN Xue-xi, WU Li-xin, etc. 3D Stratum Modeling Method Based on Generalized Tri-Prism Volume Element[J]. Journal of Liaoning Technical University, 2006. 25(1): 36-11

[2] Chunhe Yang, Daemena, Jian-Hua Yin. Experimental Investigation of Creep Behavior of Salt Rock[J]. International Journal of Rock Mechanics and Mining Sciences, 1999, 36(2): 233-242

[3] Consenza P H. In Situ Rock Permeability Measurement for Long Term Safety Assessment of Storage[J]. International Journal of Rock Mechanics and Mining Sciences and Geomechanics Abstracts, 1999, (4): 509-526

[4] James E B. Execution and Results of a Sonar Surey in a Horizontal Leached Cavern[C]. Florida, USA: Orlando, 2001. 22-25

[5] Liu Guanling, Lin Guoming, Yan Qi. Review and Prospect of Active Sonar Detection Techniques[J]. Technical Acoustics, 2007, 26(2): 335-340

[6] Sehalge R A. Electric Compressed Air Storage Cavern From Planning to Completion[C]. //USA: Illinois Institute of Technology, 1993. 998-1003

[7] Thomas R L, Gehle R M. A Brief History of Salt Cavern Use[C].// The 8th World Salt Symposium, 2000, 207-214

[8] Tian Zhonglan. Application of Sonar Logging Technology in Evaluating Cavity in Salt Deposit[J]. China Well and Rock Salt, 2009, 40(4): 16-18, 44

[9] Wang Yongzhi, Sheng Yehua, Zhou Liangchen, etc. Intersection Detection Algorithm for 3D Objects Utilizing Spatial Sweeping Strategy[J]. Journal of Image and Graphics, 2011, 16(7): 1262-1268

[10] WANG Yong-zhi, SHENG Ye-hua, ZHOU Liang-chen, etc. Generation Method of Three-Dimensional Computing Model of Wall Rock of Salt Rock by 3D GIS[J]. Rock and Soil Mechanics, 2012. 33(05): 1557-1564

[11] Wei Donghou, Qu Dan'an. The Construction of Salt Cavern Gas Storage and Sonar Measuring Technique[J]. Oil & Gas Storage and Transportation, 2007, 26(8): 58-60

[12] Wu Li-xin, SHI Wen-zhong, Christopher Gold. Spatial Modeling Technologies for 3D GIS and 3D GMS [J]. Geography and Geo-Information Science, 2003, 19(1): 5-11

[13] Xu Xinqiao, Yang Chunhe, Li Yinping. Review on compressed air energy storage abroad and its

feasibility application to Hubei Province[J]. Chinese Journal of Rock Mechanics and Engineering, 2006, 25(supplement 2): 3387-3392

[14] YANG Chun-he, LIANG Wei-guo, WEI Dong-hou, etc. Research on the Feasibility of Energy Storage in Rock Salt in China[J]. Chinese Journal of Rock Mechanics and Engineering, 2005, 24(24): 4409–4417

[15] YANG Chunhe, LI Yinping, CHEN Feng. Mechanics Theory and Engineering of Layered Salt Rock[M]. Beijing: Science Press, 2009:1–6(in Chinese)

[16] YANG Qiang. Global Stability Analytical Method of Cavern Group in Salt Rock Based on Energy Criterion, 2011

IOP Conf. Series: Earth and Environmental Science **57** (2017) 012023 doi:10.1088/1755-1315/57/1/012023

Estimation of the surface displacement of ore mining quarry using Satellite Radar Interferometry

Maira G. Razakova

Joint-Stock Company "National centre of space researches and technologies", Almaty, Kazakhstan

mairarazakova@gmail.com

Abstract. In this study, the TerraSAR-X InSAR data acquired from August 2011-October 2015 are employed to monitor the surface deformation of the mining quarry in the Central Kazakhstan. Our results show that there is a 0.5 - 1.0 cm motion per a month. Our InSAR measurement also indicate that one side has stopped moving. This differential motion may be due to the end of mining ore. This method should be useful for estimating the dynamic of the ore mining regions based on the geo-information technology combined with ground measurement.

1. Introduction

The ore deposits in the Central Kazakhstan have been extracted for many years having the purpose of its industrial processing. It is difficult to ensure safety during the ore extraction process due to the structure of geological and petrophysical characteristics as well reservoir physical, mechanical and collecting properties. In view of the foregoing, monitoring of the earth surface in the mineral's extraction territory is of critical importance

In this work, we discuss our experience with interferometer SAR techniques that have been used to monitor dynamics of earth surface subsidence in ore mining regions of Kazakhstan

2. Processing of radar data and the results

To monitor the mining quarry terrain deformation, we have analyzed eight images obtained from TerraSAR-X and five images obtained from COSMO-SkyMed satellites from August 2011 to October 2015. Interferometric analysis of synthetic aperture radar (SAR) images [1] provides the deformation of the earth's surface. SAR images acquired by radar satellites at different times (master and slave images) can be combined into interferograms that reveal the change in range from ground to satellite expressed as interferometer fringes. Each fringe corresponds to change in range of where is the radar wavelength (3 cm for the TerraSAR-X satellites). This assumes that fringes due to topography have been removed with usage of a digital elevation model (DEM).

The resulting difference of phases in each element of the interferogram is determined by the relief of the exposed surface (the "topographic" phase component) and surfaces of displacement within the period between the surveys. The relief model is an important factor for accurate calculation of phase difference [2]. So we calculated first a digital model of the relief based up on the COSMO-SkyMed radar survey data obtained by interferometer methods. To ensure the required accuracy of the relief model elements, we chose three interferometer pairs, selecting the temporal and spatial base, and used them for creation of three DEMs by interferometer method. To make it optimal, the resulting relief was obtained by averaging three interferometer pairs. It made the resulting DEM more accurate. Finally, we obtained a digital model of the relief with a high space resolution (9 m), very close to the date of terrain displacement's calculation, which is extremely important for identification of current changes. The use of this DEM for differential interferometry calculation purpose has produced good results in the form of a high coherence of the radar interferometer pairs and enabled us to exclude the topographic component of the phase difference, see Table 1, Figure 1,2.

Table 1. Characteristics of the radar images.

No	Satellite	date of interf.pair (d/m/y)	coherence
1	TerraSAR-X	18 10 2011 - 10 08 2012	0.64
2	TerraSAR-X	9 10 2015 - 28 09 2015	0.713
3	TerraSAR-X	9 10 2015 - 17 09 2015	0.7

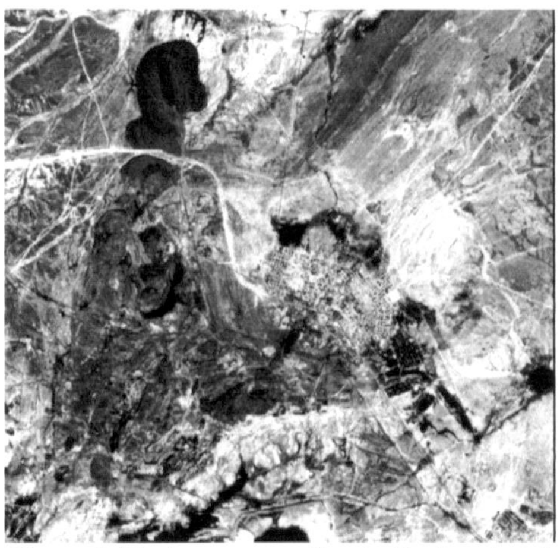
Figure 1. Test area according to the Landsat 8 data.

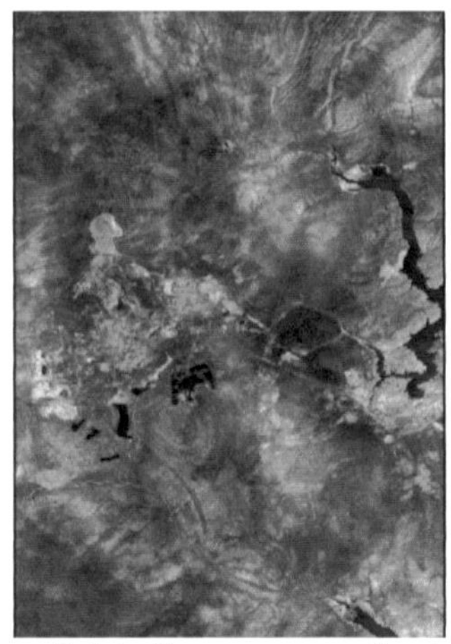
Figure 2. Interferogram according to the TerraSAR-X radar data.

The displacement values vary from -1 cm to -5 cm within a 9-month period in 2011 and 2012; the displacement area is 795 m x 1182 m. The interferogram for 2015 gives us displacement values from -0.3 cm to -0.5 cm for 22 days; the displacement area is 230 m x 268 m. If we focus on maximum displacement values, we will obtain the following displacement process rates: 0.6 cm and 0.5 cm in 2012 and 2015 correspondingly. The displacement scale was reduced by 2015. The vertical displacement area and the site location also were reduced (marked green in Figure 3). It is logical to assume that the ore extraction works were stopped in the maximum displacement area -5 cm in 2012 (marked red in Figure 3).

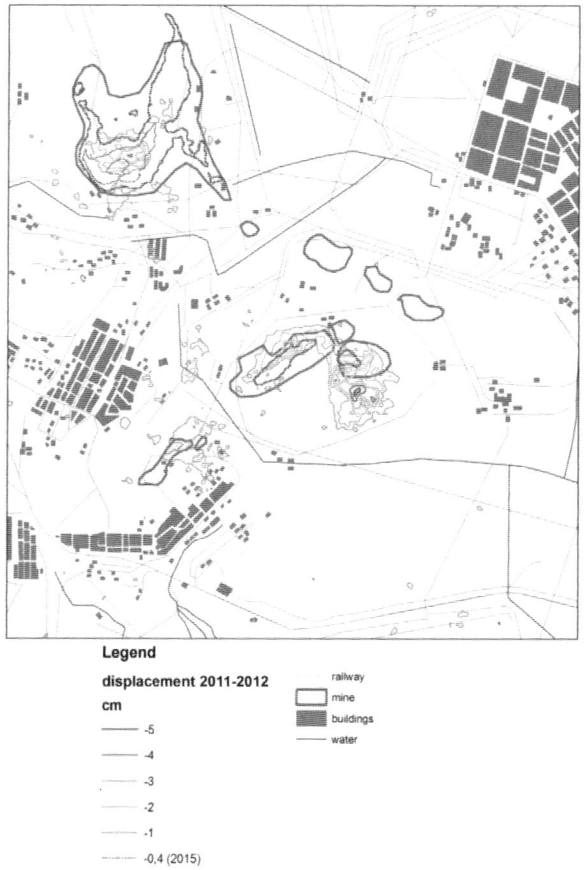

Figure 3. Map of measured changes in the terrain according to radar interferometer data.

The obtained detailed radar images enabled us to implement a possibility of tracing any changes occurred in a definite terrain for a certain period, i.e. coherent detection. Analysis of the image magnitude characteristics providing the main visual information and the phase characteristics are matched together at that. It enables us to identify changes imperceptible for the human eye.

Synthesis of such radar survey parameters as coherence, mean magnitude, and different magnitude enables us to estimate the amount of the terrain change. After conversion of the radar data interferometer pair data to one geometry with a sub-pixel accuracy and estimation of their coherence, we calculated a composite image based upon the coherence map, mean and different magnitudes (Figure 4). The prevailing red color in this image corresponds to high coherence, which is typical for unchanged areas of the Earth surface. Low and average coherence (from blue to yellow) is typical for changes related to watering basins, vegetation, and soil displacements caused by industrial processes. Vegetation exhibits seasonal changes, while water basins reflect the sent radio signal almost completely. Large rivers always have black shades on radar images. Therefore, changes took place only in the ore extraction areas.

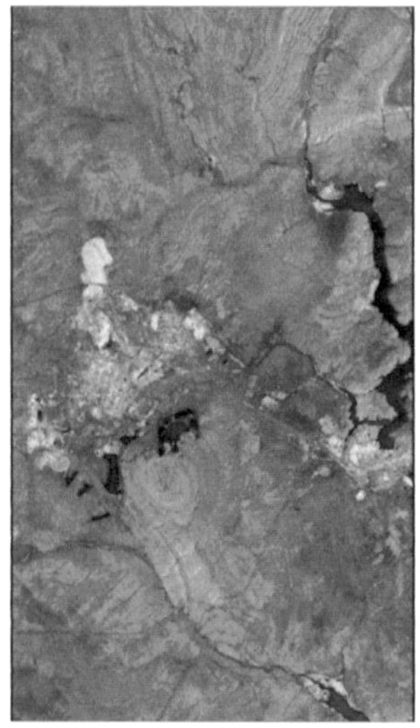

Figure 4. Composite image of radar data based upon the coherence map, mean and different magnitudes.

Nevertheless, the radar interferometry possibilities are much wider and can be used for estimation of dynamics or stabilization of terrain changes. The analyzed interferograms in this case are used for determination of the displacement process dynamics in the event of irregular radar surveys. The purpose is to find a way to recover the missing data in the intermediate uncontrolled zone of the observation interval proceeding from the data obtained at the reference points or otherwise to use analytical methods or to apply some algorithm setting a sequence of operations to calculate the parameter value. In our case, the radar interferometer methods provide us with the area parameters of the terrain changes having a local character; moreover, the displacement process turns out to be slow (at an average of several centimeters per year). Therefore, we were forced to interpolate the missing data with the moving mean of the displacement value total amount from 2011 to 2015. Using the calculated results, we created a composite image showing the dynamics of earth surface changes (see Figure 5). Terrain deformations are clearly visible in locations where mining works are performed. Such deformations are caused by formation of cavities in the process of open pit mining, elevations and dumps.

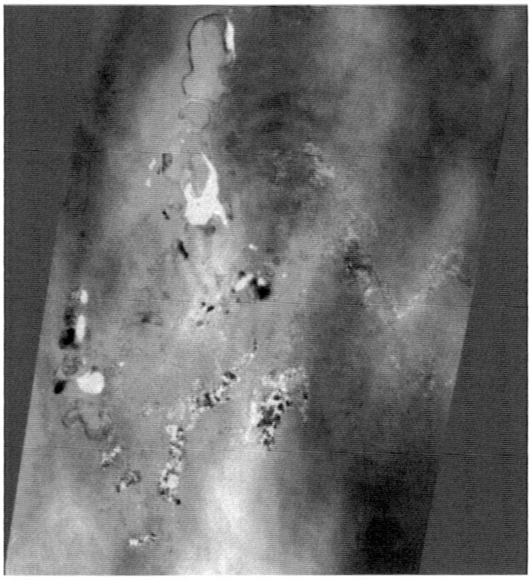

Figure 5. Composite image of displacement maps.

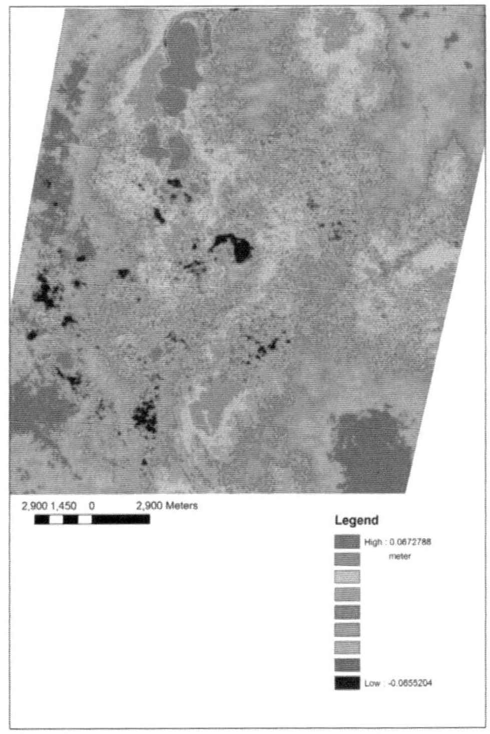

Figure 6. Dynamic of land motion.

Proceeding from the analytical results and the obtained parameters of displacements on the mining territory, we observe changing contour's dumps, stabilization of individual sections of the basin subsidence, and the manifestation of progress within the area of field development. At pits, displacement values ranged between -8.5cm to 6.7 cm (see Figure 6). It is possible to state that the movement manifestation at the ore development site is provoked by some man-caused factors such as mineral extraction and processing.

3. Conclusion

Our research shows that InSAR techniques can be used to monitor dynamics of surface deformation in the ore mining regions of central Kazakhstan. Therefore, the monitoring of active movements using InSAR techniques will be useful for estimating the underground mining.

References

[1] Fielding E.J., Blom R.G. and Goldstein R.M., 1998. Rapid subsidence over oil fields measured by SAR interferometry. *Geophysical Research Letters*, Vol. 25, No. 17, pp. 3215 3218, Earth Science Community. Washington, DC: American Geophysical Union (AGU).

[2] Rosen P., Hensley S., Joughin I., Li F., Madsen, S., Rodriguez, E. and Goldstein, R., 2000, March. Synthetic Aperture Radar Interferometry. *Proceedings of the IEEE*, 88(3), pp. 333-379.

IOP Conf. Series: Earth and Environmental Science **57** (2017) 012024 doi:10.1088/1755-1315/57/1/012024

Landslide susceptibility mapping based on GIS modle on Shicheng Jiangxi province, china

Wufu Qi[1], Yu Chen[2,3], Xianfeng Cheng[1,4], Qinjun Wang[2,3], Yongmin Wei[3]

1) Yunnan Land and Resources Vocational College, Kunming 652501, China
2) Key Laboratory of Earth Observation Hainan Province, Sanya Institute of Remote Sensing, Hainan 572029, China
3) Key Laboratory of Digital Earth Sciences, Institute of Remote Sensing and Digital Earth, Chinese Academy of Sciences, Beijing 100094, China
4) China University of Geosciences, Beijing100083, China

corresponding author: Yu Chen(chenyu@radi.ac.cn)

Abstract. A GIS model-information index model was presented for landslide susceptibility mapping on Shicheng, Jiangxi province, China.140 landslides were identified from SPOT6 fusion image with 1.5 meters resolution, and they were verified by field investigation. Application of the information index model showed that the landslides more likely occur in areas nearby the road, the river and the lower vegetation covery. The high elevation accuracy of 71% was reached using a receiver operating characteristic (ROC). The result indicates that the northeast and parts of the south of Shicheng County are highly susceptible to damages from landslides, which provides useful information for disaster management and decision making.

1. Introduction

Landslide is an important natural hazard that causes a great amount of damage around the world every year, especially during the rainy season[1]. The Shicheng County is located in the middle of China's southern mountainous area, east of Jiangxi Province, and is known to be an area prone to landslides. Remote sensing data are widely used for landslides including three main stages, namely a) detection and identification; b) monitoring; c) spatial analysis and hazard prediction[2].In this study, we firstly identified landslides distribution based on SPOT6 image (fusion image with 1.5 meters resolution) on Shicheng Jingxi province, China. Then we quantitatively analyzed landslide impact factors and their contributions to the hazards based on the information index model.

2. Study area

The study area is located about 270 km southeast of Nanchang city, the capital of Jiangxi province in Southern China, and is bounded by longitudes of 116° 05′ E and116° 38′ E, and latitudes of 25° 57′

N and 26° 36′ N. It covers 11 adjacent 1:50,000 topographic sheets of the National Geographic Institute of China and has an extent of about 1583 km^2 including 131 villages. The population of Shicheng is about278 thousand (by 2010).

3. Landslide susceptibility mapping

3.1. Landslide identification

Landslide locations were identified from interpretation of SPOT6 fusion images and field surveys in the study area. In total 140 landslides were mapped (as fig.1 shown). These data were collected and constructed into a spatial database using GIS for modeling and analysis.

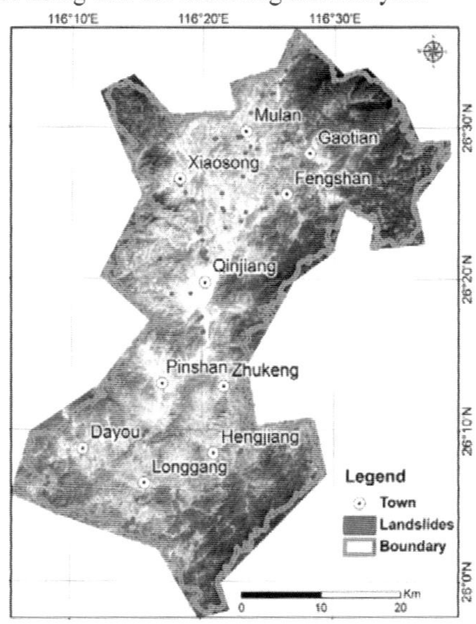

Fig.1: Landslide identification result

3.2. Impact factors retrieve

Determining the suitable impact factors that affect the landslides is crucial to analyze slope instability and failures [3].In this paper, the factors that influence landslide occurrence, such as slope, aspect, and elevation of topography, were calculated from the topographic database. Lithology, faults were extracted from the geological database. Hydrology, road and vegetation cover were extracted from the remote sensing image and geographical database.

3.3. Information index calculation

Landslide occurrence is determined by landslide impact factors. Future landslides occur under the same conditions as past landslides [3]. The information index is a bivariate statistical method that allows to evaluate the relationship between the landslide occurrence and a set of predisposing factors. The information index was calculated by the function[4]:

$$I_i = \log_2 \frac{S_i / S}{A_i / A}$$

here A is the whole area of the surveyregion; S is the whole landslides area of the survey region; A_i is the area of one class range in impact factor i; S_i is the landslides area of the class range in impact factor i. The information index of each location can be calculated as:

$$I = \sum_{i=1}^{n} I_i = \sum_{i=1}^{n} \log_2 \frac{S_i / S}{A_i / A}$$

While n is the numbers of impact factor.

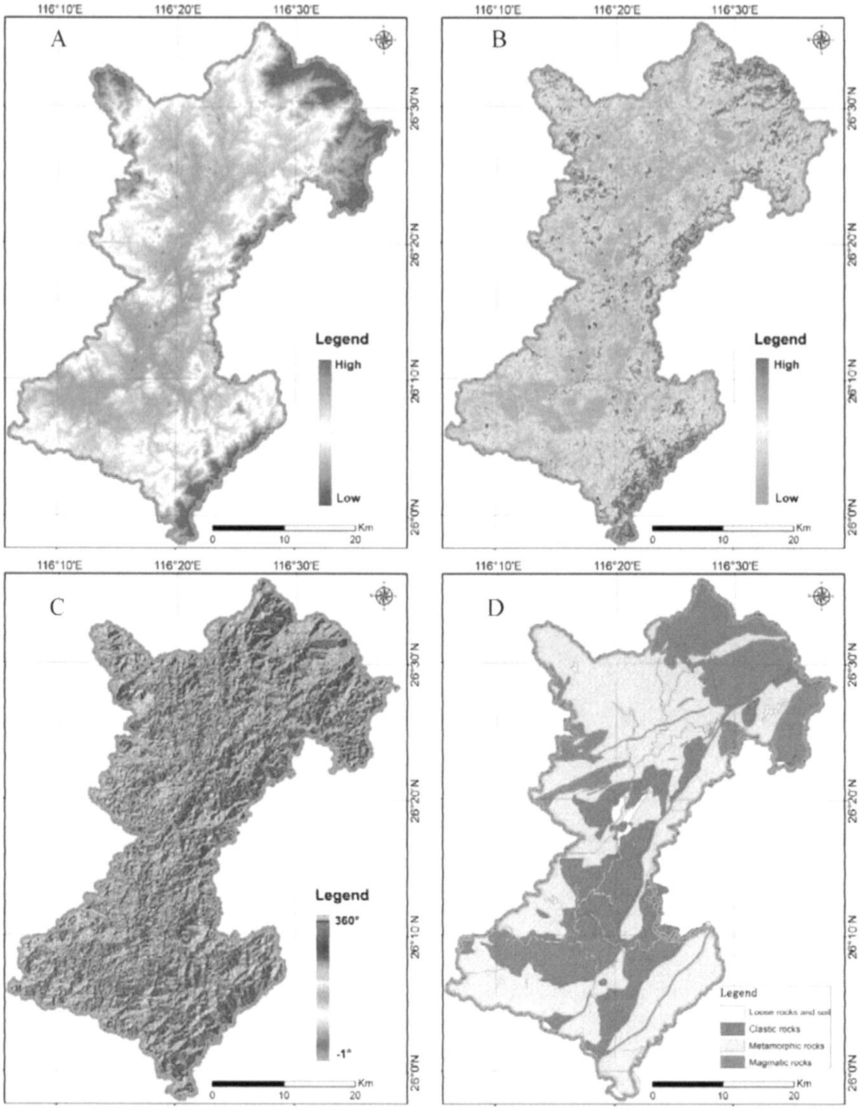

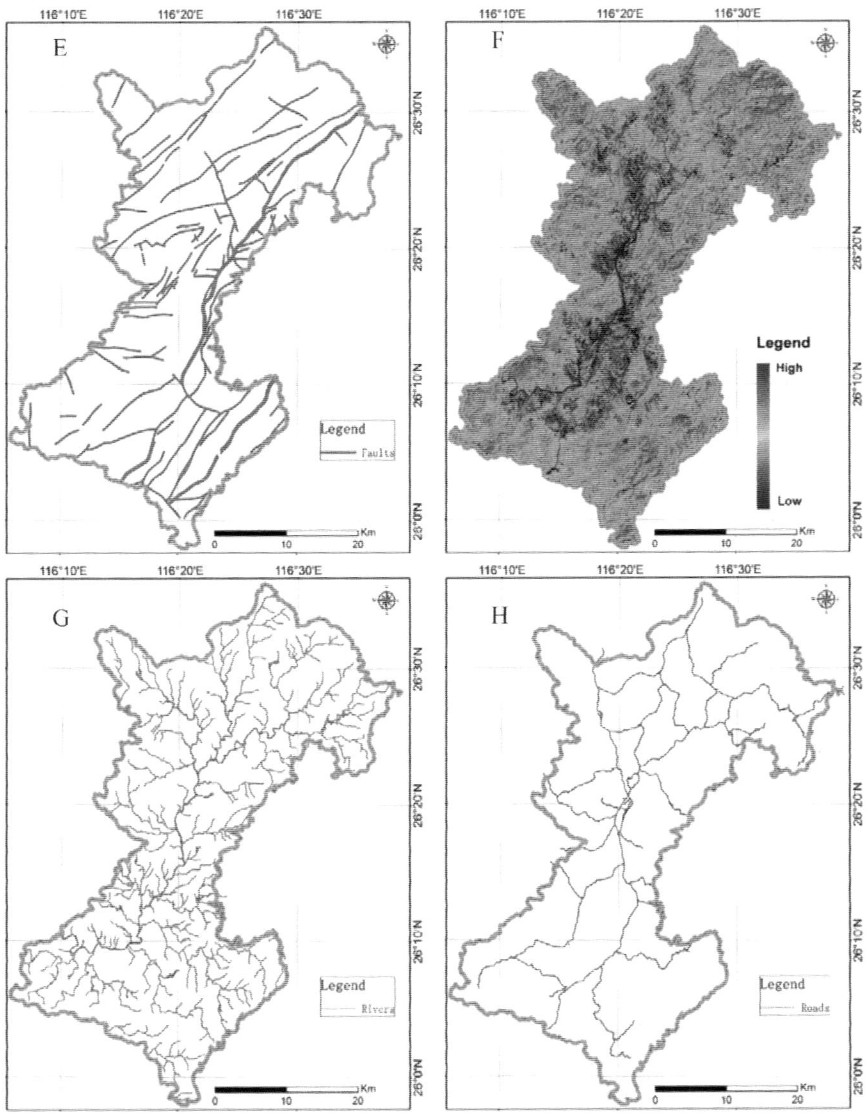

Fig.2: Impact factors retrieve result(A:DEM; B:Slop; C:Aspect; D:Lithology; E:Faults; F:NDVI; G:Rivers H:Roads;)

Table 1-Procedure results for information index calculation

Impact factors	Class range	Area(m^2)	Ratio of Landslides	Information index	Rank
Elevation（meters）	190 Less	5.78E+06	0.44%	0.18	12
	190-289	5.36E+08	43.99%	0.25	9
	467-785	3.00E+08	20.18%	0.05	18
	289-467	6.58E+08	34.18%	-0.21	29
	785 More	6.45E+07	1.21%	-1.22	38
Slope（°）	0-5.7	4.15E+08	35.61%	0.3	8
	5.7-7.9	1.97E+08	14.44%	0.14	14
	7.9-13.6	4.36E+08	27.56%	-0.01	22
	13.6-29.0	4.67E+08	21.61%	-0.32	33
	29.0 以上	5.06E+07	0.77%	-1.43	39
Aspect（°）	76 Less	3.05E+08	19.51%	0	21
	76-151	3.15E+08	21.17%	0.05	17
	151-221	3.12E+08	18.30%	-0.09	23
	221 More	3.02E+08	22.27%	0.14	13
Lithology	Loose rocks and soil	2.56E+07	2.21%	0.32	7
	Clastic rocks	3.13E+08	23.48%	0.18	10
	Metamorphic rocks	8.60E+08	54.91%	0.02	20
	Magmatic rocks	4.01E+08	19.40%	-0.26	32
Distance to fault（meters）	177 Less	2.36E+08	16.32%	0.08	16
	177-539	3.72E+08	21.28%	-0.11	25
	539-1285	5.16E+08	29.88%	-0.1	24
	1285-2814	3.87E+08	27.01%	0.09	15
	2814 More	5.58E+07	5.51%	0.44	5
NDVI	-0.19 Less	1.32E+07	0.66%	-0.24	31
	-0.16	8.98E+07	9.68%	0.52	3
	-0.07	1.43E+08	17.16%	0.63	2
	0.04-0.20	6.89E+08	52.81%	0.18	11
	0.20 More	6.31E+08	19.69%	-0.72	36
Distance to river（meters）	90 Less	2.26E+08	24.15%	0.51	4
	90-276	3.44E+08	32.97%	0.41	6
	276-657	5.53E+08	28.78%	-0.21	28
	657-1440	4.03E+08	12.46%	-0.73	37
	1440 More	3.91E+07	1.65%	-0.41	34
Distance to road（meters）	370 Less	2.73E+08	33.19%	0.64	1
	370-1065	4.06E+08	26.57%	0.03	19
	1065-2370	4.98E+08	26.02%	-0.2	27
	2370-4819	3.27E+08	11.03%	-0.64	35
	4819 More	6.20E+07	3.20%	-0.21	30

4. Results and conclusions

The result of susceptibility mapping is shown as figure 3. It indicates that the northeast and parts of the south of Shicheng County are highly susceptible to the damage from landslides. The result provides useful information for disaster management and decision making. The contribution of each impact factor (table 1 column rank) indicates that the landslides more likely occurred nearby the road (with the information index of 0.64), which means that the slope failure mostly caused by the road construction (excavating the bottom of the hillside makes the slope unstable). The other factors include the less vegetation cover (with the information index of 0.63 and 0.52) and the closing to rivers (with the information index of 0.51). The high elevation accuracy of 71% is reached using a receiver operating characteristic (ROC, figure 4).

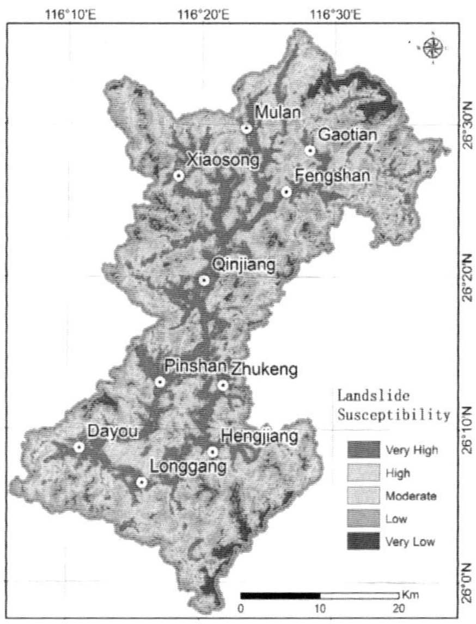

Fig.3:*Susceptibility mapping result*

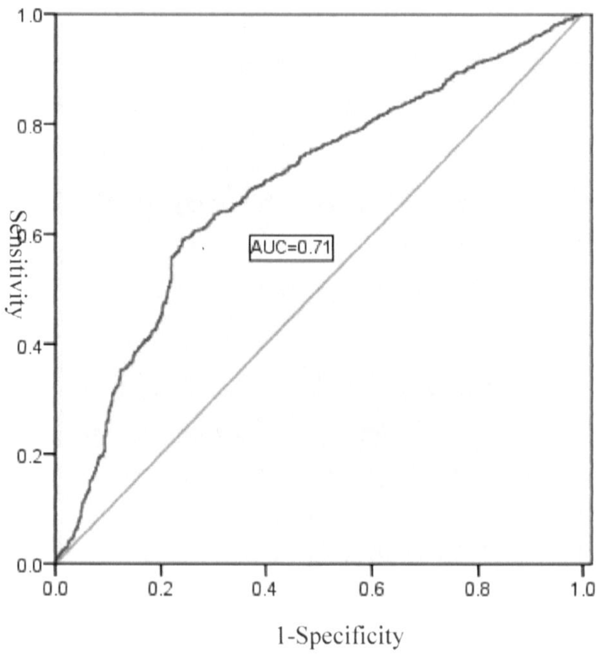

Fig.4: The result of elevation accuracy using ROC

Acknowledgement
This research was financially supported by the project of Natural Science Foundation of Hainan Province (No. 20154188), Hainan Provincial Department of Science and Technology (No. ZDKJ2016021) and Sanya Institute of Science and Technology Cooperation projects (No.sy14ys01132).

References

[1] H. Hong, H. R. Pourghasemi, and Z. S. Pourtaghi, "Landslide susceptibility assessment in Lianhua County (China): A comparison between a random forest data mining technique and bivariate and multivariate statistical models," *Geomorphology,* vol. 259, pp. 105-118, 4/15/, 2016.

[2] G. Metternicht, L. Hurni, and R. Gogu, "Remote sensing of landslides: An analysis of the potential contribution to geo-spatial systems for hazard assessment in mountainous environments," *Remote Sensing of Environment,* vol. 98, no. 2-3, pp. 284-303, Oct 15, 2005.

[3] A. M. Youssef, B. Pradhan, M. N. Jebur *et al.*, "Landslide susceptibility mapping using ensemble bivariate and multivariate statistical models in Fayfa area, Saudi Arabia," *Environmental Earth Sciences,* vol. 73, no. 7, pp. 3745-3761, Apr, 2015.

[4] Y. Chen, Q. Wang, Y. Wei *et al.*, "Application of information index model in landslide susceptibility mapping on Tonggu Jiangxi PROVINCE, China," in 2014 IEEE Geoscience and Remote Sensing Symposium, Beijing, 2014, pp. 4307- 4310.

Comparison of SMMI, PDI and its applications in Shendong mining area

Ying Liu, Hui Yue, Haoren Wang and Wei Zhang

College of Geomatics, Xi'an University of Science and Technology, Xi'an 710054, Shaanxi, China

E-mail: liuying712100@163.com

ABSTRACT. Soil moisture has become the dominant factor in the Shendong mining area for vegetation growth and recovery. Soil moisture monitoring index (SMMI) and the Perpendicular Drought Index (PDI) were widely used to monitor soil moisture, which were all obtained from the NIR-Red spectral space. To validate whether SMMI and PDI were all can be applied to evaluate the soil moisture conditions, SPOT-6 images on October 5, 2015 in Shendong mining area are used to calculate the SMMI and PDI over ground measuring points. The SMMI and PDI are then compared to ground-measured soil moisture data with the satellite overpass. This study concludes that the SMMI and PDI provide quite similar results for soil moisture monitoring and they are quite suitable to assess the soil moisture statue in 0-5-cm-depth. The highest correlation of R2=0.54 for the SMMI and R2=0.52 for the PDI is obtained when compared with relative soil moisture from 0 to 5 cm soil depth. It is evident from the results showing the spatial distribution of drought in Shendong mining area that the SMMI and PDI are highly accordant with landform types and underlying.

1.Introduction

Soil moisture is one of the most direct and important indicators of drought events, moreover, drought is also a complex phenomenon which related to regional climate change and water cycle. Therefore, an understanding of soil spectral behaviour is critical to the drought monitoring and estimation. Ghulam [1] established the Perpendicular Drought Index (PDI) which developed on the basis of spatial characteristics of moisture distribution in NIR–Red space that proved to be effective in large-scale applications over western China using MODIS data. The modified perpendicular drought index (MPDI) is based on the combination of two important indicators of drought soil moisture and fraction of green vegetation. MPDI showed potential advantages for regional surface dryness estimation as reported by Ghulam [2]. Liu [3] according to the distribution characteristics of soil moisture in the spectral space and using the distance of any point to origin indicating the status of soil moisture, constructed soil moisture monitoring index (SMMI) of TM/ETM+ which was not rely on soil line. To validate the drought indices of PDI and SMMI, SPOT-6 images in Shendong Mining Area with various drought conditions is used to calculate PDI and SMMI over ground measuring points. The PDI and SMMI are then compared to an in-situ drought index obtained from field measurements made synchronously with the satellite overpass, including the bulk soil moisture content at different soil depths.

2.Study area and data

2.1 Study area

Shendong mining area (Figure 1) is located in southeast of the Ordos Plateau, northern edge of the Loess Plateau in northern Shaanxi and the South-East edge of Mu Us Desert (110°18′30″E, 39°11′30″N). This mining area is one of the major coal production bases in China, belonging to the arid and semi-arid desert mine area. The ecological environment of Shendong mining area is fragile and the climate is characterized by less precipitation, uneven seasonal distribution, strong evaporation, and scarcity in surface water resources. Soil moisture is the leading factor of vegetation growth and recovery in this area.

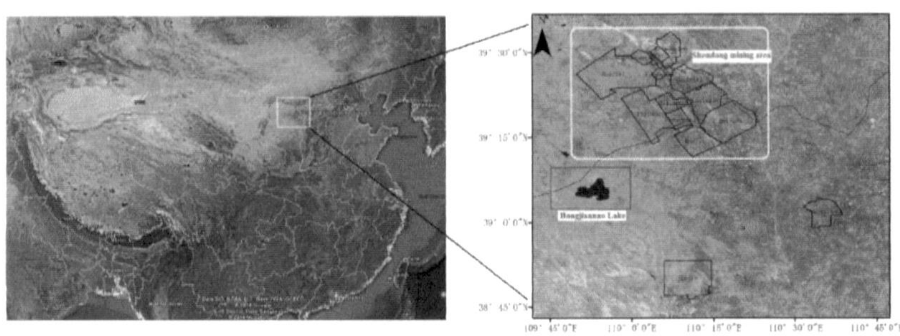

Figure 1. Location of Shendong mining area

2.2 Data

Satellite data used in this study is SPOT-6 image with 6m resolution registered on October 5, 2015 over Shendong mining area and its surrounding region in China. After the geometric correction, digital numbers (DNs) values were converted into spectral radiance and FLAASH in the ENVI software was carried out to eliminate the atmospheric perturbation and obtain the reflectance at ground level. Data on relative soil moisture to compare with SMMI and PDI were obtained by field sample on October 5, 2015. Soil samples from the Shendong site were taken back to the laboratory, and a traditional weighing method was used to obtain relative soil moisture. The number of field sample points was 31.

3.Method

3.1 Perpendicular drought index (PDI)

PDI was defined as line segment that is parallel with the soil line and perpendicular to the normal line of soil line intersecting the coordinate origin in the two-dimensional scatter plot of red against near infrared (NIR) wavelength reflectance. In the NIR-Red space, PDI can be defined as the vertical distance from a random point E (ρ_{NIR},ρ_{Red}) to line L and it can be expressed as

$$PDI = |EF| = \frac{1}{\sqrt{M^2+1}}(\rho_{Red} + M\rho_{NIR}) \quad (1)$$

Where ρ_{NIR} and ρ_{Red} are the atmospherically corrected surface reflectance of Red and Near Infra-Red (NIR) bands of data, M represents the slope of the soil line BC in the NIR–Red space. PDI values vary between 0.0 and 1.0. A pixel which locates closer to the coordinate origin O with a smaller PDI means less water stress or wet surface while it locates further away from the coordinate origin O with a larger PDI value indicates more severe water stress. PDI assumes land cover and soil types are homogeneous over the area of study which is based on the concept of a fixed soil line assumption. Some studies indicated that PDI was an effective index for soil moisture monitoring in bare soil area [4-5].

3.2 Soil Moisture Monitoring Index (SMMI)

Soil moisture monitoring index (SMMI) is defined as the distance of random point E (ρ_{NIR},ρ_{Red}) to coordinate origin O in the two-dimensional scatter plot of red against near infrared (NIR) wavelength reflectance, which was not rely on soil line. Generally speaking, the distance of a black body is almost equals zero which is located at the origin O. Any other objects with some reflectance, it will be closer to the origin O due to the higher soil water content it possesses. Meanwhile, the farther the distance, the less the soil moisture or vice versa. The SMMI can be expressed as following.

$$\text{SMMI} = \frac{|OE|}{\sqrt{2}} = \frac{\sqrt{\rho_{Red}^2+\rho_{NIR}^2}}{\sqrt{2}} \quad (2)$$

Where ρ_{NIR} and ρ_{Red} are the surface reflectance of the NIR and red bands, respectively. Values of SMMI vary between 0 and 1; lower SMMI means less water stress and higher PDI represents more severe water stress. Figure 2 shows that in ΔOEF, the right-angled side of EF was used to represent the PDI and the hypotenuse OE means the SMMI.

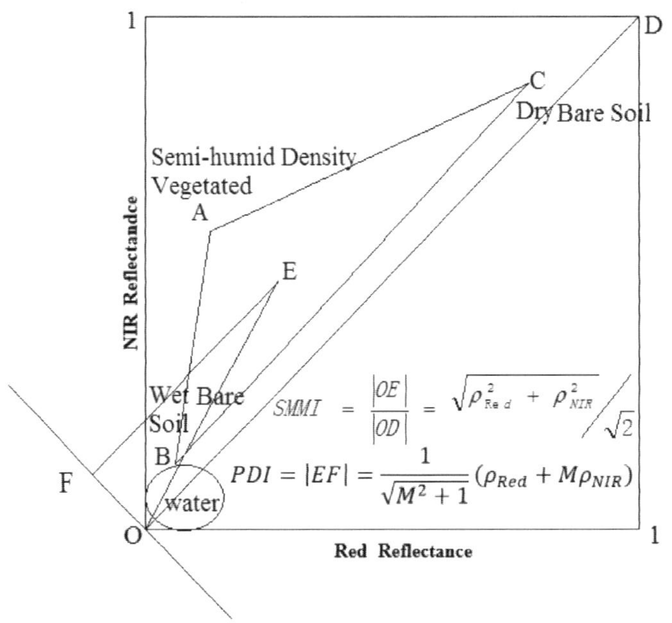

Figure 2. Sketch map of Soil Moisture Monitoring Index (SMMI) and Perpendicular Drought Index (PDI)

4.Results and Discussion

4.1 Relationship between soil moisture and the SMMI and the PDI

Available soil moisture can directly be linked to the drought conditions. To further understand the performance of SMMI and PDI in monitoring the soil moisture status, they are compared with in-situ measurements of 0-5-cm-depth, 10-cm-depth and 20-cm-depth relative soil moisture data (SM). Taking SM as the abscissa and SMMI and PDI as ordinates, respectively, SM-SMMI and SM-PDI scatter plots (Figures 3) were constructed. Figures 3 shows that higher soil moisture values correspond to lower SMMI and PDI values. There are significant negative linear correlations (reliable at 1% significance level) between the SMMI and PDI and different depths average soil moisture.

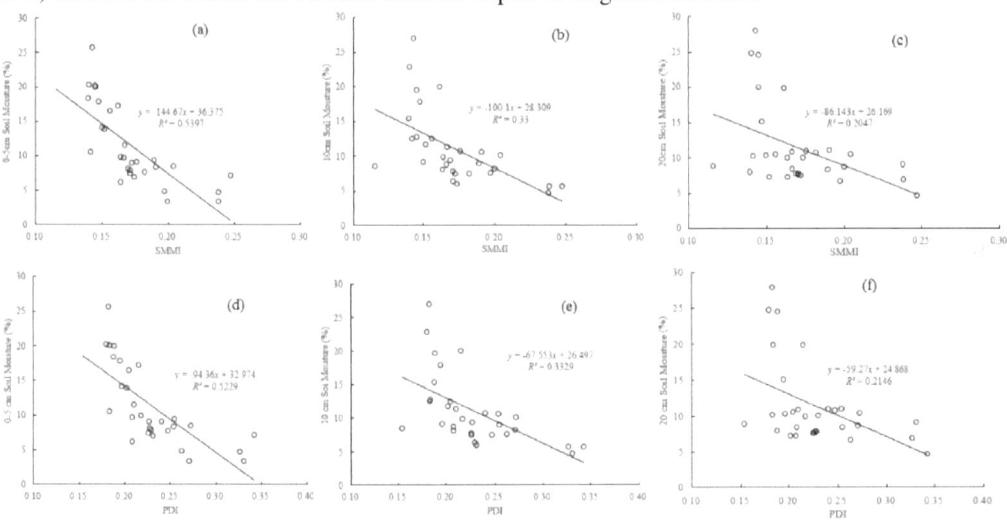

Figure 3. Relationship between soil moisture at different depths and SMMI and PDI in the Shendong mining area

The coefficients of determination between SMMI, PDI and 0-5-cm-depth relative soil moisture are the highest, with R^2=0.54 and 0.52, respectively. The coefficients of determination between SMMI, PDI and 10-cm-depth relative soil moisture is higher than the 20-cm-depth relative soil moisture, with R^2>0.33 and 0.20, respectively. The results showed that SMMI and PDI all can reflect the surface soil moisture and they are more suitable for assessing 0-5-cm-depth soil moisture conditions. However, SMMI is slightly better than the PDI for monitoring 0-5-cm-depth soil moisture conditions, while the latter are slightly better than the former in reflecting 10-cm-depth and 20-cm-depth soil moisture status.

4.2 Surface drought assessment with the SMMI and PDI in Shendong mining area

The SMMI and PDI are all constructed on the basis of spectral behaviour of soil moisture in NIR–Red space. They can indicate soil moisture status which directly links to the drought, therefore, we can be used the two indexes assess the drought conditions of the study area (Figure 4). Figure4 shows that the drought condition is quite similar between SMMI and PDI and it is consistent with landform types and underlying in mining area. In the surrounding of Wulanmulun mine and the northern and eastern parts of Shigetai mine, the value of SMMI and PDI is higher than the other parts of the two mines and the drought conditions is more severe. The SMMI and PDI values are lower than other parts of the study area and the soil water content is higher on the two sides of the River Kuye. Compared to the underlying of the mining area (Figure 4(a)), the drought status is normal in most area of the mining area where with better vegetation cover. However, the surface is bare, open pit mining or cover by sand (like samples A, B, C, D), the drought is more severe with the higher SMMI and PDI values. Samples E and F is water with lower value of SMMI and PDI (Figure 4). It is obvious from the results that both the SMMI and PDI are effective methods for monitoring drought events in study area.

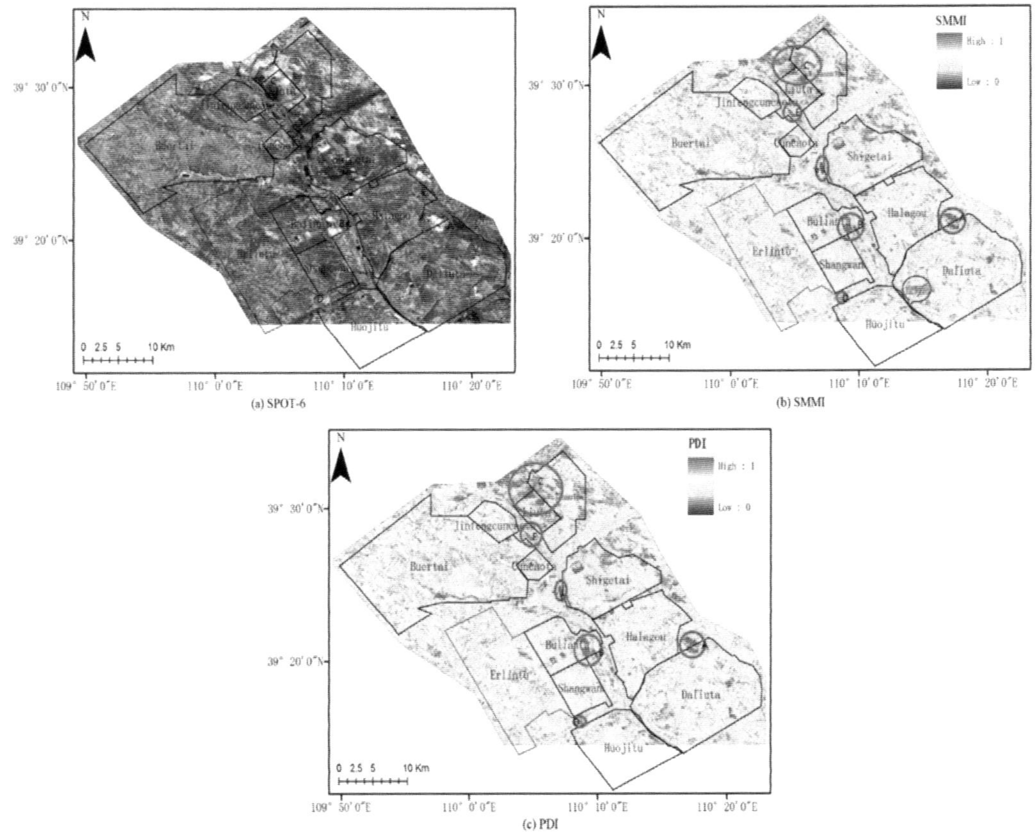

(a) SPOT-6 (b) SMMI (c) PDI

Figure 4. Comparison of SMMI and PDI over spatial distribution of drought in Shendong mining area

5.Conclusion

This paper has explored the performance of SMMI and PDI in soil moisture assessing. It was revealed from the results that SMMI and PDI derived from NIR-Red space all demonstrated a strong negative linear correlation with different depths field measured soil moisture. The correlation between SMMI, PDI and soil moisture where is highest with R^2=0.54 and 0.52, respectively, which passed the 0.01 significant test. The drought conditions are quite similar which derived from SMMI and PDI. The paper concluded that SMMI and PDI are simple and effective methods for monitoring soil moisture status and drought events.

However, PDI is based on a fixed soil line assumption. However, the shape of the soil line changes with the variation of soil type and soil fertilization conditions and, therefore, affect the PDI by changing the slope parameter M. SMMI was only inferred to Red and NIR reflectance which was not rely on a fixed soil line. Both SMMI and PDI were not consider the effects of vegetation greenness on the index, yet further work needs to be done to discriminate the effects of vegetation greenness on the index.

6.References

[1] Ghulam A, Qin Q, Teyip T 2007 *ISPRS J PHOTOGRAMM*, **62** 150-164.
[2] Ghulam A, Qin Q, Zhan Z 2007 *Environment Geology*, **52** 1045-1052.
[3] Liu Y, Wu L X, Ma B D 2013 *Journal of China University of Mining & Technology*, **42** 296-301.
[4] Qin Q, Ghulam A, Zhu L, Wang L, Li J, Nan P 2008 *Int. J. Remote Sens.* **29** 1983–1995.
[5] Shahabfara A., Ghulamb A., Eitzingera J., 2012. *Int. J. Appl. Earth.* **18** 119-127.

Acknowledgements

This work was jointly supported by The National Natural Science Foundation of China (Grant No. 41401496), China Postdoctoral Science Foundation Funded Project (Grant No. 2016M592815) and Student's Platform for Innovation and Entrepreneurship Training Program (201610704120).

IOP Conf. Series: Earth and Environmental Science **57** (2017) 012026 doi:10.1088/1755-1315/57/1/012026

Emergency Broadcast System Based On GIS

Kang Junfeng*, Wang Baoyu, Jiang Xueying, Liu Xiaosheng

Jiangxi University of Science and Technology, Ganzhou, China, 341000

jfkang@sina.com, wangby3214@foxmail.com

Abstract. Nowadays, disaster occurs more and more frequently, which brings huge losses to the society. Presented a GIS based emergency broadcast system (GIS-EBS), to provide rapid, accurate information to the public, can reduce the people's losses maximum. Different from the traditional radio data system (RDS), by adding broadcasting terminal control instruction on the FM signal, GIS-EBS extended the RDS protocol, which can control each radio terminal's state. And through the extended RDS protocol, GIS-EBS can broadcast and release the emergency information to the right place and the people most in need in the shortest time.

1. Introduction

More and more extreme natural disasters continue to emerge, which put forward a higher request that the emergency management needs to be strengthened and be improved. In the process of occurring M9.0 earthquake in Japan[1], many people successfully managed to escape the danger in a short time because they heard the alarm broadcast before the earthquake. Similarly, broadcast played a significant role in Wenchuan the M8.0 earthquake as well[2]. It had proved that the broadcast is the quickest and most effective approach of signal transmission in the critical situation.

Wang[3] proposed to build the national emergency broadcast architecture, which focusing on clarify the level of emergency, and combining emergency response and prevention. Through researching the Japanese natural disaster emergency broadcasting system, Guo[4] analyzed the problems and deficiencies of Chinese disaster broadcasting system. Jiang[5] developed a township's emergency broadcast system based on Radio Digital System(RDS) technology. Wen[6] designed a shortwave remote terminal waken method, that used in the development and application of country's emergency broadcast system.

Adding digital signal to the FM has become the trend of the construction of the emergency broadcasting system. By extending the RDS protocol, proposed a GIS-based emergency broadcast system (GIS-EBS), GIS-EBS's main characteristics including: Addressable FM radio data system, GIS-EBS operates a set of extended FM-RDS instructions on both PC and encoder, which can combine users' inputting text and voice and the radio terminal control command to FM signal; Radio terminal breakdown monitor method, in order to manage a large number of distributed radio terminals, developed an active access method by using embed GPRS module, so GIS-EBS can query the terminals' real-time state automatically. Weak signal processing based on spatial analysis, by combining with GIS spatial analysis method[7], GIS-EBS can calculate the most reasonable location to place the radio terminal, which can receive the optimal broadcast signal.

2. GIS-EBS 's key technology

As showed in Figure 1, when the unexpected events occurred, GIS-EBS could automatically determine the scope of broadcasting, and convert the emergency signal to radio signal at the first time, according to the type of emergency signal's type and location. The process of emergency signal network transmission was as follows: firstly, emergency signal needs to convert into IP data flow, then it would be sent into the Internet or intranet network by IP Protocol Transmit Device, secondly, RDS Protocol Encoder would receive the IP data flow, and convert it to the RDS data, finally, RDS Protocol modulator would modulate RDS data into FM signal, sent to the broadcasting terminal on CATV network or unwire FM signal.

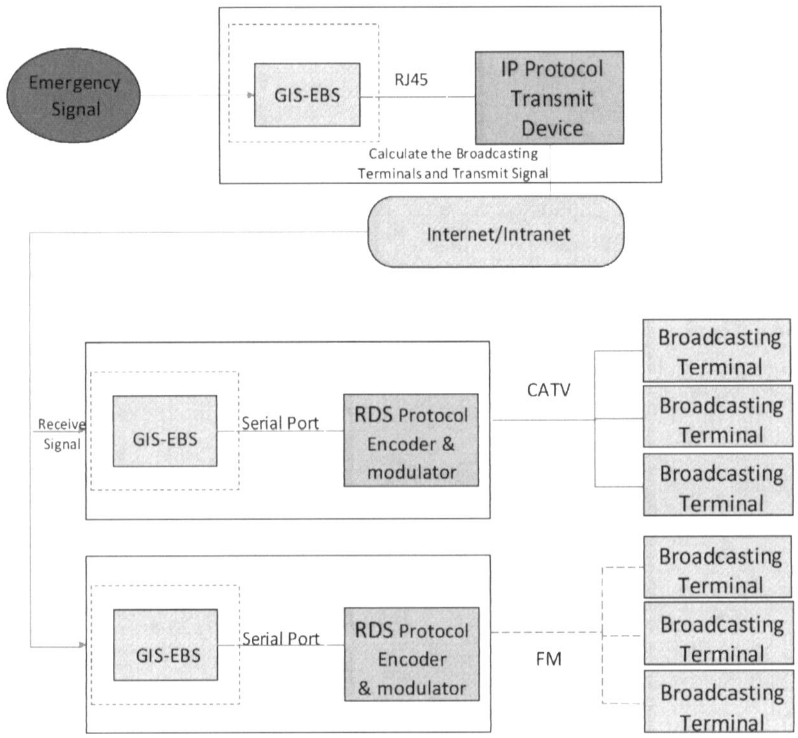

Fig.1 GIS-EBS Topo Framework

2.1 Extended RDS Protocol

According to the characteristics of the emergency broadcast, GIS-EBS designs three kinds of communication protocol type: extended RDS protocol, IP protocol, and GPRS protocol. And the extended RDS protocol was the core of GIS-EBS, and it was the key to broadcasting radio accurately. GIS-EBS's Extended RDS protocol was showed in table 1.

Tab. 1 GIS EBS Extended RDS Protocol

	Head	Command type	protocol type	Frame Length	Data	CRC check	End
Length (byte)	1	1	1	2	n	2	1
Default Value	0x24	CMD					0x0A

Taking data "04 01 01 11 10 10 10 10 11 02 02 01 01 0A 00 01 01 92 BC" as an example:

04 (representative data level): county level, 0x01 - Central ,0x02 - province, 03 -city, 04 –county, 05 –country, 06- village.

01(representative data type): Emergency.

01(represents a group of broadcasting terminal): 0x01 - one group, maximum group number - 0x16.

11 10 10 10 10 11: (broadcasting terminal address): length - 6 bytes, on behalf of the province, city, county, country, village, terminal address. Each level's length is a byte.

02(represents a type directive): 0x01---start, 0x02---stop.

02 (on behalf of event level): 0x01--- particularly serious, 0x04---normal serious.

01 01 0A 00 01 (represents the emergency type): 0x01- flood,0x02-debrisflow, 0x03-Blizzard, 0x04-Typhoon.

GIS-EBS's extended RDS signal could add digital information on the FM signal, such as warning area (national\provincial\city (county, district)\township); warning type (earthquake\ tsunami\air strikes\Typhoon\debris flow\flood and other emergency situation), and etc. When the Broadcasting Terminal received the FM signal, it would judge whether or not to response according to the added digital information.

2.2 Broadcasting Terminal Failure Detection method based on GPRS

Broadcasting Terminal is usually sprawled across in various areas. The traditional way to checking the Broadcasting Terminal's state was by manual inspection, which was really time-consuming[8]. GIS-EBS implemented an automatic monitor service based on GPRS. The real-time data of Broadcasting Terminal would send its' heartbeat data to the internet network on GPRS protocol, and GIS-EBS's GPRS monitor service can automatically detect the broadcast terminal's heartbeat data, the Broadcasting Terminal's statues would show in the Map Viewer intuitively. If Broadcasting Terminal communication failure, GIS-EBS would alarm the user in real time, so it can make the Broadcasting Terminal's maintenance easier.

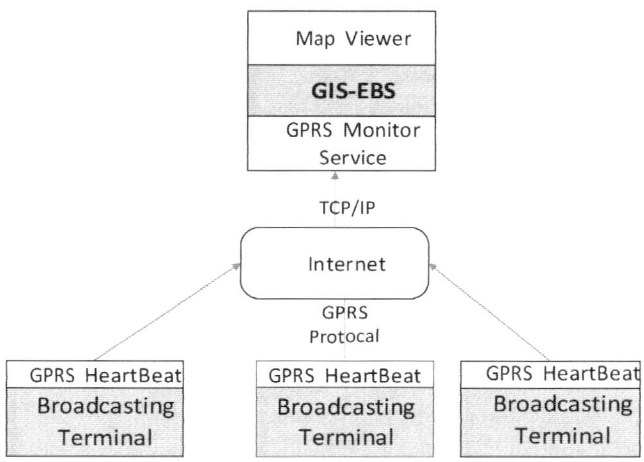

Fig 2 Broadcasting Terminal Failure Detection

2.3 Broadcasting Terminal's Location choice based on GIS

Broadcasting Terminal could be received radio signal through CATV network or wireless FM signal, in the complex mountain village area, Broadcast Terminal was usually using FM wireless way to set up. But FM signal suffered some signal attenuation during the long distance transmission; the quality of the FM signal will depend on the receiving environment, frequency shift, time delay, modulation type and other factors [9].

In the complex mountain village area, for the special path, using the following formula (1) [10] to estimate the standard variance of location variability σL.

$$\Delta h/\lambda < 3\ 000 \quad \sigma_L = 6 + 0.69\left(\frac{\Delta h}{\lambda}\right)^{1/2} - 0.0063\left(\frac{\Delta h}{\lambda}\right) \text{ dB} \tag{1}$$
$$\Delta h/\lambda > 3\ 000 \quad \sigma_L = 25 \text{ dB}$$

The parameters in the formula:

Δh: Decile height (meters)

λ: wavelength (meters)

$\lambda = 300/f$

f: frequency (MHz)

Generally, the standard deviation would decrease with the increase of the distance between the transmitters, and the main factors that affected the h included elevation, vegetation, buildings and so on. Not considering ocean and other special terrain, the received signal strength would increase with the height of the antenna. The experience height of the antenna (heff) was between 3 and 15 km. According to the path length, the optimum height of the transmitting station antenna (h1) could be calculated according to the follow formula.

$$h_1 = h_a \quad \text{for} \quad d \leq 3 \text{ km} \tag{2}$$

$$h_1 = h_a + (h_{eff} - h_a)(d-3)/12 \quad \text{for} \quad 3 \text{ km} < d < 15 \text{ km} \tag{3}$$

ha refered to the antenna height above the earth.

According to the FM signal attenuation equation, by using remote sensing image and DEM data[11], GIS-EBS can analyze the location and get the best height of the antenna in the area, and achieve the reasonable location of Broadcasting Terminals, so that can get the balance between the broadcast signal and construction resources.

3. Design and Implementation of GIS-EBS

GIS-EBS's system architecture is showed in figure 1, divided into the hardware layer, protocol layer, business logic layer and user presentation layer.

The hardware layer included IP Protocol Transmit Device, RDS Protocol Encoder, modulator and Broadcasting Terminal. IP Protocol Transmit Device was responsible for converting received emergency signals to IP data, and sending the IP data to Internet or Intranet network; RDS Protocol Encoder and modulator were responsible for convert RDS data to FM signal; Broadcasting Terminal would convert it to RDS signal when receiving the FM signal, and based on the address information in FM signal to determine whether or not execute the command. Besides, broadcasting Terminal would encode its' state to IP data, then send to the wireless network through GPRS module, GIS-EBS would judge broadcasting Terminal whether or not online through decoding the received IP data.

Protocol Layer has three protocols, including RDS Protocol, IP Protocol, GPRS Protocol.

Business Layer has divided into Addressing module and Broadcasting module. Broadcasting module included Emergency Voice Broadcasting, Emergency Text Broadcasting, Daily Voice and Text Broadcasting and Plan Broadcasting. Broadcasting module is enabled to the real-time broadcast of emergency voice signal and text signal. Addressing modules included IP Address Module and GPRS Address Module. IP Address Module was responsible for sending an emergency signal to the specified IP Protocol Receive Device; GPRS Address Module was responsible for sending the terminal state to GIS-EBS.

Presentation Layer included the function of Map Viewer, Boarding and Terminal Fault Detection.

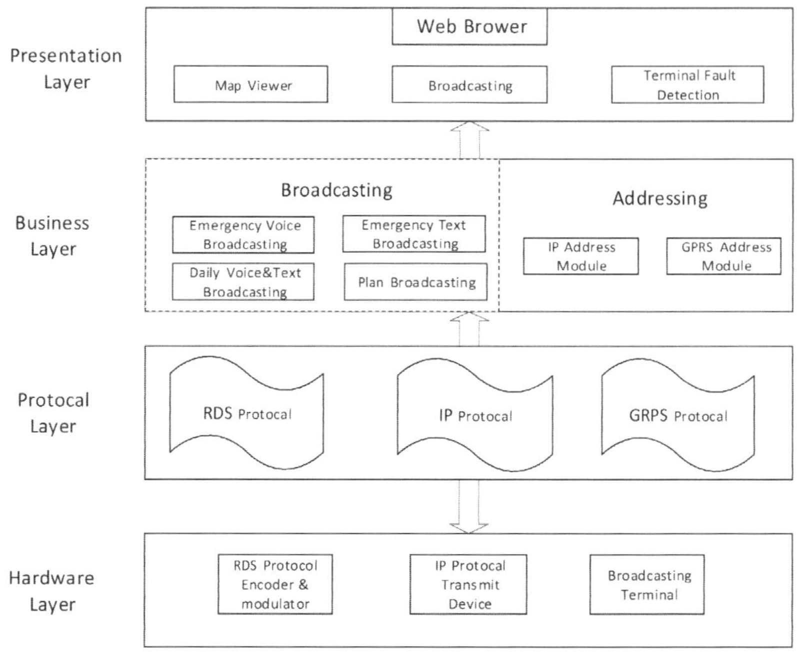

Fig3 GIS-EBS System Architecture

The interface of GIS-EBS was showed in Figure 4. The main function included daily broadcast, text broadcast, emergency broadcast, timing broadcast and fault detection. According to the broadcast plan, GIS-EBS would broadcast on the schedule automatically.

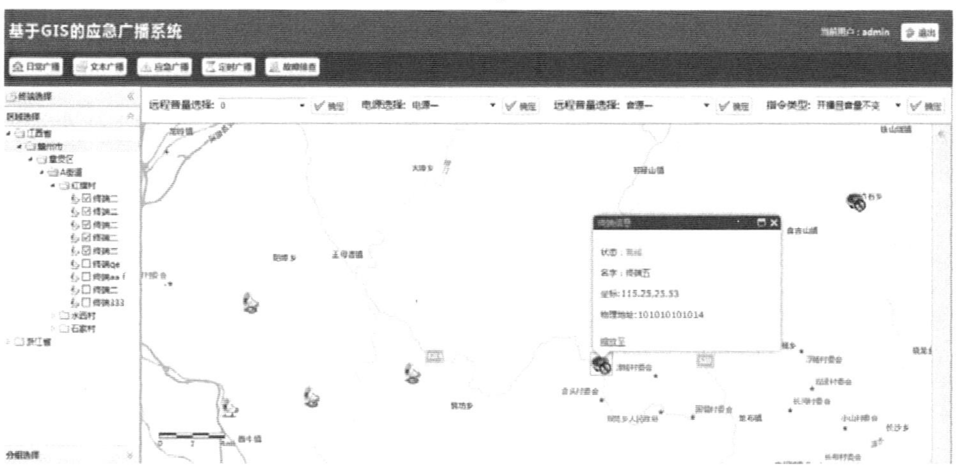

Fig 4 GIS-EBS's running interface

If emergency occurred, GIS-EBS would suspend the current program, and gave the priority to the emergency broadcast information, in addition, the system supports broadcast the text and voice information. When the Broadcasting Terminal was failure, GIS-EBS could display the fault equipment in the map viewer, which provided convenience to quick retrieval and maintenance.

4. Summarize and Future work

The frequent occurrence of natural disasters in recent years makes people aware of the value of

emergency broadcasting, GIS-EBS not only could provide real time and accurate emergency radio broadcasting, but also provide a convenient method to maintenance broadcasting terminal.
In the future, we will combine with the Internet of things (IOT) into GIS-EBS, to provide real-time warning broadcast signal, such as automatic monitoring and information of landslides, earthquakes, and floods, to provide the greater value to the disaster prevention and mitigation.

5. Acknowledgments

The project is partially funded by the National Natural Science Foundation of China No. 41561091, and the Natural Science Foundation of Jiangxi No 20142BBE50024, 20132BAB211024, and Jiangxi Provincial Department of Education Science and Technology Research Projects No.GJJ13414.

References

[1]. Cui, J.F., Overview of emergency broadcast system implementation. Radio & TV Broadcast Engineering, 2011(1 2): p.40-43.
[2]. DenisNurmel, W., Emergency Broadcast System. PRWeb, 2003(10).
[3]. Wang, C.P., Analysis and Reflection on the construction of national emergency broadcasting system. China Emergency Management, 2012: p. 26-31.
[4]. Guo, Y.W., The Establishment and Improvement of Natural Disaster Emergency Broadcast System. Value Engineering,2014. 106(19): p. 319-320.
[5]. Jiang, Z.B., Development and application of township emergency broadcasting system based on RDS Technology. Radio & TV Journal, 2015(4): p. 20-21.
[6]. Wen, H.J., Application of Emergency Broadcast in County-level and Remote Terminal Awakening. Radio & TV Broadcast Engineering, 2014. 41(6): p. 114-117.
[7]. Liu, H., et al., Dynamic visualization of disaster information in GIS. Journal of Natural Disasters, 2012. 21(6): p. 19-24. [8]. Yu, R.W., Fault analysis of DF_100A type broadcast transmitter modulator. Southeast Communication, 2014(10): p.143-144.
[9]. Guo, S.H., J.T. Sun and R.H. Xie, Attenuation characteristics of electromagnetic wave penetrating walls. High Power
Laser and Particle Beams, 2009. 21(1): p. 113-117.
[10]. Hai, X., Correlation correction method of ITU_RP_1546_3 proposal. Radio & TV Broadcast Engineering, 2008(9): p.124-126.
[11]. Liu, C., W.W. Sun and H.B. Wu, Terrain Complexity Factor and Its Relationship with Accuracy of DEM Terrain Representation. Geomatics and Information Science of Wuhan University, 2009. 34(9): p. 1014-1019.

Digital reconstruction of the Song Dynasty Ganzhou drainage system based on AR technology and its' application in the new urban area planning and revision

H L. Cao[1], Y L. Chen[2] and T H.Tao[3]

School of Architectural and Surveying Engineering, Jiangxi University of Science and Technology, 341000 Ganzhou, China

[1]634211660@qq.com, [2]34564594@qq.com, [3]250904570@qq.com

Abstract. Water-logging problem is a common problem in modern city. The urban built-up area of Zhangjiang new district in Ganzhou has the same water-logging problem, however, the old urban area of Ganzhou was praised as "Millennium no flood".The drainage system of the old urban area of Ganzhou—Fushougou, which is not flooded for hundreds years because of the perfect drainage system. It's valuable to be referenced to the modern city drainage and waterproof comprehensive planning. In order to explore the mystery of "Millennium no flood" of old urban area of Ganzhou, at the same time to provide directive opinion to the sustainability of Zhangjiang new urban area drainage system , this paper attempts to digital reconstruct the drainage system in old urban area of Ganzhou by augmented reality(AR). It will provide a new technological means and ways to evaluate the sustainability of urban underground drainage system under the surface feature changes in the landscape. On the basis of digital reconstruction of the drainage system in the old urban area of Ganzhou, the sustainability evaluation index of drainage system is studied by analyzing and contrasting with Zhangjiang new urban area drainage system, to guide the revision of comprehensive planning about city drainage and water-logging in the new urban area of Zhangjiang.

1. Introduction

In 2015, a total of 20.79 million residents in the 20 provinces (including autonomous regions and municipalities) suffered flood, which has caused 108 dead, 21 missing and 13461 displaced. The worst flood, which has ever seen in past years, brought about the direct economic loss of 35.3 billion Yuan, based in fact that it affected 1.717 million hectares of crops and left 44000 collapsed houses. The Intensity is rare throughout the years. According to the special investigation from Ministry of Housing to 351 domestic city,a total of 170 cities including 12 important cities, 33 secondary cities and 125 general city haven't developping the urban flood control planning. There are still 340 cities (53%) did not reach the flood control standards set by the state. The design of more than 70% of the urban drainage system construction in China heavy rains return period is less than 1 year. According to the national wide range of urban water-logging problem, the government has issued <The town drainage and sewage disposal ordinance>(No.641 decree of the state council), national standard <regulations on the urban drainage and wastewater treatment>, national standard <specifications for design of outdoor drainage >(2014 revision).[1]The Ministry of Water Resources published <the urban flood control planning outline> (revised). According to the flood control standards and specifications which have been revised, the local government are writing and developing the planning of flood control and revision. How to solve the problem of urban water-logging, which is one of the important subjects to be studied and discussed.[2]

Ganzhou is located in the southern part of Jiangxi province ,where the new and old districts present totally different environment features .That is, the new district which is on the southern of the Zhangjiang river is often water-logging, while the old district which is on the northern of the Zhangjiang river is reputed for its stable environment conditions . This phenomenon has attracted the attention of scholars at home and abroad. In order to explore the mystery of old district of Ganzhou which is not suffering from a flood for a thousand year and provide guidance for the sustainable evaluation of the drainage system of the new district, this paper attempt to use the augmented reality technology(AR) to get digital reconstruction of drainage system of the old districts of Ganzhou .On the basis of digital reconstruction of the drainage system of the old districts, this paper studies the safety evaluation index of urban drainage system. A comparison between the drainage systems of Ganzhou old district and Zhangjiang new district is made, which provides favourable guidance for urban water-logging planning.

2. Drainage system of the old district of Ganzhou-fushougou

2.1 Fushougou Overview[3]

Ganzhou located in the southern part of Jiangxi province was built in 201 BC. Ganzhou surrounded by water on three sides. Zhangjiang and Gongjiang River flow across the city and they merge into Ganjiang River in the northern of the city. Ganzhou often suffered from floods from 201 BC to the Song Dynasty. Until AD 1068 - 1077 Song dynasty, Liuyi who is served as chief of a prefecture planned and constructed city street. According to the layout of the streets and topographic features, his team adopt the principle of partition drainage and then built two main drainage system. What they do make Ganzhou exempt from the flood for a thousand years and protect this city. The two main gutter ways are named as Fushougou because they look like "Fu" (good fortune)and "Shou" (longevity) of Chinese seal script.

2.2 Planning and Design Features of Fushougou

(1) Districted Draining System Adaptive to Terrain

"The Shou Gutter way takes in water from the northern city and the Fu Gutter way from the southeastern city." Starting from the South Gate, the Fu Gutter way passes by the middle part of Jianguo Road, Junjing Alley, Pangaopu Street and stretches to Bajing Road. The Shou Gutter way starts from the new Gannan Road and goes along the two Xinkai Roads, Xijin Road, reaching the Weat Gate. After the completion of the two main ditches, a number of sub-ditches have been built, presenting a systematic and ordered draining system in the ancient Ganzhou city.

(2) Natural Draining Availing of Elevation

The Fushougou makes full use of urban terrain elevation, using the natural flow to make the city rainwater and sewage goes into the river.

(3) Reasonable Selection of Material and Large Cross Sections

The cross sections of the two drainage systems are rectangular in large size, made of masonry. According to historical records, the gutter way is "about 0.6 to 1 meter wide and 1.5 to 2 meters deep, built firstly by bricks and then stones". Although having been maintained and improved for several times, the gutter way still keeps its structure of brick walls and stone shields.

(4) Intact Draining System

The old urban district of Ganzhou city has a drainage system dating back to the Song Dynasty. It is composed of five parts: the Fushougou, reservoir, city walls, ditches and water windows, consisting an intact drainage system. The Fushougou is connected with the dozens of ponds inside the city, constituting a reservoir with large capacity. This system can regulate the flow quantity of rainstorm and reduce the flow of sewer system. Once the rainfall intensifies and the water in the Fushougou floods, the water will go into the ponds to be regulated. Twelve water windows are built at the outlet to prevent the river water in Zhang and Gong Rivers rises beyond the outlet and thus floods into the inner city. Water windows consist of inner gate at the inlet, ditch and outer gate at the outlet. Because of sharp gradient, water will form a strong flow after entering water windows, taking away dirt and other solids, opening the gate and flowing into the river. The rule underlying here is that when river water is below water window, it opens windows with the force of sewer water and then drains out; when river water is above water windows, it closes windows with the force of river water in case of backward flow.

3. Augmented reality technology in Fushougou digital protection application

3.1 Digitization of Fushougou

The walls were demolished and the water windows were ruined during the Cultural Revolution in China (1967-1977). Then the pond and moat were filled after the Reform and Opening up. This paper is based on the platform of ArcGIS 10.1. According to the urban map of Ganzhou city in Tong Zhi dynasty and recorded Fushougou condition in combined with field research, we get digital reconstruction of Fushougou and draw the Ganzhou Fushougou digital map.See Figure 1 for details. This study attempts to restore the Song dynasty's Fushougou drainage system in Ganzhou city with the use of modern technology, which provides reference for today's drainage planning.[4]

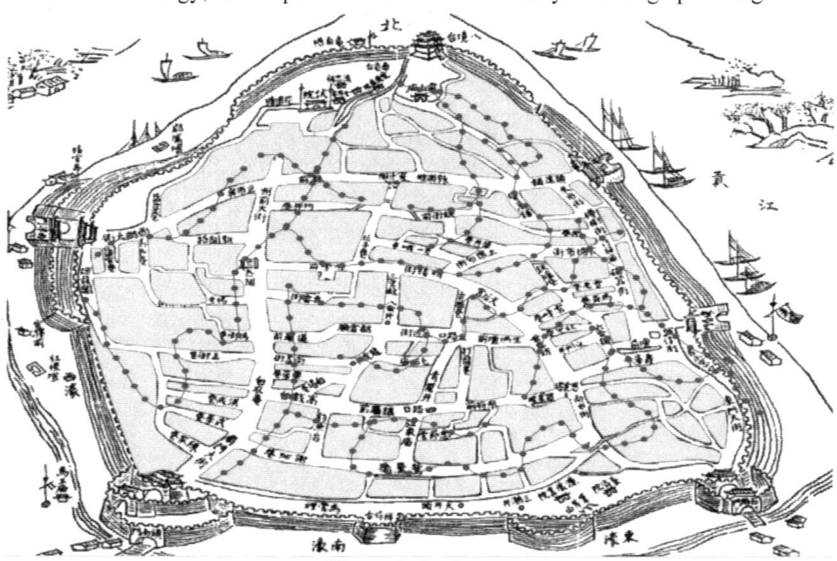

Figure 1. Study area

By using ArcGIS statistical analysis function and combining with historical data, we can obtain the statistical data of the Song dynasty's Fushougou drainage system in the old districts of Ganzhou. These data include watershed area(2.7 km^2),main channel length (12.78km),average cross section area of drainage ditch(1.4m^2),sub channel length(44.56km), average cross section area(0.032m^2),the number of the pond for water storage(79), capacity(110600 m^3),city wall length(8.126km),ditch and pond length (3.78km),average cross section area(18.67 m^2) and water windows' number(12). [5]

3.2 The Application of Augmented Reality Technology in the Digital Protection of Fushougou

The basic idea of AR Technology is to combine the 3D model of the virtual object generated by the computer with the real world.AR technology enables virtual and real images to become an organic whole and enhances the users' perception and understanding of the real world. In augmented reality system, the main task is to match the virtual objects around the users with the real world and display them in real time. Therefore, virtual reality alignment and display are the key technology in the Augmented Reality Technology.[6]

(1) System framework

We use ARToolKit to develop the augmented reality system of outdoor 3D pipe network. The video tracking database of ARToolKit can calculate the position and direction of the camera relative to the mark point in real time because it is based on computer vision. While it cannot be directly applied to the development of the outdoor system, this paper has improved the ARToolKit. The system frame diagram is shown in Figure 2.

(2) The main functions of the system

The system functions are as follows:

a) Virtual information enhancement
The users can see that the images generated by the computer are overlapped on the video image in reality.
b) Query of attribute information
Pop up the corresponding attribute information of the pipe ditch when a pipe ditch was selected.
c) Navigation
The two-dimensional digital map can be embedded into the system and switched at any time. Users can get real-time observation of their current location. The system can display the relevant information automatically according to the user's line of sight direction.
d) Spatial analysis
Three-dimensional pipe ditch can be carried out buffer analysis in order to determine the distance between the different types of pipe trench meets the safety requirements. Besides, tube bursting analysis and plug tube analysis of pipe network can also be carried on.[7]

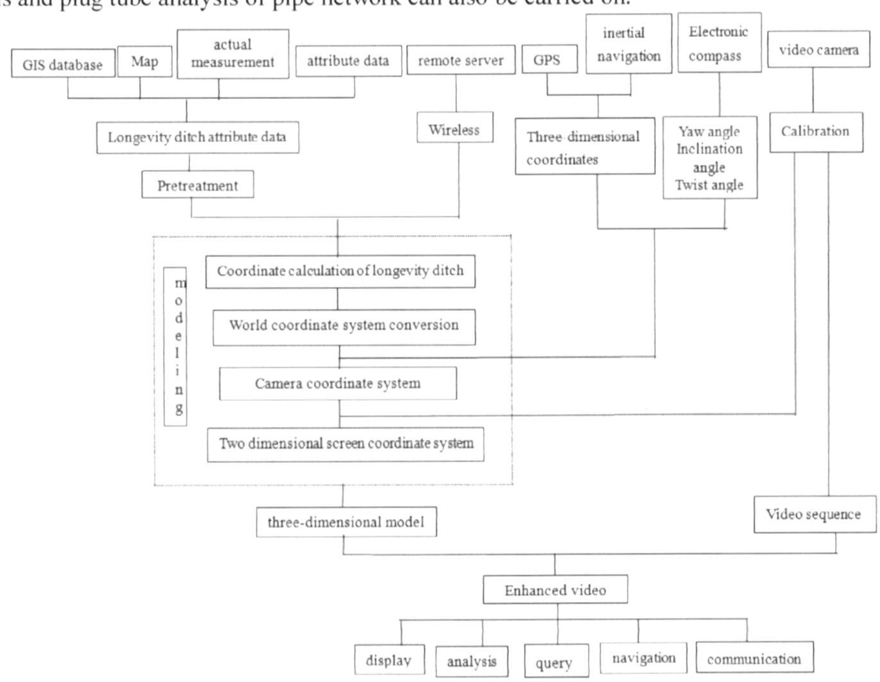

Figure 2. Based on ARToolKit 3D augmented reality system diagram

4. Drainage system of the Ganzhou old district in the Song dynasty and safety evaluation of storm-water system of the new district

4.1 Construction of index
The security of the system includes physical security and social security according to the theory of VPM (vulnerability of place model) model proposed by Cutter.[8] The reason why we focus on the physical security index is because this paper mainly analyzes the significance of the Song Dynasty drainage structures to the modern drainage system planning. On the basis of the safety evaluation index, this paper optimized the actual situation of the song drainage system in the old district of Ganzhou constructed the following 9 indexes as the safety evaluation index of the drainage system:

a) The rain mouth relative height(*ha*), b) Grid spacing(*qa*), c) Diameter of pipe(*da*), d) Unit of the catchment area of pipeline capacity(*ra*), e) Pipe slope(*ia*), f) The unit of catchment area building storage capacity(*ga*), g) Water outlet(*Oa*), h) Runoff coefficient(*aa*),i) Emergency facilities(*ja*).

4.2 The Determination of the index weight
Because each index is different in the security of the rain water system and has different importance, the weight of each evaluation index should be determined before the evaluation.According to the

composition and operation characteristics of the rainwater system, this study used the analytic hierarchy process (AHP) to determine the weight of each index (W).
see Table 1.

Table 1. Water drainage system safety evaluation index weight value

W_1	W_2	W_3	W_4	W_5	W_6	W_7	W_8	W_9
0.1081	0.0624	0.1513	0.1697	0.1249	0.1009	0.0729	0.1649	0.0448

4.3 Security evaluation model

We use the Delphi method and direct calculation method to evaluate the performance of the above mentioned safety evaluation index in the drainage system. According to the weight value of safety evaluation index in the system, the mathematical model for calculating the safety of drainage system is established:[9]

$$V = haW_1 + qaW_2 + daW_3 + raW_4 + iaW_5 + gaW_6 + OaW_7 + \alpha aW_8 + jaW_9 \quad (1)$$

The V in formula (1) represents security degree.

4.4 Determination of safety level

Data so in this study, we use the return period P and the safety degree V as the basic index, the water depth and the water time as the apparent index in view of the return period P which is the most direct indicator of urban storm water system .This article made urban rainwater safety rating scale (Table 2) based on the characteristics of China's city water-logging in recent years. The water depth, water time and the safety evaluation level of the rain water system may be obtained under different rainfall intensity according to the P value of the rainwater safety degree.[10] See Table 2.

Table 2. Safety assessment of urban storm water system

Safety degree V	Characterization description	system performance	Security level
[0.9,1.0]	P≥50 year,h≤0.10m,t<0.5h,	very good	Very high
[0.7,0.9]	50>P≥10 year,h≤0.15m,t<1.0h	good	high
[0.5,0.7]	10>P≥5 year,h≤0.20m,t<1.5h	secondary	commonly
[0.3,0.5]	5>P≥1year,h≤0.30m,t<2h	difference	low
<0.3	P<1year,h>0.30m,t≥2h	Poor	Very low

4.5 Drainage system of the Ganzhou old district in the Song dynasty and safety evaluation of storm-water system of the new district

According to the above mathematical model, we can calculate the token value of each index element and comprehensive safety evaluation value of drainage system of the old district and storm-water system of the new district. See Table 3

Table 3. Token value of drainage system of the Ganzhou old district in the Song dynasty and the new district

Serial number	index	weight	Drainage system of the Ganzhou old district in the Song dynasty		Zhangjiang new rainwater system	
			Token value	Comprehensive value	Token value	Comprehensive value
1	h.	0.1081	0.91	0.098	0.61	0.066
2	q.	0.0624	0.86	0.054	0.35	0.022
3	d.	0.1513	0.96	0.145	0.65	0.098

4	r.	0.1697	0.95	0.161	0.32	0.054
5	i.	0.1249	0.91	0.114	0.58	0.072
6	g.	0.1009	0.97	0.098	0.25	0.025
7	O.	0.0729	0.92	0.067	0.42	0.031
8	α.	0.1649	0.93	0.153	0.47	0.078
9	j.	0.0449	0.81	0.036	0.73	0.033
Total		1.0		0.927		0.479

By comprehensive evaluation, the safety degree of drainage system of the old district was 0.927 and the safety degree of the rainwater system in the new urban area was 0.479.The system performance of the old district is "very good" and the safety level is "very high", but the system performance of the Zhangjiang new urban rainwater system is "poor" and the level of safety is "low". We can get the interpretation of the mystery "Millennium no flood" of the old district and "the reality of frequent water-logging" of the new district.

5. The advice of rainwater system planning of the new district of Ganzhou

In order to ensure the safety of the rainwater system planning of the Zhangjiang new district, we need to optimize the existing planning from the following aspects based on the the digital results of Fushougou and safety assessment analysis:

(1) Increasing the design return period of drainage system of the new district of Ganzhou

At present,the new district of Zhangjiang often emerges water-logging phenomenon as drainage pipe network design return period of drainage system of the new district is one year.Therefore,the thesis proposed that the design return period should increase up to 30 years.

(2) Rational use of pipe materials

Reasonable selection of pipeline materials can ensure the effective long-term operation of the urban drainage system. Standard reinforced concrete structure was widely used in pipes of rainwater drainage system in Zhangjiang district. Without preservative treatment, the endurance of the reinforced concrete structure is about 50 years old, and culvert pipes with such structure can rarely be replaced. While the use of masonry structure is extensively available in drainage pipeline in Fushougou, the structure owns its merits: it is easy to repair and its economic endurance is longer than that of reinforced concrete structure. As a result we recommend that masonry structure should be applied to the rainwater system in Zhangjiang district of Ganzhou city.

(3) The determination the reasonable section form

At present, the main water pipe diameter is basically above 1000mm in the new district of Ganzhou, while the branch pipe diameter is 150-400mm.It is too small to collect and discharge.Therefore, this paper suggests that we should increase the diameter of the branch pipe.

(4) Rational use of paving materials

In the construction process of the new district of Ganzhou, an army of concrete structures covered the ground, which increase the pressure of urban drainage system by reducing the ground water permeability and increasing the urban rainwater runoff.

(5) Zoning plan according to terrain

The new district of Ganzhou which has 1833.27 hectares set up six water drainage area to drain away water, but merely four partitions were carried out at the old district which only has 2.7 square kilometers. Obviously the setting number of the water drainage area of the new district is too small. So, the new district rainwater system planning should be based on the actual situation of the gap between high and low terrain area and subdivises drainage area scientifically and reasonably.

(6) Increasing the outlet

The new district of Ganzhou which has 1833.27 hectares set up six water drainage area to drain away

The catchment area of the new district is 1800 hectares and the number of outlet is 4 at existing planning. While the old district is only 2.7 square kilometers, which has built 12 water outlet. So it is necessary to greatly increase the outlet to reduce the emission intensity of the existing water outlet.

(7) Strengthening the combination of water storage and drainage

Strengthening the combination of urban rainwater storage and drainage can enhance the ability of city water-logging. At present, the new district has built 1002 acres of Urban Central Ecological Park, of which the lake area is more than 600 acres. The lake is connected with Zhangjiang River through a 2 kilometers water Creek. The lake can play a role in flood storage and flood drainage by saving a lot of rainfalls and discharged into the Zhangjiang River through the river system. Reference Fushougou design concept, this paper aims to form ecological system which combines water storage and drainage by proposing a substantial increasement for the storage pond and connecting the pond to the urban underground rainwater system under construction data source.

Acknowledgements
This work is supported by science and technology projects from Jiangxi Provincial Office of Education Grant #GJJ10483.

References

[1] Shi G W, Wang Y T, Liu Y and Zheng W. Digital Conservation of Cultural Heritage Using Augmented Reality. *Journal of System Simulation*, 2009,21(7) :2090-2093.
[2] Peng Q Y, Xu H F and Lu F .Application of Geographic information System to Shao xing Urban Drainage Network. *China Water & Waste Water*, 2010, 26(18) : 13-17.
[3] Han G F and Huang Y R . Study on the Construction of Drainage System of Historical City in Perspective of Urban Safety: Case Study of Fushou Ditch, *Modern Urban Research*, 2013 ,(12): 72-76.
[4] Liu J H, Xia G K and Zhang Y M On the Research about the Application of Augmented Reality Technology in Cultural Heritages Conservation. *Journal of Xiangtan Universitg(Philosophy and social Sciences)*, 2015 ,(2): 124-126.
[5] The 15th General Assembly of ICOMOS,2005. XI'AN Declaration on the Conservation of the Setting of Heritage Structures, Sites and Areas[EB/OL]. Xi'an, CHINA.
[6] Chen J Z, Liu Y L, Lin N Y and Peng Q S . Photorealistic Visualization of Underground Pipelines to Enhance the Renlity. *Journal of computer-Aided Design& computer Graphics*, 2012, (9):1164-1170.
[7] Chang Y and He Z Y . Research on Underground Pipeline Augmented Reality System Based on ARToolKit. *Computer engineering and application*, 2005 ,(29):195-199.
[8] Du Q Y and Liu T . Design and Implementation of a Prototype Outdoor Augmented Reality GIS,*Geomatics and Information Science of Wuhan University*, 2007,(11):1046-1049.
[9] Chang Y and Xue L M. Research on Constructing Outdoor Augmented Reality System Based on Improved ARToolKit, *Journal of System Simulation*, 2009,21(8):2273-2276.
[10] Ren B , Li L J , Cao W M and Wang C .Interactive Urban Design Based on Augmented Reality, *J. of HUST. (Urban Science Edition)*, 2006, 23(2):32-34.

Historical ruins of remote sensing archaeology in arid desertified environment, northwestern China

N. K. Hu [a, *], **X. Li** [b]
[a] College of Tourism and Environment, Shaanxi Normal University, 710000 Xi'an, China
[b] Northwest Institute of Eco-Environment and Resources, Chinese Academy of Science, 730000 Lanzhou, China

Abstract. Silk Road is an important exchange channel for human communication and culture propagation between ancient China and the West during historical periods. A lot of human activities performed in Silk Road and many historical ruins leave behind to present. Archaeological ruins can play a significant role in studying and restoring the past human activities, as well as understanding regional environmental changes. There were many flourishingly human activities during different historical periods that were developed in ancient Juyan Oasis in the downstream of the Heihe River Basin. A large number of historical ruins that reflect past human activities preserved between numerous of the nebkhas and sand dunes. In this study, combined high-resolution remote sensing imageries with in situ truths investigated during the fieldwork, certain unknown ruins were identified according to the image features of historical ruins that appear in remotely sensed data, which were undiscovered during the previous field archaeological investigations and unreported in the past public literatures. Almost all of the newly discovered ruins that were identified using remote sensing images are distributed in the Lvcheng and BJ2008 surroundings. Newly findings supplement the missing gaps that were not taking into account during the previous field surveys.

1. Introduction

Archaeology is named to explore and study the relics and ruins that were left behind through a variety of past human activities, in order to research and understand the conditions of history and human activities that happened in those ancient society. Remote sensing archaeology is namely the use of modern remote sensing techniques to explore various marks that represent the human activities of past thousand years in the legacy of the Earth's surface from the air. Archaeology with remote sensing was firstly used to the study of airborne archaeological sites with balloon by British scientists in the early 20th century, and it was further attention and development in heritage conservation during the World War II and the post-war rebuilding and economic development (Nie, 2009). From the 1950s of the artificial satellite successfully launch to the early 21st century, the rapid development of a variety of satellites for earth imaging and information technology provide a new platform for space remote sensing archaeology. Plentiful of remote sensing technology for all types of archaeology provides a rich source of data and multiple research approaches, which play an important role in remote sensing archaeology at home and abroad (McCauley, 1982; Clark, 1998; Philip, 2002; Goossens, 2006; Rowlands, 2007). Compared with traditional field archaeology, remote sensing technology can be achieved to conduct research from whole to part, but also can supply information for some inaccessible or difficult to access regions. It will play a more important role in future archaeological due to its many advantages.

Silk Road is an essential communication channel of many human activities and cultural exchanges between ancient China and the West. Most part of the land Silk Road is covered with arid climate environment. As the most important water resources in arid region, inland river basin give birth to oases civilization, and where is the primary place for human activities and agricultural production from ancient to modern (Zhang, 2001; Fan, 2002; Xie, 2009; Qin, 2012). However, many oases, with good ecological environment and suitable for human survive in the past, had become to desert due to many factors, and a large number of ruins of ancient human activities stay. These archaeological ruins is the direct evidence and first-hand documents when carrying out the research on historical human activities and environmental change (Fan, 2002; Xie, 2009; Qin, 2012; Hou, 1964; Zhu, 1983; Li, 1990; Liu, 2010; Turner, 1974; Alizadeh, 2004; Yu, 2012). Because of the extreme climatic characteristics with drought and less precipitation and large evaporation, a large number of ruins still keep in the desert after thousands of years. These ruins is the most important data for the study on the historical human activities and the process of oasis environmental change.

Heihe River Basin is one of the major arid inland rivers of the western China. Ejin Oasis is the downstream for the modern river, where the principal place to serving human's lives and production. While the ancient Juyan Oasis is the river terminal during historical period, which is located southeast of the Ejin Oasis (Figure 1). Historical and archaeological discoveries indicated that there was lush vegetation and good environment in ancient Juyan Oasis. And there were also prosperity human activities during different historical periods. However, this oasis is completely desert, and plentiful of ruins and relics are the witness of the past human flourishing activities. It can do nothing for traditional field archaeology to some smaller targets hidden in countless sand dunes and nebkhas due to the oasis severe desertification. In addition, it is difficult for traditional archaeological field surveys to reach those areas that intensive sand dunes and nebkhas stopping the flow into the vehicle due to the serious sand erosion in the sub-region of the oasis. There are some advantages for remote sensing application in archaeology: observation from aerial avoiding the influence by terrain and topography, taking into account anywhere of the study area, and identifying the small ruins hidden in the sand dunes and nebkhas. Several of remote sensing images are comprehensively used to identify the historical ruins that were difficult to find in previous traditional field investigation. They enrich the archaeological information to a certain extent. It confirms that remote sensing technology can play a significant role in distinguish the small ruins in desertified environment. This study provides new quantitative evidence to further understanding of the relationship between human beings and nature in arid desert environment.

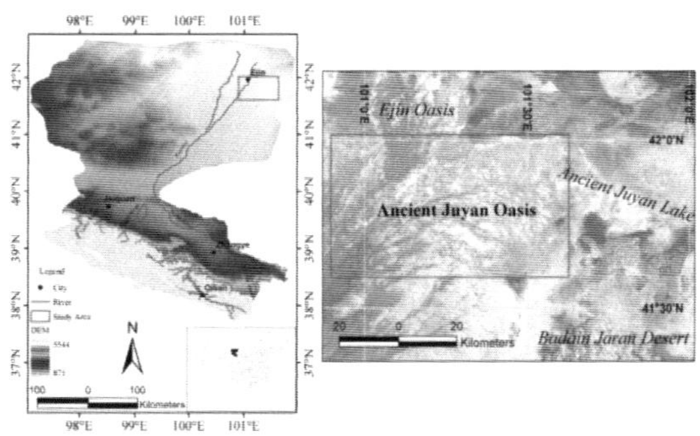

Figure 1. Geographical location of the study area

2. 2. MATERIALS AND METHODS

2.1 Data sources

Besides collecting the archaeological reports officially published of the study area, two kinds of remote sensing imageries with high resolution were also used in the present study. As the first space satellite reconnaissance program, CORONA was deployed in 1958. The camera, which was equipped with KH-4A, obtained early high-resolution images at a resolution of approximately 3 m for the covered parts of China from August 1963 to September 1969. As access to an earlier era, the historical ruins is preserved better due to their less influence by early environment. These images served as an earlier important data source in this study. The CORONA images from September 29, 1969 were ordered and acquired from the USGS for the study and were primarily used to identify the unknown ruins. Because of its black and white photo, the appearance for some ruins are not enough outstanding, which features are not obvious. GeoEye images are very high resolution satellite data, and they were obtained from the Google Earth, which is a virtual global map service that contains several types of satellite images with different spatial resolutions. In the current study, GeoEye images from Google Earth for the study area have a resolution of 0.5 m, and the acquisition dates were March 28, 2010 and April 28, 2010. These images were primarily used as supplement to distinguish and extract the locations of the historical ruins. GeoEye images become the most important source to discriminate the ruins from the desertified environment, which lie in their false color composition and higher spatial resolution. The Landsat-Enhanced Thematic Mapper Plus (ETM+) images from May 28, 2002 were acquired and incorporated into this study, which were used as background for mapping, and analysing the environmental changes in the study area.

Several remotely sensed data were used in the current study, and preprocessing was required for these images to distinguish the ruins more accurately. Geometric corrections to the images were carried out using ground control points (GCPs) from the field investigations. The GCPs includes selected important sites and obvious targets (e.g., crossing points of rivers or discovered ruins). Next, the mosaicked images from the CORONA and GeoEye data were obtained using image processing software that applied the multi-scene images. The images used for this study were clipped with the boundary of the study area. Image enhancement was the last operation executed in order to improve the visibility of archaeological features on satellite data. Standard deviations and histogram equalizations were calculated for all of the images to highlight the image features of the ruins, particularly for the early black and white satellite images from CORONA..

2.2 Methodology

Archaeological documents, remote sensing data and field investigation are comprehensively used to distinguish the historical ruins in desertified environment. Previous archaeological reporters are primarily used to confirm and analyze the spatial distribution of the discovered ruins during the past surveys, which can help determining the target area where unknown ruins probably existing in the current oasis. The unknown ruins are primarily identified and extracted by indoor remote sensing data. Field investigation is for validating the accuracy of the interpreted results by indoor remotely sensed imageries, and obtaining other relevant information about the unknown ruins.

Historical ruins almost belong to artificial structures, which have fixed shapes and contour features. There are different conditions for the historical ruins in the desert: some are completely covered by sand dunes or nebkhas, while others expose to wind eroded land and surround by sand dunes and nebkhas. All the above mentioned information in remote sensing images are performed with regular geometric texture features, e.g. square or polygon. Additionally, shadow features will perform for some building with walls. These geometric textures and shadow features are the most important basis for the archaeological application with remote sensing data.

After analyzing the spatial distribution of the discovered ruins, we determined the approximate range of unknown ruins, where key areas are to conduct remote sensing archaeology. According to the typical imagery features of the historical ruins, we identify and extract the possible ruins by visual interpret. Then, we carry out the field investigation to validate the indoor interpreted results and observe the environment of the ruin surroundings.

3. Results and discussion

3.1 Spatial distribution of historical ruins

3.1.1 Distribution of discovered ruins

There were many outdoor adventures and archaeological investigations in Juyan Oasis during latest one hundred years since the 20th century to the 21st century, and a large number of historical ruins of human activities had been found (Figure2).

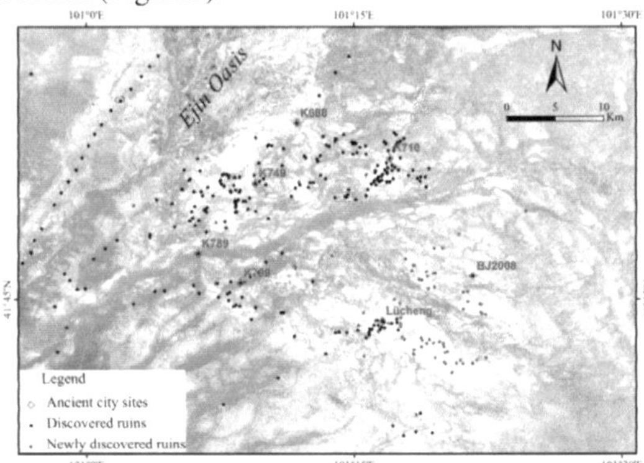

Figure 2. Spatial distribution of the historical ruins

As seen from Figure 2, mostly of the historical ruins that were discovered previously are mostly distributed in the northern region of the oasis, and some of them are distributed in the southern region.

In addition, a small number of the ruins are dispersed around the Gobi desert. In the view of spatial distribution of the surrounding ruins for each important city site, there are a large number of ruins distributed around the city sites of the northern oasis, such as K710, K688, and K749, and many ruins are dispersed around the city sites of K799 and Lvcheng that located in the southern oasis. However, the historical ruins around the Lvcheng are mostly distributed in the west, while there is nothing in the eastern and southern part of the city. And, there is also no ruins found around the BJ2008 site. Generally, a large number of small sites should distribute around the important city site because that many human activities performed around the large city site during the historical periods that were dominated with agricultural events. According to the archaeological surveys, BJ2008 site was built in the Han Dynasty, which is one of the largest city sites in the present discoveries (Sohma, 2009). It is an important area for Lvcheng site to have inhabited human activities at multiple historical stages, and there was prosperity agricultural settlements (Li, 1991; Li, 2003). According to the above mentioned evidence, we confirm that some unknown ruins have still undiscovered around BJ2008 and Lvcheng sites, in where the interested sub-area to conduct archaeological application with remote sensing.

3.1.2 Spatial pattern of newly discovered ruins

As is seen from Figure 2 about the spatial pattern of the newly discovered ruins, most of the ruins that were identified by remote sensing data are distributed in the northeastern part of the Lvcheng and BJ2008 surroundings, which both located in the south oasis. Furthermore, a few of newly discovered ruins are scattered around the K799 site and other region of the northern part of the oasis. The newly discovered ruins by remote sensing data supplement the spatial distribution of the missing remnants that were not discovered during those previous traditional archaeological surveys. These ruins ensure that many human activities are distributed around each important city sites. In particular, they update the missing information of the historical human activities around the Lvcheng and BJ2008 sites.

3.2 Field investigation and validation

Desertification results in complex surface shapes and complicated landforms. Image information appeared in remote sensing data is incomplete due to dilapidated ruins by sand dunes and nebkhas. Thus, the distinguished results based on imagery features need to be validated through field and in site observation in order to determine the accuracy and reliability of the newly discovered ruins. In addition, some apparent characteristics of the detected ruins can be displayed on remote sensing images, e.g. shape, texture, shadow, size, etc. But we do not know other attributes by remote sensing data, such as building materials, construction mode, ground relics and current conditions, which need to be obtained from field investigation.

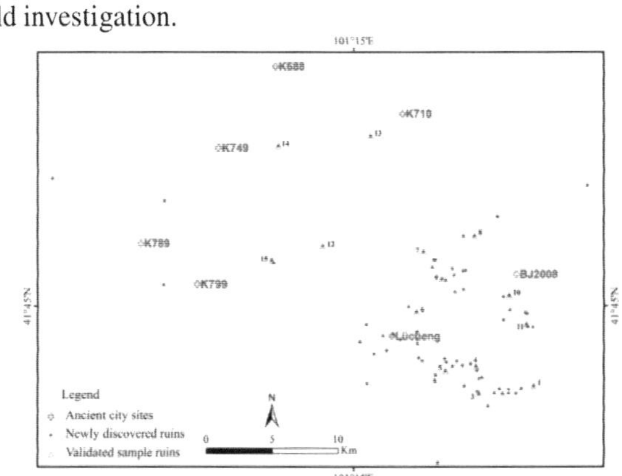

Figure 3. Spatial distribution of the validated sample ruins

In this study, 15 samples of the newly discovered ruins are selected to conduct the field investigation and validation (Figure 3). By detailed comparison with the results of the in site truths, we found all the newly discovered ruins using remotely sensed data are built by man-made structures. Furthermore, according to the building materials and structured skills that were investigated during the fieldwork, as well as the ground relics surroundings, it is speculated that almost all of the newly discovered ruins were primarily built and used in the Western Xia and Yuan Dynasties (A.D. 1028-1375), while only a few of them should be the Han Dynasty.

There are different construction methods and area as different types and usage for the historical ruins. The current conditions of the ruins vary with their different locations and environments. According to the image features exhibited in remote sensing images, current size, ruin types, and site surroundings, 3 samples from the field investigated ruins are selected to show and compare their image features with in site characteristics (Figure 4).

Figure 4. Image features and in site truths of the ruins

4. Conclusion

In this study, some small historical ruins that were not discovered previously were identified and interpreted using remote sensing data in ancient Juyan Oasis. About 70 historical ruins were newly distinguished by remotely sensed data, and almost all of them are distributed in the Lvcheng and BJ2008 city sites of southern part of the oasis. These newly discovered ruins were primarily built and used in the Western Xia and Yuan Dynasties according to their constructed methods investigated during the field surveys, expect for a few of them are the Han Dynasty. Additionally, the newly discovered ruins supplement the missing relic information around the Lvcheng and BJ2008 city sites. Furthermore, newly discovered ruins greatly enrich the archaeological information that represent the ancient human activities in Juyan Oasis. It further proved that remote sensing technology can play a significant role in identifying small targets in desertified environment. In future, microwave remote sensing data with high resolution is expected to play a more important role to explore the historical ruins in the desertified environment.

Acknowledgements

Acknowledgements of support for the Fundamental Research Funds for the Central Universities (No. 1301020479).

References

Alizadeh, A., Kouchoukos, N., Wilkinson, T. J., Bauer, A. M., Mashkour, M., 2004. Human-environmental Interactions on the Upper Khuzestan Plains, Southwest Iran, Recent Investigations. Paleorient, 30(1), pp. 69-88.

Clark, C. D., Garrod, S. M., Parker, P. M., 1998. Landscape Archaeology and Remote Sensing in Southern Madagascar. International Journal of Remote Sensing, 19(8), pp. 1461-1477.

Fan, Z. L., Mu, G. J., Ma, Y. J., Ma, Y. J., 2002. Formation and Development of the Artificially Irrigated Oases in the Northern Piedmont of the Tianshan Mountains. Scientia Geographica

Sinica, 22(2), pp. 184-189.

Giardino, M. J., 2011. A History of NASA Remote Sensing Contributions to Archaeology. Journal of Archaeological Science, 38(9), pp. 2003-2009.

Goossens, R., Wulf, A. D., Bourgeois, J., Gheyle, W., Willems, T., 2006. Satellite Imagery and Archaeology: the Example of CORONA in the Altai Mountains. Journal of Archaeological Science, 33(6), pp. 745-755.

Hou, R. Z., 1964. Desertification Changes of Eastern River from Remnants of Human Activities in Ningxia. Chinese Science Bulletin, 3, pp. 226-230.

Li, Y. Y., 1991. Manuscripts Unearthed in Khara Khoto (in Chinese Volume). Beijing, Science Press, pp. 18-19.

Li, B. C., 1990. Investigation on the Ancient Ruins in the Western Sandy Land of Minqin County. Journal of Desert Research, 10(2), pp. 35-42.

Li, B. C., 2003. Study of the Desertification in Hexi Corridor in Historical Period. Beijing, Science Press, pp. 32-47.

Liu, W., Cao, S. K., Xi, H. Y., Feng, Q., 2010. Land Use History and Status of Land Desertification in the Heihe River Basin. Natural Hazards, 53(2), pp. 273-290.

McCauley, J. F., Schaber, G. G., Breed, C. S., Grolier, M. J., Haynes, C. V., Elachi, C., Blom, R., 1982. Subsurface Valleys and Geoarchaeology of the Eastern Sahara Revealed by Shuttle Radar. Science, 218, pp. 1004-1020.

Nie, Y. P., Yang, L., 2009. Applications and Development of Archaeological Remote Sensing Technology in China. Journal of Remote Sensing, 13(5), pp. 940-951.

Philip, G., Donoghue, D. N. M., Beck, A. R., Galiatsatos, N., 2002. CORONA Satellite Photography: an Archaeological Application from the Middle East. Antiquity, 76(291), pp. 109-118.

Qin, X. G., Liu, J. Q., Jia, H. J., Lu, H. Y., Xia, X. C., Zhou, L. P., Mu, G. J., Xu, Q. H., Jiao, Y. X., 2012. New Evidence of Agricultural Activity and Environmental Change Associated with the Ancient Loulan Kingdom, China, around 1500 Years Ago. Holocene, 22(1), pp. 53-61.

Rowlands, A., Sarris, A., 2007. Detection to Exposed and Subsurface Archaeological Remains Using Multi-sensor Remote Sensing. Journal of Archaeological Science, 34(5), pp. 795-803.

Sohma, H., Tian, R., Wei, J., Moriya, K., Iguro, S., Ito, T., 2009. Unrecognized Ruins Interpreted from the High-resolution Satellite Images and Their Significance, in Case of the Lower Reaches of the Heihe River, Inner Mongolia, China. The Association of Japanese Geographers, 7, pp. 106.

Turner, B. L., 1974. Prehistoric Intensive Agriculture in the Mayan Lowlands. Science, 185(4146), pp. 118-124.

Xie, Y. W., Chen, F. H., Qi, J. G., 2009. Past Desertification Processes of Minqin Oasis in Arid China. International Journal of Sustainable Development and World Ecology, 16(4), pp. 260-269.

Yu, Y. Y., Guo, Z. T., Wu, H. B., Finke, P. A., 2012. Reconstructing Prehistoric Land Use Change from Archaeological Data: Validation and Application of a New Model in Yiluo Valley, Northern China. Agriculture. Ecosystems and Environment, 156(1), pp. 99-107.

Zhu, Z. D., Liu, S., Gao, Q. Z., Hu Z. Y., Yang, Y. L., 1983. The Environmental Changes and Desertification Processes in Historical Period in the Areas of Ancient Juyan-Heicheng Region in Western Inner Mongolia. Journal of Desert Research, 3(2), pp. 1-8.

IOP Conf. Series: Earth and Environmental Science 57 (2017) 012029 doi:10.1088/1755-1315/57/1/012029

Regional specific features and temporal evolution of the quasi-biennial oscillation of total ozone (TO) in the atmosphere

M D Orozaliev[1] **and A S Abdyldaev**[1]

[1]Institute of Innovative Professions, Kyrgyz State University of Construction, Transport and Architecture named after N Isanov, Bishkek, Kyrgyzstan.

E-mail: omdi1952@mail.ru

Abstract. This work presents results of the research on features of quasi-biennial oscillation (QBO) of total ozone content over the North Tien-Shan based on 35 year old (1980-2015) experimental database of Issyk-Kul high mountain monitoring station (lat. 42.62; long. 76.98; 1640 m.a.s.l.). It has been established that regional specific features and temporal evolution of QBO in total ozone (TO) are largely defined by interrelation - synchronizations of quasi-biennial and annual (seasonal) variations.

1. Introduction

The atmospheric ozone not only protects the biosphere from harmful Sun's ultraviolet radiation, but also plays one of the major roles in energy exchange processes between the different parts of the atmosphere. However, inter influence of atmospheric ozone evolution and global and regional climate changes in different time scales remains as one of the most important and not fully understood problems of atmosphere physics, meteorology and climatology. That is why experimental study of spatiotemporal ozone variations has scientific and applied value.

Monitoring of ozone and main greenhouse gases in Central part of the Eurasian continent has been conducted for more than 35 years at Issyk-Kul Station, located on the coast of the mountain lake Issyk-Kul (lat. 42.62; long.76.98; 1640 m.a.s.l.). There were received temporal series of TO, having unique duration for this region, which significantly complimented the measurement results of the monitoring network managed by the World Meteorological Organization (WMO).

The ozone content measurements are performed with the help of a spectrophotometer scanning set (SPS). Determination of a systematic measurements error and fitting TO values to the world ozonometric scale were conducted by simultaneous verification of TO measurements on SPS and spectrophotometer Dobson No. 108, located at Main Geophysical Observatory named after Voeikov A I, which is the national standard of Russia, as well as with SPS and spectrophotometer Brewer No. 44 (SPA "Typhoon"). The deviations in measurement results, obtained by the SPS, do not exceed 2 % from the world ozonometric scale. Comparative analysis of TO measurements at "Issyk-Kul" station and satellite measurements showed that the average relative deviation of TO between "Issyk-Kul" station and SBUV/SBUV2 satellite data for the period 1980-2013 made less than 1%.

Detailed description of measurement methods, set of measurement devices used at the research station Issyk-Kul, measurement results processing, analysis methods and detailed bibliography list on this subject are given in the works [1, 2, 3, 4].

Research in quasi-biennial oscillation (QBO) of total ozone were carried out by many scientists, particularly in works [5, 6]. This work contains the results of the research on features of quasi-biennial oscillation (QBO) development of total ozone over the North Tien-Shan based on 35 year old (1980-2015) experimental database of Issyk-Kul high mountain monitoring station (ISK 347). The interest in this research is due to the fact that powerful mountain ecosystems of Middle and Central Asia (the

Himalayas, Tien-Shan, Pamir-Alai, the Tibetan Plateau), making significant influence on the high-altitude jet streams, play an important role in spatial and temporal distribution of total ozone in the region [4].

Due to the large volume of material presented in the article, informative and most significant results are shown in the form of graphs.

2. Results of total ozone content measurements at Issyk-Kul Station for the period 1980-2015

Figure1(a) shows the variations of average monthly TO values (curve 1) received from the data measured at "Issyk-Kul" Station (ISK 347). It also depicts the smoothed curve 2, that reflects inter annual TO variations, as well as the parabolic trend (curve 3):

$$(X(t) = 0.00044319 * t^2 - 0.249 * t + 347.09)$$

Figure 1(b) presents deviations of average monthly TO values (columns 1), expressed as a percentage to the average annual cycle of TO for year 1980. Curve "smooth" represents inter annual variations of the TO deviations, curve 3 is a polynomial trend of the second order, showing the TO evolution during the monitoring and coming periods. It shows that the depletion in the ozone layer was 11% on average during 1980-2005 and in the last decade its recovery has been observed. The data analysis revealed that the rates of the ozone layer depletion and recovery have seasonal dependence. It was established that the highest ozone depletion rates V= -1.79 DU/year (on linear trend for 1980-2005) and the highest recovery rates V= 5.13 DU/year (2005-2015) were observed in May.

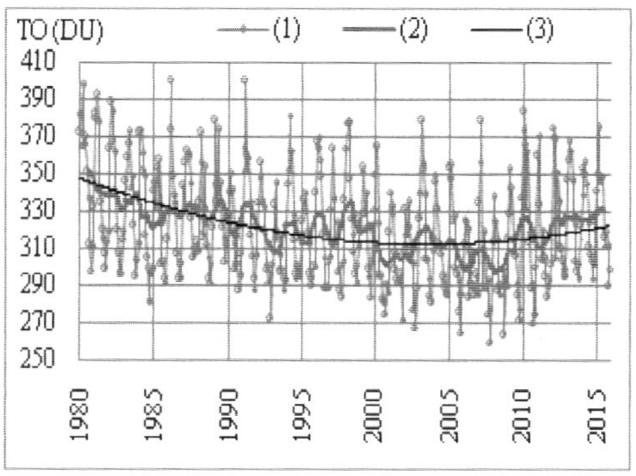

(a). Average monthly (1), inter-annual (2) TO variations, parabolic trend (3)

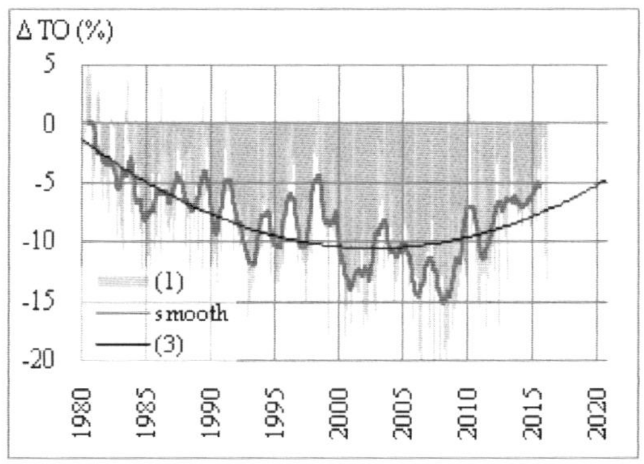

(b). TO value deviations in % (1) relatively to TO values in 1980 and smoothed(smooth) and parabolic trend (3)

Figure 1. Temporal variations of total ozone content according to data measured at Issyk-Kul station (ISK 347).

Figure 2 shows the variations of inter-annual (1) and average annual (2) TO values with linear (3) and parabolic trends (4). The slopes of the linear trends ($X = X_0 + b \cdot t$) reflect the rates of the ozone layer depletion and recovery at different time intervals: for 1980 - 1985, value of b = -0.4 DU/mon.; 1984 - 1990, b = 0.18 DU/mon.; 1989 - 1994, b = -0.26 DU /mon.; 1993 - 1998, b = 0.14 DU /mon.; 1997 - 2006, b = -0.16 DU/mon.; 2006 - 2015, b = 0.29 DU/mon.

Note that in the observed long time TO variations (figure 2), there are deposits of natural (solar activity cycles, gravitational tidal forces, volcanism, the flux of galactic cosmic rays, etc.) and anthropogenic factors (Montreal protocol acts 1987), differentiation of which is a complex urgent problem that requires thorough research

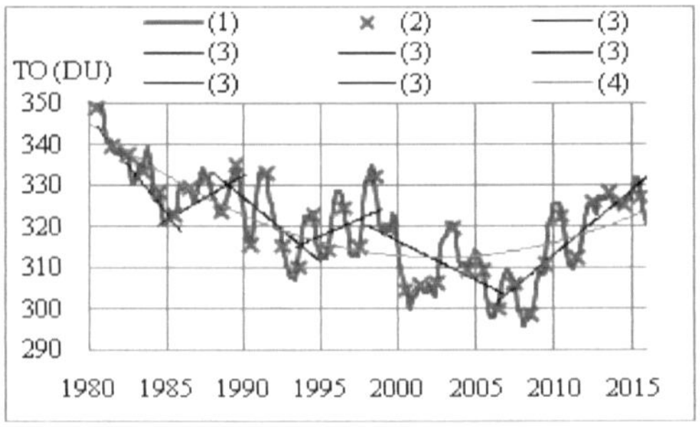

Figure 2. Variations of inter-annual (1) and average annual (2) TO values with linear trends(3) and parabolic trend(4)

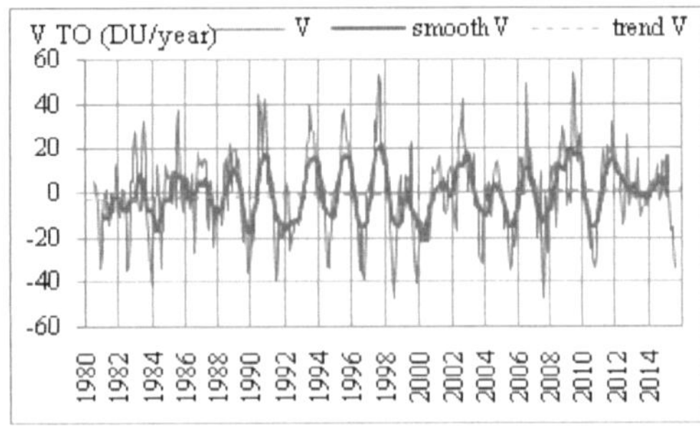

Figure3. Fluctuation of TO change rate (V) and its smooth (smooth V) for 13 months.

Figure 3 presents fluctuations of the TO change rate (V), obtained as a derivative from the inter-annual TO variations (figure 2). The linear trend (figure 3) shows that the TO change rate had negative values till 2002-2003, i.e. depletion of the TO was observed, and further it has positive values, i.e. the TO has been recovering.

Figures 1, 2, 3 show that TO change is a complicated oscillatory process with distinct quasi-biennial oscillation. Oscillation amplitude of TO change rates (figure 3) varies within wide limits, and increase in magnitude of the amplitudes is observed at the beginning and its reduction at the end of the observation period (1980-2015) like the beating of amplitudes.

3. Methods and results of analysis of quasi-biennial oscillation of total ozone (TO)

To study the quasi-biennial oscillation of total ozone, the following methods were used in this work:

1. The average annual cycle norm of TO (figure 4a) for the period 1980-2013, caused by the orbital motion of the Earth around the Sun, was subtracted from the temporal series of average monthly TO (figure 1). To improve the accuracy, unlike previously used method [1, 2, 3] in determining the "norm" of TO, instead of the linear trend, we subtracted a non-linear trend (inter-annual variations – figure 2) of TO.

2. The cleaning of temporal series of TO from long periodic oscillations with periods of T> 36 months was carried out in order to suppress influence of helio-geophysical and other processes, that have oscillations T more than three years, on QBO TO.

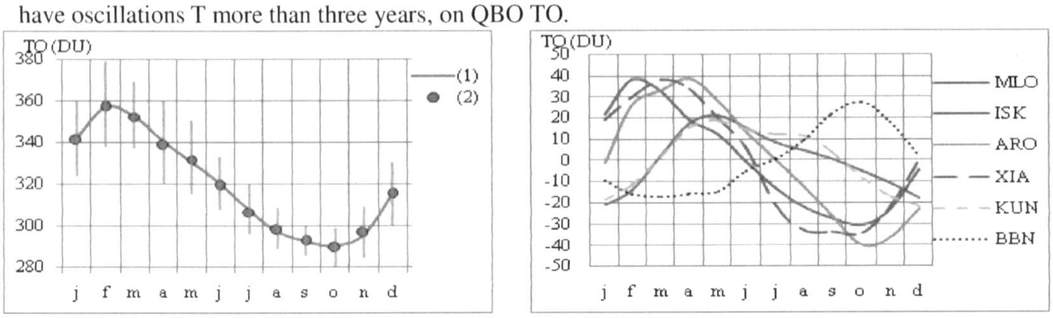

(a). TO norm for ISK 347 (2) versus the model (1)

(b). The relative norms: TO ISK 347; MLO 3; ARO 35; XIA 208; KUN 209; BBN 27.

Figure 4. (a) The TO norm (2) with standard deviations versus model (1) for ISK 347; (b) Comparison of relative norms of TO ISK 347, MLO 31, ARO 35, XIA 208, KUN209, BBN 27.

Figure 4(a) presents average annual (1980-2013) variations-norm of TO (2) in comparison to the model (1), received by us as a result of spectral Fourier analysis:

$$X(t) = \sum A_i \cos(\omega_i t + \varphi_i) + X_{mean} \qquad (1)$$

where: $\omega_i = 2\pi/T_i$; $T_1=12$, $A_1=31.57$, $\varphi_1=1.05$; $T_2=6$, $A_2=7.10$, $\varphi_2=0,86$; $T_3=4$, $A_3=2.95$, $\varphi_3=0.97$; X_{mean} = 320.81DU; T, A, φ correspondingly months, DU and rad.

Figure 4 presents the comparison of relative norms of TO, received with above mentioned method for the observation stations: ISK 347, MLO 31, ARO 35, XIA 208, KUN209, BBN 27 (table 1) (http://www.woudc.org/).These data further will be used for analysis of regional features of QBO of TO.

Table 1

Platform name	Platform id	Latitude	Longitude	Elevation (m)	Country	WMO region
ISSYK-KUL (ISK)	347	42.62	76.98	1640	KGZ	II
XIANGHE (XIA)	208	39.88	116.65	62	CHN	II
KUNMING (KUN)	209	25.03	102.68	1917	CHN	II
MAUNA LOA (MLO)	31	19.53	-155.57	3405	USA	V
AROSA (ARO)	35	46.78	9.68	1840	CHE	VI
BRISBANE (BBN)	27	-27.42	153.12	3	AUS	V

Figure 5 presents averaged (smoothed) values (remainders) of TO for three and thirteen months, received after subtraction of the annual cycle (norm) and non-linear trend (correspondingly curves 3 and 13). In the same figure, in order to demonstrate the features of interrelation between quasi-biennial oscillation of TO and quasi-biennial oscillation of equatorial zonal wind, the diagrams of temporal TO variations, zonal wind velocities V (QBO ZW) at the level of 20 hPa (13) (www.fu-berlin.de), as well as of temporal variation of solar activity index (F 10.7 – radio waves length 10.7 cm) were made (http://www.ngds.noaa.gov/stp/solar/flux.html/).

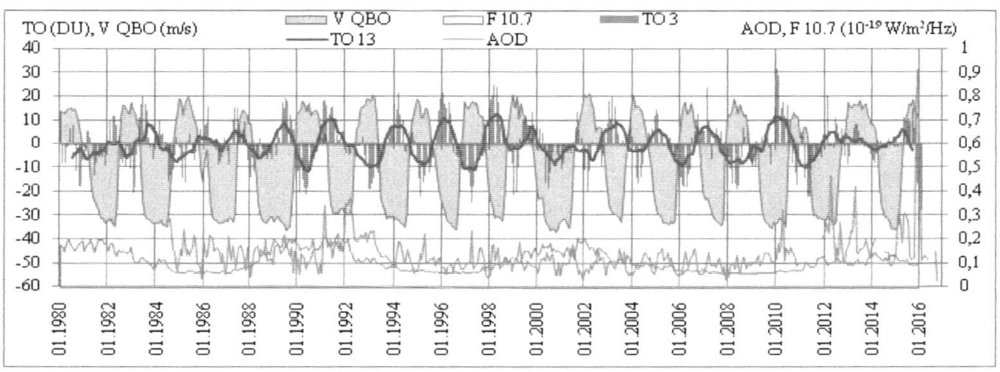

Figure 5. Correlation of quasi-biennial oscillation of TO (smoothed in three months -TO 3 and in thirteen months -TO 13) in DU with quasi-biennial oscillation of velocity (m/c) of Equatorial Zonal Wind (QBO) at the level of 20 hPa, with Solar Activity (Index F 10.7)and aerosol optical depth (AOD).

Correlation coefficient value, determined for QBO TO and QBO ZW, are equal to r= - 0.51 (for smooth 13, figure 5). The negative value of r means that when ZW changes its direction from western to eastern, TO increases and when oppositely, TO decreases. Changes in determined values of correlation coefficients between QBO TO (ISK) and QBO ZW year by year are presented in figure 8(a).

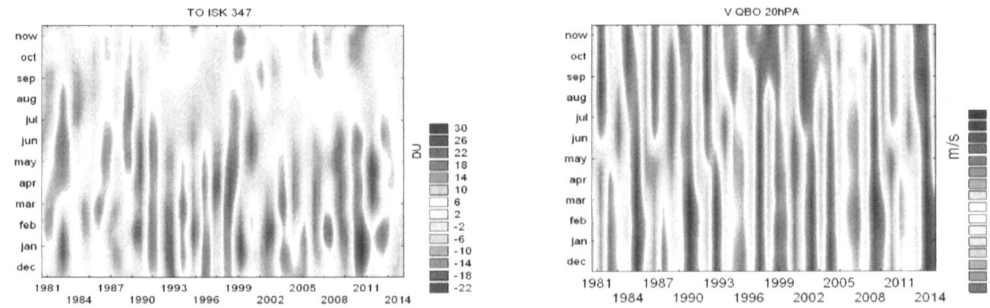

(a) TO ISK 347 **(b)** V QBO

Figure 6. Dynamics of seasonal dependence: a) QBO TOISK 347; b) QBO ZW

In figure 6, there are diagrams showing the dynamics of seasonal dependence of QBO TO (a) and QBO ZW (b). It can be seen in the figure that the effect of QBO ZW on QBO TO in winter and spring period is more significant than in summer and autumn period.

Analysis of TO and ZW correlation in comparison with the solar activity data of NOAA (http://www.ngds.noaa.gov/stp/solar/flux.html) and measured data on atmosphere's aerosol optic depth at Issyk-Kul station [7] (http://aeronet.gsfc.nasa.gov/) showed that pattern of interrelationship between TO and ZW gets destroyed in the years when aerosol content in the atmosphere is high (volcanic activity), as well as during the periods of magnetic polarity change of the Sun and its flare activity (up to r sign change). Excluding such periods (1991, 2000 and 2013) from the data processing, the relationship ratio increases and r reaches -0.68. Moreover, anomalies in correlation of TO and QBO were observed within the interval of 1986-1988 (figure 5), the reason of which can be related to the explosion of the nearest supernova (SN 1987A). Exclusion of this anomaly from time series results in increase of r up to - 0.78. It must be noted that such an anomaly of r for this time period was observed at other monitoring stations too (figures 5, 6, 7, 8).

3.1. Comparative analysis.

The data of Xianghe (XIA 208), Kunming (KUN 209), Arosa (ARO 35), Mauna Loa (MLO 31) and Brisbane (BBN 27) stations (http://www.woudc.org/data/products/totalozone/), which are located at different latitudes and longitudes, were processed using above mentioned method (table 1).

Figure 7 presents peculiarities of quasi-biennial oscillation of total ozone according to the data of stations ISK (347), MLO (31), ARO (35), XIA (208), KUN (209),BBN (27) (DU), quasi-biennial oscillations of Zonal Wind velocity (V QBO) at a level of 20 hPa (m/s) and solar activity (F 10.7 cm) for the period of 1980-2015. One can see that QBO TO observed at various stations differ by values of oscillation parameters and reaction to solar activity.

Figure 8, as an example, presents distribution of correlation coefficients for V QBO (20 hPa) and QBO TO arranged by the years for the stations ISK 347, MLO 31, ARO 35, XIA 208. Arithmetic average correlation coefficients of TO with V QBO made: a) <r> = - 0.6 for MLO 31; b)<r> = - 0.39 for ARO 35; c)<r> = - 0.52 for XIA 208; d) <r>= - 0.57 for ISK 347.

Thus, presented analysis results show significant correlation of QBO in total ozone with quasi-biennial oscillations of zonal wind (ZW) average velocity in the equatorial stratosphere, as well as peculiarities of narrowness and dynamics manifestation of this relationship of QBO in total ozone depending on Solar variability and latitude and longitude coordinates of monitoring stations. Relationship patterns of TO and ZW get destroyed in the years when aerosol components in the atmosphere have high values (volcanic activity), as well as during the periods of magnetic polarity change of the Sun and its flare activity, up to r sign change.

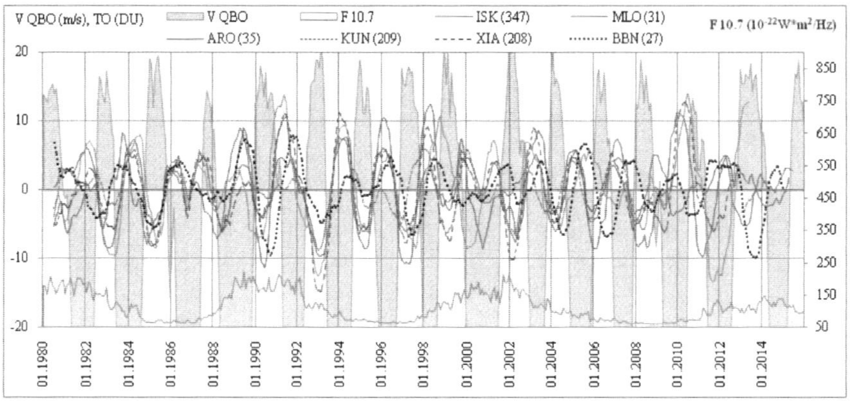

Figure 7. Comparison of quasi-biennial oscillations of TO of the stations ISK 347, MLO 31, ARO 35, XIA 208, KUN 209, BBN 27 in DU with quasi-biennial oscillations of the zonal wind velocity (m/s) (V QBO) at a level of 20 hPa.

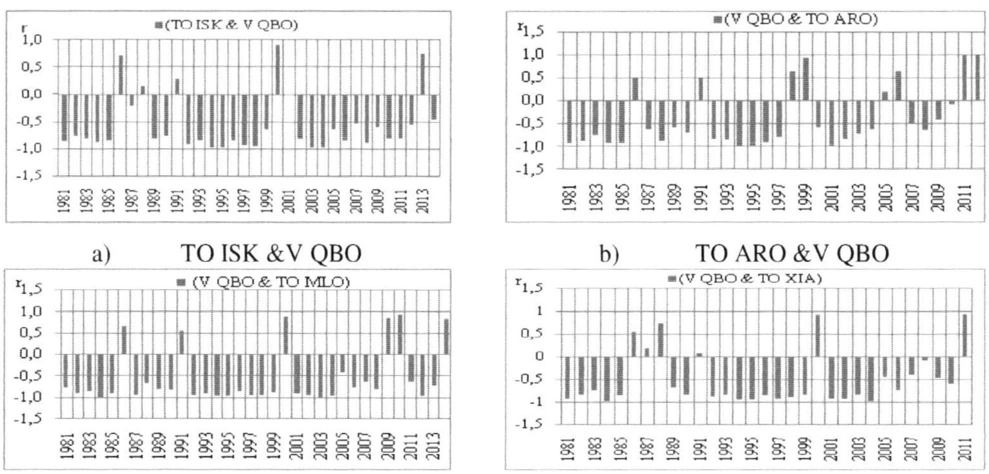

a) TO ISK &V QBO

b) TO ARO &V QBO

c) TO MLO & V QBO

d) TO XIA & V QBO

Figure 8. Distribution of correlation coefficient values for V QBO (20 hPa (m/c)) and QBO in total ozone arranged by the years: (a) ISK 347 (<r>= -0.57); (b) ARO 35 (<r>= -0.39); (c) MLO 31 (<r>= -0.6); (d) XIA 208 (<r>= -0.52).

4. Regional specific feature of quasi-biennial oscillation of total ozone (QBO TO)

Features of significant harmonics of QBO TO and QBO ZW components were found out by using Fourier and wavelet analysis for time series of TO and ZW. The relation of periods, amplitudes and phases of significant harmonics of the analyzed oscillations are presented in figure 9. The comparison of harmonics parameters of QBO TO and QBO ZW for Issyk-Kul station revealed significant harmonics (main) in fluctuation of QBO TO with periods of T_m= 34.0; 31.4; 29.1; 27.2; 25.5; 24.0 (months), where T values match harmonics periods of quasi-biennial oscillations of zonal wind velocity (QBO ZW). Moreover, QBO TO also had "additional" significant harmonics with periods of T_{ad}= 21.47; 20.40; 19.43; 18.55; 22.60 (months).

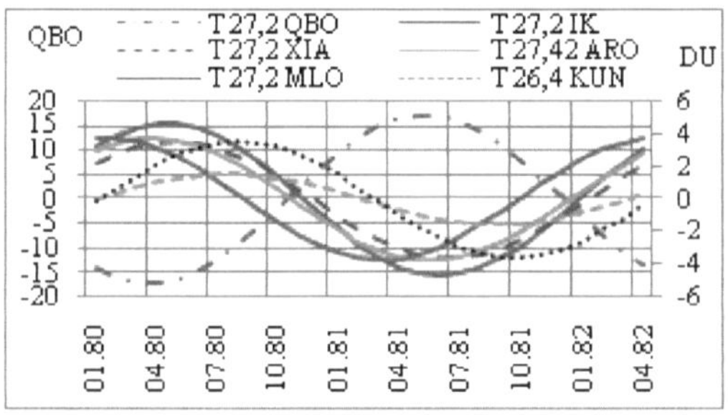

a)T 27.2

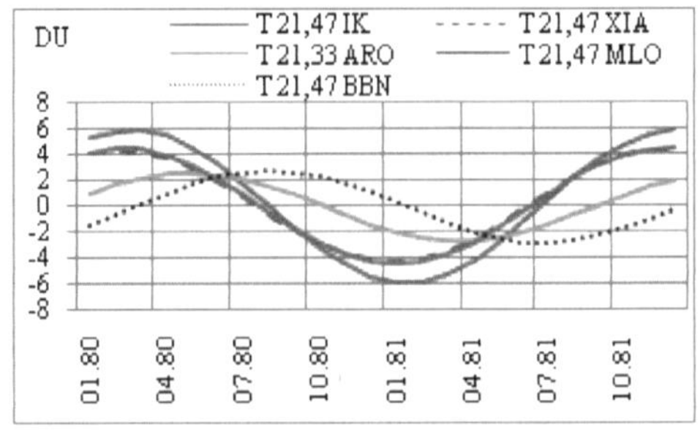

b)T 21.47

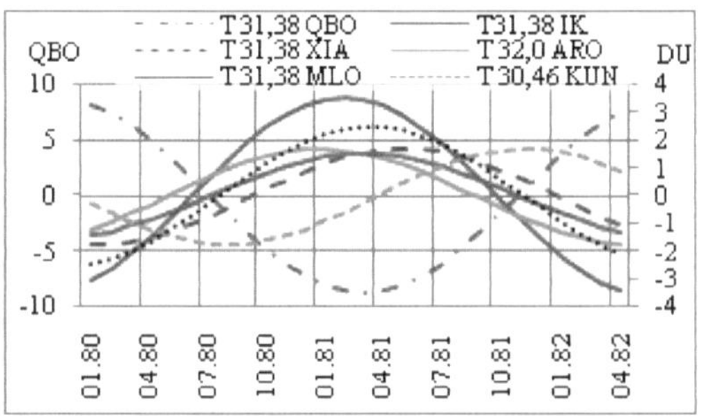

c)T 31.38

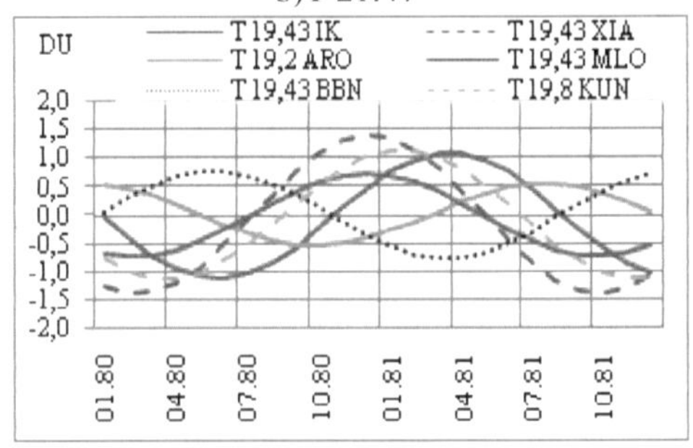

d)T 19.43

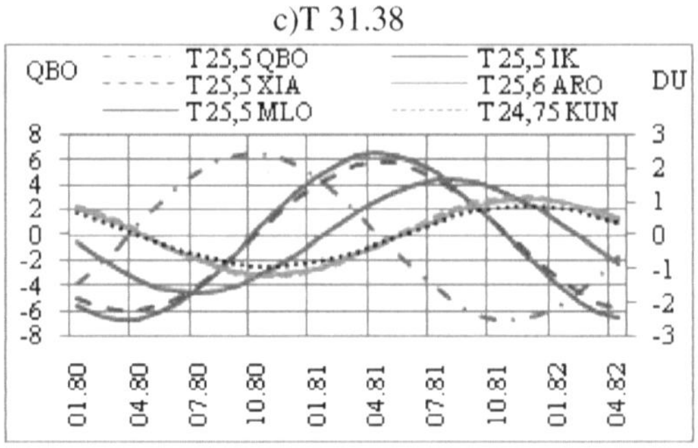

e)T 25.5

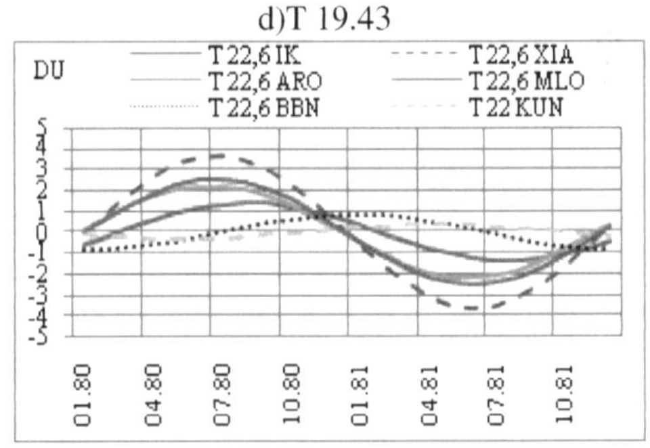

f)T 22.6

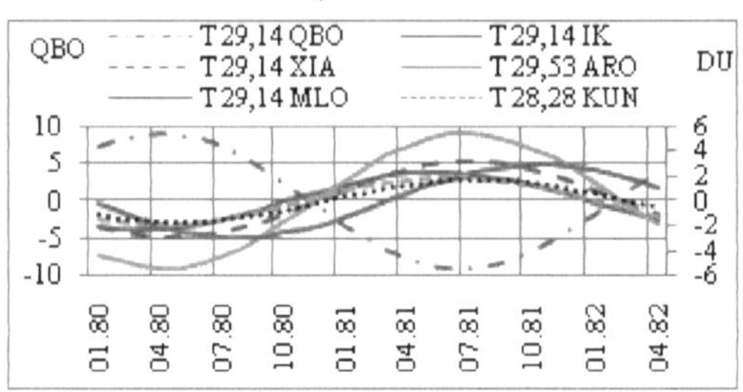

g)T 29.14

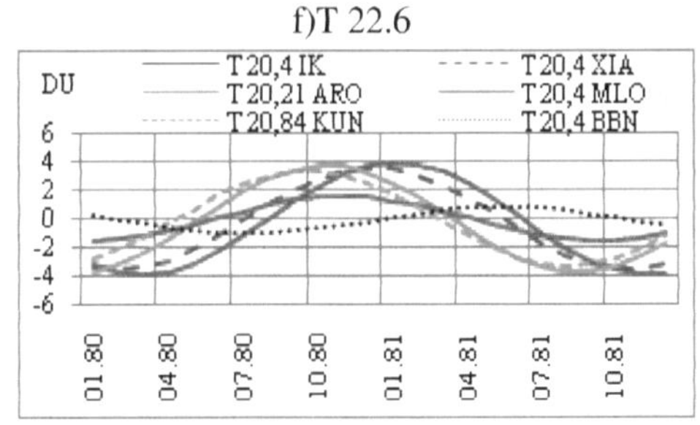

h)T 20.4

Figure 9. Relationship of QBO TO harmonics' amplitudes and phases with periods, coinciding with the harmonic periods of ZW: a) T = 27.2; c) T 31.38; e) T= 25.54; g) T= 29.17; and "additional" harmonics of QBO TO with periods of b) T = 21.47; d) T 19.43; f) T = 22.6; h) T= 20.4

Analysis of received data gives grounds to assume that the "additional" harmonic arises as a result of mutual synchronization of the "main" harmonics in QBO with annual harmonics (12 months). Then, the frequency of the resulting oscillation at mutual synchronization of two oscillations, as known [8], can be determined using equation: $\frac{1}{T_T} = \left(\frac{1}{T_1} - \frac{1}{T_2}\right)$

For our case: $\frac{1}{T} = \left(\frac{1}{12} - \frac{1}{27.2}\right) = \frac{1}{21.47}$ $\frac{1}{T} = \left(\frac{1}{12} - \frac{1}{29.14}\right) = \frac{1}{20.40}$

$$\frac{1}{T} = \left(\frac{1}{12} - \frac{1}{31.38}\right) = \frac{1}{19.43} \quad \frac{1}{T} = \left(\frac{1}{12} - \frac{1}{34.0}\right) = \frac{1}{18.55} \quad \frac{1}{T} = \left(\frac{1}{12} - \frac{1}{25.5}\right) = \frac{1}{22.60}$$

Comparison of the above calculated T_T with the periods of T_{ad}, obtained from the spectral analysis of experimental data, showed that they match to the third decimal point. Such a good match between the theoretically calculated T_T and T_{ad} values, determined from experimental data of Issyk-Kul station, confirms our assumption about the nature of "additional" harmonics, resulting from mutual synchronization of QBO with the annual cycle.

As one would expect, "additional" harmonics in QBO TO with similar values of T periods, but with different values of the amplitudes (power) and phases were identified for other monitoring stations too: Xianghe (XIA 208), Kunming (KUN 209), Arosa (ARO 35), Mauna-Loa (MLO 31). These differences, obviously, are due to the fact that TO annual cycle (figure 4(b)), as well known [9], has a longitude and latitude dependence, which is caused by the orbital motion of the Earth around the Sun.

The analysis of QBO TO at other stations, including Brisbane (BBN 27) station, located in the southern hemisphere, has also shown that regional and evolutional specifics of QBO TO are mostly defined by synchronization of quasi-biennial oscillation with annual oscillations. Figure 9 shows, as an example, relationship ratio of QBO TO harmonics' amplitudes and phases with periods, coinciding with the harmonic periods of ZW: a) T = 27.2; c) T 31.38; e) T= 25.54; g) T= 29.17; as well as "additional" harmonics in QBO TO with periods: b) T = 21.47; d) T 19.43; f) T = 22.6; h) T= 20.4 for some monitoring stations.

It is established that some observed particularities of temporal evolution of QBO in total ozone (figure 3 and 5.) are due to the superposition of "additional" harmonics. For example, for the data of Issyk-Kul station (ISK 347), a superposition of harmonics with periods of 21.47 and 20.4 months gives the beating with a period of 34.1 years: $\frac{1}{T} = \left(\frac{1}{21.47} - \frac{1}{20.4}\right) = \frac{1}{409.3}(mon.) = \frac{1}{34.1}(years)$.

Figure 10 (a) shows the resulting oscillations, received during superposition of "additional" harmonics with periods: T=21.47; 20.40; 19.43; 18.55; 22.60 months, and resulting oscillations, obtained during superposition of "main" harmonics with periods:T=34.0; 31.4; 29.1; 27.2; 25.5;24.0 months of QBO TO (figure 10 b).

Figure 11 shows a good match between the experimental data ISK 347 (figure 5) and the model, obtained by adding "main" and "additional" harmonics of QBO TO.

$$X(t) = \sum A_i \cos(\omega_i t + \varphi_i)$$

where

Table 2 Harmonic parameters

#	1	2	3	4	5	6	7	8	9	10	11
T	34,00	31,38	29,14	27,20	25,50	24,00	22,67	21,47	20,40	19,43	18,55
A	1,22	3,64	2,33	4,58	2,60	2,85	1,25	5,80	3,89	1,21	0,34
φ	1,04	0,21	0,96	1,97	0,79	2,42	1,31	3,03	0,07	0,84	0,90

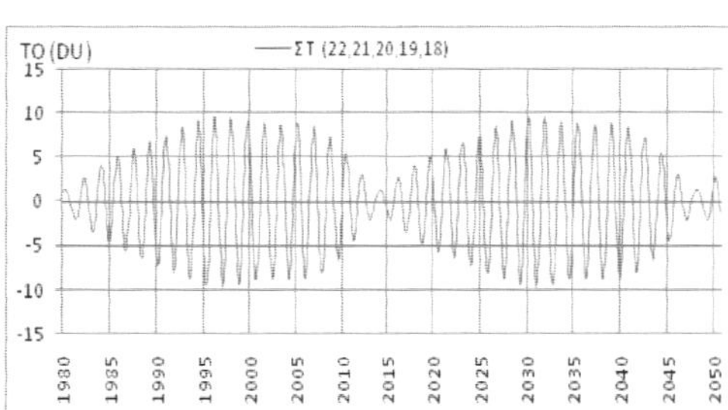

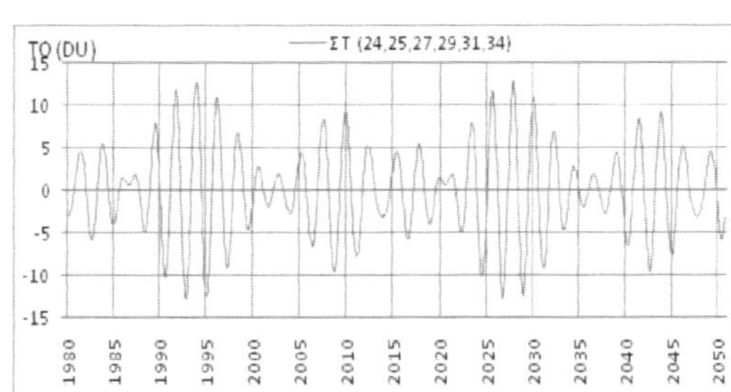

a) The oscillation amplitude beats with period of b) The superposition of "main" harmonis of TO

T~34 years, received by adding "additional" harmonics of TO with T=21.47; 20.40; 19.43; 18.55; 22.60 (mon.)

with T=34.0; 31.4; 29.1; 27.2; 25,5; 24.0 (mon.)

Figure 10. The model of quasi-biennial oscillation beats in TO for ISK347 in comparison to:
(a) The oscillation amplitude beats with period of T~34 years, received by adding "additional" harmonics of TO with T_{ad}=21.47; 20.40; 19.43; 18.55; 22.60 (mon.).
(b) The superposition of "main" harmonis of TO withT_m=34.0; 31.4; 29.1; 27.2; 25.5; 24.0(mon).

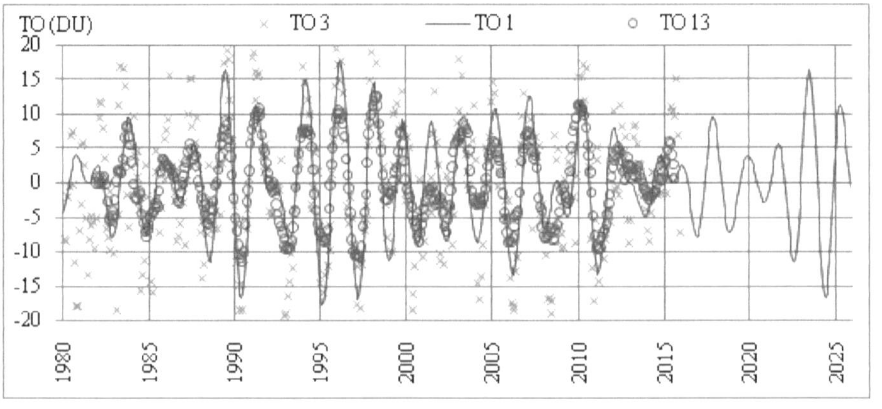

Figure 11. The QBO TO amplitude beats and comparison of experimental data of ISK 347 with the model.

The given data shows that the observed decrease in the amplitude excursion of QBO TO in recent years is due to the transition to the new long-period (34 year) regime of atmospheric circulation.

The period and beating mode of total ozone time series for other stations are different from the data obtained at ISK station. This is due to the dependence of the value of "additional" harmonics of total ozone on latitude and longitude.

Thus, the model calculation results allow describing the existing experimental data with sufficient accuracy and make forecast estimates for the expected TO changes in the atmosphere (in the future).

On the basis of detailed analysis of above mentioned experimental data, we found out, that regional and evolutional specific features of QBO TO, to a significant degree, are defined by mutual interaction of synchronization of quasi-biennial and annual (seasonal) oscillations.

5. Conclusion

Based on analysis of long-term (1980-2015) experimental data of Issyk-Kul station, in comparison with the data from other stations, there were defined specific features of quasi-biennial oscillation (QBO) of total ozone content over the North Tyan-Shan.It is established that the regional specific feature and temporal evolution of QBO in total ozone, to a significant degree, are resulted by the synchronization of quasi-biennial and annual (seasonal) oscillations.

It was found, that experimentally observed QBO TO amplitude beats are based on superposition of "additional" harmonics of QBO TO, produced as a result of mutual synchronization of quasi-biennial and annual (seasonal) oscillations. Calculations revealed that the minimum in QBO TO amplitude excursion over North Tien-Shan was reached in 2015 and new 34-years period of ozone layer variations started.

It is shown that the pattern of the relationship between TO and ZW breaks down in the years when the value of aerosol content in the atmosphere is high (volcanic activity), as well as during the periods of Magnetic Polarity change of the Sun and its flare activity, up to the change of the correlation coefficient r sign, and moreover, the degree and duration of the disturbance have regional features.

Acknowledgements.
The authors are grateful to the think-tanks of NASA, NOAA, WOUDC, SIDC for the opportunity to access the database created by them.

References

[1] Arefyev V N, Kamenogradskiy H E, Semenov V K and Sinyakov V P 1995 Ozone and nitrogen dioxide in the atmosphere over the North Tien-Shan *Izvestiya Atmospheric and Oceanic Physics*1 pp 20-25
[2] Toktomyshev S J, Semenov V K, Amanaliev M K, Orozaliev M D and Sinyakov V P 2009 *Regional monitoring of atmospheric ozone* (Bishkek, Kyrgyzstan, L L Color) p 316
[3] Toktomyshev S J and Orozaliev M D 2011 *The ozone holes over Central Asia* (Bishkek, Kyrgyzstan) p 316
[4] Visheratin K N, Nerushev A F, Orozaliev M D, Xiangdong Z, Shumen S and Li L 2016 Temporal variability of total ozone in the Asian region inferred from ground-based and satellite measurement data *Space Study of the Earth* (in the press).
[5] Gruzdev A N and Bezverkhny V A 2004 *About sources of quasi-biennial variations in the Northern Hemisphere extratropical atmosphere* Proc. XX Quadrennial Ozone Symposium 1-8 June 2004, (Kos, Greece) pp 341-342.
[6] Sitnov S A 2004 QBO effects manifesting in ozone, temperature, and wind profiles *Annales Geophys* vol 22 ‹ 5 pp 1495–1512.
[7] Sizov N I , Akimenko R M, Arefev V N, Kashin F V, Orozaliev M D, Sinyakov V P and Sorokina L I 2015 Aerosol optical depth veriability of the atmosphere over the North Tien-Shan *Meteorology and hydrology* № 3 pp 51 – 57
[8] Pikovsky A, Rosenblum M and Kurts Y 2003 *Synchronization* A universal concept in nonlinear sciences. (Moscow, Technosphere)
[9] Perov S P and Hrgian A H 1980 Current atmospheric ozone problems *Hydrometeoizdat* (Leningrad, Russian Federation) p 288

Detect, map, and preserve Bronze & Iron Age monuments along the pre-historic Silk Road

Timo Balz[1,2], **Gino Caspari**[3], **Bihong Fu**[4]

[1] State Key Laboratory of Information Engineering in Surveying, Mapping and Remote Sensing, Wuhan University, Wuhan 430076, China
[2] Collaborative Innovation Center for Geospatial Technology, Wuhan 430076, China
[3] Department for Cultural Studies, Hamburg University, Hamburg, Germany
[4] Kashgar Research Center, Institute of Remote Sensing and Digital Earth (RADI), Chinese Academy of Sciences, Beijing 100101, China

balz@whu.edu.cn

Abstract. Central Asia is rich in cultural heritage generated by thousands of years of human occupation. Aiming for a better understanding of Central Asia's archaeology and how this unique heritage can be protected, the region should be studied as a whole with regard to its cultural ties with China and combined efforts should be undertaken in shielding the archaeological monuments from destruction. So far, international research campaigns have focused predominantly on single-sites or small-scale surveys, mainly due to the bureaucratic and security related issues involved in cross-border research. This is why we created the *Dzungaria Landscape Project*. Since 2013, we have worked on collecting remote sensing data of Xinjiang including IKONOS, WorldView-2, and TerraSAR-X data. We have developed a method for the automatic detection of larger grave mound structures in optical and SAR data. Gravemounds are typically spatially clustered and the detection of larger mound structures is a sufficient hint towards areas of high archaeological interest in a region. A meticulous remote sensing survey is the best planning tool for subsequent ground surveys and excavation. In summer 2015, we undertook a survey in the Chinese Altai in order to establish ground-truth in the Hailiutan valley. We categorized over 1000 monuments in just three weeks thanks to the previous detection and classification work using remote sensing data. Creating accurate maps of the cemeteries in northern Xinjiang is a crucial step to preserving the cultural heritage of the region since graves in remote areas are especially prone to looting. We will continue our efforts with the ultimate aim to map and monitor all large gravemounds in Dzungaria and potentially neighbouring eastern Kazakhstan.

1. Introduction

During the Iron Age, large parts of the Eurasian steppe belt from Western China to Eastern Europe were inhabited by nomadic pastoralists. Many of these cultures built grave mounds (so called kurgans) to bury their dead and thus left distinctive marks in the landscape. There are tens of thousands of these monuments distributed over the steppe zones of Eurasia and not even a rough overview has been established yet. Iron Age grave mounds are also a primary target for looting, since they are known to contain valuable objects. A large part of these monuments has already been robbed and destroyed by looters. In order to effectively protect and study the Iron Age in Xinjiang and neighbouring Kazakhstan, the area should be studied as a whole and combined efforts should be undertaken to protect these extraordinary archaeological monuments. Until now, international research campaigns have focused predominantly on single-sites or small-scale surveys confined to today's borders, mainly due to bureaucratic and security related problems in cross-border research.

With remote sensing, large areas can be searched for material remains. The rather large and distinct burial mounds are suitable for remote sensing based landscape archaeological analysis and for creating an inventory. Since the acquisition of large chunks of optical data is expensive, we jointly analysed the possibility of using synthetic aperture radar (SAR) data for the detection of burial mounds. To analyse large areas, automated or semi-automated detection methods are required. Certainly so, if we want to get a complete overview over the burial mounds in Dzungaria and eastern Kazakhstan. The size of a burial mound has commonly been interpreted as a measure for social status. Simplistically speaking, the larger the mound the more important the person buried within it and the richer the artefacts and treasures buried with the dead. Therefore, our focus is on the rather large burial mounds which would equal to the most substantial loss of cultural heritage if robbed. The main objective of our *Dzungaria Landscape Project* is to establish a method which allows for efficient large-scale detection of big grave mounds with diameters of 20 m and more. It has been shown that very often larger Bronze and Iron Age monuments form nuclei for smaller archaeological structures [1],[2]. The automatic classification of kurgan structures, especially of dense structures of kurgans forming archaeological 'hotspots' is essential to get an overview over the regional spatial distribution of the sites. The detection of so-called princely graves can thus reveal interesting archaeological hotspots in a given landscape as a first step for further, more detailed archaeological inquiries and measures for heritage protection.

In section 2, we describe our previous work using remote sensing for the detection of burial mounds in the Altai region. In section 3, we describe the planned extension of the work on a larger scale. Finally, we draw the conclusions.

2. Previous work in the Altai Mountains

2.1. Selecting an area of interest: Hailiutan valley, Chinese Altai

As a first test site we selected an archaeologically rich landscape in the foothills of the Southern Chinese Altai Mountains based on an automated large-scale survey. The area around the village Hailiutan close to the airport of Kanas is a broad valley with an extensive alluvial plain as its centre. It is part of the territory of both Bu'erjin and Habahe County. Framed by mountain ranges to its North and its South the valley lies on 1150-1300 m a. s. l. Steep mountain passes allow access to the plain from the Dzungarian steppe. The area lies on the border of two climatic zones: the Dzungarian steppe to its South and the alpine climate of the high Altai to its North and thus encompasses two major living environments of antique Central Asian nomadic pastoralists.

2.2. Selecting an area of interest: Hailiutan valley, Chinese Altai

For the detection of burial mounds covering a wide area, automated or semi-automated detection methods are preferred. In a previous study, we tested three different approaches for the automatic detection of grave mounds in IKONOS-2 datasets [3]. We used convolution with ring templates exploiting the shape characteristics of the gravemounds and implemented and tested class-specific Hough Forests for object detection. We achieved the best results with class-specific Hough Forests (see Figure 1).

In this experiment, we showed that archaeological features that are clearly visible with a strong contrast against the background can be detected quite well and are relatively robust to changes in threshold value using class-specific Hough Forests. Monuments that are well identifiable, but for example have been changed in their appearance by ploughing or vegetation, are still detected at a reasonable rate; however, changes in the parameters for a higher detection rate of these structures lead to an increase in the false positives. Burial mounds that are hardly identifiable visually in the remote sensing data are also not identified by the algorithm (see [3]).

Generally, large burial structures are clearly identifiable and automatically detectable in IKONOS-2 and WorldView-2 data. Because the structures are often spatially clustered with smaller mounds being in proximity to larger mounds, the detection of larger mounds is sufficient to detect larger clusters of archaeologically important structures.

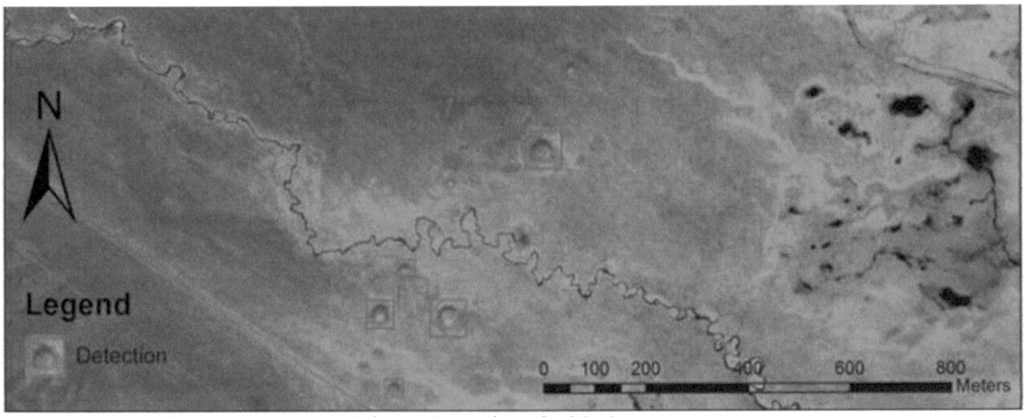

Figure 1. Detections on IKONOS-2 data (Caspari et al., 2014)

2.3. Detecting burial mounds from Synthetic Aperture Radar (SAR) satellite data

Based on the promising results for detecting burial mounds from high-resolution optical data, we tested the detectability of such mounds in very high resolution SAR data. SAR has some advantages over optical data, especially regarding continuous monitoring and detecting relative small height differences using interferometry. Getting larger data coverage from SAR can also be more economical than optical data acquisition, making it an interesting alternative for covering large areas.

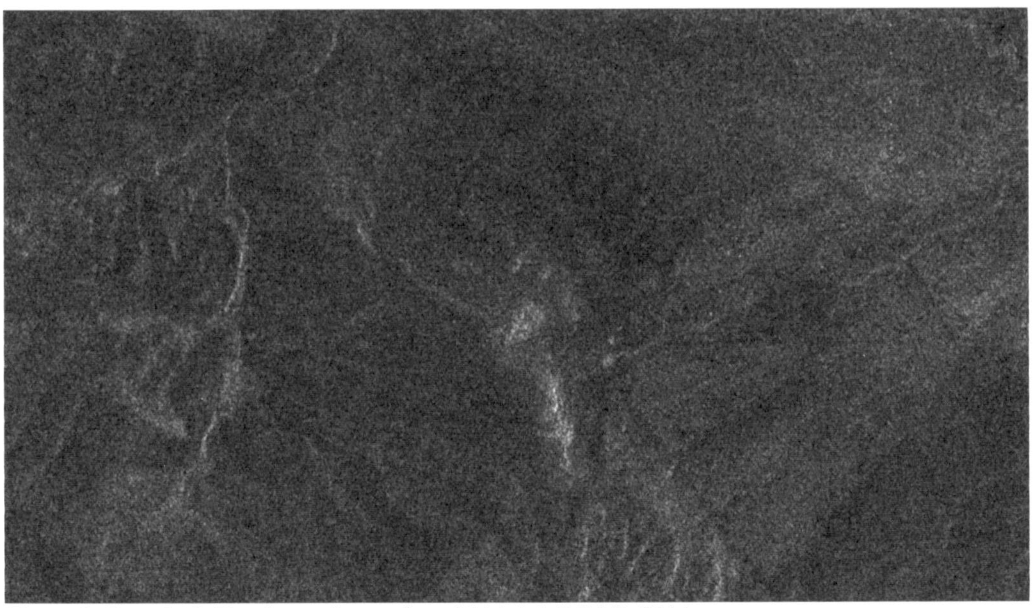

Figure 2. TerraSAR-X stripmap image from 2014-6-5 © DLR, 2014

For our experiments we used TerraSAR-X data acquired with different SAR modes [4]. First, we acquired TerraSAR-X stripmap data with approximately 3 m spatial resolution. Furthermore, we acquired TerraSAR-X spotlight data with approximately 1.6 m resolution. Finally, we used the new TerraSAR-X staring spotlight mode [5] with around 0.25 x 0.9 m resolution. As the dimension of the larger burial mounds we are interested in exceed 10 m, they should be identifiable in the test data.

However, as we can see in Figure 2 even large burial mounds are hardly identifiable in the stripmap images with around 3m resolution. This situation improves when using spotlight images as shown in Figure 3 However, the detectability even of larger mounds is still very limited in the spotlight images.

Figure 3. TerraSAR-X spotlight image from 2015-7-14 © DLR, 2015

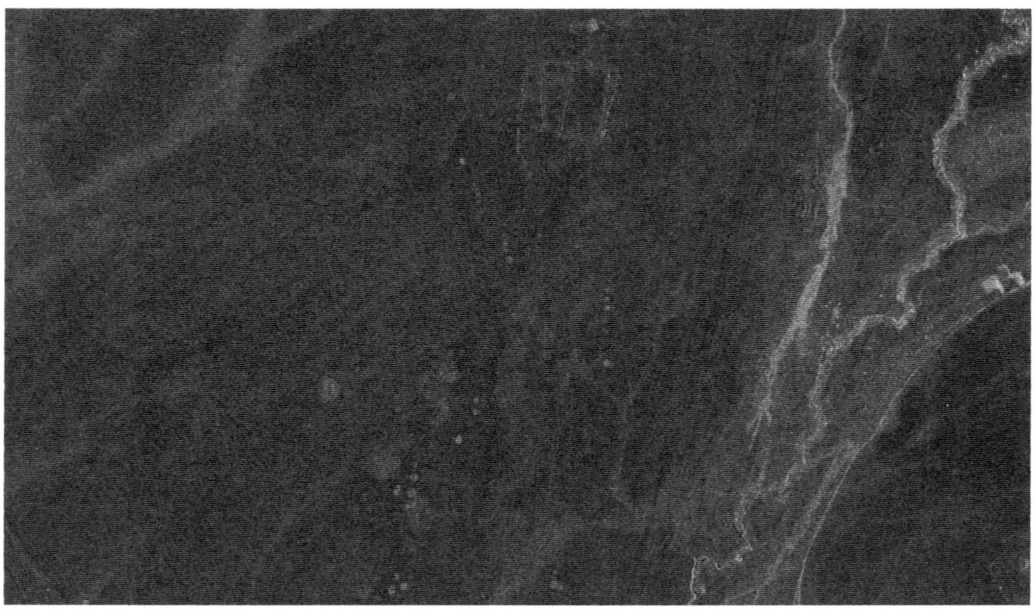

Figure 4. TerraSAR-X staring spotlight image from 2015-9-5 © DLR, 2015

The situation is different though when we are using TerraSAR-X staring spotlight images with very high spatial resolution. As we can see in Figure 4, burial mounds of different sizes are clearly identifiable in the staring spotlight data. There are several reasons for this. First, the higher spatial resolution alone already allows for a better identification. Additionally, the speckle effect disturbs the clear identification of mounds in the SAR image and we can use the additional spatial resolution in azimuth for a four time multi-looking of the TerraSAR-X staring spotlight image for a reduction of the speckle in the staring spotlight image. This tremendously increases the detectability. Unfortunately, with the stripmap and spotlight data, multi-looking would reduce the spatial resolution even further. Therefore, the increase in radiometry from multi-looking is rendered useless due to the loss of spatial resolution.

Using staring spotlight data, automated detection methods can also be applied. Using Hough-Forests, good detection rates for larger mounds can be achieved. As shown below in Figure 5, larger mounds can be detected in both cases. Increasing the detection rate for the smaller mounds would increase the false detection rate though, so that we cannot expect a complete detection with automated methods, but instead a good detection of clusters of burial mounds.

Figure 5. Example of classification results from TerraSAR-X staring spotlight data (left) and IKONOS-2 data (right)

Unfortunately, although we demonstrated that burial mounds are detectable in very high-resolution staring spotlight data, this does not help us regarding our goal of detecting the burial mounds on a larger, regional scale. Staring spotlight images only cover a small area and are, by design, not suitable for large area coverage. To cover the Dzungarian steppe, stripmap data or even ScanSAR / TOPS data would be better suited. But, with that data we cannot detect even larger burial mounds.

2.4. Ground survey

In summer 2015, we conducted a field survey in the Hailiutan valley, based on the area of interests detected with optical and SAR remote sensing as described above. Our survey was based on field walking with the aim of mapping all visible sites in the area. We worked with small and very mobile teams usually consisting of 3-4 people. In this way, we were able cover an area of well over 140 km^2 in just about three weeks of intensive fieldwork. The remote sensing data helped us immensely with planning and allocation of our efforts. In total, the team mapped close to 1'000 structures.

We found that the construction of grave mounds appears to be constrained by certain environmental factors which – if taken into account – reduce the amount of data which has to be processed considerably. Our survey showed that the centre of the basin was not favoured for the construction of funerary monuments since it is rife with rivulets, the soil therefore has a high moisture content and is suboptimal for construction. Thanks to the ground truth we were also able to improve the accuracy of our remote sensing survey. Especially short-lived circular soil features like the one's resulting from herding activities and the placement of yurts led to false positive marks before the field survey (see Figure 6.)

Figure 6. Temporary soil marks that looked like archaeological features in remote sensing data.

Already through a rough analysis of the archaeological features we were able to divide them according to our dating hypotheses and assign them to specific periods. We analysed the archaeological landscape of the valley of Hailiutan which has been occupied at least since the Bronze Age. During the Early Iron Age, we see an upsurge of anthropogenic building activity in the area. During this time the southern foothills of the Altai Mountains seem to have been an important passage way and the home base of a highly developed culture with intensive ties to both modern day Kazakhstan and the northern Altai.

3. Extension to a regional scale

Extending the *Dzungaria Landscape Project*, we aim to map the complete plain for large archaeological features, especially Iron Age grave mounds. Special attention will be given to the passage ways into the west, since these have been early on places of intensive cultural interaction that connect eastern Chinese cultural influences with Central Asia. Our aim is to integrate the archaeological research in Dzungaria with the efforts of our colleagues in Kazakhstan and create research interfaces through remote sensing by providing a standardized methodology.

As a first step we need to complete several case studies through remote sensing. This will help us document the variety of monuments in northern Xinjiang and establish a system for classification. Throughout these case studies we will be using both optical and SAR imagery and continue to assess the validity and accuracy of our predictions through on-ground surveys. The case studies will follow a north-south line from the foothills of the Altai Mountains to the northern flank of the Tianshan. Through this methodology we might be able to relate specific cultural phenomena to environmental circumstances, thus increasing the specificity of our survey results.

In a second step we need to conduct small scale excavations at a number of monuments in order to obtain dating material and further develop our methodology for protection. Knowing the date and the preservation circumstances of individual monuments will enable us to make recommendations for protection measures from environmental factors and anthropogenic destruction like looting, agricultural activities or construction.

The third step will then consist of building up a geographic information system which points out areas of high archaeological density. Such a dataset will help local heritage administration to consistently monitor, protect and - if not possible - excavate the cultural treasures of northern Xinjiang and eastern Kazakhstan.

4. Conclusion

Over the past three years we have built a strong network of remote sensing specialist and archaeologists, which are able to perform the tasks described above. From the large-scale automatic scanning of thousands of square kilometres of data to landscape archaeological analysis as well as excavation and survey-based ground truth we have developed a process that will help further our understanding of the cultural heritage of Xinjiang and Kazakhstan. In order to show the cultural connection this part of China established during prehistoric times with Central Asia and to protect the valuable heritage of this region we need to continue our research within the frame of the Dzungaria Landscape Project.

We have obtained good results in detecting large Iron Age gravemounds from both optical and SAR data. Due to the valuable objects contained in those mounds they are especially prone to looting activities and – as a matter of fact – a large proportion of those mounds has already been destroyed. Mapping all large gravemounds in Dzungaria will provide a cultural heritage management tool in order to preserve these magnificent monuments for future generations. Furthermore, it will increase our understanding of the archaeology in Xinjiang and neighbouring Kazakhstan thus showing the culturally strong connection these regions had in the past. But further effort is needed to accomplish this aim.

Relations to the local cultural heritage administration and the Academy of Social Sciences have to be built so we can establish further ground truth. We need to obtain excavation data of a small number of gravemounds to be able to give accurate advice on how to preserve and monitor monuments of specific types. We also need to continue to acquire data and ground surveys. But, step-by-step we will complete the largest archaeological survey China has ever seen.

Acknowledgements
The work presented in this manuscript was supported by the Natural Science Foundation of China (Grant No. 61331016 and 41174120) as well as the Swiss National Science Foundation (Grant No. P1SKP1_158714) and the ArchaeoCare Foundation. The TerraSAR-X data was provided by DLR via the MTH2240 project. The IKONOS high-resolution optical data was provided through an imagery grant of the Digital Globe Foundation.

References

[1] Gheyle W 2009 *Highlands and Steppes: An Analysis of the changing archaeological landscape of the Altay Mountains from the Eneolithic to the Ethnographic period* PhD diss., University of Ghent.
[2] Caspari G, Plets G, Balz T, Fu B 2016 Landscape Archaeology in the Chinese Altai Mountains *Archaeological Research in Asia*, forthcoming.
[3] Caspari G, Balz T, Liu G, Wang X.Y, Liao MS 2014 Application of Hough Forests for the detection of grave mounds in high-resolution satellite imagery *Proc. IGARSS 2014* Quebec, Canada, pp. 906-909.
[4] Balz T, Caspari G, Erasmi S, Fu BH, Rosenbauer R, Liao MS 2015 Analyzing TerraSAR-X Staring Spotlight Mode Data for Archaeological Prospections in the Altai Mountains *Proc. APSAR 2015*, Singapore
[5] Mittermayer J, Wollstadt S, Prats-Iraolo P, Schreiber R 2014 The TerraSAR-X staring spotlight mode concept *IEEE Transactions on Geoscience and Remote Sensing* 52 pp. 3695-3706.

Radar remote sensing for archaeology in Hangu Frontier Pass in Xin'an, China

A H Jiang[1,2], F L Chen[2,3], P P Tang[2,3], G L Liu[1], W K Liu[1], H C Wang[4], X Lu[4] and X L Zhao[5]

[1] Geomatics College, Shangdong University of Science and Technology, Qingdao 266590, China;
[2]Key Laboratory of Digital Earth Science, Institute of Remote Sensing and Digital Earth, Chinese Academy of Sciences, No. 9 Dengzhuang South Road, Haidian District, Beijing 100094, China;
[3]International Centre on Space Technologies for Natural and Cultural Heritage under the Auspices of UNESCO;
[4]Institute of conservation and management of Xin'an Hangu Pass (Han Dynasty), Cultural relic bureau of Xin'an County, Luoyang, 471800, China;
[5]Institute of cultural relics and archaeology of Zhengzhou city, Zhengzhou, 450052, China.

E-mail: jiang.ai.hui@163.com

Abstract. As a non-invasive tool, remote sensing can be applied to archaeology taking the advantage of large scale covering, in-time acquisition, high spatial-temporal resolution and etc. In archaeological research, optical approaches have been widely used. However, the capability of Synthetic Aperture Radar (SAR) for archaeological detection has not been fully exploded so far. In this study, we chose Hangu Frontier Pass of Han Dynasty located in Henan Province as the experimental site (included into the cluster of Silk Roads World Heritage sites). An exploratory study to detect the historical remains was conducted. Firstly, TanDEM-X SAR data were applied to generate high resolution DEM of Hangu Frontier Pass; and then the relationship between the pass and derived ridge lines was analyzed. Second, the temporal-averaged amplitude SAR images highlighted archaeological traces owing to the depressed speckle noise. For instance, the processing of 20-scene PALSAR data (spanning from 2007 to 2011) enabled us to detect unknown archaeological features. Finally, the heritage remains detected by SAR data were verified by Ground Penetrating Radar (GPR) prospecting, implying the potential of the space-to-ground radar remote sensing for archaeological applications.

IOP Conf. Series: Earth and Environmental Science **57** (2017) 012031 doi:10.1088/1755-1315/57/1/012031

1. Introduction

Remote sensing was applied in archaeology almost since the beginning of aerial photography owing to the synoptic observation of large-scale landscapes. This technology together with the geophysical prospection have been indispensable tools for the investigation of ancient cities and surrounding landscapes without causing a loss of archaeological evidence. Moreover, it is beneficial for the schedule of archaeological excavations [1, 2].

Nowadays, Synthetic Aperture Radar (SAR) systems have not been widely used in archaeological prospecting. This is mainly due to the rather complex nature of SAR data processing. However, the advantages of SAR remote sensing for archaeology are obvious. First, SAR systems are capable of acquiring data in all-weather, all-day conditions. Second, digital surface models with a high height accuracy can be generated by using interferometric SAR, and in turn the terrain data can be used for the ortho-rectification of SAR images as well as very high-resolution (VHR) optical data [2, 3].

Up to now, there are several studies have been undertaken to assess the potential of satellite SAR data [3, 4, 5]. Archaeological investigations conducted by SAR data have moved forward to another stage focusing on archaeological detection and monitoring by developing technologies related to the polarimetric enhancement, backscattering change detection, motion detection and diagnosis. The second-generation spaceborne SAR data, represented by multi-band, polarization as well as high-resolution, e.g., ALOS PALSAR-2, TerraSAR/TanDEM-X and Cosmo-SkyMed, are favorable in archaeology. For instance, the X-band COSMO-SkyMed data have been used for the detection of archaeological marks in diverse environmental conditions [6].

2. Experimental site and data

2.1. Experimental Site

Hangu Frontier Pass of Han Dynasty, in Xin'an County of Henan Province, China, was built in the Western Han Dynasty (114 B C). Hangu Frontier Pass is the gate as well as one of most important defences of Xi'an that was the capital of Western Han. When Luoyang was chosen as the capital in Eastern Han Dynasty, the Pass was even more important and nominated as the head of Eight Great Passes. After that, as a military pass, it held a responsible position until Sui and Tang Dynasties. It was also the first-round pass of the Silk Road Corridor. The pass was not only a military fortress but also the monument that had witnessed the prosperity of the Silk Trade. As a heritage site of "Silk Road Routine", it was listed as a World Heritage Site in 2014. History records that the Hangu Frontier Pass of Han Dynasty stretched from Yellow River to Luohe River, with a dimension of approximately 60 kilometers (Figure 1). The storehouse of Hangu Frontier Pass of Han Dynasty was excavated in 1998, and the core section in Xin'an was excavated in 2012. However, most sections of Hangu Frontier Pass are still unknown, in particular the lost ruin of San-kuan Pass (the southern end of Hangu Frontier Pass).

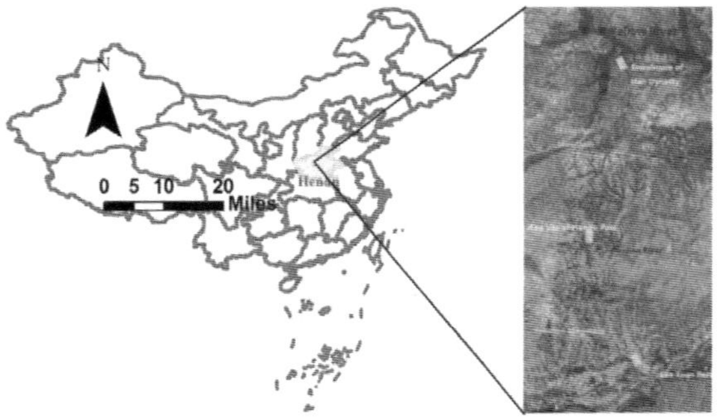

Figure1. Location of Experimental Site

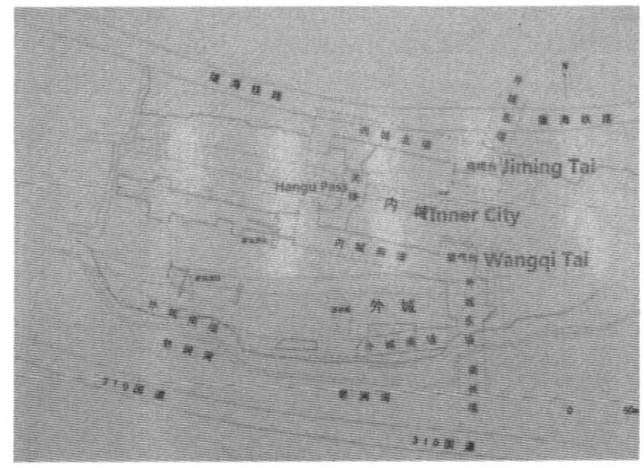

Figure 2. Location of the sub-area for the GPR prospection (highlighted by the green rectangle)

2.2. Data

Satellite SAR data have played an important role in archaeological prospecting. As an active survey method by transmitting microwave frequency signatures, SAR has several advantages such as all-weather and all-day acquisitions [7]. Three categories of microwave remote sensing data were acquired, including X-band TanDEM, L-band PALSAR-1 satellite images as well as GPR prospecting data from the field.

2.2.1. TanDEM-X. TanDEM-X is a SAR system which allows not only an independent operation from TSX-1 in mono-static mode but also supports synchronized operations [8]. The main objective of the TanDEM-X mission is to generate a high resolution global DEM by SAR interfereometry (InSAR). The TanDEM CoSSC datasets for the DEM generation were listed in Table 1. In addition, the SRTM DEM (1 arc-second) of Hangu Frontier Pass has been used as the complementary data for the generation of TanDEM-X DEM.

Table 1. TanDEM-X CoSSC data specifications

Date	Pol./Path	Pixel size	Angle of Incidence	Baseline(perp)	Height of Ambiguity
08 January 2014	HH/Desc	2.26×2.03m	37.1°	112.9	52.6m

2.2.2PALSAR. L-band ALOS PALSAR-1 data applied in this study were acquired with two modes, including the fine-beam dual-polarization (HH and HV) and the single-polarization (HH). In total, 20 scenes of PALSAR-1 were collected. There are 9 scenes acquired in single-polarization with the pixel spacing of 4.68m and 3.15m in range and azimuth direction, respectively, and the other 11 scenes acquired in dual-polarization with the pixel spacing of 9.37m and 3.15m in range and azimuth direction, respectively(see Table 2).

Table 2. Parameters of multi-temporal PALSAR-1 images applied in this study

No.	Acquisition(Day Month Year)	Pol.	No.	Acquisition(Day Month Year)	Pol.
1	03 January 2007	HH	11	23 February 2009	HH

2	18 February 2007	HH	12	11 July 2009	HH+HV
3	06 July 2007	HH+HV	13	26 August 2009	HH+HV
4	21 August 2007	HH+HV	14	11 October 2009	HH+HV
5	06 October 2007	HH+HV	15	11 January 2010	HH
6	06 January 2008	HH	16	26 February 2010	HH
7	21 February 2008	HH	17	29 May 2010	HH+HV
8	23 May 2008	HH+HV	18	14 July 2010	HH+HV
9	08 July 2007	HH+HV	19	14 October 2010	HH+HV
10	08 January 2009	HH	20	01 March 2011	HH

2.2.3Ground-Penetrating Radar. There are several geophysical methods that can be used for the detection of archaeological features under beneath. One of them is the Ground-Penetrating Radar (GPR). When thousands of radar reflections are measured as antennas moving along transects within a grid, a three-dimensional imagery related to physical properties of soils, sediments and feature changes can be derived [1,9]. In this study, a sub-area with the occurrence of outer-south city-wall is measured by the GPR prospecting, as marked by a green rectangle in Figure 2.

3. Data processing

3.1. TanDEM-X Data Processing

TanDEM-X is the first mission which provides interferometric data in bistatic mode. There is a standard InSAR procedure for the interferometric processing of both monostatic and bistatic data [10]. The whole process for the high resolution DEM generation was done using the SARscape software. Figure 3 provides the flowchart adopted for the TanDEM-X DEM generation. The co-registration procedure was not needed for the co-registered CoSSC image pairs. Then the ridge lines were extracted using terrain analysis tools in ArcGIS software by utilizing TanDEM-X DEM data.

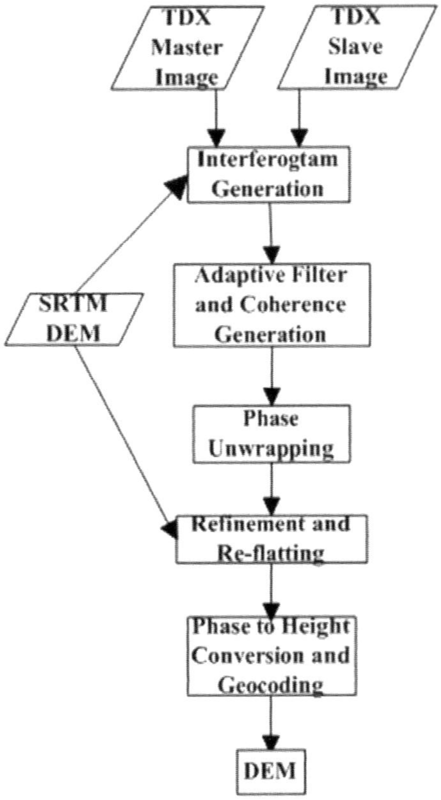

Figure 3. Flow chart for the DEM generation using TanDEM CoSSC data

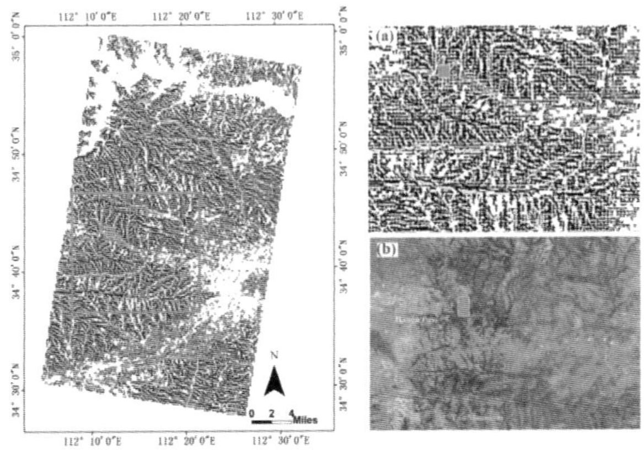

Figure 4. Spatial correlation between ridge lines and the stretching direction of outer walls

3.2. PALSAR Data Processing

The pixel-based image analysis and interpretation is generally prevailing in archaeological applications. Consequently, when multi-temporal SAR data were available, co-registration was an essential preprocessing step not only significant for the enhancement of features (e.g. temporal averaging) but also for the implementation of the change detection. The co-registration of PALSAR-1 data was conducted by using GAMMA software tools. In this procedure, it is generally preferable to register the lower-resolution image (dual-polarization data) into the higher-resolution image (single-polarization data) to preserve the spatial details. The following 1×3 multi-looking (looks in range and azimuth, respectively) generated amplitude SAR products with a pixel spacing of 4.7m× 9.4m. In order to enhance the spatial anomalies, a temporal stacking technique was applied. In such a way, the emerging wall-remains become more evident that was beneficial for the identification of archaeological traces using SAR images.

3.3. GPR Data Processing

For the selected sub-area (highlighted by the green rectangle in Figure 2), 15 GPR profiles were carried out using the RIS K2-Fastwave (K2FW). Taking the occurrence of buried remains as well as the penetration capability of radar signals, 100MHz antenna with the monostatic type was applied. The spatial dimension of the sub-area is 60m×30m. A 2m step-size was selected for the next profile in the data acquisition. The GRED software has been used to process the data. Image filtering and the antenna gain correction were jointly used to for the signal enhancement, including the processing steps of background removal, vertical bandpass filter, linear gain and smoothed gain.

4. Results and analysis

4.1. Topography Analysis

Ridge line is the boundary of terrain changes; therefore, it is an effective factor for the analysis of topography. Generally, there is a close relationship between the occurrence of ancient passes and the direction of ridge lines. Owing to the high-resolution of DEM data (with a relative height accuracy of 2m), the primary and thin ridge lines were all extracted (see Figure 4). Considering the collapse phenomenon, the identification of ram-earthed outer walls (extending from the core zone to the Yellow River or to the San-kuan Pass) by ridge lines was challenging. Nonetheless, the spatial correlation could be observed, such as the direction of several outer-wall sections followed the trend of ridge lines, particularly for the section as marked by the green rectangle in Figure 4.

4.2. Temporal Stacking

The processing of PALSAR imagery was aimed at the enhancement of features linked to the occurrence of frontiers. For the test site selected (marked by green rectangle in Figure 5), the capability of multi-temporal PALSAR imagery in identifying spatial anomalies of emerging remains was assessed. The comparison of temporal-stacked imagery and the single-acquisition imagery was illustrated in Figure 5. A more remarkable visualization of the latter demonstrated that multi-temporalstacking provides a better performance for the detection of archaeological traces due to the suppressed speckle noise in SAR data. Furthermore, one significant feature high-probably linked

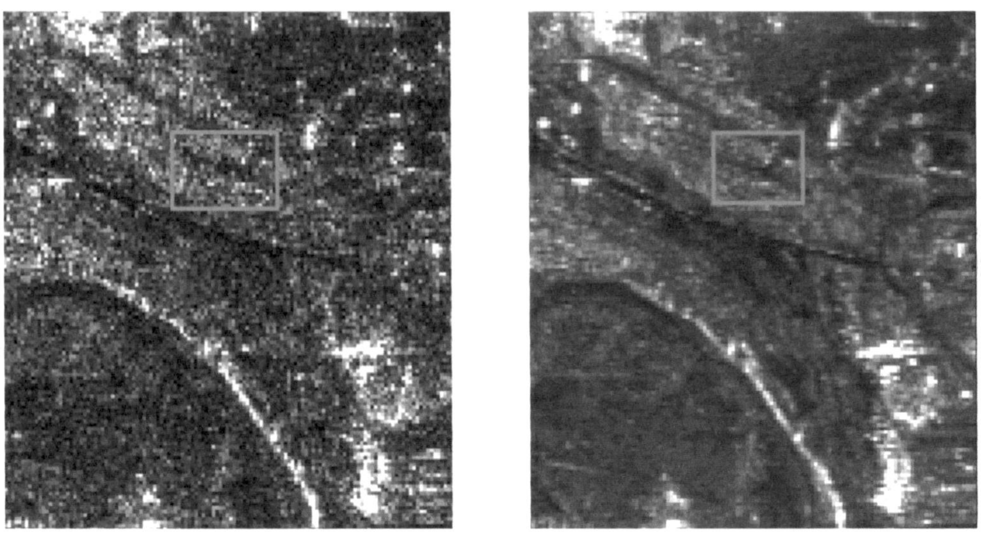

a: Single-acquisition of PALSAR-1 b:Multi-temperal stacked PALSAR-1 imagery

Figure 5. Comparison between Single-date and Multi-temporal SAR data

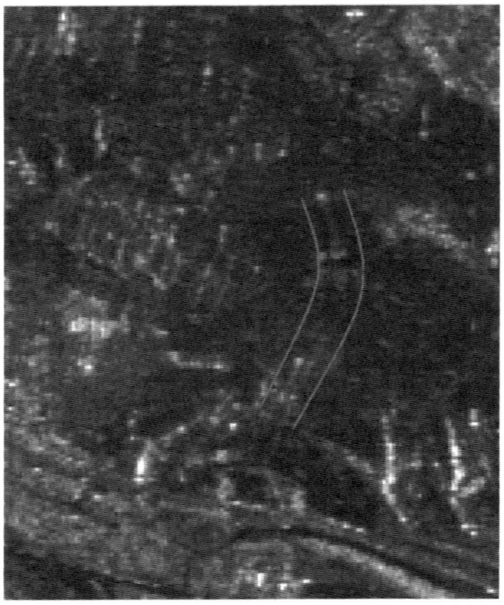

Figure 6. Backscattering anomalies on the temporal-averaged imagery of 20-scene PALSAR-1 data (marked by red polylines). The phenomenon was high-probably linked to the outer-wall of the pass stretching from the core zone to Yellow River

to the stretching of Hangu Frontier Pass was identified in the temporal-averaged imagery (Figure 6), as marked by red polylines. According to the historical document, this archaeological trace was coincidence with the trend of Hangu Frontier Pass approaching to the Yellow River.

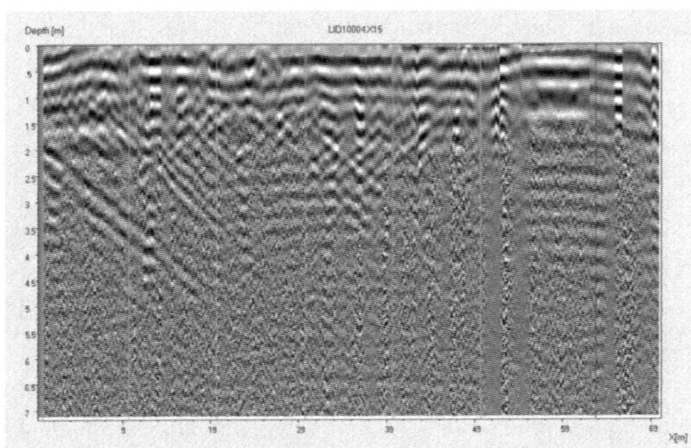

Figure 7. GPR profile of L4

4.3. GPR Prospection

GPR profiles of the investigated area show reflections that could be associated with the occurrence of buried ancient walls. Figure 7 showed the vertical profile of L4 with a 60m length in T direction. The penetration depth of radar signatures is approximately 5m. This profile pointed out reflection anomalies with a depth of 3.5m, which could be related to the south outer-wall of the ancient city. Moreover, several other hyperbolas were also visible in the L4 profile caused by the reflection from the stones and broken tiles buried.

5. Summary

In this study, data jointly from satellite SAR images and GPR prospecting have been analyzed in order to assess their capability in detecting archaeological remains in Hangu Frontier Pass of Han Dynasty. The spatial correlation of ridge lines with the occurrence of outer walls was rather preliminary and thus further investigations were required. Temporal-stacking of PALSAR-1 imagery enabled us to detect unknown archaeological remains. Finally, buried archaeological remains, particularly the south outer wall, were detected by the profile of the GPR prospecting. Consequently, it is feasible and the time to develop a methodology framework of radar remote sensing in archaeology by integration the space and ground observations.

Acknowledgements
This work was supported by funding from Hundred Talents Program of the Chinese Academy of Sciences (CAS) (Y5YR0300QM), CAS-TWAS Centre of Excellence on Space Technology for Disaster Mitigation "Joint Program on Space Technology for Disaster Mitigation in Asia", and Scientific Research Foundation Project for the Talents of Shandong University of Science and Technology (2014RCJJ047).

References

[1] Balz T, Liao M, Caspari G, Fu B, Rosenbauer R and Erasmi S 2015 Analyzing TerraSAR-X Staring Spotlight Mode Data for Archaeological Prospections in the Altai Mountains *IEEE 5thAPSAR* 528-532

[2] Rowlands A and Sarris A 2007 Detection of exposed and subsurface archaeological remains using multi-sensor remote sensing *J. Arc. Sci.* **34(5)** 795-803

[3] Linck R, Busche T, Buckreuss S, Fassbinder J W E and Seren S 2013 Possibilities of Archaeological Prospection by High-resolution X-band Satellite Radar - a Case Study from Syria *Arc. Pro.* **20(2)** 97-108

[4] Chen F, Lasaponara R and Masini N 2015 An overview of satellite synthetic aperture radar remote sensing in archaeology: From site detection to monitoring J. Cul. Her.

[5] Chen F, Masini N, Yang R, Milillo P, Feng D and Lasaponara R 2014 A Space View of Radar Archaeological Marks: First Applications of COSMO-SkyMed X-Band Data *Rem. Sen.* **7(1)** 24-50

[6] Moreira A, Krieger G, Hajnsek I, Hounam D, Werner M, Riegger S and Settelmeyer E 2004 TanDEM-X A TerraSAR-X add-on Satellite for Single-pass SAR Interferometry *IEEE IGARSS* **2** 1000-1003

[7] Gruber A, Wessel B, Huber M, Breunig M, Wagenbrenner S and Roth A 2012 Quality Assessment of First TanDEM-X DEMs for Different Terrain Types *EUSAR. 9th Eur. Conf.* 101-104

[8] Krieger G, Fiedler H, Hajnsek I, Eineder M, Werner M and Moreira A 2005 TanDEM-X Mission Concept and Performance Analysis *IEEE Int. Geo. Rem. Sen. Sym.* **7** 4890-4893

[9] De Reu J, Bourgeois J, De Smedt P, Zwertvaegher A, Antrop M, Bats M and Crombé P 2011 Measuring the relative topographic position of archaeological sites in the landscape, a case study on the Bronze Age barrows in northwest Belgium *J. Arc.Sci.* **38(12)** 3435-3446.

[10] Pandit A Ramsankaran R and Rao Y S 2014 Generation and validation of the interferometric SAR DEMs from TanDEM-X data for Gangotri and Hamtah Glaciers of Indian Himalayas *Pro. Tech.* **16** 793-805.

Spatial information management platform for Dunhuang Global Geopark

YU Yan-long[1], WU Fa-dong[1], HAN Jin-fang[1], WANG Yan-Jie[1], CHU Hao[1]

1. *School of Earth Sciences and Resources, China University of Geosciences(Beijing), Beijing 100083, China*

yuyanlong_cugb@126.com

Abstract. As a member of UNESCO Global Geoparks, Dunhuang Global Geopark has developed a great quantity of landforms formed under special geological background and extremely droughty climate, which integrate together with specific geographic location and cultural relics on the "Silk Road Economic Belt". The main geoheritage in Dunhuang Global Geopark is Yardang landform, which is formed by loose Quaternary sediments. According to different shapes, the Yardang landform were divided into five types, namely, ridge-shaped Yardang, wall-shaped Yardang, tower-shaped Yardang, column Yardang and Yardang monadnock. In order to monitor and protect the unique morphological features of Yardang landforms , a spatial information management platform is established, using SPOT 6 remote sensing image, with object oriented approach and manual interactive interpretation. Study shows that the maximum area, perimeter, length and width of Yardang were 324843.1 m2, 3447.52 m, 1508.41m, and 285.81 m, respectively. Additionally, the aspect ratio of Yardang has a certain positive correlation, with the coefficient of correlation being 0.675. Furthermore, the relationship between length and width of Yardang is calculated using formula Y=2.546X, where Y = length, X= width.

1. Introduction

Located in the western of Gansu Province, as a key position of the ancient Silk Road, Dunhuang City will play an irreplaceable role in constructing the Silk Road Economic Belt, realizing the communication of policies, transportation, trades and currency among countries along the ancient Silk Road. Situated in the extremely arid climatic region, there has developed very typical geoheritage in Dunhuang Geopark. Being a member of GGN will not only open up a new way for ecological civilization construction, ecotourism development and geoheritage protection simultaneously, but also bring Dunhuang City a great opportunity to develop worldwide. Geoparks play significant role in promoting geoconservation and realizing sustainable development of geoheritage (Gizzi & Lazzari, 2012; Dowling & Newsome, 2006).

The Yardang landform are widely distributed in Dunhuang Geopark, including different development stages. They are more densely distributed than their counterparts anywhere else in the world. Therefore, the Geopark showcases the most typical Yardang landforms in China and even in the world. Their formation was a result of the combined action of tectonic uplift, wind, running water and gravity on incompletely consolidated Quaternary fluvial and lacustrine sediments (Cooke et al. 1993; Lin et al. 2012; Ren et al.2011; Dong ,2013). The Dunhuang Yardang landform has high aesthetic value and also serves as the best place to study their formation, development and evolution.

2. Materials and method

2.1 Yardangs

"Yardangs" were used by the Swedish explorer Sven Hedin(1903) meaning "steep bank". They are streamlined forms up to 150 km long and 75 m in height resulting from a number of formative processes, including wind abrasion, deflation, fluvial incision, desiccation cracks, slumping, weathering and mass movement(Goudie, 2007; Mccauley et al., 1977b;Ward & Greeley, 1984). The Dunhuang Yardang landforms are typical wind-erosion landforms.

2.2 Study area

Yardang area is located in the western part of Dunhuang Global Geopark, 160 kilometers away from Dunhuang city, covering an area of 346.35 square kilometers. Yardang Landform is the typical landform under the extremely arid climate, mainly formed by the wind and water erosion on the Quaternary sediments during the past 100,000 years. The major landscapes of this geo-area include mound-shaped, wall-shaped, tower-shaped and columnar Yardang landforms, which are very lifelike and has become the real ghost city in western China, and associated with the vast Gobi and desert.

Fig. 1 Transportation Sketch Map of Dunhuang Geopark of China

2.3 Materials and method

SPOT 6 satellite, launched on September 9, 2012, is the first satellite of SPOT 6 and SPOT 7 constellation that will assure continuity of 1.5 m high spatial resolution, 60×60 km wide-area image data through to 2024. It provides 6m multispectral image with blue, green, red and NIR bands in addition to 1.5m panchromatic image(Alganci ,et al. 2013）. The image covering the study area was acquired on

February 18, 2014.

With the development of remote sensing technology, remote sensing image has a higher spatial resolution and makes us obtain very abundant information of nature. In addition to the spectral information outside of itself, it also contains a large number of shapes and texture information. The traditional image classification method that based on spectral of pixels, can work well for the image that have middle or low resolution. But this method cannot be fully utilized other information such as texture, not well adapt to the development of today's remote sensing data and classification accuracy. In this context, the classification of object-oriented methods have emerged. In object-oriented classification, different features have different best scale for image segmentation. Using visual methods to determine best segmentation scale and rules for extracting as below.

Refine: 9053.30 to 13784.0

Attributes Computed:

Spatial

Spectral

Texture

Classification: Rule-Based

Rule Set:

#1 (1.000): If area [867.4445, 17887.1465] AND length < 1500.0000 AND numholes < 0.0000 AND holesolrat > 1.0000 AND maindir > 60.0000, then object belongs to "Feature_1".

Export Options:

Vector Output Directory: C:\Users\Administrator.PC-20121124JOLJ\AppData\Local\Temp\

Output Class Image: G:\yadan20140116\ FX\ class result

Output Rule Image: G:\yadan20140116\ FX\ rule result

Feature Info:

Feature_1 Type: Polygon

Smoothing: Threshold of 1

Feature Name	Feature Count	Total Area	Mean Area	Min Area	Max Area
Feature_1	958	5580675.00	5825.34	450.00	60187.50

Through this we got Yardangs classification and established a Spatial information management platform in Arcgis software to calculatedata of all the Yardang bodies of area, perimeter, length and width.

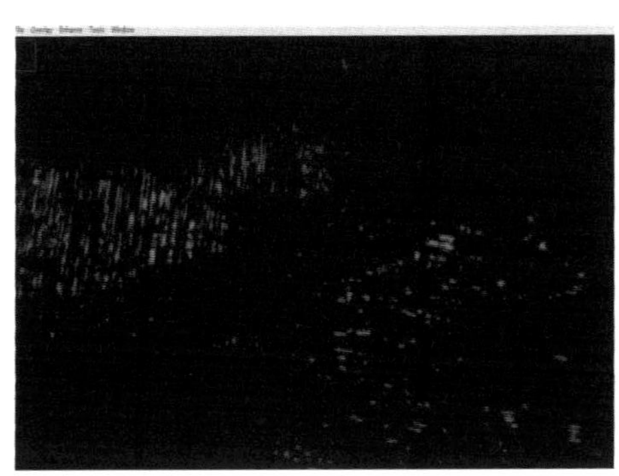

Figure 2. Yardangs got through rule-based classification

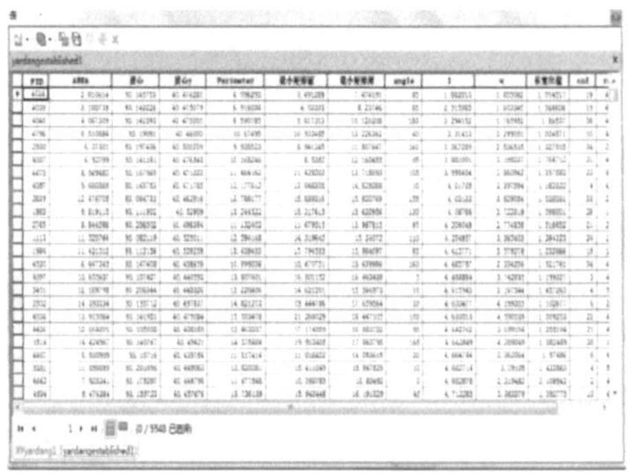

Figure 3. Date base of Yardangs in Spatial information management platform

3. Conclusion and discussion

3.1 Morphological features of Yardang landforms

By calculation and statistical analysis of all the Yardang bodies such as area, perimeter, length and width, with the maximum area, perimeter, length and width were 324843.1 m^2, 3447.52 m, 1508.41m, and 285.81 m, respectively. Additionally, the aspect ratio of Yardang has a certain positive correlation, with the coefficient of correlation being 0.675. Furthermore, the relationship between length and width of Yardang is calculated using formula Y=2.546X, where Y = length, X= width.

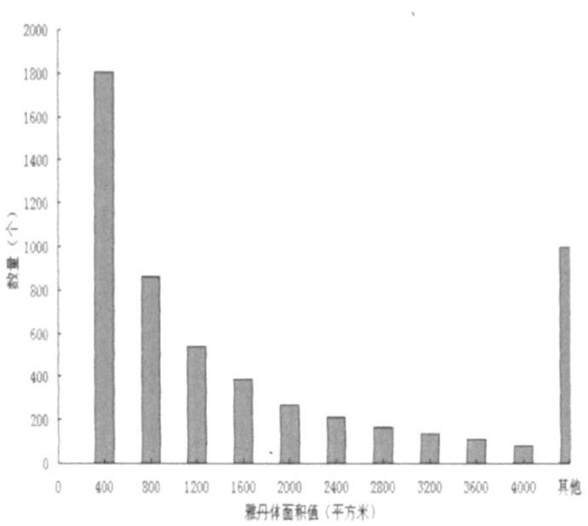

A

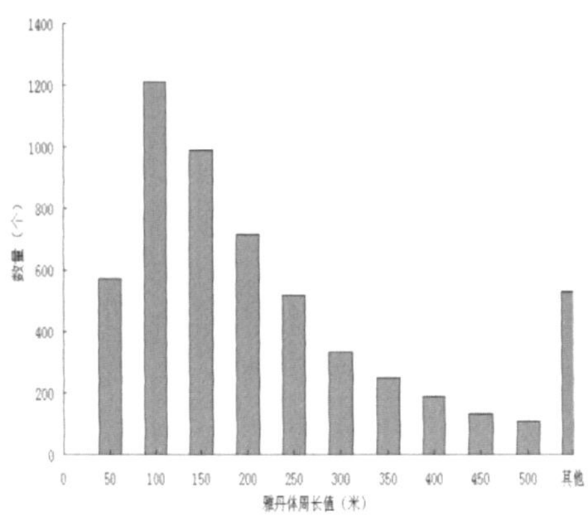

B

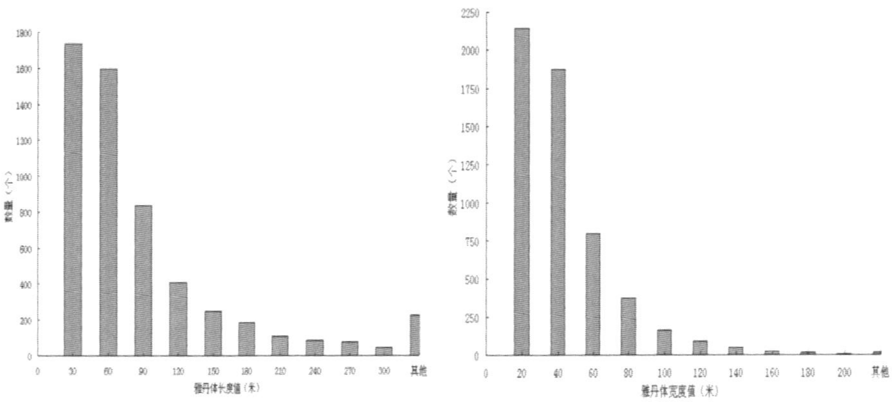

C D

Figure 4. Morphological features of Yardang landforms（A area, B perimeter, C length, D width）

3.2 Yardang types based on morphological features

According to the research data and field observations, combined with Yardang body growth process: incubation period, childhood, youth, maturity, old age and death period, Yardangs were divided into five types: ridge-shaped Yardang, wall-shaped Yardang, tower-shaped Yardang, column Yardang and Yardang Monadnock.

Table 1. Typical Yardang body of five type in Dunhuang Geopark with brief descriptions

No.	Type	Description	Evolution stage
1	Mound-shaped Yardangs, wall-shaped Yardangs (Fleet out to Sea)	The mound-shaped Yardangs run from south to north, with a distribution range of about 5 km2. They cover an area of 50-200 m long and 5-30 m wide, and rise 15-25 m high, with an interval of 20-80 m between adjacent mounds. The stratum is basically level, with vertical joints. There are water erosion gullies along the joints. Some of them have collapsed.	The mound-shaped Yardangs are lathy. Mound-shaped Yardangs can be considered the adolescents of the Yardang landforms. They are of great importance for studying the development of Yardang landforms.
2	Mound-shaped Yardangs (Western Sea Fleet)	This area has hundreds of south-to-north, mound-shaped Yardangs and wall-shaped Yardangs. It is like a neatly arranged fleet sailing grandly on the sea. The joints which developed in the Yardangs run at 185°-194°.	
3	Wall-shaped Yardang (Grotto)	Wall-shaped Yardangs about 10 m high and 100 m long are distributed in a nearly south-north direction. Differential weathering formed a concave-convex Yardang surface, which looks like a cliff cave, giving you a solemn and serene feeling.	Wall-shaped Yardangs represent mature Yardang landforms, which have diversified shapes and a complete range of types. Home to a wide range of

4	Wall-shaped Yardang (Yardang Castle)	Surface erosion changed the tall wall-shaped Yardangs into magnificent, towering castles measuring 170 m in length and 40 m in height. The castle shape comes from the development of 12 vertical joints and gullies on the surface.	mature Yardang landforms, the Geopark has become the optimal site for the study of Yardang landforms and for geotourism for both national and international visitors because of its great international and domestic significance.
5	Wall-shaped Yardang (Gold Turtle Playing with Toad)	This is a kind of mature Yardang landform between wall-shaped Yardangs and tower-shaped Yardangs, measuring 50-100 m in length, 20 m in height, on average, and 10 -15 m in width. Co-existing with the Yardangs are desert wetlands, fixed dunes, etc.	
6	Wall-shaped Yardang and columnar Yardang (Intimacy)	Due to erosion along vertical joints, formerly wall-shaped Yardangs were split into columnar Yardangs. Various forces caused the Yardangs to collapse and lean towards each other.	
7	Tower-shaped Yardang (Lotus Throne)	This kind of tower-shaped Yardang looks like a lotus throne and measures 15 m in height and 60 m in diameter. It consists of alternating layers of sand and clay.	
8	Tower-shaped Yardang (Golden Lion Greeting Guests)	This is a tower-shaped Yardang, about 20 m high. Its base is about 50 m long and 30 m wide. It looks like a golden lion lying with its face toward the east.	Tower-shaped Yardang marked the beginning Yardang landform development towards to old age. It has important significance for evolution of studying Yardang landform.
9	Tower-shaped Yardang (Lonesome Boat on the Black Sea)	This is the only oval Yardang in the vast black gobi, measuring about 30 m in diameter and 10 m in height, and looking like a lonesome boat on the black sea.	
10	Tower-shaped Yardang (North Gate)	Two parallel tower-shaped Yardangs, each measuring 15 m in height and 15 m in diameter, stand on each side of the road, forming a natural gate.	
11	Tower-shaped Yardang (West Gate)	Located in the west of the Yardang Geo-area, these Yardangs, measuring 15 m in height and 40 m in diameter, form a natural gate to the Yardang Geo-area.	
12	Columnar Yardang (Peacock Standing Gracefully)	This kind of Yardang looks like a graceful peacock standing about 10 m high. The clay layer has stronger resistance to wind erosion than the arenaceous layer. It is this differential weathering that helped shape the lifelike peacock standing gracefully.	As a kind of old-age Yardang landform, it marks an important step in Yardang landform development and has important research value.

13	Monadnock-shaped Yardang	This was formed in the late period of Yardang landform development. Gravity and other forces caused avalanches along the vertical joints and fractures, resulting in the collapse of the Yardang.	The Yardangs evolved into the decline phase, collapsing along the joints, resulting in Yardang monadnocks piling up in a disorderly fashion.

3.3 International significance of Yardangs in study area

Mostly developed in arid regions, the Yardang landforms are distributed in deserts on every continent except Australia. In China, they are distributed mainly in the Qaidam Basin in Qinghai, in the middle and lower reaches of the Shule River, and around Lop Nur in Xinjiang. Compared with other typical Yardang landforms in the Iranian Lut Desert, Kuwait, the Chad Basin in Africa, and the Mojave Desert in California, and other major Yardang landforms in China, the Dunhuang Yardang landforms feature complete and typical variety and have high scientific value (Mccauley ,et al. 1977a; Parsons & Abrahams,2009; Desir , et al. 2002).

Great diversity, dense distribution, graceful shapes, large quantities and obvious changes in trend make the Dunhuang Yardangs unique in the world. The north part of the Yardang Geo-area features nearly SN-trending Yardangs, while the south is characterized by nearly EW-trending Yardangs, which shows that the major agents creating Yardang landforms vary significantly. This is important for the study of the formation of Yardang landforms and for comparative studies of the regional environment. The great variety of Yardang landforms reveals multiple evolution stages in their formation, which is of great significance in the study of the evolution of landforms, the paleo-geography and paleo-climate in arid regions, and distinct wind erosion (Dong,2013; Cao &Yang,2009; Huang ,et al. 2006).

References

[1] Alganci U, Sertel E, Kaya S and Berkustundag B.2013. A research on agricultural mapping capabilities of the SPOT 6 satellite images. International Conference Onagro-Geoinformatics (Vol.59, pp.93-96).

[2] Cao J, Yang G. Characteristics of tourism resources in Xinjiang Lop Nur [J]. Acta Geologic Sichuan, 2009, 29:235-241.

[3] Cooke R U, Warren A and Andrew S G. Desert Geomorphology [M].UCL Press, 1993.

[4] Dong L. the Sedimentary Characteristics and Cause Analysis of Yardangs in Lop Nur [D]. Xinjiang Normal University, 2013.

[5] Desir G, Gutierrez-Elorza M and Gutieriez-Santolalla F. Yardangs in the semiarid central sector of the ebro depression (NE Spain) [J]. Geomorphology, 2002, 44(1):155-170.

[6] Dowling R, Newsome D. Geotourism: Geotourism's Issues and Challenges. Elsevier, London, pp. 2006, 242-254.

[7] Gizzi F T, Lazzari M. Geological risks and cultural heritage safeguard, DisasterAdv. 2012, 5: 3–4.

[8] Goudie A S. 2007. Mega-Yardangs: A Global Analysis. Geography Compass 1, 65–81.

[9] Huang Y L, Wei X H and Li F. Discuss on Problems Existing in Desert-Tourism Resources Development of North China [J]. OALib Journal, 2006, 26(5):739-744.

[10] Lin Z K, Cai L P and Luo T. Analysis of characteristics of dunhuang basin formation development [J]. Nei Jiang Science & Technology, 2012, (3):149-149.

[11] Mccauley J F, Breed C S and Grolier M J. Yardangs [A].In: Doehring D O. Geomorphology in Arid Regions[C]. London: Allen and Unwind, 1977, 233-269.

[12] Mccauley J F, Grolier M J and Breed C S .1977. Yardangs in D. O. Doehring (Ed.), Geomorphology in Arid Regions (pp. 233–269). London: Allen and Unwin.

[13] Parsons A J, Abrahams A D. Geomorphology of desert environments [M]. Berlin: Springer, 2009:597-627.

[14] Ren S M, Meng J Y and Lin J Y. Grain Size Analysis and 36Cl Dating of Early Pleistocene Sediments of Dunhuang Basin, Tectonic Uplift Implication[J]. Journal of Jilin University (Earth Science Edition), 2011, 41(5):1380-1388.

[15] Ward A W, Greeley R.1984. Evolution of the Yardangs at Rogers Lake, California .Geological Society of America Bulletin, 95, 829–837.

Long term spatio-temporal analyses of snow cover in Central Asia using ERA-Interim and MODIS products

J P Liu[1,2] **and W C Zhang**[1,3]

[1] Academy Key Laboratory of Digital Earth, Institute of Remote Sensing and Digital Earth, CAS, Beijing 100094, China
[2] University of Chinese Academy of Sciences, Beijing 100049, China

E-mail: zhangwc@radi.ac.cn

Abstract. In this study, an approach for generating a long term series of snow-cover products from 1979 to 2015 was proposed by integrating the data of ERA-Interim snow-depth and 8-day cloud-free MODIS snow-cover derived by removing cloud from MOD10A2/MYD10A2 product. On the basis of the spatio-temporal analyses and evaluation of snow-cover duration (SCD) during the hydrological year from 1979/1980 to 2014/2015 over Central Asia, the average start and melt date of snow-cover（SCS and SCM, respectively）were estimated using the long term snow-cover product. The results suggested that the snow-cover product derived by this approach is fairly satisfactory with the mean bias error (MBE) of -0.55%±5.03%. The SCD, SCS and SCM all presented an apparent north-south towards gradient as long as mountainous regions and waterbodies were avoided. The mean SCD over the high-latitude and high-mountainous regions were all beyond 122 days, however, it gradually became shorter with a significant level of a < 0.05 or even a < 0.001 from 1979/1980 to 2014/2015. In contrast, the SCD over low-latitude and low-altitude regions, like Turkmenistan and Uzbekistan, were evidently shorter than the former, but it became significantly longer with a significant level of a < 0.05 or even a < 0.001 in the southwestern, northern Turkmenistan and most of Uzbekistan in the same duration. Notably, most of Turkmenistan and Uzbekistan, where snow-cover usually appeared late and melting out early, even always stay snow-free throughout the year.

1. Introduction

Snow is one of essential factors in cryosphere and is very sensitive to global climate change, especially in the arid and semi-arid Central Asia [1]. ERA-Interim is the latest global atmospheric reanalysis dataset produced by European Centre for Medium-Range Weather Forecast (ECMWF) [2] under the project following ERA-15 and ERA-40. The goal of this project is to resolve several difficult problems mainly related to representation of hydrological cycle, the consistency of the reanalyzed geophysical fields on time and optimization of various reanalysis techniques, such as data selection, quality control, bias correction and performance monitoring, which seriously affected the quality of the reanalysis production of ERA-40 in data assimilations [2]. Previous study shows that ERA-Interim has made substantial progress in producing global estimates of the basic dynamical fields [3]. Temperature variations from 1979 to 2011were studied in Central Asia, the result suggested that temperature variations revealed from the ERA-Interim reanalysis matched well with the local

[3] To whom any correspondence should be addressed.

climate records [4]. The reanalysis dataset of ERA-Interim ozone was verified by the ERA-Interim TCO (total column ozone) only about 2% lower than the ozone monitoring instrument measured between 50°S and 50°N [5]. Nevertheless, to our best knowledge, long term analysis on spatio-temporal variability of snow-cover by using ERA-Interim reanalysis dataset of snow-depth (m of water equivalent) was not found in literatures, which is essential to understand dynamics of long-term snow-cover variations in Central Asia from 1979 to 2015.

Aiming to generate long term snow-cover product in Central Asia by using ERA-Interim snow-depth reanalysis and cloud-free MODIS 8-day snow-cover datasets, this study proposed an approach by integrating ERA-Interim snow-depth reanalysis and cloud-free MODIS 8-day snow-cover datasets to investigate spatial distribution of mean SCD (snow-cover duration) in each hydrological year in the period from 1979/1980 to 2014/2015 over Central Asia. In order to examine and evaluate the spatio-temporal variability of SCD, Mann-Kendall rank statistical test was used, and the average beginning (SCS) and melting out (SCM) date of snow-cover for each hydrological year were estimated by the produced long time series of snow-cover product to investigate the variations of SCD in the same duration in Central Asia.

2. Materials and methods

2.1. Study region

Located at mid-latitudes of the north-western hemisphere (35~55°N, 46~85°E), Central Asia, as the target area of this study, consists of five Central Asian countries without the Caspian Sea: Kazakhstan, Kyrgyzstan, Tajikistan, Uzbekistan and Turkmenistan, covering an area of nearly 4×10^6 km^2 [6, 7] as illustrated in Figure 1. Mountains and plateau (*i.e.* Tianshan, Pamir plateau), vast deserts (*i.e.* Kara Kum, kyzyl Kum and Taklamakan) with less forest (1%), lots of grasslands and shrubs (48%) [8] characterize its diverse and complex underlying surface with elevation ranging from -227 m a.s.l. to 7441 m a.s.l. of decreasing tread in topography from southeast to northwest of this region. The temperate continental climate predominates in Central Asia with characterization of warm and arid summer months and cold winter seasons [9], which is primarily caused by the rain-shadow effect of the Tianshan Mountains and Pamir plateau situated in the southeast of this region, isolating almost entire Central Asia from warm and humid air masses originating from the Indian Ocean [10]. Global climate changes pose heavy effects on eco-environment condition of the region, especially on glaciers and snow covers in this climate change sensitive region [11].

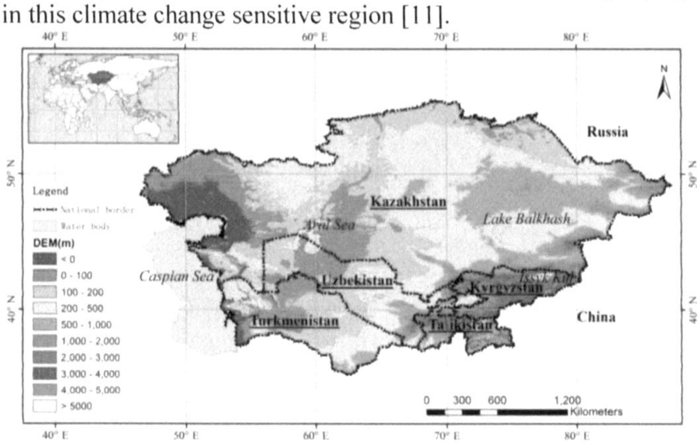

Figure 1. Overview of Central Asia including topographic feature and major water body.

2.2. Data utilized

2.2.1. 8-day cloud-free MODIS snow-cover dataset. The 8-day snow products MOD10A2 (Terra, V005) and MYD10A2 (Aqua, V005) obtained from the National Snow and Ice Data Centre (NSIDC) were utilized to calculate the threshold of snow-depth for generating cloud-free and accurate time series of snow-cover dataset. For covering the entire Central Asia, eight MODIS tiles (h21v03, h22v03, h23v03, h21v04, h22v04, h23v04, h22v05 and h23v05) were selected for the period from the 1st of September, 2000 to the 31st of August, 2015, spanning totally fifteen complete hydrological years during 2000/2001 and 2014/2015. Spatial resolutions of MOD10A2/MYD10A2 are both 500m with projection of Sinusoidal (SIN). The cloud removal approach in five steps ensured the satisfied quality of cloud-free images being generated with accuracy ranging from 90% to 94%, the overall accuracy of post-processed snow-cover dataset between 2000/2001 and 2014/2015 was between 92.79% and 93.34% over Central Asia [8, 12].

2.2.2. ERA-Interim snow-depth data. The ERA-Interim, covering the period from January 1979 to present, was initiated in the year of 2006 by the ECMWF, which was derived based on a 12-hourly four-dimensional variational analysis (4D-Var) including the adaptive estimation of biases in satellite radiance data [13]. Two analyses per day were implemented at 00:00 and 12:00 UTC (universal time coordinated) in ERA-Interim, serving as initial data fields for the subsequent forecasts [14]. Snow-depth (m of water equivalent) product with the highest spatial resolution of 0.125° derived from the ERA-Interim reanalysis with WGS-84 geodetic coordinate provided by ECMWF (http://apps.ecmwf.int/) was selected in this study. In order to match the MODIS 8-day snow-cover product in time, snow-depth at 12:00 UTC was adopted. According to [13], some locations of coastal sea and lake area were falsely identified as snow-free due to some pronounced errors were introduced in the pre-processing of ice mapping system (IMS) snow-cover derived from NOAA/NESDIS at ECMWF, a possible solution was to remove all the major water body from the original daily snow-depth reanalysis. In present study, we just adopt this suggestion as shown in Figure 1.

2.3. Methods

2.3.1. The algorithm for generating long-term snow-cover product. In order to accurately generate a long time series of snow-cover product, it is necessary to determine the threshold of SD derived from ERA-Interim to determine the dates when the snow-cover started to accumulate as well as the snow-cover began to vanish. The algorithm for generating long-term snow-cover product involves four steps. The first step is to generate 8-day maximum SD product through compositing 8 days of SDs derived from ERA-Interim. Note duration of 8 days of SDs should be in the same time duration of 8 days for MOD10A2 datasets from 2000/2001 to 2014/2015. The second step is to determine the threshold of each 8-day maximum SD according to the snow coverage rate of MOD10A2 for the same period with the method of iteration. The third step is calculating the mean value of each 8-day thresholds from 2000 to 2015. The fourth step is to apply the mean 8-day threshold to the corresponding 8 days of daily SD data derived from ERA-Interim for the period from 1979 to 2015.

2.3.2. Calculation of snow-cover start and melt Date. The produced daily snow-cover data were then aggregated to generate full time series of hydrological years (1 September through 31 August) for all years from 1979/1980 to 2014/2015 according to Equation (1):

$$SCD = \sum_{i=1}^{n} (s_i) \tag{1}$$

where SCD represents snow-cover duration; n stands for the number of observations, which is beginning with 1st September and ending with 31st August of the next year; and the s_i is the value of pixel , recoded as 0 or 1, derived from produced daily produced snow-cover data sets.

The snow-cover start (SCS) and snow-cover melt (SCM) date were considered as important indicators of trends and lasting changes [8]. The algorithms described in [15] for computing SCS and SCM were adopted in this study to calculate the SCS and SCM during the period from 1979/1980 to 2014/2015 in Central Asia with some parameters being modified as shown in Equation (2) and (3), respectively.

$$SCS = Fd - SCD_{bFd} \tag{2}$$

$$SCM = Fd + SCD_{aFd} \quad (3)$$

where SCD_{aFd} and SCD_{bFd} stand for after and before Fd refering to SCD, respectively. The Fd was determined to consist with the maximum snow-cover extent throughout all the study period, which is on the date of 26 January. Notably, snow-cover was conceived to remain throughout a year once it appears when the Equation (2) and (3) are applied, which means that we ignore any transient snow events during a hydrological year. This assumption may not directly reflect the actual conditions, however, it can present the quantitative variation of SCS and SCM, which was also the best way to display SCS and SCM in only single chart, respectively.

3. Results and discussions

3.1. *Accuracy of the algorithm for generating snow-cover data*

As shown in Figure 2, each mean 8-day threshold of SD throughout the year was derived for snow-depth of ERA-Interim between each individual in successive during hydrological years from 2000/2001 to 2014/2015. After applying these thresholds to 8-day ERA-Interim snow-depth products from 2000/2001 to 2014/2015, mean bias error (MBE) of the estimated results derived by this algorithm was only -0.55%±5.03% (MBE ± Standard Deviation), which was much better than that (16.42% ±13.98%) derived by original algorithms from ERA-Interim data. Finally, these mean 8-day thresholds of SD were applied to daily snow-depth data of ERA-Interim correspondingly, which can generate a long time series of daily and high-spatial-resolution (0.125 degree) snow-cover products from 1979 to 2015 over the Central Asia.

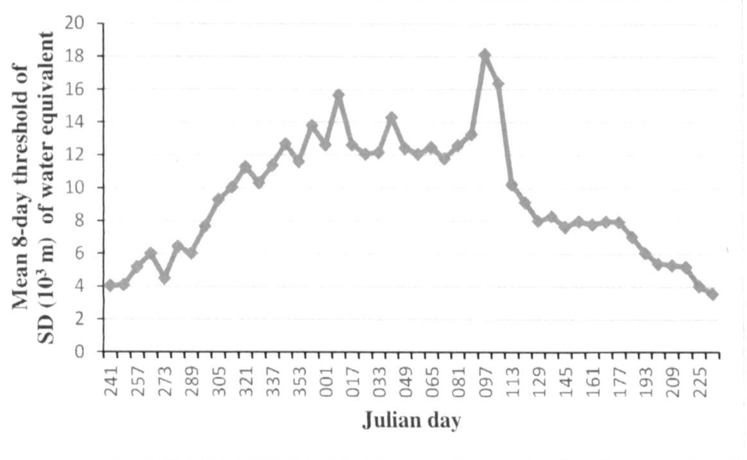

Figure 2. The mean 8-day threshold of SD throughout the year.

3.2. *The spatial distribution of mean snow-cover days*

Spatial distribution of snow-cover is essential for better understanding the snow-runoff processes. Using the produced daily snow-cover data sets, SCD map are generated during the period from 1979/1980 to 2014/2015. The mean SCD derived from the entire time series is better suited for a general overview of snow-cover conditions. The annual mean SCD between 1979/1980 and 2014/2015 were illustrated in Figure 3. From Figure 3, we can find that SCD tended to decrease from north to south with decreasing latitude, except these alpine regions of southeastern Pamirs and Tianshan Mountains and northeastern Altai Mountains over Central Asia. SCD below 50°N was less than 122 days in this north-south gradient. And furthermore, it has been snow- cover free for many years in the southern Turkmenistan. SCD in the Tianshan Mountains and northern Kazakhstan above 50°N are all beyond 122 days. Specially, in the alpine regions of southeastern Pamirs and northeastern Altai Mountains, some locations were snow covered longer than 244 days. Generally, the SCD presented an increase trend from south to north over Central Asia during the period in hydrological year from 1979/1980 to 2014/2015, besides that the altitude of terrain is also an important factor affecting the spatial distribution of SCD.

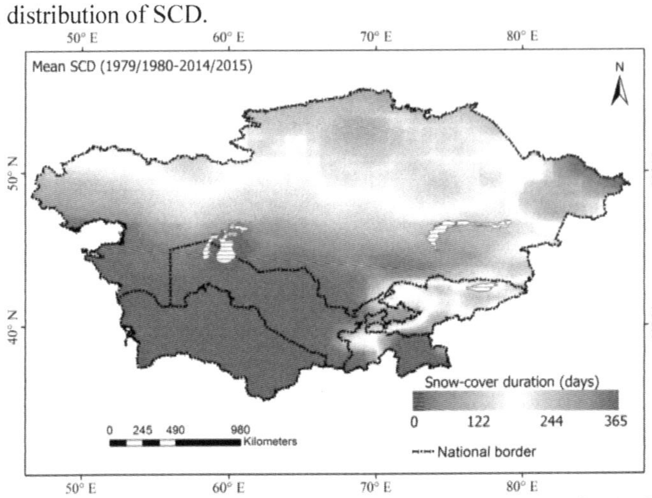

Figure 3. Annual mean snow-cover days from 1979/1980 to 2014/2015 in the Central Asia.

3.3. Detection of the spatio-temporal variations of snow-cover days

The method of Mann-Kendall test was applied in this study to detect the spatial and temporal variation trend of SCD during the period from 1979/1980 to 2014/2015 in Central Asia. As illustrated in Figure 4, Mann-Kendall test for spatial variation of SCD between 1979/1980 and 2014/2015 in Central Asia suggested that SCD decreased in most part of Central Asia, i.e. Kazakhstan, Pamirs, Tianshan Mountains and Altai Mountains, significant decrease appeared in the central and northwestern Kazakhstan, eastern Pamirs and southern Tianshan,. In the meanwhile, the first EOF mode (EOF-1) of temperature experienced a drastic increase trend in these regions from 1979 to 2011(Hu et al., 2014). However, the SCD became significantly longer with a significant level of a < 0.05 in the southwestern Kazakhstan, northern Turkmenistan and most of Uzbekistan. In the most of Turkmenistan, especially in the southeast, SCD almost kept no change during the period from 1979/1980 to 2014/2015. As the snow-cover in Central Asia is affected by a variety of climatic factors, such as temperature, precipitation, sea-level pressure and so on, unfortunately, a few studies on the correlation between climatic factors and snow-cover in Central Asia was conducted. The following the sensibility of snow-cover in Central Asia to climatic factor will be analysed.

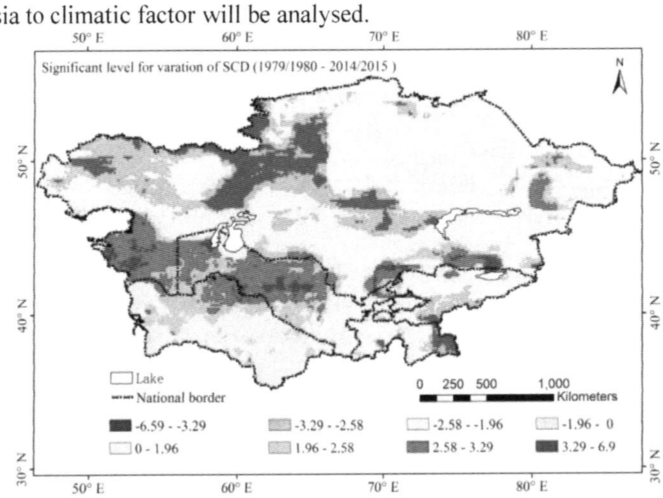

Figure 4. Mann-Kendall test for variation of SCD from 1979/1980 to 2014/2015.

3.4. The spatial distribution of mean SCS and SCM

The start and melt of snow-cover duration (SCS and SCM) are considered the most important parameters except SCD. Mean SCS and SCM, derived from single-year SCS and SCM during the period from the hydrological year of 1979/1980 to 2014/2015, were illustrated in Figure 5 and 6.

From Figure 5 we can find that mean conditions of SCS conform to the mean SCD shown in Figure 4 with earliest snow-cover in the alpine regions of southeastern Pamirs and northeastern Altai Mountains. Conversely, the latest snow-cover is located in the Turkmenistan and parts of Uzbekistan in the southwest, where usually stay snow-free. Like the spatial distribution of SCD, an apparent north-south gradient of SCS emerged in Central Asia except the alpine regions and waterbodies. In these alpine regions the topography predominated the main influential factor on snow-cover but not latitude. Figure 6 illustrated the mean SCM for all hydrological years from 1979/1980 to 2014/2015, which coheres with mean SCD (Figure 3) and mean SCS (Figure 5). The mean SCM was also characterized by north-south gradient as long as mountains and water bodies were avoided. In these alpine regions the key factor affected the SCM is topography but not latitude.

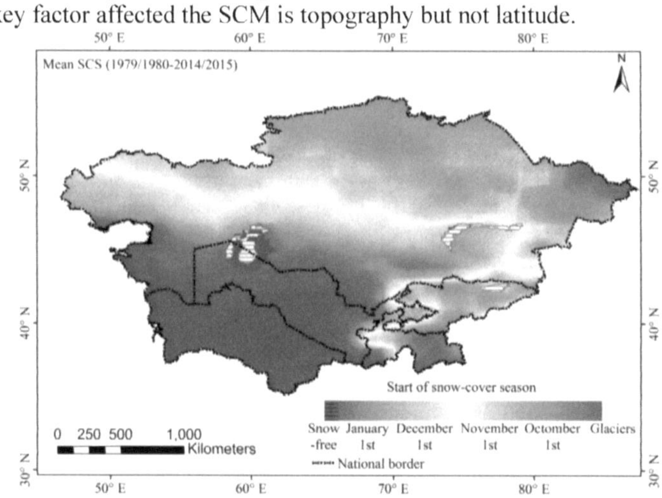

Figure 5. Mean SCS for all the hydrological years from 1979/1980 to 2014/2015.

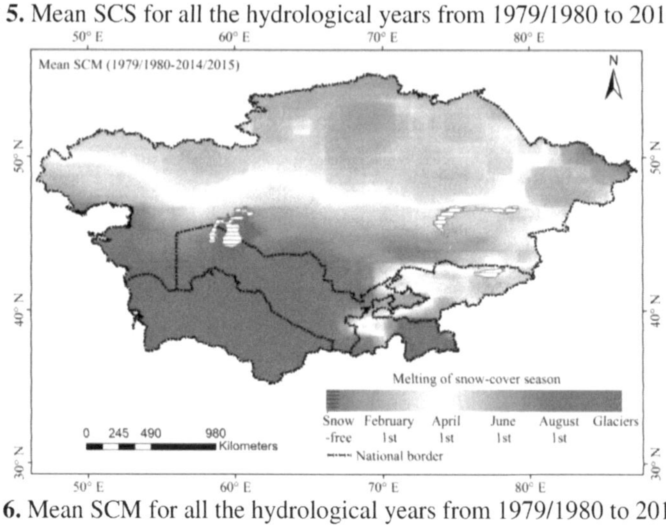

Figure 6. Mean SCM for all the hydrological years from 1979/1980 to 2014/2015.

4. Conclusions

In this study, an approach for generating a long time series of snow-cover product from 1979 to 2015 was proposed by integrating the data of ERA-Interim snow-depth and 8-day cloud-free MODIS snow-cover derived by removing cloud from MOD10A2/MYD10A2 product. Based on the spatio-temporal variations and evaluation of snow-cover duration (SCD) during the hydrological year from 1979/1980 to 2014/2015 over Central Asia, the average start and melt date of snow-cover（SCS and SCM, respectively）were estimated using the long time series of snow-cover product. The results suggested that:

(1) The mean bias error of the proposed algorithm is only -0.55% ± 5.03%, which is much better than the one unprocessed procedure (16.42% ±13.98%). It suggests that the algorithm combined with ERA-Interim snow-depth and cloud-free MODIS snow-cover is significantly effective and the snow-cover results have high accuracy.

(2) The spatial distribution of mean SCD presented obviously north-south gradient as long as alpine regions and waterbodies were avoided. The mean SCD in high-latitude regions were all beyond 122 days, however, it became significantly shorter with significant levels of $a < 0.05$ or even $a < 0.001$ from 1979/1980 to 2014/2015. In contrast, the SCD in low-latitude and low-elevation regions, like Turkmenistan and Uzbekistan, were relatively very shorter than other regions. Furthermore, the trends of their SCD became notably longer with a significant level of $a < 0.05$ or even $a < 0.001$ in the southwestern, northern Turkmenistan and most of Uzbekistan.

(3) The spatial distribution of both mean SCS and SCM conforms to mean SCD, presenting an obvious north-south gradient except the alpine regions and waterbodies. In the high-latitude or high-mountains regions, including Pamirs, Tianshan Mountains, Altai Mountains and the Kazakhstan northward 50°N, where the snow-cover appeared early and melting late. However, in the other regions, such as Turkmenistan and Uzbekistan, snow-cover appeared late and melting early, even stayed snow-free throughout the year.

Acknowledgements
This study is funded by CRSRI Open Research Program (CKWV2015226/KY), and the National Key Technologies R&D Program of China (No. 2016YFA0602302 and No. 2016YFB0502502). We are thankful to Professor Tie Liu (Xinjiang Institute of Ecology and Geography, CAS) for his helpful suggestions.

References
[1] Hall D K and Riggs G A 2007 Accuracy assessment of the MODIS snow products *Hydrological Processes* **21** 1534-47
[2] Dee D P and Uppala S 2009 Variational bias correction of satellite radiance data in the ERA-Interim reanalysis *Quarterly Journal of the Royal Meteorological Society* **135** 1830-41
[3] Bao X and Zhang F 2013 Evaluation of NCEP–CFSR, NCEP–NCAR, ERA-Interim, and ERA-40 Reanalysis Datasets against Independent Sounding Observations over the Tibetan Plateau *Journal of Climate* **26** 206-14
[4] Hu Z Y, Zhang C, Hu Q and Tian H Q 2014 Temperature Changes in Central Asia from 1979 to 2011 Based on Multiple Datasets* *Journal of Climate* **27** 1143-67
[5] Dragani R 2011 On the quality of the ERA-Interim ozone reanalyses: comparisons with satellite data *Quarterly Journal of the Royal Meteorological Society* **137** 1312-26
[6] Chen X, Bai J, Li X Y, Luo G P, Li J L and Li B L 2013 Changes in land use/land cover and ecosystem services in Central Asia during 1990–2009 *Current Opinion in Environmental Sustainability* **5** 116-27
[7] Klein I, Gessner U and Kuenzer C 2012 Regional land cover mapping and change detection in Central Asia using MODIS time-series *Applied Geography* **35** 219-34
[8] Dietz A J, Kuenzer C and Conrad C 2013 Snow-cover variability in central Asia between 2000 and 2011 derived from improved MODIS daily snow-cover products *International Journal*

of Remote Sensing **34** 3879-902

[9] Kottek M, Grieser J, Beck C, Rudolf B and Rubel F 2006 World Map of the Köppen-Geiger climate classification updated *Meteorologische Zeitschrift* **15** 259-63

[10] Guo H, Chen S, Bao A, Hu J, Gebregiorgis A, Xue X and Zhang X 2015 Inter-Comparison of High-Resolution Satellite Precipitation Products over Central Asia *Remote Sensing* **7** 7181-211

[11] Sorg A, Bolch T, Stoffel M, Solomina O and Beniston M 2012 Climate change impacts on glaciers and runoff in Tien Shan (Central Asia) *Nature Clim. Change* **2** 725-31

[12] Gafurov A and Bárdossy A 2009 Cloud removal methodology from MODIS snow cover product *Hydrology and Earth System Sciences* **13** 1361-73

[13] Foppa N, Wunderle S, Hauser A, Oesch D and Kuchen F 2004 Operational sub-pixel snow mapping over the Alps with NOAA AVHRR data *Annals of Glaciology* **38** 245-52

[14] Tang Z G, Wang J, Li H Y and Yan L L 2013 Spatiotemporal changes of snow cover over the Tibetan plateau based on cloud-removed moderate resolution imaging spectroradiometer fractional snow cover product from 2001 to 2011 *Journal of Applied Remote Sensing* **7** 2431-44

[15] Wang X and Xie H 2009 New methods for studying the spatiotemporal variation of snow cover based on combination products of MODIS Terra and Aqua *Journal of Hydrology* **371** 192-200

Plant conservation priorities of Xinjiang region, China

L P Li[1], W H Cui[1], T Wang[1], S Tian[2], W J Xing[1,3], L K Yin[4], N Abdusalih[5] and Y M Jiang[1,]*

[1] Institute of Remote Sensing and Digital Earth, Chinese Academy of Sciences, 100101 Beijing, China

[2] South China Sea Branch, State Oceanic Administration, 510310 Guangzhou, China

[3] College of Geoscience and Surveying Engineering, China University of Mining and Technology, 100083 Beijing, China

[4] Xinjiang Institute of Ecology and Geography, Chinese Academy of Sciences, 830011 Urumqi, China

[5] College of Resources and Environmental Sciences, Key Laboratory of Oasis Ecology of the Ministry of Education, Xinjiang University, 830046 Urumqi, China

jiangym@radi.ac.cn

Abstract. As an important region in the Silk Road, Xinjiang is getting a good chance of developing economy. However at the same time, its natural environment is facing a big challenge. To better protect the plant diversity, it is urgent to make a thorough conservation plan. With a full database of vascular and medicinal plant distributions and nature reserve plant lists and boundaries in Xinjiang of China, we analysed the plant diversity hotspots, protection gaps and proposed the plant conservation priorities of this region. Differed from the widely accepted viewpoints that lots of plants were not included in nature reserves, we found that most of the plants (> 90%) were actually included in the current nature reserves. We believe that compared with establishing more nature reserves, improving the management of the existing ones is also important. Furthermore, the very few unprotected plants (< 10%) were distributed mostly in the regions of Aletai, Tacheng, Zhaosu, Manasi, Qitai and Hetian which could be the future conservation priorities.

1. Introduction

1.1. Biodiversity hotspots and conservation

Biodiversity is a great treasure for human being. While with the fast economic development, biodiversity is destroyed severely by resource over-exploitation, habitat destruction and other human activities. One good way to protect biodiversity is to build nature reserves. Prior to this, a scientific planning is important, *e.g.*, to have a comprehensive plan on where to build nature reserves and how big the nature reserves should be. In order to support most biodiversity with relatively least cost, nature reserves are better built in regions with high biodiversity, *i.e.* in biodiversity hotspots (Myers, 1988; Myers *et al.*, 2000).

With the development of GIS (Geographic Information System) method, biodiversity hotspots are more easily defined to guide the nature conservation. Through the intensive species distribution data collection, via the application of GIS, people could define the high biodiversity regions fast and reliably. GAP (Geographic Approach to Protection of biodiversity, Burley, 1988; Scott *et al.*, 1993) analysis is also a broadly used method in biodiversity conservation. A fundamental assumption of the method is that the best time to decrease the probability of a species' extinction due to human activities is well before its population is diminished to the point of endangerment (Jennings, 2000). So GAP analysis tries to find species that are not well protected whether the species are meanwhile endangered or not. Through all these methods, the destination of conservation is to keep common species always common.

With some newly developed methods in macroecology and some good quality regional plant checklists, distribution patterns of vascular plants were being acquired, *e.g.*, in Xinjiang, China (Li *et al.*, 2011, 2013). With these data available, combined with the nature reserve data, currently it would be possible to investigate the biodiversity hotspots, conservation gaps and conservation priorities.

1.2. Medicinal plant conservation in Xinjiang

Species having economic values is usually over-used which makes huge biodiversity loss. Medicinal plants are among the species facing over-exploitation. In 1970s, paclitaxel in *Taxus chinensis* var. *mairei* was found an active constituent of cancer-combating, and then this species was severely logged to a very low stock. It was then listed as endangered species in *China Plant Red Data Book* (Fu & Jin, 1991). *Sinopodophyllum hexandrum* has limited use in traditional Chinese medicines, and it has always been a common species without many disturbances. But all of a sudden, when podophyllotoxin was found high in this plant, it was very soon severely depleted and listed as a rare species in *China Plant Red Data Book* (Fu & Jin, 1991; Li *et al.*, 2005). If there was basic research on normal species, when they were even not threatened, and some reserves could be set beforehand, stocks of these species probably could not be so easily going down to extremely low, and status of these species could not be so quickly going from common to rare or endangered.

Xinjiang is a region with many minorities. There is more possible plant utilizing ways in different cultures. Besides traditional Chinese materia medica (CMM), Uygur medicine (UM) and Kazak medicine (KM) were also very popular in Xinjiang region. As the unplanned exploitation of these medicine source plants, vegetation in the region was highly destroyed, for example, the high mountain ecosystems including *Saussurea involucrate*, desert ecosystems including *Cynomorium songaricum*, *Cistanche deserticola*, and piedmont regions with *Ferula fukanensis*, *Ferula sinkiangensis* were all damaged due to the specific medicinal plant gathering.

The distribution patterns of vascular and medicinal plants were mapped and basic conclusions were given, for example, high plant richness in mountainous areas (Li *et al.*, 2011, 2015). While for conservation, some important issues are still not known, for example, what plants are protected in nature reserves, what plants are not, and where are the unprotected plants. Here in this research, with data of plant distributions and current nature reserves, we comprehensively analysed the vascular plant and medicinal plant diversity hotspots, protection gap species and protection gap areas in Xinjiang to give suggestions for plant conservation planning of this region. With this case study, we also would get a systematic method to obtain plant conservation strategies in arid regions.

2. Materials and Methods

2.1. Study area

Xinjiang was anciently the main region of "Xiyu" (Western Regions) where there were various minorities. It was an important place in ancient Silk Road, especially the regions around north and south of Tianshan mountains. In 2014, jointly submitted by China, Kazakhstan and Kyrgyzstan, "Silk

Roads: the Routes Network of Chang'an-Tianshan Corridor" was inscribed on the World Heritage List. The ecosystems in Xinjiang are also very important. Tianshan mountains, including areas of Bogda, Tomur, Kalajun-Kuerdening, and Bayinbuluk, were announced as world natural heritage in 2013.

With the great development chance under the "Belt & Road" economic policy of Chinese government, more infrastructure projects are going which makes the environment protection planning urgently needed in this region. There are currently more than 30 nature reserves (national and provincial level) in Xinjiang region (Figure 1). They are distributed very unevenly with some nature reserves targeted for wild animals occupying very large areas, for example, Lop Nur Wild Camel National Nature Reserve (4.5×10^4 km^2) and Altun Mountain National Nature Reserve (6.7×10^4 km^2).

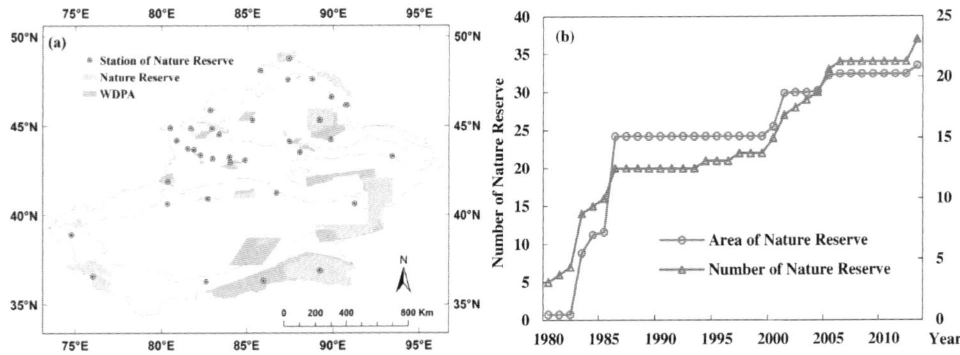

Figure 1. Current nature reserve distribution patterns (a) and the changing patterns in 30 years (b) in Xinjiang, China. In (a), dots show the location of the nature reserve management administrations, green area is the nature reserve extent digitalized from literatures, and blue area is the nature reserve extent downloaded from the IUCN World Database on Protected Areas (WDPA, http://www.protectedplanet.net/); the data of (b) was mainly from Statistic Bureau of Xinjiang Uygur Autonomous Region (2006, 2013) and Du & Zhang (2006); the division of Xinjiang in (a) is according to Li *et al.* (2011).

2.2. *Data sources*

The distributions of vascular plants were compiled from *Florae Xinjiangensis* (Commissione Redactorum Florae Xinjiangensis, 1992-2011), and online database (Xinjiang Ecological Resources and Environment Database, http://www.csdb.cn/). The medicinal plant checklists of Chinese Materia Medica (CMM) were from database *Classification and Codes of Traditional Chinese Medicine Resource* (Wang *et al.*, 2012) and *Records of Chinese Medicinal Resources* (China National Corporation of Traditional & Herbal Medicine, 1994). The UM and KM checklists were mainly from *Chinese Materia Medica: Uighur Cartridge* (State Administration of Traditional Chinese Medicine -- *Chinese Materia Medica* Editorial Board, 2005) and *Materia Medica of Kazak Nationality in China* (Wang, 2008-2012). All these data were formerly digitalized into 0.1° × 0.1° equal area grid cells and systematically recorded in Li *et al.* (2011, 2013, 2015). Please see the three references for detailed processing methods.

From an intensive literature survey, we collected scientific reports (or project planning) of 29 nature reserves (Table 1), including most of the national and provincial level reserves and part of the reserves in planning. They were mostly managed by State Forestry Administration. The boundaries and plant lists of each reserve were acquired from these literatures.

Table 1. Nature reserves with plant species lists or ranges collected

. No	Nature Reserve (Abbreviated)	Latitude	Longitude	Area	Year	Level	Management

1	Altun Mountains	36.91	89.24	45000	1983	National	MEP
2	Lop Nur Wild Camel	40.64	91.25	67000	1986	National	MEP
3	Kanas	48.78	87.47	2500	1980	National	SFA
4	Bayin bulah Swan	42.95	84.00	1487	1980	National	SFA
5	Tomur Peak	41.87	80.37	2376	1980	National	SFA
6	Westen Tianshan Mountain	43.15	82.98	312	1983	National	SFA
7	Ganjiahu *Haloxylon ammodendron*	44.52	83.35	1040	1983	National	SFA
8	Tarim *Populus euphratica*	43.09	84.88	3954	1983	National	SFA
9	Ebnur Lake Wetlands	44.85	82.93	2671	2000	National	SFA
10	Barleh Mountain	45.90	82.83	1150	1980	National	SFA
11	Tianchi Lake Bogda Peak	43.52	88.10	381	1980	Provincial	SFA
12	Burgun Beaver	46.17	90.73	88	1980	Provincial	SFA
13	Karamayle Mountain Ungulate	45.34	89.25	17330	1982	Provincial	SFA
14	Taxhorghan	36.57	75.75	15000	1984	Provincial	SFA
15	Altay Two River Headstream	46.63	89.90	6759	2001	Provincial	SFA
16	Middle Kunlun Mountains	36.34	86.00	32000	2001	Provincial	SFA
17	Koksu Wetlands	47.59	87.40	307	2001	Provincial	SFA
18	Pamir Plateau Wetlands	38.92	74.75	1256	2005	Provincial	SFA
19	Ertish River Koktoghay Wetlands	48.10	85.77	990	2005	Provincial	SFA
20	Xarxel	44.88	81.70	314	2000	Provincial	SFA
21	Yining *Fraxinus sogdiana*	43.65	81.88	4	1983	Provincial	SFA
22	Huocheng *Testudo horsfieldii*	44.20	80.89	350	1983	Provincial	SFA
23	Shaya Upper Reaches of Tarim River Wetlands	40.91	82.70	2568	2004	Provincial	SFA
24	Altay Xebarkol				2014	In planning	SFA
25	Yutian Keriya River				2014	In planning	SFA
26	Kashhar *Juniperus jarkendensis*				2014	In planning	SFA
27	Qarqan River Wetlands				2014	In planning	SFA
28	Changji Laolong River				2014	In planning	SFA
29	Ulughqat *Ammopiptanthus nanus*				2014	In planning	SFA

Note: *Latitude* and *Longitude* is the geographic location of the management administration; *Area*: km^2; *Year*: the initial founding year of the reserve; In the *Management* column, MEP: Ministry of Environmental Protection of the People's Republic of China, SFA: State Forestry Administration of the People's Republic of China.

2.3. *Data analyses*

Based on the plant distributions (Li *et al.*, 2011, 2013, 2015), with some 400 more species altitude ranges, species distributions were re-mapped and 5% of the highest plant richness regions were selected as the diversity hotspots. These were conducted for vascular plants, CMM, UM and KM, respectively.

In this research, we would deem species not appear in nature reserves the protection gap species, *i.e.* the species not being protected. While for a species appearing in a nature reserve, whether it is practically well protected or not, we deem it protected species.

The ranges of some nature reserves were digitalized and species lists of the nature reserves were carefully structured and checked. Then two methods were used to get the protection gap species.

Method 1 uses mainly the nature reserve species lists as basic data and method 2 uses mainly the species richness data clipped by the nature reserve ranges.

Method 1: Merge the plant lists from these reserves and remove the duplicates, then species appearing in nature reserves of Xinjiang were obtained. Compared with the species of Xinjiang (Commissione Redactorum Florae Xinjiangensis, 1992-2011), plants not appearing in nature reserves were considered protection gap species.

Method 2: Firstly, digitalize the nature reserve ranges according to the range maps in scientific reports (Figure 1a). For the nature reserves without range maps in the scientific reports, their ranges were simulated with buffer analysis in GIS. In particular, according to the area of the specific nature reserve, the range was drawn as an equal area circle with the management station as circle center. Secondly, by overlaying the ranges of each plant, we got the species that fell into the ranges as species in nature reserves. Then we overlaid the data of all nature reserves, removed the duplicates and the species lists of nature reserves could be obtained. Lastly, compared the lists with the species of Xinjiang (Commissione Redactorum Florae Xinjiangensis, 1992-2011), plants not appearing in nature reserves were considered protection gap species.

Combined the two results above, and after an expert review, we got the final list of protection gap species. Then the distributions of these gap species were mapped. Regions with high richness of these gap species were defined conservation priority regions for the future nature reserve construction.

3. Results

3.1. Vascular and medicinal plant biodiversity hotspots

High medicinal plant diversity regions were mainly in Altay mountains, Yili valley, part of mountains in west Jungar, and part of northern slopes of Tianshan mountains (Figure 2a,b,c). High vascular plant diversity regions were also concentrated mainly in mountains (Figure 2d). The hotspot regions for vascular plant, CMM, UM, and KM were concordant (Figure 2e).

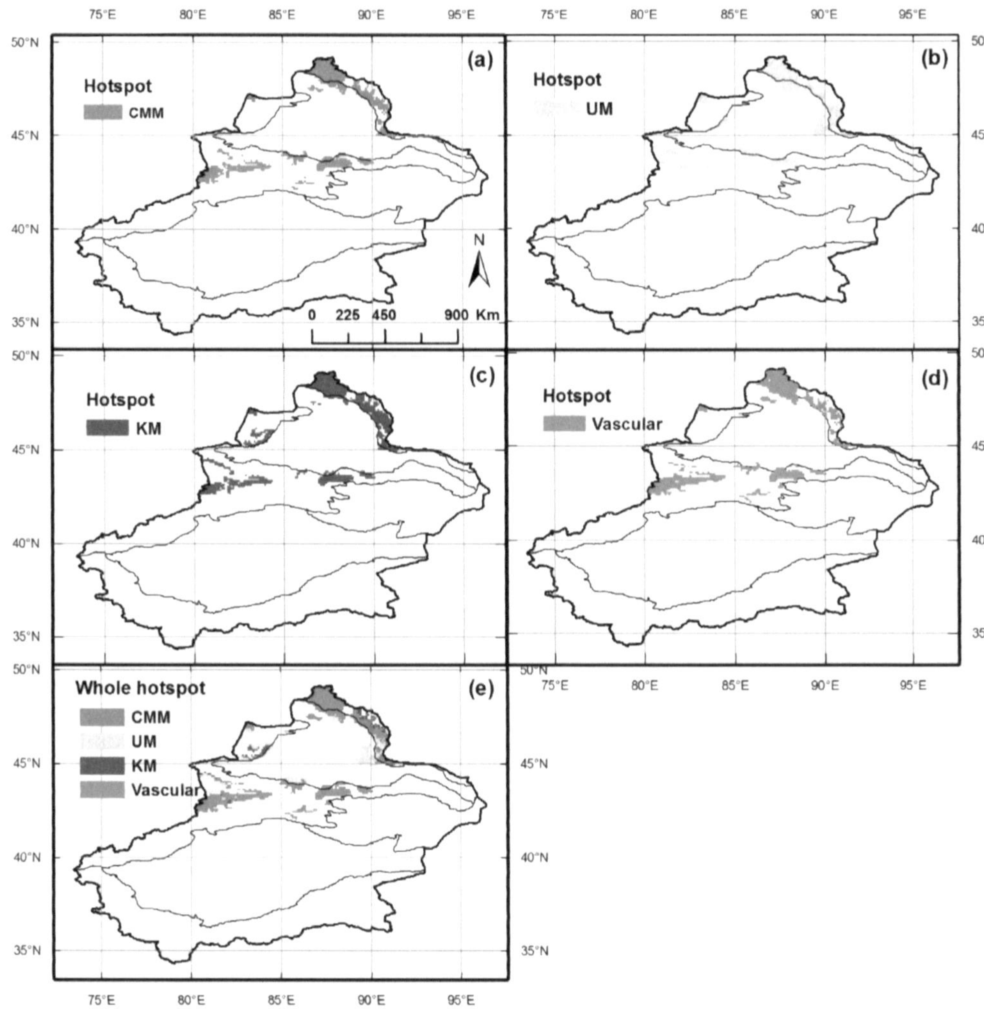

Figure 2. Biodiversity hotspots of vascular and medicinal plants in Xinjiang, China (a, CMM; b, UM; c, KM; d, vascular plants; e, the combination of the above four). Vascular plant layer was at the bottom in (e). With more species altitudinal distribution range data in this study, there were slight differences between (d) and hotspots in Li *et al.* (2011).

3.2. Vascular and medicinal plant protection gaps

After comparing the species list of nature reserves and *Florae Xinjiangensis* (Commissione Redactorum Florae Xinjiangensis, 1992-2011), we found that about 24% of vascular species were not under the protection of nature reserves. From this method, we further found that about 105 species of CMM (12% of whole CMM) not appearing in nature reserves. Luckily, most species of UM and KM were in the ranges of nature reserves (Table 2). By sending the results to experts, we modified the list of species not protected. The experts went over the lists and removed the ones that were ever seen in a nature reserve with their field experiences. The experts eventually removed 694 vascular species from the not protected plant lists. From this method, there were finally 175 of vascular plant species, including 73 of medicinal plant species, identified not appear in reserves.

Table 2. Plant species numbers in or not in nature reserves based on the species lists of nature reserves

Items	Vascular	CMM	UM	KM
Total	3638	846	63	266
Species in reserves	2769	741	59	253
Species not in reserves	869	105	4	13
Ratio of species in reserves (%)	76.1	87.6	93.7	95.1

Note: Based on the species list comparison, the not protected species were, for example, *Ferula fukanensis*, *Lycium chinense*, *Ferula sinkiangensis*, *Althaea nudiflora*, *Saposhnikovia divaricate*, *Hyssopus cuspidatus* var. *albiflorus*, *Mentha spicata*, *Campanula fedtschenkoana*, *Gnaphalium affine*, *Inula salicina*, *Achillea impatiens*, *Allium sinkiangense*, *Stellaria dichotoma* var. *lanceolate*, *Lotus corniculatus*, *Aconitum barbatum* var. *puberulum*, *Orchis roborovskii*, and *Digitaria ischaemu*.

With the second method, we differentiated the species from the plant distribution maps by using the digitalized nature reserve ranges. We found 419 vascular plants, including 40 CMM (*e.g.*, *Physochlaima physaloides, Lycopodium japonicum, Matricaria recutita*), 1 UM (*Hyssopus cuspidatus* var. *cuspidatus*), and 9 KM (*e.g.*, *Echinops sphaerocephalus, Sonchus palustris, Daphne altaica*) species were not conserved in nature reserves. These results were also sent for an expert review.

Combining the results of the above two methods, and after the expert review, we preliminarily concluded that there were totally about 262 species of vascular plants, including 124 species of medicinal plants (CMM, UM, and KM) not appearing in nature reserves. From this result, we could conclude that most of the native vascular plants (93%) and medicinal plants (90% of the whole medicinal plants) were distributed in nature reserves.

By mapping the distributions of these plants, we found that northern Xinjiang had more gap species than southern Xinjiang. After the careful consideration, we defined six regions waiting for more attentions, including northwest of Tacheng basin (belonging to Tacheng city), southwest of Yili valley (belonging to Zhaosu county), northwest of Altay mountains (belonging to Aletai city), middle Tianshan Mountains in the northern slope (mainly in Manasi county and Qitai county) and part of Kunlun mountains (especially Hetian area) (Figure 3). The diversity of gap species in Hetian area was lower than the regions in northern Xinjiang, but it was relatively high in Pamir Plateau and Kunlun Mountains, so we listed this area also as high gap species region.

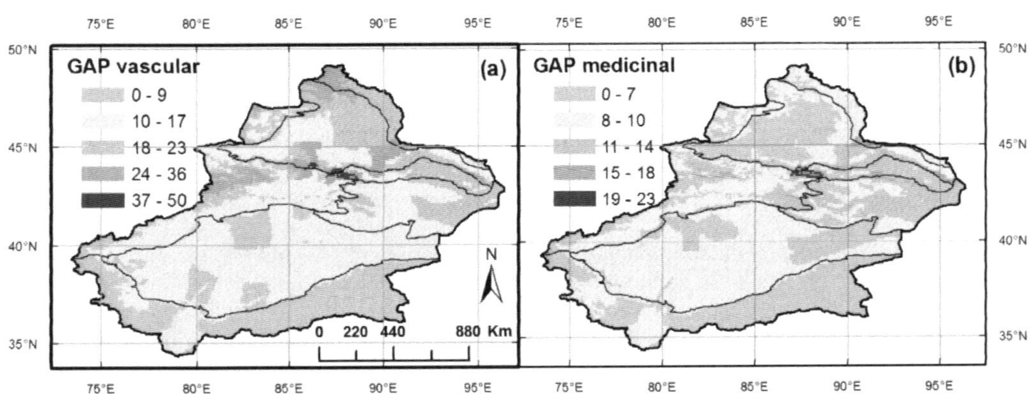

Figure 3. Vascular and medicinal plant protection gap species distribution patterns in Xinjiang region.

4. Discussion

4.1. Plant protection status evaluation

Vascular and medicinal plant biodiversity was formerly mapped and the distribution patterns were clear in Xinjiang region (Li *et al.*, 2011, 2015). Some conclusions were obtained with concrete data, for example, the high biodiversity regions were concentrated in mountainous areas. This could guide the regional conservation policy. But how the species was protected in this region was not evaluated, and what species was not protected was still not clear. Based on the former research, by further collecting the nature reserve species lists and boundaries, this study comprehensively analysed the presence of species within and out nature reserves, and mapped the locations of species not in reserves. Analysis was conducted for plants and medicinal plants, respectively. This is beneficial for the future conservation plan of vascular and medicinal plants.

Some important plants were evaluated in Chinese national level nature reserves, for example, Yuan *et al.* (2009) found that 20% of the Chinese state key protected plant species was not included in national nature reserves. Protection of orchid species in China were also evaluated (Qin *et al.*, 2012). Although being a highly concerned family, orchids were very poorly protected, with nearly 50% of the orchid species being not well protected in nature reserves (Qin *et al.*, 2012). The large scale methods were not very precise, for example, data from scientific reports was not of good quality, not having enough new references and materials included in the analysis, and without specimen identification (Jin 2012). New protection status evaluation of orchid species in China was conducted recently (Zhang *et al.*, 2015b) and found that 154 of 1449 orchid species were not included in nature reserves. This means about 11% of orchid species was till now still without any protection in China. Similar research found that about 12% of threatened species in China was not covered by reserves (Zhang *et al.*, 2015a).

Comparing species lists of nature reserves with targeted protection species is the most frequently used method for evaluating species protection status. With this method, most studies got high percentages of species not protected, for example, the above mentioned studies, all higher than 10% even for threatened and mostly concerned species, *e.g.*, orchids. We also used this method as method 1 in this study. After the expert review, we found that this method under-estimated the species protection status. Caused possibly by data problems, including not enough reserve information and the species lists of reserves not containing the entire species in the reserves, some more species would be listed as unprotected species. We suggested that an expert review was needed after comparing the species list.

Not as imagined, with low nature reserve numbers in Xinjiang, there are actually only low percentages of species never appear in reserves in Xinjiang. It includes the species of some specific variations in taxonomy, *e.g.*, *Hyssopus cuspidatus* var. *albiflorus* and *Artemisia sacrorum* var. *messerschmidtiana*; and some important medicinal plant species, *e.g.*, *Ferula* spp. We concluded that most species in Xinjiang are currently included in reserves, but whether they were well protected still lacks evaluation.

4.2. Vascular and medicinal plant conservation

In the last six decades, Chinese government built more than 3000 nature reserves (according to the database of NSII, http://www.papc.cn/). These reserves played very important role in species conservation. But some of the reserves were not well managed, for example, there were not any investigation data, there were not workers and they even did not have fixed boundaries. Because of data deficiency, there has not been comprehensive research on all these nature reserves till now. The very few comprehensive research included only national nature reserves (Zhao *et al.*, 2013).

Xinjiang government has built more than 30 nature reserves (national and provincial level) since 1980. The reserves cover about 15% of the total region area. The areas of the reserves are quite uneven with five reserves (> 10,000 km^2, Lop Nur Wild Camel, Altun Mountains, Middle Kunlun Mountains, Karamayle Mountain Ungulate, and Taxkorgan) cover 75% of the total reserve area. From the *Statistical Yearbook* (Statistic Bureau of Xinjiang Uygur Autonomous Region, 2006, 2013), there were few reserves established after 2005. There is still big development space for reserve construction. Because of the climate change and also human activities, the landscapes and ecosystems are changing in Xinjiang, for example, the ecological functions of wetlands are reducing in Bayin bulah (Shi *et al.*, 2015).

Using as medicines is an important reason for the species loss. So some specific measures for protecting medicinal plants are necessary. The same as China, India is a country with traditional herbal medicines broadly used. Although without systematic analysis, medicinal-plant conservation areas were tentatively put forward 10-15 years ago in India (Dhar *et al.*, 2000; Kala, 2000, 2005). In Pakistan, Important Plant Areas (IPAs) for medicinal and economic plants were identified (Sher *et al.*, 2012). Here in this research, we found that some medicinal plants were not well protected, including the broadly used species *Ferula fukanensis* and *Ferula sinkiangensis*. Some more attentions should be given to these species.

4.3. Data quality and method problem

The gap analysis is based on data of regional species lists / distributions and nature reserve species lists / boundaries. Regional species data were formerly published (Li *et al.*, 2011, 2015) while the nature reserve data were not easy to be acquired. In this study, we included part of the nature reserve information of Xinjiang. Our current nature reserve dataset included the all national ones and most of the provincial ones, while the county level nature reserves were mostly not included. This may increase the number of species not in reserves but we believe the impact from this is few. According to our investigation, the county level nature reserves were not very effectively managed. Species in these county level nature reserves were actually very possibly not very different with species out of the nature reserves. So our dataset highly represents the situations of nature reserves in Xinjiang.

During the analysis process, we found that the data quality of species lists in Xinjiang reserves was very poor. Species in the checklists of nature reserves were very frequently not the accept names. This makes mistakes on the judgement of species appears in a reserve or not. Through the expert review, nearly 80% of the gap species obtained from the species list comparison (method 1) was found actually in nature reserves. Further, the data quality problems lead to some conflict results for gap species analysis with different methods (for example, method 1 and method 2 in this research). Thus, more investigations of these nature reserves are needed.

This research defined plants in the nature reserves well protected but actually it might not. Species in the nature reserves could also be severely damaged. So, further methods are still needed in the evaluation processes, for example, detail investigations of the plant protection status in the nature reserves.

5. Conclusion

We preliminarily found that most plants in Xinjiang (> 90%) appeared in nature reserves. So we concluded that besides building new reserves, it should be very important improving the management status of current nature reserves and more resource investigation should be conducted in nature reserves. And further, more attentions should be given to the 262 unprotected species. The gap species were mostly distributed in Aletai, Tacheng, Zhaosu, Manasi, Qitai and Hetian. These areas could be the possible target region for future new reserve construction.

Acknowledgements

The project is supported by National Natural Science Foundation of China (81503183, 81560660). We thank Hui Zhang for supplying the national level nature reserve boundaries. We consulted Y Hai and G P Wang for the modification of the gap species. The method of buffer analysis for nature reserve borders was from G H Zhao's master thesis.

References

[1] Burley F W 1988. Monitoring biological diversity for setting priorities in conservation. In:

Wilson E O (Ed) *Biodiversity* (National Academy Press: Washington DC) pp 227-230.

[2] China National Corporation of Traditional & Herbal Medicine 1994. *Records of Chinese Medicinal Resources* (Science Press: Beijing). (in Chinese)

[3] Commissione Redactorum Florae Xinjiangensis 1992-2011. *Florae Xinjiangensis* (Vol 1-6). (Xinjiang Science & Technology Publishing House: Urumqi). (in Chinese)

[4] Dhar U, Rawal R S and Upreti J 2000. Setting priorities for conservation of medicinal plants — a case study in the Indian Himalaya. *Biological Conservation,* **95**(1): 57-65.

[5] Du N and Zhang P 2006. *Xinjiang Nature Reserves*. (Xinjiang Science & Technology Publishing House: Urumqi). (in Chinese)

[6] Fu L G and Jin J M 1991. *China Plant Red Data Book: Rare and Endangered Plants* (Vol 1). (Science press: Beijing). (in Chinese)

[7] Jennings M D 2000. Gap analysis: concepts, methods, and recent results. *Landscape Ecology*, **15**(1): 5-20.

[8] Jin X H 2012. A comment on "Assessment of in situ conservation of 1,334 native orchids in China". *Biodiversity Science,* **20**(2): 235-237. (in Chinese)

[9] Kala C P 2000. Status and conservation of rare and endangered medicinal plants in the Indian trans-Himalaya. *Biological Conservation,* **93**(3): 371-379.

[10] Kala C P 2005. Indigenous uses, population density, and conservation of threatened medicinal plants in protected areas of the Indian Himalayas. *Conservation Biology,* **19**(2): 368-378.

[11] Li L P, Wang Z, Zerbe S, Abdusalih N, Tang Z Y, Ma M, Yin L K, Mohammat A, Han W X and Fang J Y 2013. Species richness patterns and Water-Energy Dynamics in the drylands of northwest China. *PLoS ONE,* **8**(6): e66450.

[12] Li L P, Yin L K and Tang Z Y 2011. Distribution patterns of the species richness of plants and animals in Xinjiang, China. *Arid Zone Research,* **28**(1): 1-9. (in Chinese)

[13] Li L P, Zhang B G, Xiao P G, Zhang Z, Qi Y D, Li X J, Wang G P and Liu H T 2015. Native medicinal plant richness distribution patterns and environmental determinants of Xinjiang, Northwest China. *Chinese Herbal Medicines,* **7**(1): 45-53.

[14] Li Z C, Wang X L and Ge X J 2005. The biology and conservation of *Sinopodophyllum hexandrum*, an endangered medicinal herb. *Guihaia,* **25**(2): 179-185. (in Chinese)

[15] Myers N 1988. Threatened biotas: "hot spots" in tropical forests. *Environmentalist,* **8**(3): 187-208.

[16] Myers N, Mittermeier R A, Mittermeier C G, da Fonseca G A B and Kent J 2000. Biodiversity hotspots for conservation priorities. *Nature,* **403**(6772): 853-858.

[17] Qin W H, Jiang M K, Xu W G and He Z H 2012. Assessment of in situ conservation of 1,334 native orchids in China. *Biodiversity Science,* **20**(2): 177-183. (in Chinese)

[18] Scott J M, Davis F, Csuti B, Noss R, Butterfield B, Groves C, Anderson H, Caicco S, D'Erchia F, Edwards T C, Jr Ulliman J and Wright R G 1993. Gap analysis: A geographic approach to protection of biological diversity. *Wildlife Monographs,* **123**: 1-41.

[19] Sher H, Ali H and Rehman S 2012. Identification and conservation of important plant areas (IPAs) for the distribution of medicinal, aromatic and economic plants in the Hindukush - Himalaya mountain range. *Pakistan Journal of Botany,* **44**: 187-194.

[20] Shi H, Yang Z P, Han F, Shi T G and Li D 2015. Assessing landscape ecological risk for a world natural heritage site: a case study of Bayanbulak in China. *Polish Journal of Environmental Studies,* **24**(1): 269-283.

[21] State Administration of Traditional Chinese Medicine -- "Chinese Materia Medica" Editorial Board 2005. *Chinese Materia Medica: Uighur Cartridge*. (Shanghai Scientific and Technical Publishers: Shanghai). (in Chinese)

[22] Statistic Bureau of Xinjiang Uygur Autonomous Region 2006,2013. *Xinjiang Statistical Yearbook*. (China Statistics Press: Beijing). (in Chinese)

[23] Wang R 2008-2012. *Materia Medica of Kazak Nationality in China* (Vol 1-4). (Xinjiang Science & Technology Publishing House: Urumqi). (in Chinese)

[24] Wang X J, Sun C Z, Chen S L, Lin Z J and Li H 2012. Design and development of network query system on "Classification and Codes of Traditional Chinese Medicine Resource" National Standard. *World Science and Technology/Modernization of Traditional Chinese Medicine and Materia Medica,* **14**(4): 1904-1908. (in Chinese)

[25] Yuan H, Zhang Y B, Qin H N, Liu Y and Yu M 2009. The in situ conservation of state key protected wild plants in national nature reserves in China. *Biodiversity Science,* **17**(3): 280-287. (in Chinese)

[26] Zhang Z J, He J S, Li J S and Tang Z Y 2015a. Distribution and conservation of threatened plants in China. *Biological Conservation,* **192**: 454-460.

[27] Zhang Z J, Yan Y J, Tian Y, Li J S, He J S and Tang Z Y 2015b. Distribution and conservation of orchid species richness in China. *Biological Conservation,* **181**: 64-72.

[28] Zhao G H, Tian Y, Tang Z Y, Li J S and Zeng H 2013. Distribution of terrestrial national nature reserves in relation to human activities and natural environments in China. *Biodiversity Science,* **21**(6): 26-33. (in Chinese)

Synergistic Use of WorldView-2 Imagery and Airborne LiDAR Data for Urban Land Cover Classification

M F Wu [1, 2], **Z C Sun** [2, 3], **B Yang** [1], **S S Yu** [2, 4]

[1] Hunan Normal University, Changsha Hunan 410006, China
[2] Key Laboratory of Digital Earth Science, Institute of Remote Sensing and Digital Earth, Chinese Academy of Sciences, Beijing 100094, China
[3] Hainan Key Laboratory Earth Observation, Sanya Hainan 572029, China
[4] Shandong University of Science and Technology, Qingdao Shandong 266510, China

Email: sunzc@radi.ac.cn

Abstract. There are lots of challenges for deriving urban land cover types for high resolution optical imagery because of spectral similarity of different objects, mixed pixels, shadows of buildings and large tree crowns. In order to reduce these uncertainties, recently, it's a trend of the classification of urban land cover from multi-source sensors in the field of urban remote sensing. In this study, a hierarchical support vector machine (SVM) classification method was applied to the urban land cover mapping, using the WorldView-2 imagery and airborne Light Detection and Ranging (LiDAR) data. The results showed that: (1) The overall accuracy (OA) and overall kappa (OK) were 72.92% and 0.66 for WorldView-2 imagery alone; while the OA and OK were improved up to 89.44% and 0.87 for the synergistic use of the two types of data source. (2) Buildings and road/parking lots extracted from fused data were more precision and well-shaped. The two classes from fused data were optimally classified with higher producer's accuracy and user's accuracy than WorldView-2 imagery alone. The trees were also easily separated from the grasslands when the airborne LiDAR data was added. (3) The fused data could reduce the phenomenon of different spectral character of the complex and detailed objects. It was also helpful to address the problem of shadows from the high-rise buildings. The results from this study indicate that the synergistic use of high resolution optical imagery and airborne LiDAR data can be an efficient approach to improving the classification of urban land cover.

1. Introduction

Sustainable and cost effective urban sprawl dynamic monitoring is desirable, but the development and implementation of accurate urban land cover classification remain a significant challenge. It is often difficult to accurately classify urban land cover from Very High Resolution (VHR) optical images alone. Some limitations include: (a) some materials of urban objects have similar spectral properties. For example, some buildings with the grey roofs share similar reflectance characteristics with the asphalt roads; and the spectral properties of high albedo objects such as bare soils, cloud, sand and the buildings with white roofs are similar [1]. (b) Because of the heterogeneity in urban landscapes, the mixed pixel has been recognized as a challenge to improve classification accuracy [2]. (c) Shadows casted by buildings or large tree crowns have become a constraint for the classification [3]. Therefore, recent advances in Earth Observation allow using multi-source high resolution data to conduct the more detailed urban land cover classification.

The fusion of Light Detection and Ranging (LiDAR) data and VHR optical imagery has been used for urban land cover mapping in the emergence. Chen et al. combined Quick Bird image with LiDAR normalized digital surface model (nDSM) data to classify the urban areas based a hierarchical object oriented classification method. After the height information from LiDAR data was integrated, high buildings and low buildings could be distinguished from the original buildings class. Roads could also be further classified into roads and crossroads [4]. Kyle et al. conducted the land cover classification and mapping by the method of the Classification and Regression Tree (CART), based on the fusion of National Agriculture Imagery Program (NAIP) multispectral data and LiDAR derived data (intensity, digital terrain model and canopy height model). They realized adding the LiDAR height information, compared with NAIP multispectral (MS) data and derived Normalized Difference Vegetation Index (NDVI) data, produced a more accurate urban land cover classification and increased the overall accuracy by 5.2% [5]. Im et al. quantified the urban impervious surface using a synthesis of artificial immune networks, based on the LiDAR nDSM data and WorldView-2 multispectral imagery. They successfully identified the urban impervious surface and the overall accuracy was greater than 90% [6]. Zhou et al. implemented the fusion of WorldView-2 imagery and LiDAR elevation, intensity layer and its pseudo-waveform for classifying image objects using the Kullback-Leibler divergence-based curve matching approach. They showed the use of the fused dataset improved the overall classification accuracy by 7.61% over the use of WorldView-2 imagery alone [7]. Langkvist et al. achieved land cover classification from WorldView-2 multispectral image and airborne LiDAR DSM data using convolutional neural networks. The overall accuracy was up to 90.02%-94.49% [8]. Chen et al. improved the accuracy of thematic classification from hyperspectral and LiDAR data using Support Vector Machine (SVM) with extended morphological attribute profiles (EMAPs) and CNN. Deep fusion of two data sources increases 3%-5% of overall accuracy (Compared with Hyperspectral image alone), 12%-15% of overall accuracy (Compared with LiDAR height alone) [9]. According to the above analyses, we can come to the conclusion that LiDAR data are useful for urban land cover classification, especially for LiDAR-derived height information.

In this study, synergistic use of the WorldView-2 image and airborne LiDAR data, a hierarchical support vector machine (SVM) classification method was used to map the urban land cover types. A compared analysis between WorldView-2 imagery alone and the fusion of this two data source was conducted. The objective of this study is to expand the application of fusion of VHR optical imagery and LiDAR data in the field of urban land cover.

2. Datasets and Method

The study area is located in the Pingdingshan City, Henan Province, China. Collected WorldView-2 imagery was acquired on 21 August 2014. The 11-bit WorldView-2 imagery consists of one panchromatic (PAN) band with 0.5×0.5 meter pixel size and eight multispectral bands across wavelengths ranging from 400 to 1040 nm, with the spatial resolution of 2 meter. Collected airborne LiDAR data was acquired on 19 August 2013 and were provided in ASCII format, including the X, Y, and Z coordinates and delivered first return intensities. The two dimensional mean point density was 23.0 points/m^2.The DSM and the DEM were derived from airborne LiDAR point clouds. The nDSM image was generated by subtracting the DSM with DEM, with 0.5 meter spatial resolution (Yan, 2015).

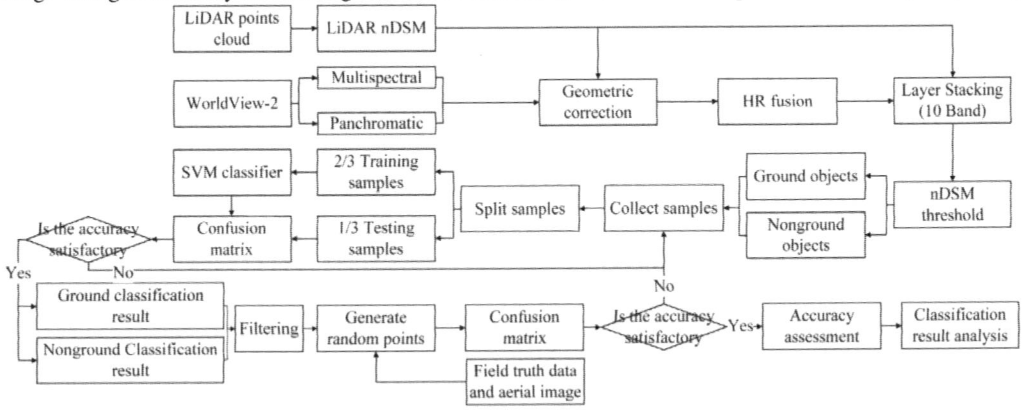

Figure 1. The flowchart of urban land cover classification in the study.

Figure 1 shows the data procession for classifying the urban land cover from WorldView-2 imagery and airborne LiDAR data, Including data pre-processing, a hierarchical SVM classifier construction, accuracy assessment and classification result analysis.

2.1 Data pre-processing

During the data pre-processing, the WorldView-2 imagery was rectified to a Universal Transverse Mercator (UTM) coordinate system with the corresponding rational polynomial coefficients (RPC) sensor camera model. The LiDAR nDSM image, with 0.5 meter resolution, was registered to the WorldView-2 imagery (UTM), and the Root Mean Square (RMS) value was less than 1 pixel

A Haze-and-ratio-based (HR) fused scheme was used by WorldView-2 MS and PAN bands. The HR fused method is a PAN modulation (PM) based fusion method that takes into account haze [10]. Equation (1) shows that $M_{i,f}$ $(i = 1..n)$ denotes the fused i th MS band, M_i is the corresponding i th MS band, PAN denotes the high-resolution PAN image. h_p and h_i are related to atmosphere effects in PAN band and in the i th MS band, respectively. P_L is a spatially degraded PAN image by averaging the high-resolution PAN pixels. A 10 bands image, including a WorldView-2 MS image with 0.5 meter spatial resolution by HR fusion method, WorldView-2 PAN image and LiDAR nDSM image, were generated to conduct following classification.

$$\frac{M_{i,f} - h_i}{M_i - h_i} = \frac{PAN - h_p}{P_L - h_p} \quad (1)$$

2.2 Hierarchical support vector machine (SVM) classifier

A pixel-based SVM method was used in the classification. SVM is supervised, non-parametric statistical learning techniques designed to solve classification and regression problems [11]. SVM has shown great capabilities to model complex multi-modal, nonlinear data distributions in high-dimensional features spaces [12]. In our study, a SVM method with the kernel of Radial Basis Function (RBF) was used to perform the classification. The kernel parameter G was set to 0.1, the reciprocal of band number (10), and the penalty parameter was set to 100 [13].

Based on the traditional SVM model, a hierarchical classification was added. An nDSM threshold determined by the Decision Tree for splitting ground and non-ground objects was 2.4 meter. After mask, the SVM classifier was used to ground and non-ground images respectively. The land cover types in our study area were classified into six classes, including buildings, trees in non-ground objects, and road/parking lots, grasslands, bare soils, and waterbody in ground objects. The waterbody was manually clipped before classification because of the proportion of area of waterbody was little. The pure samples were manually selected by cross-validating geo-referenced field sites and aerial photos. Selected samples polygons were randomly split to training sample datasets and test sample datasets according to 2:1 ratio. The training samples were used to the classification processing of SVM, and the test samples were used to build the confusion matrix and validate SVM modal accuracy. A "Majority/Minority" filtering for each class, with 3×3 window, was employed to reduce the "salt and pepper" phenomenon in pixel-based classification.

2.3 Accuracy validation

As for accuracy validation, the random points of every class (the number of random points was Buildings: 200, Trees: 150, Roads/Parking lots: 200, Grasslands: 100, Bare soils: 100) were selected based on aerial photos and field truth data and the producer's accuracy (PA), the user's accuracy (UA), the overall accuracy (OA) and the overall kappa coefficient (Kappa) were calculated. The confusion matrix was built to analyse the classification accuracy.

3. Results and analysis

Table 1. The classification accuracy of the study area land cover/land use

WorldView-2	Buildings	Trees	Roads/Parking lots	Grasslands	Bare soils	Total	UA（%）
Buildings	114	11	60	3	10	198	57.58
Trees	0	128	0	22	0	150	85.33
Road/Parking lots	47	5	133	10	4	199	66.83
Grasslands	2	10	2	84	1	99	84.85
Bare soils	4	0	11	0	85	100	85.00
Total	167	154	206	119	100	746	
PA（%）	68.26	83.12	64.56	70.59	85.00		
OA（%）	**72.92**						
OK	**0.66**						
WorldView-2+LiDAR nDSM	**Buildings**	**Trees**	**Roads/Parking lots**	**Grasslands**	**Bare soils**	**Total**	**UA（%）**
Buildings	169	6	16	4	4	199	84.92
Trees	0	141	2	5	1	149	94.63
Road/Parking lots	4	9	181	4	2	200	90.50
Grasslands	1	6	3	90	0	100	90.00
Bare soils	1	1	8	2	88	100	88.00
Total	175	163	210	105	95	748	
PA（%）	96.57	86.50	86.19	85.71	92.63		
OA（%）	**89.44**						
OK	**0.87**						

PA: the producer's accuracy; UA: the user's accuracy; OA: the overall accuracy; OK: the overall kappa coefficient

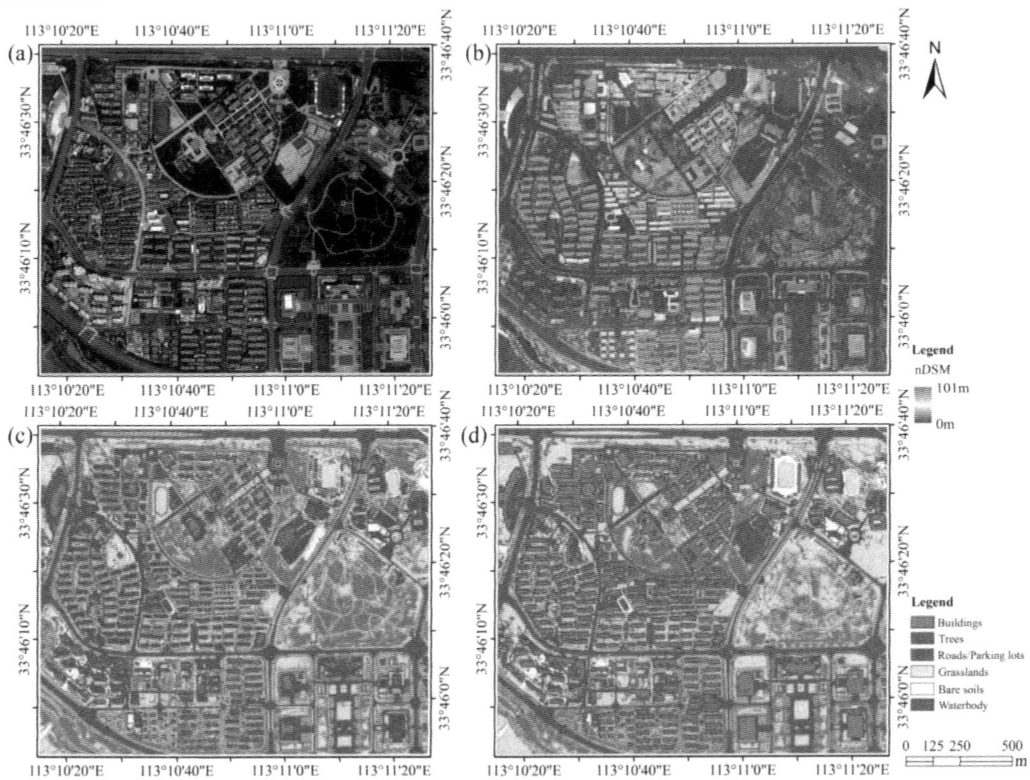

Figure 2. The land cover classification results in the development area.
(a) WorldView-2 imagery; (b) LiDAR nDSM image; (c) classification result of WorldView-2 image alone; (d) classification result of the synergistic use WorldView-2 image and LiDAR nDSM data

The hierarchical classification results, of our study area, from fusion of WorldView-2 imagery and LiDAR nDSM data were generated to compare with WorldView-2 imagery alone using a non-hierarchical classification (Figure 2). Then the results of accuracy evaluation are listed in Table 1.

According to the Figure 2 and Table 1, we generated the finer classification result from fusion of WorldView-2 imagery ad LiDAR nDSM data, compared with WorldView-2 imagery alone. Buildings and roads/parking lots, extracted from fused data, were more precision and well-shaped. Only 68.26% of buildings could be classified correctly from WorldView-2 imagery alone. 96.57% of buildings could be extracted accurately after integrating LiDAR height information. Similarly, the producer's accuracy and user's accuracy of roads/parking lots improved from 64.56% to 86.19%, and from 66.83% to 90.50%. What's more, adding height information, the PA of grasslands was from 70.59% up to 85.71%, and the UA of trees was from 85.53% up to 96.63%. The fused data reduce the misclassification between grasslands and trees. In summary, for the synergistic use of WorldView-2 imagery and airborne LiDAR data, the overall accuracy improved from 72.92% to 89.44%, overall Kappa coefficient improved from 0.66 to 0.87.

A visual comparison is showed in Figure 3. The height information improved the buildings extraction. For WorldView-2 imagery alone, much grey roofs misclassified as roads and parking lots (Figure 3(a)). Then as for the spectral difference in an individual roof, the buildings extracted were well-shaped from fused data (Figure 3(b)). Furthermore, due to the non-orthographic phenomenon in WorldView-2 high spatial resolution data, the roofs and floors of tall-buildings were classified as buildings and roads respectively, and the shadows on the roads, from high-rise buildings, were misclassified as buildings. After adding nDSM data, the high-rise buildings were extracted more precision and the misclassification between shadow and buildings was reduced (Figure 3(c)). In addition, hierarchical fused data classification reduced the mixture of some grasslands and trees. Therefore, height information can contribution to classification (Figure 3(d)).

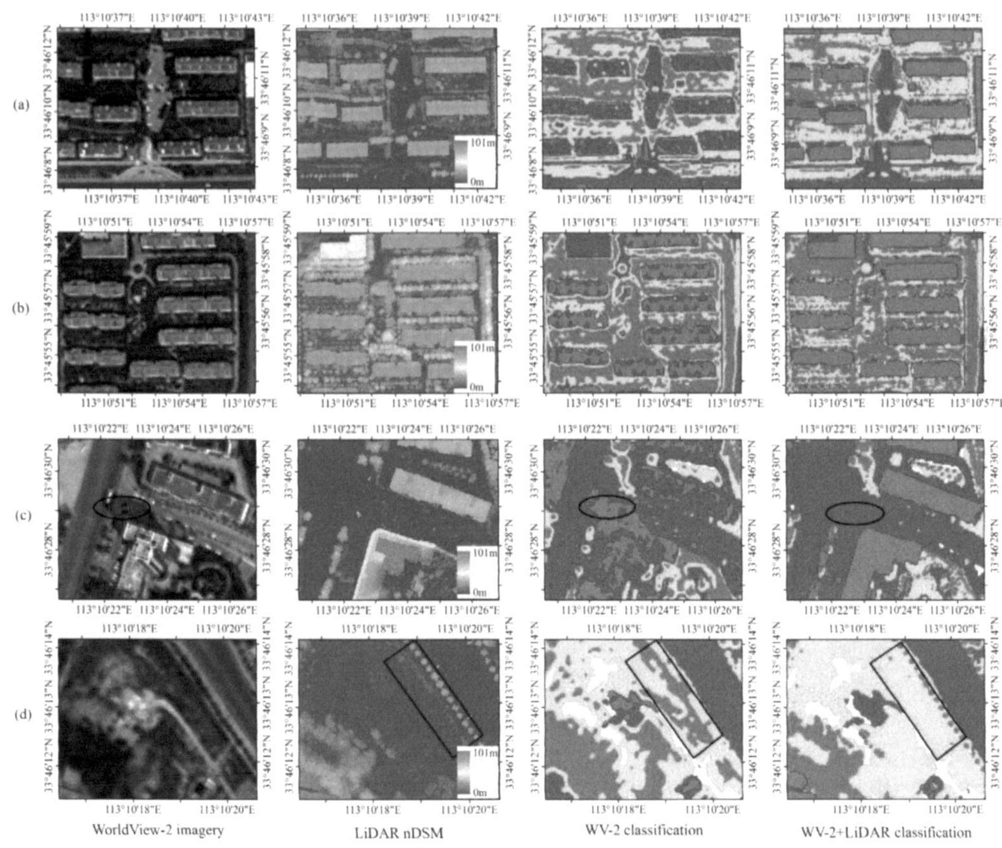

Figure 3. The detail of land cover classification results in study area.

(a) shows WV-2+LiDAR classification reduced the mixture between grey roofs and asphalt roads; (b) presents WV-2+LiDAR classification was valuable for solving the problem of spectral difference in individual roof; (c) illustrates WV-2+LiDAR classification improved the extraction of high-rise buildings and shaded roads; (d) explains WV-2+LiDAR classification classified trees and grasslands more accurately.

4. Conclusions

In the study, based on the fusion of WorldView-2 imagery and airborne LiDAR data, a hierarchical SVM classification method was applied to urban land cover classification, and compared with the non-hierarchical classification using WorldView-2 imagery alone. Three results were obtained:
(1) Compared with WorldView-2 image alone, the overall accuracy and overall Kappa coefficient of classification from fused data improved from 72.92% to 89.44%, and from 0.66 to 0.87 respectively. The major reason was improving the classification accuracy of buildings and roads/parking lots.
(2) Due to adding the height information, fused data reduced the confusion of white roofs with high albedo roads, and grey roofs with low albedo roads. Fused data also reduced the imprecise buildings extraction from spectral difference in a roof due to complex detail of urban objects.
(3) Moreover, fused data could extract the high-rise buildings more precisely due to the LiDAR nDSM orthographic image, and fused data was helpful for addressing the shadows from buildings to some extent.

Acknowledgments

The paper is supported by the National Natural Science Foundation of China (No.41201357). The authors wish to acknowledge the experts providing the WorldView-2 imagery and airborne LiDAR data used for the study, and the anonymous reviewers for providing helpful suggestions that greatly improved the manuscript.

References

[1] Hangqiu X and Meiya W 2016 Remote sensing-based retrieval of ground impervious surfaces J. Remote Sens. **20** 1270-1289
[2] Huadong G, Huaining Y, Zhongchang S, Xinwu L and Cuizhen W 2014 Synergistic use of optical and PolSAR Imagery for urban impervious surface estimation Photogramm. Eng. Remote Sens. **80** 91-102
[3] Dengsheng L, Scott H and Emilio M 2011 Impervious surface mapping with Quickbird imagery. Int. J. Remote Sens. **32** 2519-2533
[4] Yunhao C, Wei S, Jing L and Zhongping S 2009 Hierarchical object oriented classification using very high resolution imagery and LiDAR data over urban areas. Adv. Space Res. **43** 1101-1110
[5] Hartfield K A, Landau K I, and Leeuwen W J D V 2011 Fusion of high resolution aerial multispectral and lidar data: land cover in the context of urban mosquito habitat. *Remote Sens.* **3** 2364-2383
[6] Im J, Lu Z, Rhee J, and Quackenbush L J 2012 Impervious surface quantification using a synthesis of artificial immune networks and decision/regression trees from multi-sensor data *Remote Sens. Environ.* **117** 102-113
[7] Zhou Y and Qiu F 2015 Fusion of high spatial resolution worldview-2 imagery and lidar pseudo-waveform for object-based image analysis *ISPRS J. Photogramm.* **101** 221-232
[8] Längkvist M, Kiselev A, Alirezaie M and Loutfi A 2016 Classification and segmentation of satellite orthoimagery using convolutional neural networks *Remote Sens.* **8** 1-21
[9] Yushi C, Chunyang L, Pedram G, Chunyu S and Yanjeng G 2016 Deep fusion of hyperspectral and LiDAR data for thematic classification IGARSS 3591-3594
[10] Linhai J and Qiuming C 2009 Two improvement schemes of PAN modulation fusion methods for spectral distortion minimization. *Int. J. Remote Sens.* **30** 2119-2131
[11] Schölkopf B, Smola A 2002 Learning with kernels-Support vector regularization, optimization, and beyond, MIT Press
[12] Huang C, Davis L S and Townshend J R G 2002 An assessment of support vector machines for land cover classification *Int. J. Remote Sens.* **23** 725-749
[13] Zhongchang S, Huadong G, Xinwu L, Linlin L and Xiaoping D 2011 Estimating urban impervious surfaces from Landsat-5 TM imagery using multilayer perceptron neural network and support vector machine, *J. Appl. Remote Sens.* **5** 1-17

Extracting built-up areas from TerraSAR-X data using object-oriented classification method

SuYun Wang [1, 2], **Z C Sun** [2, 3]

[1] China University of Geoscience (Beijing), Beijing 100083, China
[2] Key Laboratory of Digital Earth Science, Institute of Remote Sensing and Digital Earth, Chinese Academy of Sciences, Beijing 100094, China
[3] Hainan Key Laboratory Earth Observation, Sanya Hainan 572029, China

Email: sunzc@radi.ac.cn

Abstract. Based on single-polarized TerraSAR-X, the approach generates homogeneous segments on an arbitrary number of scale levels by applying a region-growing algorithm which takes the intensity of backscatter and shape-related properties into account. The object-oriented procedure consists of three main steps: firstly, the analysis of the local speckle behavior in the SAR intensity data, leading to the generation of a texture image; secondly, a segmentation based on the intensity image; thirdly, the classification of each segment using the derived texture file and intensity information in order to identify and extract build-up areas. In our research, the distribution of BAs in Dongying City is derived from single-polarized TSX SM image (acquired on 17th June 2013) with average ground resolution of 3m using our proposed approach. By cross-validating the random selected validation points with geo-referenced field sites, Quick Bird high-resolution imagery, confusion matrices with statistical indicators are calculated and used for assessing the classification results. The results demonstrate that an overall accuracy 92.89 and a kappa coefficient of 0.85 could be achieved. We have shown that connect texture information with the analysis of the local speckle divergence, combining texture and intensity of construction extraction is feasible, efficient and rapid.

1. Introduction

Urban expansion has been served as one of the most important dynamic change processes in the context of global land use changes. The detrimental consequences of urban sprawl result in various urban environmental and ecological challenges, especially in most developing countries [1]. Therefore, in order to reduce the negative consequences, it is valuable for urban sustainable development to estimate and map accuracy built-up areas (BAs).

Radar satellite systems can acquire data at day and night without the consideration of the weather or environmental conditions. Consequently, radar data are more reliably available compared to optical imagery [2]. And the high spatial Resolution (HR) satellite images can offer a great potential for the extraction of information for urban areas [3]. In order to derive useful thematic maps from HR satellite images, traditional methods to extract information from high resolution satellite data include visual interpretation or pixel-based classification methods. Thomas et al. presented an overview and comparison of various classification methods for HR satellite data [3]. Until now, various advanced approaches have been developed to extract urban information from HR satellite images. Regarding the identification of settlements, based on IKONS data, H. Taubenbock et al. proposed an object-based, multi-level, hierarchical classification framework combining shape, spectral, hierarchical and contextual information for the extraction of urban features [4]. Guo et al. utilized C5.0 decision tree algorithm to select features extracted to build a decision tree for urban impervious surface classification [5]. Ban performed urban detection using a generalized Kittler-Illingworth minimum error thresholding algorithm [6]. D. Marmanis implemented a deep learning technique for classifying above-ground objects within urban environments by using a multilayer perception model and VHSR DEM data [7]. Mourad Bouziani et al. proposes a new method for change detection of buildings in urban environments from very high spatial resolution images (VHR) and using existing digital cartographic data [8]. Andreas R. et al. demonstrated and discussed the potential of very high resolution radar imaging of urban areas by means of SAR and interferometric imaging [9]. Fauvel et al. proposed a technique called decision fusing, combining several individual classifiers [10]. Meher et al. presented a wavelet feature based classification method analysing the spatial and spectral characteristics of a pixel along with its neighbours [11]. T. Esch and Michael Thiel et al. presented an approach toward the semi-automated detection of BAs based on single-polarized TSX images [12]. There are several attempts to detect urban areas, utilizing medium-resolution images to map urban areas has solely focused on the settlements, industrial, and commercial zones. However, HR images can produce more detailed urban land cover such as individual roads and buildings in the urban environment. Based on segmentation and object-oriented classification method, we can take advantage of texture and intensity information to acquire more accurate result.

In this paper, based upon method presented by T. Esch et al [12], we improved this method to quickly extract BAs from single-polarized TerraSAR-X (TSX) stripmap model (SM) image. Firstly, accuracy true texture image is calculated by analysing local speckle characteristics. Secondly, the multi-resolution segmentation is performed using the original intensity image based on the software eCognition Developer. Finally, the objects of BAs are derived by combining information on the local true texture and local backscatter intensity.

2. Study area and data collection

Dongying City located in the north of Shandong province with longitude $118^{\circ}7'$ to $119^{\circ}10'$ and latitude $37^{\circ}20'$ to $38^{\circ}10'$. Land use of this area is dominated by settlements, roads, vegetation and water body. The data is based on single-polarized TerraSAR-X(TSX) images acquired in stripmap mode on June 11,2013. The average ground resolution of the image is 3m$\times$3m, with incidence angle of 44.47°.

3. Methodology

3.1. The principle of modelling build-up radar footprints

There are three main characteristics of buildings in SAR: layover, double-bounce, and shadowing effects, which are caused by the side-looking and ranging properties of radar sensor. In order to throw light upon this, Fig. 1 shows a schematic view of the scattering profile of a simplified gable-roof building model, which is imaged by a sensor with incidence angle θ.The annotation a refers to backscattering from the ground surface. acd denotes the layover area where scattering from the ground, from the vertical building front wall and from parts of the flat roof are superimposed since these parts have the same distance to the sensor [13]. b stands for double bounce effect that due to the ground-wall compose a corner reflector. d denotes the roof-scattering only. There is shadow effect in e. In Figure. 2(a) shows an example of a flat-roof building in 3m resolution TerraSAR-X data with viewing direction from left. The double-bounce line is highlighted with a red arrow, (b) is the same building in Google Earth. For gable-roof buildings as shown in Fig. 3, besides the main three scattering mechanisms, metallic structures lead to bright spots. acd is at the sensor close side resulting from direct backscattering from the roof. In Figure. 4(c) gives an example of gable-roof buildings with small aspect angles in 3m resolution TerraSAR-X data with viewing direction from left. The double-bounce and roof scattering lines are highlighted with red and yellow arrows, respectively, (d) is the same buildings from (c) in Google Earth.

IOP Conf. Series: Earth and Environmental Science **57** (2017) 012036 doi:10.1088/1755-1315/57/1/012036

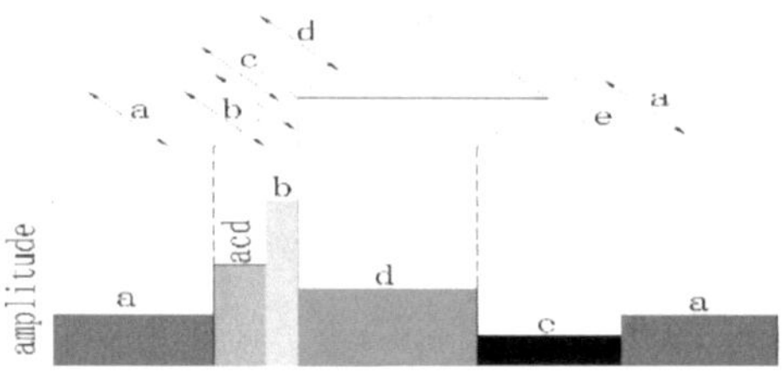

Fig.1 Scattering model for a flat-roof building

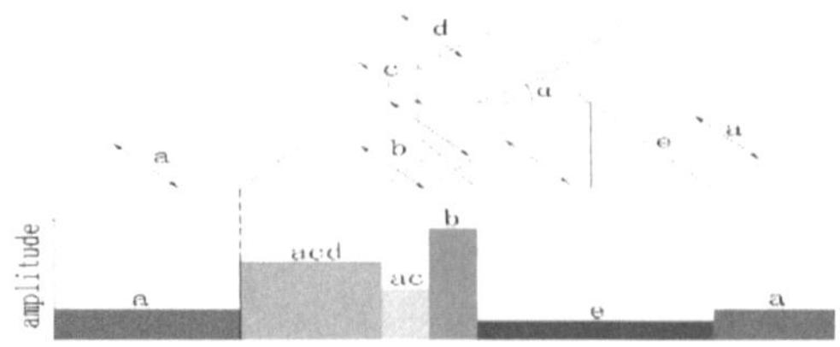

Fig.3 Scattering model for a gable-roof building

Fig.2 (a) Building in 3m resolution TerraSAR-X data. The double-bounce line is highlighted with a yellow arrow. (b) The same building in Google Earth.

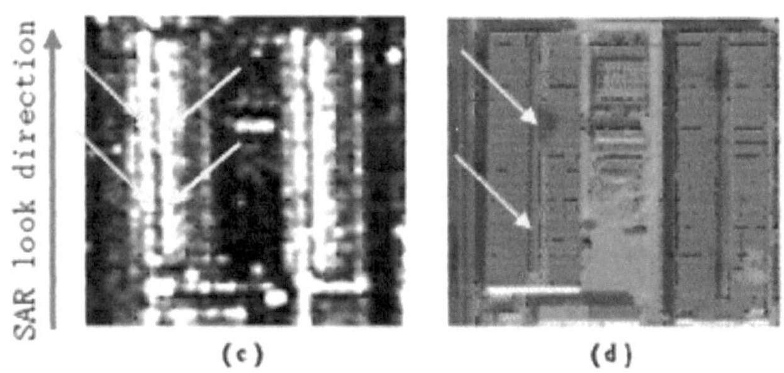

Fig4. (c)Buildings in 3m resolution TerraSAR-X data with viewing direction from left. The double-bounce and roof scattering lines are highlighted with yellow arrows, respectively. (d)The same building from (c) in Google Earth.

3.2. Methodology

The workflow of the method is described (as showed in Fig.5). Firstly, accuracy true texture image is calculated by analysing local speckle characteristics. Secondly, the multi-resolution segmentation is performed using the original intensity image based on the software eCognition Developer. Finally, the objects of BAs are derived by combining information on the local true texture and local backscatter intensity.

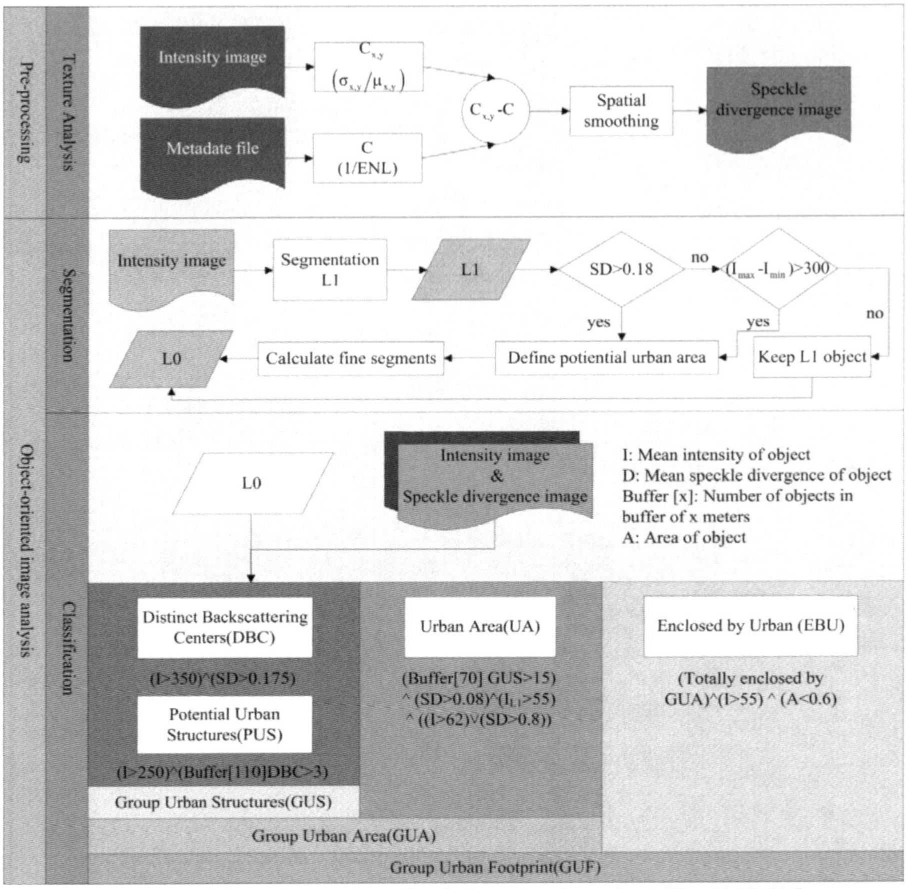

Fig.5 The workflow of extracting urban information from single TerraSAR-X data

3.2.1. Generation of texture image based on speckle analysis

Speckle appearing in SAR images is due to the coherent interference of waves reflected from many elementary scatterers [14]. This effect causes a pixel-to-pixel variation in intensities, and the variation manifests itself as a granular noise pattern in SAR images. A reasonable representation of fully developed speckle is a multiplicative noise model assuming a Gaussian distribution, and the standard deviation of data is related to mean of signals, thus multiplicative. However, this assumption is not applicable for urban sceneries since the multiple-scattering processes within a resolution cell frequently show a directional behaviour instead of a random distribution.

A conventional parameter to quantify the multiplicative noise in SAR data is the coefficient of variation C

$$C = s / m \tag{0}$$

Where σ is standard deviation and μ is mean value.

The equivalent number of looks (ENL) is a good indicator of the speckle noise level in SAR images.

$$ENL= L_a + Lg= 1/ C \tag{2}$$

With L_a and Lg defining the effective number of looks in the azimuth and range.

The local speckle divergence $D_{x,y}$,which means the difference between the local estimated coefficient of variation and the calculated scene-specific coefficient of variation. It can be quantified by

$$D_{x,y} = C_{x,y} - C \tag{3.1}$$

$$C_{x,y} = s_{x,y} / m_{x,y} \tag{3.2}$$

Where C denotes the theoretical heterogeneity due to the fully developed speckle, $C_{x,y}$ providing the local coefficient of variation defined by the local standard deviation $\sigma_{x,y}$, and the local mean $\mu_{x,y}$ calculated via the

IOP Conf. Series: Earth and Environmental Science **57** (2017) 012036 doi:10.1088/1755-1315/57/1/012036

maximum-likelihood estimated in a defined local neighbourhood. In this paper, $D_{x,y}$ was calculated with the best compromise window size of 15 pixels for the study area.

3.2.2. Object-oriented mapping and region-growing algorithm of urban extraction

The first step of an object-oriented image analysis is image segmentation, which is a way to separate the image into simple regions with homogeneous behaviour. The multi-resolution segmentation algorithm is a bottom-up region-growing technique within the eCognition Developer software, starting at the seed points, in numerous threshold-based which shows completely in Fig. 5, then discriminate regions on the basis of intensity value difference between pixels, merging into larger ones. There are three main parameters dominate the object generation during segmentation: scale parameter, shape and compactness. At first, a coarse segmentation level (L1:150, 0.3, 0.8) was created, acquiring large-scale features such as settlement bodies. The more detailed objects can get from subdivided level (L0:25, 0.3, 0.8), which was focused on potential urban areas, through some threshold like high value $D_{x,y}$ or high intensity $I_{x,y}$.

Based on scattering properties of man-made structures, the distinct backscattering centres (DBC) are composed of direct reflection and double- or multiple-bounce effects in urban sceneries, which characterised by an extremely high $I_{x,y}$ and $D_{x,y}$. In order to identify further detailed settlements----the potential urban structures (PUS), we focus on the neighbourhood of the DBC objects. Then, DBC and PUS are combined to the urban structure (GUS). PUS features a very high $I_{x,y}$. After these threshold, the potential location and shape of the build-up areas are achieved. By using the GUS segments as seed objects, the surrounding area of 25 pixels are classified as UA, aiming at the water bodies and rivers, which shows a high intensity or a very high speckle divergence. All objects are enclosed by segments classified as GUA, and enclosed by urban (EBU). Finally, GUA and EBU are merged into one.

3.2.3. Validation

We generated 957 randomly distributed reference points by cross-validating QuickBird data and geo-referenced field sites, confusion matrices with statistical indicators are calculated and used for quantitatively assessing the classification results.

We adopted two statistics, overall kappa coefficient (OK) and overall accuracy (OA) based on confusion matrices, to evaluate the classification performances, and the two formula are listed below:

$$Kappa = \frac{N \sum_{i=1}^{r} X_{ii} \quad \sum_{i=1}^{r} (X_{i+} X_{+i})}{N^2 - \sum_{i=1}^{r} (X_{i+} X_{+i})} \tag{4}$$

$$OA = \frac{\sum_{i=1}^{r} X_{ii}}{\sum_{i=1}^{r} X_{+i}} \tag{5}$$

4. Results and analysis

Based on Object-based and region growing delineation of urban footprints from Fig. 6 (a) TSX intensity image and (b) speckle divergence image starting with (c) the identification of DBC and followed by (d) the definition of PUS and (e) the supplementation of UAs and (f) regions completely EBU objects before (g) finishing with the reconstruction of the urban footprint (GUF) and smoothing the border line of the extracted settlement object. And the extraction result of Dongying city is showed in Fig. 7.

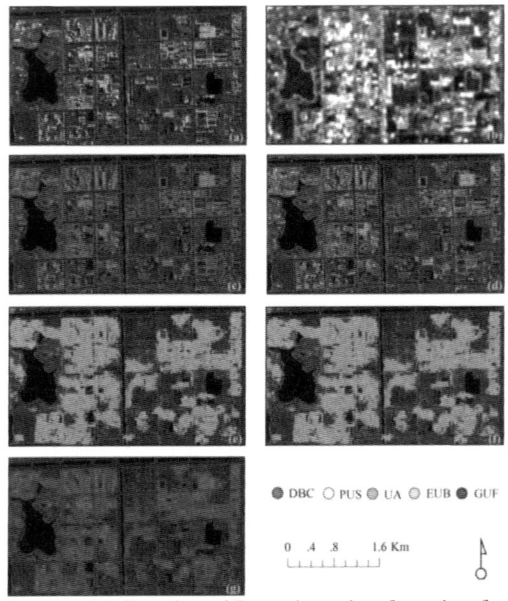

Fig. 6 the detail of Object-based delineation of Dongying urban footprints from TerraSAR-X image

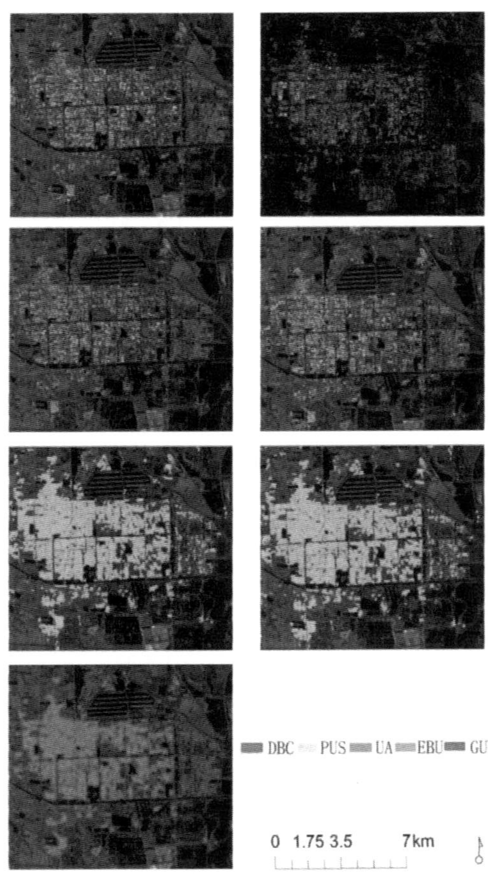

Fig.7 Object-based delineation of Dongying urban footprints from TerraSAR-X image

The method was applied to single-polarized TSX scene of Dongying city. The result of the object-oriented classification procedure is shown in Fig. 7, the figure illustrates the derived urban footprints provide accurate information on both the actual location and shape of the settlements. We validated the classification result generating 957 randomly distributed reference points by cross-validating QuickBird data and geo-referenced field sites, confusion matrices with statistical indicators are calculated and used for quantitatively assessing the classification results. The accuracy assessment for classification is summarized in Table1. From the Table1, we found that the overall accuracy 92.89% and a kappa coefficient of 0.85 can be achieved.

Table1 Accuracy assessment for classification based on Object-oriented

Category	Build-up	No Build-up	Sum-up
Build-up	492	43	535
No Build-up	25	397	422
Sum-up	517	440	957
OK (%)	92.89		
OA (%)	0.85		

The Fig. 8 shows (a)-(c) the TSX intensity data; (d)-(f) the derived speckle divergence; (g)-(i) the result of the classification process; and (j)-(l) an HR optical image for comparison. Potentials and limitations of the developed approach in terms of detecting Bas such as (c), (f), (i), and (l) industrial sites; (a), (d), (g), and (j) farmsteads; and (b), (e), (h) and (k) high-density housing. Analysing the most frequent errors of omission and commission, in countryside area (a), (d), (g) and (j), some isolated homesteads can be derived, but the hedges and grasses will deteriorate the result, leading to overestimated build-up areas. In the fringe between countryside and country, due to bare soil exists, natural land cover types are located close to settlements and some build-up areas are under constructions, misclassifications are unavoidable. In commercial zones, errors are often emerged in the cause of foreshortening or layover in mountainous areas. We could verify that the large-area specular reflection at asphalted surfaces such as squares or broad streets and large flat roofs of factory buildings. The corresponding regions appear as zones with a very low $I_{x,y}$ and $D_{x,y}$ [Fig. 8 (b),(e),(h) and (k)]. Due to lake of distinct corner reflection, the buildings are strongly interspersed with trees and hedges [Fig. 8 (a),(d),(g) and (j)]. Based on $I_{x,y}$, the woodland and bare soil can hardly be discriminated from the settlement, resulting in the overestimate of urban area[Fig. 8 (c),(f),(i) and (l)].

The advantage of using $D_{x,y}$, which is supplemental to $I_{x,y}$, is highlighted by Fig.8(a),(b),(d),(e),(g),(h),(j) and (k)l Fig. 8(a), (d), (g), and (j) illustrates a group of homesteads which can be differentiated from and delineated much better from the surrounding by considering the $D_{x,y}$ image. We came to the conclusion that the object-oriented classification method with intensity and divergence information can improve accuracy in urban area.

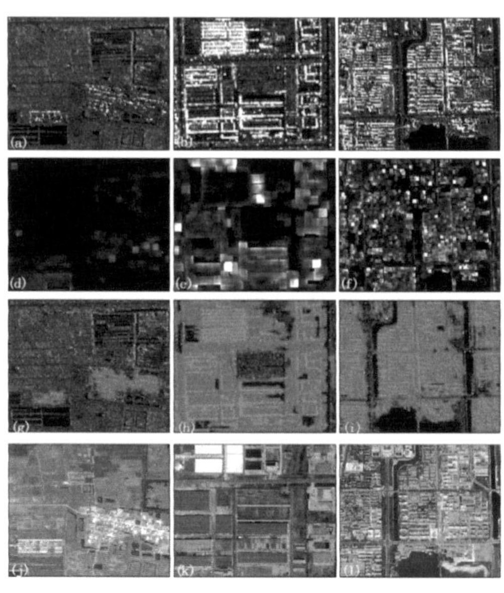

Fig.8 Potentials and limitations of the developed approach in terms of detecting BAs.

5. Conclusion

This paper presented an object-oriented and region growing classification method to extract build-up areas from TSX image. On the basis of image segmentation, combined local true texture and local backscatter intensity, we applied the method in Dongying city. The result illustrated that the overall accuracy was up to 92.89%, manifesting the classification with texture proposed by $D_{x,y}$ and single-polarized intensity information had a high accuracy. However, the error of commission still existed, which may result from the complexity of land covers themselves such as the confusion between detached houses in the outskirts of the settlements and heterogeneous landscapes composed of trees and rocks. Therefore, the future work will focus on the development of the classification rules and automated determination of the thresholds used for the classification.

Acknowledgement

The paper is supported by the National Natural Science Foundation of China (No.41201357). The authors would like to thank the DLR for providing the SAR data for this paper, and the anonymous reviewers for providing helpful suggestions that greatly improved the manuscript.

References

[1] M. P. Johnson,. 2001. Environmental impacts of urban sprawl: A survey of the literature and proposed research agenda. Environ. Plan. A. 33(4), pp. 717–735.

[2] Thomas, N., Hendrix, C., Congalton, R. G., 2003. A comparison of urban mapping methods using high-resolution digital imagery, Photogrammetric Engineering and Remote Sensing. 69(9), pp. 963-972.

[3] F. Tupin and M. Roux,. 2003. Detection of building outlines on the fusion of SAR and optical features. ISPRS J. Photogramm. Remote Sens. 58(1/2), pp. 71–82.

[4] Taubenbock, T. Escha, M. Wurm, A. Roth and S. Dech,. 2010. Object-based feature extraction using high spatial resolution satellite data of urban areas. Journal of Spatial Science. 55(1) pp.117-132.

[5] Guo, H. D., et al,. 2014. Synergistic Use of Optical and PolSAR Imagery for Urban Impervious Surface Estimation. Photogrammetric Engineering and Remote Sensing 80(1): 91-102.

[6] Ban, Y. F. and O. A. Yousif., 2012. Multitemporal Spaceborne SAR Data for Urban Change Detection in China. Ieee Journal of Selected Topics in Applied Earth Observations and Remote Sensing, 5(4), pp.1087-1094.

[7] Marmanis, D et al,. 2015. Deep Neural Networks for above-Ground Detection in Very High Spatial Resolution Digital Elevation Models. ISPRS Annals of Photogrammetry, Remote Sensing and Spatial Information Sciences. II-3/W4, pp.103-110.

[8] Mourad Bouziani, Kalifa Goïta, Dong-Chen He,. 2010. Automatic change detection of buildings in urban environment from very high spatial resolution images using existing geodatabase and prior knowledge. ISPRS Journal of Photogrammetry and Remote Sensing. 65, pp. 143-153.

[9] Andreas R. Brenner and Ludwig Roessing., 2008. Radar Imaging of Urban Areas by Means of Very High-Resolution SAR and Interferometric SAR. IGASS, 46(10), Oct 2008. pp. 2971-2982.

[10] Fauvel, M,. 2008. Decision fusion for the classification of urban remote sensing images. IGARSS, 44(10) pp. 2828–2837.

[11] Meher, S.K., Shankar, B.U., & Ghosh, A,. 2007. Wavelet-feature-based classifiers for multispectral remote sensing images. IGASS. 45(6), pp. 1881–1886.

[12] T. Esch, Michael Thiel, Andreas Schenk, et al., 2010. Delineation of Urban Footprints From TerraSAR-X Data by Analyzing Speckle Characteristics and Intensity Information. IGARSS, 48(5), pp. 905-916.

[13] Guo, H. D., 2000 Radar Earth Observation on Theory and Applications[M].Beijing: science press.H.

[14] Jong-Sen Lee. et al,. 2012. On polarimetric SAR speckle filtering", IEEE Trans. Geosci. IEEE International, pp. 111– 114.

Suitability evaluation of urban terrain in China

Shunguang Hu, Z X Zhang, X L Zhao, X Wang, B Liu, L Yi, Q K Wen, L J Zuo, F Liu, J Y Xu

Institute of Remote Sensing and Digital Earth, CAS, 100094 Beijing, China

zhangzx@radi.ac.cn

Abstract. To clarify the terrain characteristics of Chinese cities, three terrain parameters that altitude, relief and slope are selected to be analysed. Based on these factors, we make a classification of urban landforms and then compare the differences of urban terrain using the classification results. In conclusion, firstly, it is in a micro scale that the terrain of built-up areas has similar characteristics. Secondly, a method to evaluate the suitability of urban terrain in China is proposed and the suitability evaluation of terrain around built-up areas is realized. Finally, statistical data shows that Chinese urban expansion potentials are all considerable except for urban areas in high altitude mountain.

1. Introduction

A city is the center of politics, economics, technology, culture and education in a country or district, where the human activity is focus and the natural environment is the most intense affected. Although urban areas account for less than 3% of the world's land, the impact of urban expansion on ecological environment is global. Urban expansion is a concentrated expression of urban spatial distribution and structural changes, which has become a hot spot in the research of urban development both at home and abroad. Domestic and foreign research about urban expansion mainly focused on monitoring, mechanism, influence, simulation prediction and optimal adjustment, etc. (Batty, 2013; Zhang & Zhang, 2012; Zhang, et al., 2014;XU, Zhao & Min, 2014). Some scholars take terrain as one of the important factors to restrict urban expansion and make researches to clarify how much influence the topographic does(Diao,1990;Zhang,2008;Zhao,et al.,2015).During the process of urban expansion, terrain from macro aspect control the shape, structure and extension direction of a city (Verstappen,1983). From the above all, the terrain is mainly restricting the intense and direction of urban expansion. Analysis on urban topography may help to estimate and forecast the city extension status in the future, which is of great significance in urban planning decision and suitable terrain selection. Moreover, research on urban terrain influence is an important factor when we study urban expansion mechanism. And the relationship between urban scope change and terrain can provide rule and data basis to simulation prediction and optimal regulation of urban expansion.

From the perspective of formation and development of a city, gentle terrain is one of the most favorable external conditions for the development of the city(Diao,1990). And from the aspect of the overall construction of the city, gentle open terrain is more favorable (Tongji university,1991). However, for urban terrain, what mentioned "gentle open" above is only a kind of qualitative description, and transferring it to quantitative characterization is more meaningful. From low to high, plain to mountain, there exist cities. The landform types in traditional landform classification system, restricted by use and division basis, are mainly based on macro landform factors. Traditional system has an important limitation. It is unable to distinguish suitable urban terrain for its classification rule is based on a macro scale. Less micro scale terrain factor is necessary to evaluate the suitability of urban terrain. In this research, to compare the similarities and differences of different topographic parameters, more than two hundred cities are analyzed. At last, a method named neighborhood relief is proposed to distinguish micro terrain characteristics. Experiments show that neighborhood relief is not only accordant with urban terrain, but has a good universality in different landform types in China.

There are great differences of urban topographic distribution characteristics between the east and west sides of "Hu Huanyong Line". And these differences directly restrict the distribution of population and buildings, which is a key factor to answer whether or not the line can be broken through in the strategy of "new urbanization" in China. Meanwhile, related problems derived from the strategy such as "how much geomorphology controls the temporal and spatial evolution of urban expansion" become increasingly concerned in the field of city geomorphology, and it is true that the research on urban topography is the foundation of study on potential evaluation of urban expansion.

In this paper, we take Chinese urban peripheral terrain as the research object. To find out how to depict the urban built landform with some universal terrain parameter, we analyze the characteristics of different terrain factors and different scales of topographic relief features in the scope of Chinese 288 urban areas using 2010 land use database of China, nearly 40 years of Chinese urban expansion database and the second edition of the Aster DEM Terrain Data. As a result, a topographical statistical rule named neighborhood relief method is built, and the suitability evaluation of Chinese urban terrain is realized at last.

2. Basic characteristics of urban terrain in China

As a human gathering place, the city's landscape environment is generally open and flat. And urban landform has a distinctive feature in the complex environment of China. In this paper, we extract the terrain information in the build-up area of Chinese prefecture-level city based on land use data in 2010 and the second edition of the ASTER DEM data. Finally, the characteristics of China city landform are described by analysis of landform factors, such as elevation, slope and topographic relief. In order to further clarify the terrain features in the East and west sides of "Hu Huanyong line", all the statistical analysis of China will be divided into three parts, the western part is located in the west of "Hu Huanyong line", and the central part and the eastern part are located in the east of "Hu Huanyong line"(Fig. 1).

2.1. Altitude Characteristics of Chinese Cities

Altitude is a basic parameter of terrain. It is the vertical distance from the sea level of a region. The altitude characteristics of Chinese cities are as followings. As a whole, the altitude of built-up areas in Chinese single prefecture-level city is between 9m and 3664m, and the average value is 202m. There are 79.61% areas at an altitude of 200m below, and 88.17% areas are 500m below. But there are only 7.56% areas at 1000m above, 0.14% areas at 3000m above. Specifically, the regional difference of altitude in Chinese cities is obvious. The altitude of built-up areas in the eastern region is between 9m and 1248m, and the average value is 77m. The middle region, as located in the second step in Chinese topographic step distribution, is distinctly higher than the eastern region. And the altitude of built-up areas in this region is between 12m and 2407m, and the average value is 588m. Cities in the western region of China are mainly located in the central and western regions of the Tibetan Plateau, the Loess Plateau and the Inner Mongolia Plateau. It is the highest in the three parts of China. The altitude of built-up areas in this region is between 390m and 3664m, and the average value is 1265m.

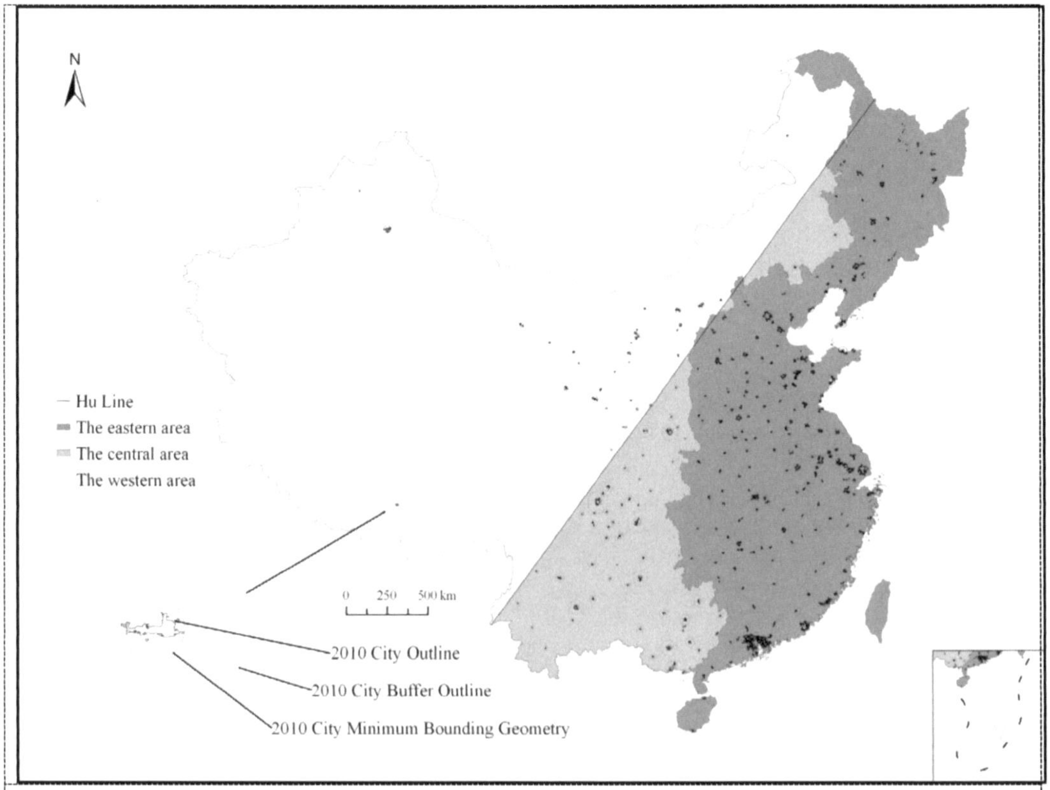

Figure 1. The distribution of Chinese prefecture-level cities.

2.2. Slope Characteristics of Chinese Cities

Slope is the ratio of vertical height and horizontal distance, and it also refers to the inverse trigonometric tangent function when showed by degree. The essence of slope is to show its steep. The slope characteristics of Chinese cities are as followings. As a whole, in 2010, the slope of built-up areas in Chinese cities is between 0.47° and 14.28°, and the average value is 2.65°. About 55.19% city areas have an average slope of 2° below, the city proportion will rise to 91.84% when the slope is increased to 6° below, 98.68% city areas have an average slope of 15° below, and 99.83% city areas have an average slope of 25° below, but only 0.17% city areas have an average slope of 25° above. Specifically, the difference of slope is small among the three parts of China. The slope of built-up areas in the eastern cities is between 0.46° and 10.29°, and the average value is 2.40°. The middle region with broken terrain is mainly located in the Yunnan-Guizhou Plateau, and the slope of built-up areas in the middle cities is between 1.41° and 14.36°, and the average value is 4.43°. The central region has the most rugged terrain in the three parts of China. The slope of built-up areas in the western cities is between 1.16° and 10.67°, and the average value is 2.98°. It is a little more than the eastern region, but less than the central region.

2.3. Topographic Relief Feature of Chinese Cities

The relief reflects the maximum height in a certain area. It is an important index to quantitatively describe the geomorphology and to divide the type of landform. According to the theory of landform development, with the increase of the statistical range, the relief becomes larger and larger, and tends to be stable after reaching a certain value. To analyze the topographic relief of Chinese cities, the best statistical unit result 13km2 based on the analysis of Chinese typical urban terrain is used in this study. As a whole, in 2010 the relief of built-up areas in Chinese cities is between 23m and 881m, and the average value is 104m. About 11.94% built-up areas are located at a relief valued 30m below, 87.59%

areas are at 200m below, and more than 80% areas are at 153m below. Specifically, Regional differences of relief in Chinese cities are obvious. In the east, the relief is between 23m and 443m, the average value is 93m, and it is the smallest relief in the three parts of China. In the middle areas, the relief is between 40m and 667m, and the average value is 164m, which is the biggest in the three parts of China, where there is large range, typical developed karst landform, broken terrain and steep surface. At last, the relief in the west is between 33m and 887m, and the average value is 153m.

2.4. Classification of Urban Terrain in China Based on Topographic Relief and Altitude

Topographic relief and altitude are important parameters for the classification of regional terrain in a macro scale and the quantification of topographic features, which is widely used in basic terrain classification. The Chinese urban terrain classification in this paper is based on the morphological classification system proposed by Zhou Chenghu et al(Zhou et al.,2009). Chinese urban terrain is divided into six types based on the average topographic parameter of its outline circle. By the relief, there are three types of cities called plain city, hilly city and mountain city. And according to altitude, urban terrain is divided into low altitude city, middle altitude city and high altitude city. More than 200 prefecture-level cities have been divided into different types(Table. 1 and Table .2) .As a result, the most types of cities in China are located in low altitude hill, and the ratio is 71.69%. The other city proportion show a descending tendency as low altitude mountain cities, low altitude plain cities, medium altitude mountain cities and high altitude mountain cities, and the proportion of the high altitude mountain cities is the least in the six types of landform, which is only 0.14%.

Table 1. Terrain classification of Chinese cities.

Altitude range	Relief range	Landform	Proportion
	<30m	low altitude plain	8.97%
<1000m	30～200m	low altitude hill	71.69%
	>200m	low altitude mountain	11.82%
1000～3500m	30～200m	medium altitude hill	4.07%
	>200m	medium altitude mountain	3.32%
>3500m		high altitude mountain	0.14%

Table 2. Terrain characteristics of Chinese cities.

Landform	Altitude (m)	Relief (m)	Slope(°)
low altitude plain	10	31	1.12
low altitude hill	92	90	2.40
low altitude mountain	220	195	4.52
medium altitude hill	1263	123	4.83
medium altitude mountain	1564	243	3.40
high altitude mountain	3664	417	3.24

The cities in different types of landform in China have distinct differences in topographic features. Above all, compared to the surrounding areas, cities tend to distribute in the low and flat terrain areas. The topographic parameters of urban built-up areas are mostly in the low value of terrain classification criteria. Taking altitude as an example, the average altitude of low altitude plain is only 10m, the value of medium altitude hill is 1263m, and the value of high altitude mountain is 3664m. Secondly, the difference of average slope among single city is small in China. Urban topography has a strong commonality in a more microscopic scale. For example, the average slope of all types of built-up areas is less than 5°(Table. 2).

3. The topographic constraints of urban expansion in China

The analysis of the terrain limitation of urban expansion is helpful to find out the suitable terrain areas of a city in the future, and it is the basis of further study on potentiality of urban expansion. According to statistics of neighborhood relief, the terrain requirement of urban expansion will be proposed in this study.

3.1. Terrain restriction parameter setting for urban expansion

The topographic restriction parameter of urban expansion should give full consideration to the terrain requirements for houses. Compared with the relief mentioned above, the neighborhood relief is more restrictive to the housing construction. Topographic relief should be analyzed based on a much closer area. Based on this, resolution of 100m pixel is used to analyze the terrain restriction condition in this paper, and the parameter is called neighborhood relief mainly due to avoiding confusion of concept understanding.

3.2. The range of urban neighbouring relief and its statistical regulation

After less than 0.01% discrete pixel value being rejected, the statistical results show that the range of urban neighbouring relief in China is from 0m to 129m, and the average value is 18m. Urban neighbouring relief with the largest proportion is 11m, its range is from 7m to 16m, and the cumulative proportion is up to 49.52%. The region where neighbouring relief is lower than 10m accounts for 32.20%, 74.50% region lower than 20m, 89.34% region lower than 30m, and 98.91% region lower than 80m(Fig. 2). In addition, the neighbouring relief range of each single city is from 3m to 79m, and there is a small regional difference(Table. 2). In the east, where there are 83.80% cities, the neighbouring relief range of each single city is from 3m to 59m, and the average value is16m. In the middle, where there are 10.06% cities, the neighbouring relief range of each single city is from 10m to 79m, and the average value is28m. In the middle, where there are 10.06% cities, the neighbouring relief range of each single city is from 10m to 79m, and the average value is 28m. In the west, there are 6.14% cities, the neighbouring relief range of each single city is from 9m to 59m, and the average value is 20m.

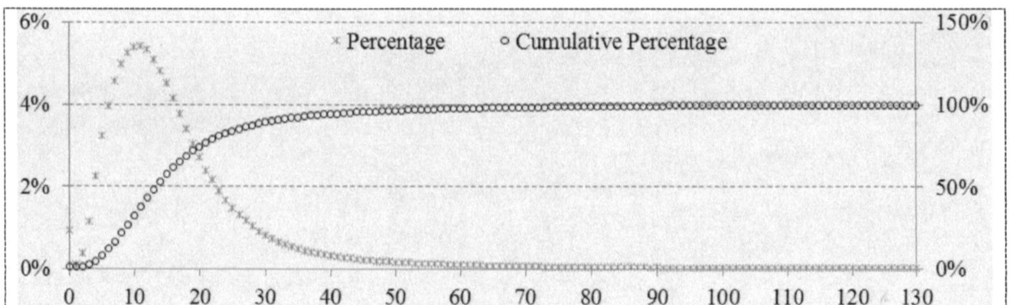

Figure 2. Scatter distribution of neighbouring relief in 288 urban built-up areas.

To find out the suitable terrain requirement of built-up areas, the first consideration is to ensure that most of the areas are in the range of built-up neighbouring relief. Taking into account the possible extreme cases that a sharp, steep, cliff terrain in the city have been packed into the neighbouring relief range, the upper height limit of neighbouring relief cannot be set as the maximum value. According to the statistical result and above analysis, 80m is chosen as the upper height limit of neighbouring relief. When the neighbouring relief is less than 80m, there exist two kinds of upper height limit. One is 30m, the neighbouring relief in these areas are less than 30m. From the statistical characteristics of neighbouring relief, the landform with this kind of height limitation is more suitable for construction of built-up areas. The other one is 80m, about 10% in proportion, here average neighbouring relief of each single city is imported to help further choosing. When the value is more than 30m, the limitation is adjusted to 80m(Fig.3).

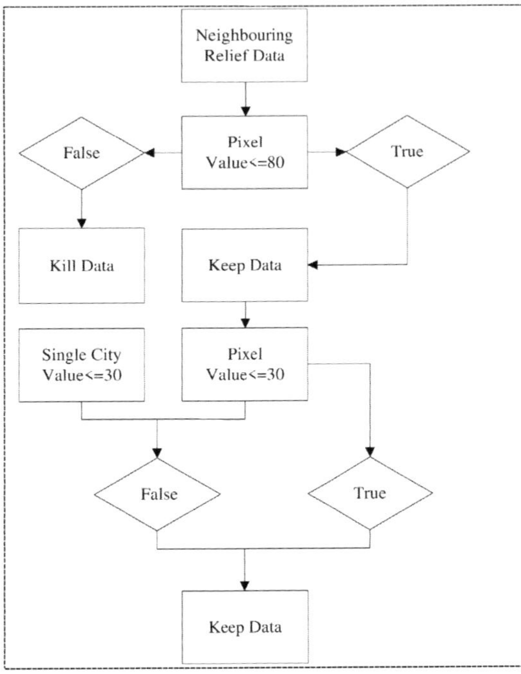

Figure 3. The extraction process of suitable terrain in urban built-up areas.

4. The distribution regularities of suitable terrain for built-up areas around Chinese cities

According to terrain limitation of urban expansion analyzed above, suitable terrain of built-up areas(Hereinafter referred to as the suitable terrain) can be found out by the elimination of the areas beyond the terrain limitation. Based on the built-up areas of 288 prefecture-level cities in China in 2010, the buffer radius is set to 2 times of the radius of each city circle, a buffer region of about 25 times of the existing urban built-up areas produces. The buffer region is the first choice for the future development of the city. The reason lies that compared with the other areas there are more regional contact and larger impact by the city in these regions. As a whole, the terrain suitability in the buffer regions around the build-up areas of Chinese cities is evaluated by the method proposed above.

4.1. Regional difference of suitable terrain around Chinese cities

In the buffers based on 2 times of the radius of Chinese cities, about 65.51% terrain is suitable for built-up areas. The proportion in the eastern, central and western parts of China is showed as 66.19%, 69.08% and 53.70%, respectively. The difference among them is not large. The proportion of urban suitable terrain in the central and west are both more than a half, and moreover, the value in the central area is even more than that in the eastern region. In conclusion, the potential of urban development in the central and west is huge from a view of microscopic scale. Otherwise, judging from the absolute amount of suitable terrain, the amount in the eastern region is still far greater than the central and western regions. And this is mainly due to the current city proportion in the eastern region of China is far greater than that in the western region(Fig. 4).

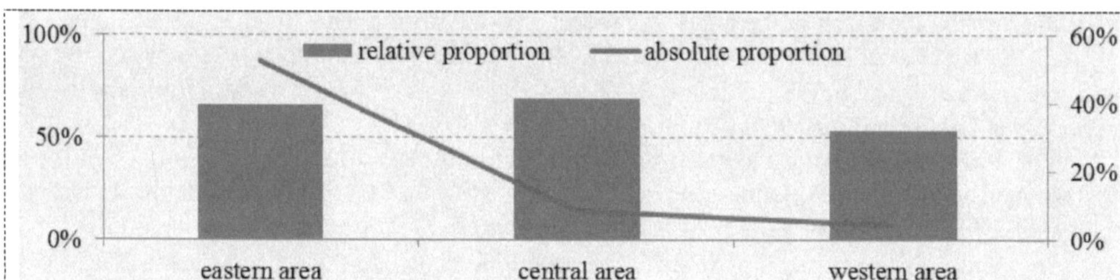

Figure 4. Suitable terrain of built-up areas in the eastern, central and west part of China.

4.2. Distribution of suitable terrain of built-up areas in different landform environment

The distribution difference of suitable terrain around Chinese cities is significant in different types of landform There is a common descending tendency of suitable terrain proportion as the altitude rises. And it shows a descending order as low altitude hill, low altitude mountain, low altitude plain, medium altitude hill and medium altitude mountain(Fig. 5). This is consis- tent with the current proportion of built-up areas in various types of landform. That is to say, greater the proportion of the built-up areas surrounding the city is, the greater proportion of suitable terrain land the type of landform has. The most prominent is the low altitude hill city, the suitable terrain areas in this type of landform are 2.45 times the total areas of the other types. From the view of the proportion of suitable terrain for built-up areas in various types of landform(called relative proportion in abbreviate form following), low altitude plain is the most suitable type of landform for built-up areas, which relative proportion is up to 78.57%. Besides, medium and low altitude hill are suitable for built-up areas, too, the relative proportion of these two types of landform are high as 71.04% and 69.22%. Moreover, each has about a half suitable terrain land in low and medium altitude mountain, the relative proportion is 50.90% and 47.37%. The terrain environment surrounding built-up areas in high altitude mountain city is the most severe, which is obviously restricted by the terrain environment, and the suitable terrain proportion is only 13.16%. In conclusion, the suitable terrain proportion of the six types of landform except for high altitude mountain is equal to or more than a half, the potential development of Chinese cities is huge.

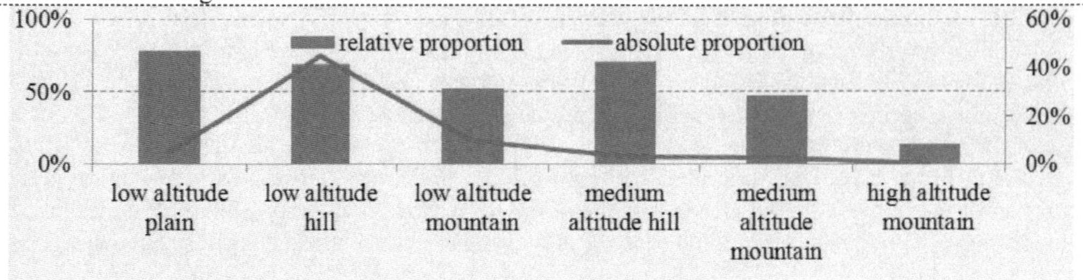

Figure 5. Suitable terrain of built-up areas in different landform environment.

5. Conclusions

Terrain characteristics of Chinese cities are analysed by a method of landform classification. Based on the terrain characteristics analysed above, suitable terrain requirement of urban expansion is built with the statistical regulation of neighbouring relief. At last, urban terrain suitability contrast in different parts of China is made. In conclusion,

1) Chinese urban areas are tend to distribute in low and gentle plain or hill, and the terrain characteristics show more unsuitable for urban areas in the west than in the east. There is a huge difference in altitude and relief but a small difference in slope among different urban areas. It is in a micro scale that the terrain in urban areas has similar characteristics. The average altitude, slope and relief in Chinese urban areas are 202m,2.65°and 104m, and the average range of these terrain factors in individual urban areas are 9～3664m, 0.47～14.28° and 23～881m respectively, and mainly concentrated in less than 500m, 6°and 200m.

2) Neighbouring relief is selected as a micro scale factor to make quantitative analysis of Chinese urban terrain characteristics. Neighbouring relief in more than 98.91% Chinese urban areas is lower than 80m, and 89.34% is lower than 30m, and the neighbouring relief range of individual urban areas is 3～79m, therefore, the upper limitation of neighbouring relief that 30m and 80m is statistically applicable to evaluate suitability of urban terrain in China.

3) In the buffers based on 2 times radius of Chinese cities, about 65.51% terrain is suitable for built-up areas. The potential of urban development in the central and west is huge from a view of microscopic terrain scale.

4) The suitable terrain around Chinese urban areas show a decline order as low altitude hill, low altitude mountain, medium altitude mountain, low altitude plain and high altitude mountain. And in the six types of landform except for high altitude mountain, the potential development of the cities is huge.

Acknowledgments

Authors wishing to acknowledge assistance from Project of National Key Technologies R&D Program of China,"Potential evaluation and scenario analysis of hollow rural settlement renovating" (No. 2014BAL01B01C).

References

[1] Batty M 2013. A theory of city size. *Science*. **340** 6139
[2] Diao C T 1990.The geomorphological environment and urban expanding in Chongqing. *Journal of Southwest Teachers University*.**15** 4
[3] Tongji University 1991.*Principles of urban planning*. Beijing:China Building Industry Press, pp. 49-68

[4] Verstappen H Th 1983.*Applied Geomorphology*.Elsevier,Amsterdam-Oxford, pp. 149-173
[5] Xu X L, Zhao M Y, Min X B 2014. Main progress and prospect of urban expansion study. *Ecological Science*.**33** 3
[6] Zhao M,Cheng W M and Huang K 2015.Spatial characteristics of urban development and its correlation with macro geomorphology in Beijing-Tianjin-Hebei region for recent 21 years. *Journal of Geo-information Science.* **17** 8
[7] Zhao K, Wang F X 2015.Landform classification from the perspective of human settlements and ecological suitability assessment about the construction of mountain settlements—a case of Quxian county in Sichuan province. *Interior Design*.3
[8] Zhang R T, Zhang X L 2012. Research progress and review of urban spatial expansion. *China Science and Technology Forum*.8
[9] Zhang Y M 2008.*The geomorphological features of the Xiaoqing river in Jinan mountain and its relationship with development of the city*. Shandong Normal University
[10] Zhang Z X, Zhao X L, Liu F, Xu J Y, Wen Q K and Hu S G 2014. *Atlas of urban expansion by remote sensing in China*. Beijing:Planet Map Publishing House, pp. 1-4
[11] Zhou C H, Cheng W M, Qian J K, Li B Y and Zhang B P 2009.Research on the classification system of digital land geomorphology of 1：1 000000 in China. *Journal of Geo-information Science*. **11** 6

Analysis of spatial pattern Change of LU/LC over the upper Tarim River region since 1990 using remote sensing data

L Y Bai[1], J Z Feng[2], Y X Ma[1], Q Y Ran[1], K Wang[1], Y Zhao[1]
[1]Institute of Remote Sensing and Digital Earth, Chinese Academy of Sciences, Beijing 100094, China;
[2]Key Lab of Agri-information Service Technology, Ministry of Agriculture of P. R. China, Agricultural Information Institute, Chinese Academy of Agricultural Sciences, Beijing 100081, China

Abstract. The upper reaches of Tarim River (URTR) is an important port of trade between China and central Asia. The development of the URTR is thus significant for the SREB initiative. The LU/LC data in the URTR from 1990 to 2015 were used to quantitatively explore the dynamics of LU/LC changes, and its driving force was discussed from two aspects of nature and human. Results showed that the unused land and grassland were the main land use types in this area, accounting for more than 79%. Compared with the data of 1990, the areas of woodland, water, farmland, and building land of 2015 increased with 3.24%, 6.53%, 10.57%, and 0.40%, respectively, and the areas of unused land and grassland decreased, which accounted for 53.25% and 26.01%, respectively. The increases of the woodland and farmland areas mainly is originated from grassland and unused land. The woodland increased sharply around 2000 due to the abundant water during the period between 1998 and 2000. Subsequently, part of the woodland was shifted into the farmland. The extension of building land wasn't obvious, but showed a salient feature of population urbanization. It was essential that the LU/LC patterns of the URTR were deeply influenced by human farming and living activities.

1. Introduction

It is crucial for land use planning, utilization of regional resources and environment management that examining the processes and trends of land use change and analyzing the primary causes. The upper reaches of Tarim River (URTR), located in the land-based Silk Road Economic Belt (SREB) (i.e., which focuses on connectivity and cooperation among main countries of Eurasia) in China, is an important port of trade between China and central Asia. The development of the URTR is thus significant for the SREB initiative.

Given that URTR is an ecological sensitive area, its land-use/land-cover changes have links to both human and nature interactions, monitoring the change of land use/land cover (LU/LC) over this region is helpful to create its harmonious coexistence pattern between man and nature and maintain its sustainable development in the future[1].

There are some existing researches focuses on the land use cover change of URTR, such as: Yongliang (2010) analyzed the changes (from 1994 to 2005) of population, land use patterns and water resource based on integrated methodology of GIS and statistic data[2].Zumrat Ebrayimland use and land cover change and trends and environmental effects of the period 1960 to 2000 were analyzed and researched quantitatively using mathematical statistical methods. Most the existing study focus on the land use cover changes before 2010[3].

In this paper the LU/LC data in the URTR from 1990 to 2015 were used to quantitatively explore the dynamics of LU/LC changes, and its driving force was discussed from two aspects of nature and human. Firstly, the LU/LC data over the URTR form 1990 to 2010 were obtained from the Data Center for Resources and Environmental Sciences, Chinese Academy of Sciences (RESDC) (http://www. resdc.cn). Furthermore, the LU/LC data over the URTR in 2015 were interpreted from Landsat satellite remotely sensed images. Then, a time series of LU/LC were built, and transfer matrix and association analysis were employed to explore the driving force of the LU/LC changes. This research provides the basic data support for decision-making so as to drive local economic development and construction of the SREB.

2. Data and methodology

2.1 Study area

Study area located between 34–43°N and 76–85°E (Figure 1) and covering more than 2.7×105 km2, including parts of hotan and aksu river basin, most of kashghar and ogen river basin, the upper reach of tarim river basin. The 495 km long upper tarim mainstream section and its sources seldom has contribution to surface runoff generation, and only fed by precipitation in the upper tributaries and snow/glacier-melt in the mountain regions, due to the extremely arid climate[4]. The unused land and grassland were the main land use types in this area, accounting for more than 79%.

2.2 Data

A 1-km gridded database of component classifiers have been completed for five stages: the late 1980s (1990), 1995, 2000, 2005, and 2010[5-7]. The database was produced under uniform quality control and integration checking, which ensured high-quality and consistency of the database[8]. There are 6 land use classes and 25 subclasses in the database, including cropland, woodland, grassland, water body, built-up land, and unused land. Additionally, we use the human–computer interactive interpretation method and the LUC information of 2010 as reference to interpret the 2015 Landsat TM digital images covering research area. Thus, 6 stages land used cover datasets (Figure 2), from 1990 to 2015, were used to spatial pattern Change analysis.

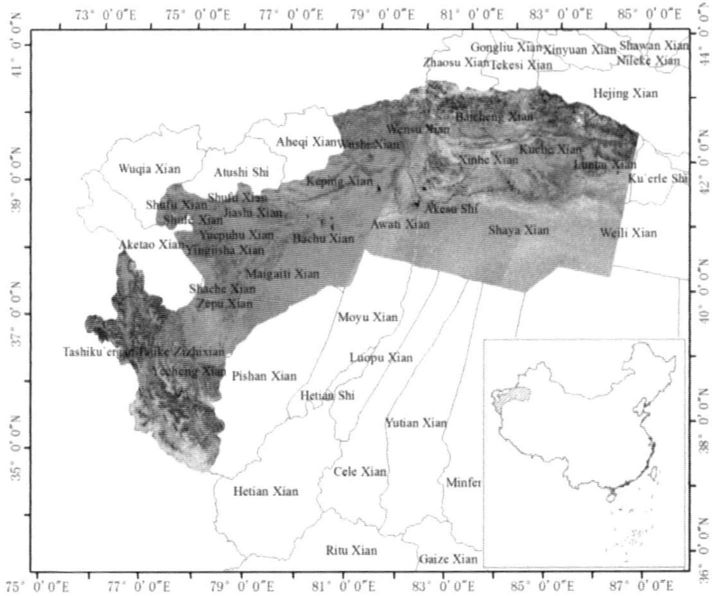

Figure 1. Position and scope of study area.

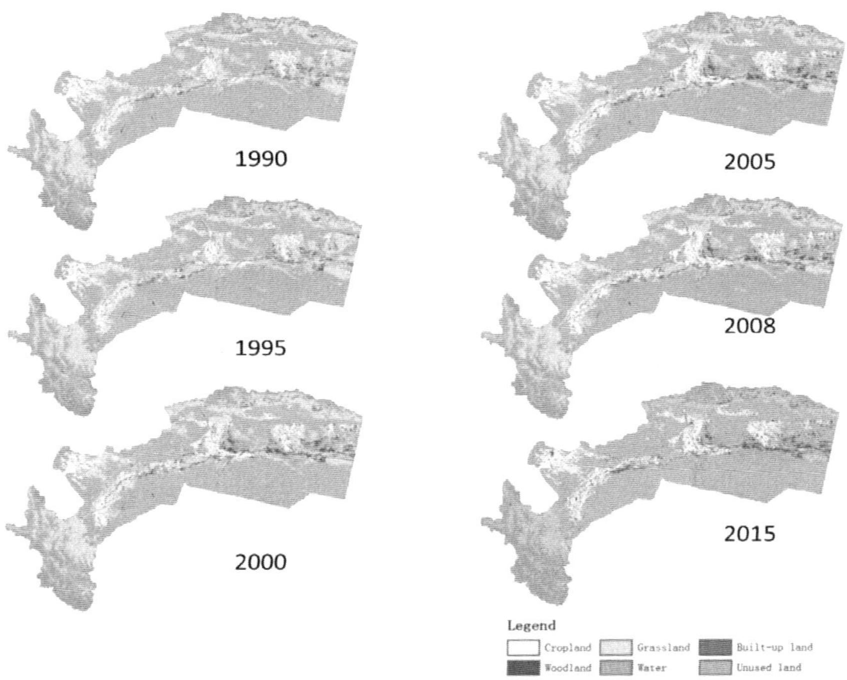

Figure 2. Land use cover of over upper reaches of Tarim River

for six stages: 1990, 1995, 2000, 2005, 2010, and 2015

2.3 Methodology

The total acreage of each land use class was calculated separately for every 5 years to analyze the quantitative land use change trend. Then transfer matrix and spatial analysis was employed to explore and mapping the dynamic transformation of land use cover. Finally, the driving force of land use transformation was discussed from two aspects of nature and human by utilizing statistics of natural resource and human society.

3. Result and discussion

3.1 Characteristics of land use acreage changes

From 1990 to 2015, land use cover in URTR indicated the significant characteristics of total acreage (figure 3) and spatiotemporal differences. Compared with the data of 1990, the areas of woodland, water, farmland, and building land of 2015 increased with 3.24%, 6.53%, 10.57%, and 0.40%, respectively, and the areas of unused land and grassland decreased, which accounted for 53.25% and 26.01%, respectively.

(1) The basic characteristics of changes in cropland can be summarized as follows. The total area of cropland increased by 11×103 km2 during past 25 years. After 2010 the cropland increased by 6×103 km2, with reclaimed cropland concentrated mainly in the farming–forestry or farming- pastoral ecotones of kashghar and ogen river basin.

(2) The basic characteristics of changes in woodland can be summarized as follows. The total area of woodland increased quickly by 2750 km2 during the period from 1995 to 2000. After then the area decreased during the following 2 stages. The increased woodland mainly distributed around/near the water, which including river, lake or marsh.

(3) The basic characteristics of other land use cover changes in grassland can be summarized as follows. The total area of grassland decreased by 14×103 km2. The total area of Unused land increased first and then decreased; the water increased in this region, and the built-up land decreased then increased.

3.2 The mainly transfer characteristics of landuse cover in URTR

The land use cover transfer matrix (table 1 to 5) of the five period, 1990 to 1995, 1995 to 2000, 2000 to 2005, 2005 to 2010, and 2010 to 2015 and the distribution of the land use change from 1990 to 2015 in URTR (figure 5) were used to primary analysis the transfer processing and its driving force.

The increases of the woodlands area mainly were originated from grassland and unused land, and the transfers mainly occurred before 2005. The woodland increased sharply around 2000 due to the abundant water during the period between 1998 and 2000 (Figure 5). Subsequently, part of the woodland was shifted into the farmland. The decreases of grassland area mainly were shifted into woodland, unused land, and cropland. During the first period most of the decreased grassland shifted into unused land or cropland. After then, also large area of grassland transfer to woodland. During 2000 to 2010 farming is the only main cause of grassland shrink. During the last period the transformation of grassland has similar characteristics with the first period. About the area of water, the increases and decreases approximately equal before 2010. However, during the last period, large area of unused land transferred into water land. The extension of building land wasn't obvious, but showed a salient feature of population urbanization. Most of the built-up lands originated from cropland.

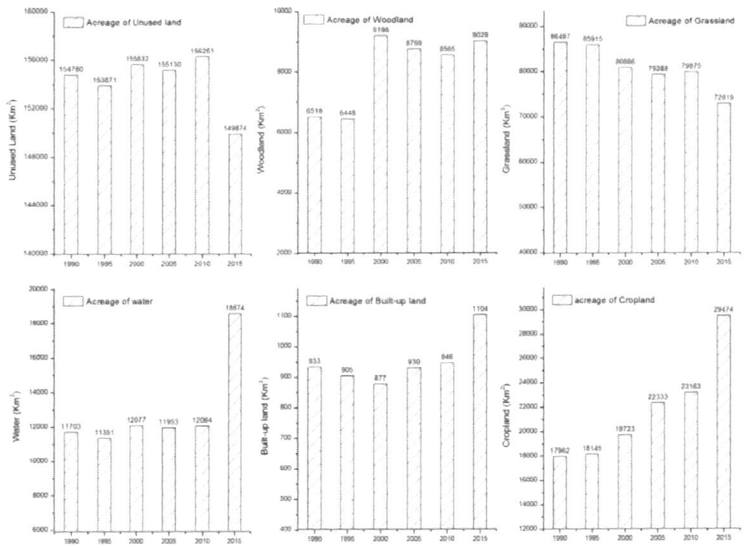

Figure 3. Acreage change trend of each land use class over upper reaches of Tarim River from 1990 to 2015

The characteristics of spatial distribution of land use cover conversions can be summarized as follows. The large area of grassland and woodland shift into cropland, meanwhile many croplands convert to built-up land, due to human activities. However, the degradations of grassland or woodland often distributed far from built-up land and water. The primary driving force is water.

Table 1. The land use cover transfer matrix from 1990 to 1995.

1995 / 1990	Woodland	Grassland	Water	Built-up land	Unused Land	Cropland
Woodland	4003	1854	108	14	370	164
Grassland	1847	74732	796	76	5983	2804
Water	103	797	8558	5	1879	128
Built-up land	3	71	5	262	36	550
Unused Land	335	6076	1919	45	145079	712
Cropland	152	2625	139	518	714	13790

Table 2. The land use cover transfer matrix from 1995 to 2000.

2000 / 1995	Woodland	Grassland	Water	Built-up land	Unused Land	Cropland
Woodland	4073	1653	111	10	346	255

Grassland	3779	68179	981	52	9037	3887
Water	115	560	8681	1	1873	121
Built-up land	11	51	4	290	29	520
Unused Land	858	7714	1804	19	142576	900
Cropland	345	2481	186	505	608	14020

Table 3. The land use cover transfer matrix from 2000 to 2005.

2000 \ 2005	Woodland	Grassland	Water	Built-up land	Unused Land	Cropland
Woodland	8703	34	16	1	5	439
Grassland	25	78940	35	8	12	1866
Water	10	151	11836	0	46	34
Built-up land	0	0	0	871	0	6
Unused Land	20	98	63	15	155067	369
Cropland	1	65	3	35	0	19619

Table 4. The land use cover transfer matrix from 2005 to 2010.

2005 \ 2010	Woodland	Grassland	Water	Built-up land	Unused Land	Cropland
Woodland	8503	95	3	1	7	150
Grassland	11	78727	6	0	88	456
Water	2	201	11625	0	86	39
Built-up land	0	0	0	929	0	1
Unused Land	20	103	2	10	154748	247
Cropland	13	72	1	6	6	22235

Table 5. The land use cover transfer matrix from 2010 to 2015.

2010 \ 2015	Woodland	Grassland	Water	Built-up land	Unused Land	Cropland
Woodland	7788	392	44	15	206	120
Grassland	619	70817	787	42	2178	5432
Water	282	514	9005	12	1901	350
Built-up land	0	18	0	764	12	152
Unused Land	150	398	8699	48	145316	1650
Cropland	190	680	39	223	261	21770

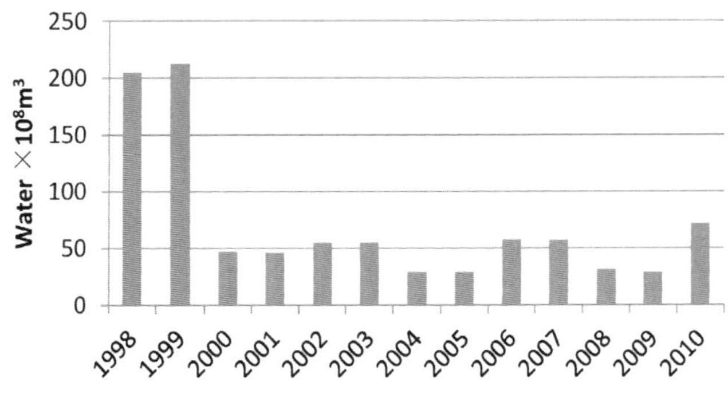

Figure 4. Annual run-off of Tarim River from 1998 to 2010.

4. Conclusion

The LU/LC data in the URTR from 1990 to 2015 were used to quantitatively explore the dynamics of LU/LC changes, and its driving force was discussed from two aspects of nature and human. Results showed that the unused land and grassland were the main land use types in this area, accounting for more than 79%. Compared with the data of 1990, the areas of woodland, water, farmland, and building land of 2015 increased with 3.24%, 6.53%, 10.57%, and 0.40%, respectively, and the areas of unused land and grassland decreased, which accounted for 53.25% and 26.01%, respectively. The increases of the woodland and farmland areas mainly is originated from grassland and unused land. The woodland increased sharply around 2000 due to the abundant water during the period between 1998 and 2000. Subsequently, part of the woodland was shifted into the farmland. The extension of building land wasn't obvious, but showed a salient feature of population urbanization.

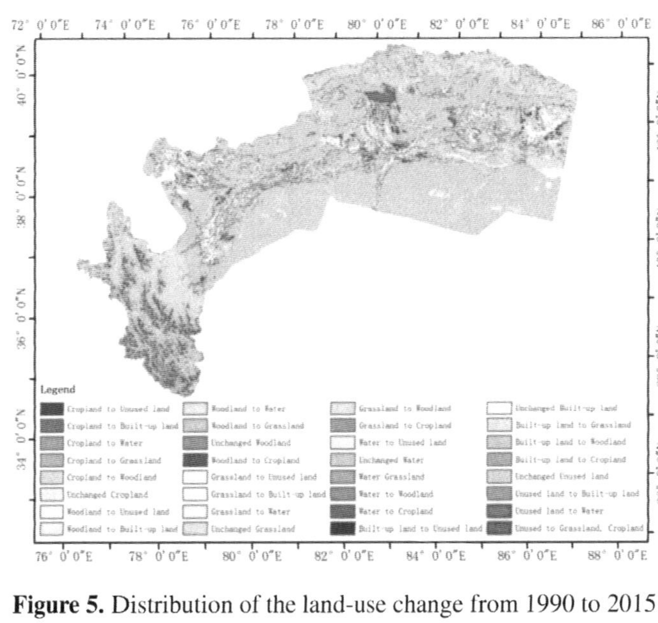

Figure 5. Distribution of the land-use change from 1990 to 2015.

Acknowledge

The Western Light Talent Culture Project (YBXM-2014-05), the Natural Science Foundation of China (NSFC; Grant Nos. 41171340 and 41101390), and the Science and Technology Innovation Program (for the Agricultural Information Institute) of the Chinese Academy of Agricultural Sciences supported this work.

Reference

[1] Wang X, Zhao X, Zhang Z, Yi L, Zuo L, Wen Q, Liu F, Xu J, Hu S and Liu B 2016 *Catena* **137** 256-68

[2] Shi Y, Wang R, Fan L, Li J and Yang D 2010 *Procedia Environmental Sciences* **21** 75-84

[3] Zumrat E, Tursun Q, Ayxam I Sum M and Han G 2011 *Journal of Arid Land Resources and Environment* **25** 29-35

[4] Su B, Wang A, Wang G, Wang Y and Jiang T 2016 *International Journal of Applied Earth Observation and Geoinformation* **48** 122-30

[5] Liu J, Liu M, Zhuang D Zhang Z and Deng X 2003a *Science in China Series D: Earth Sciences* **46** 373-84

[6] Liu J, Zhang Z, Zhuang D 2003b *Geographical Research* **22** 1-12

[7] Liu J, Zhang Z, Xu X, Kuang W, Zhou W, Zhang S, Li R, Yan C, Yu D, Wu S et al. 2010 *Journal of Geographical Sciences* **20** 483-94

[8] Zhang Z, Zhao X 2012 *Remote sensing monitoring of land use in China* (Beijing: Star Maps Publishing)

Early Stage Evolution of Nourished Beach under High-energy, Macro-tidal Environment

J H Liu[1], F Cai[1], Z W Zhang[1] and B Li[1]

[1]Island Research Center, State Oceanic Administration, 350400 Pingtan, China

E-mail: liujianhuiirc@126.com

Abstract. Beach planform evolution, profile equilibration and sediment grain size change have been studied during the first 4 months from 4th September to 24th December 2011 after the construction of beach nourishment project at Longfengtou Beach, Haitan Bay. Monthly beach profiles, shoreline surveys, sediment sampling and nearshore wave measurements were carried out after implementation of the 1.3km long nourishment project which was completed on 20th August 2011. This study indicates that: (1) rapid beach profile equilibration occurred in the early stage after the construction of the project. A null point was observed, which is equal to the height of mean high tide, basically kept dynamic stable during the process of profile evolution. Shoreface sediment accumulated beneath the height of this point while erosion happened above it, the slope between the beach berm and the landward edge of low tidal zone became more gradual accompanied with seaward transportation of beach sediment. The velocity of beach slope adjustment in earlier period is faster than later. (2) Beach planform adjustment initiated simultaneously with the combination of the process of profile equilibration and longshore sediment transport. Shoreline retreated with an average distance of 11.1m and maximum of 31.02m from 4th September to 24th December, erosion in the south part was more serious than in the north, and 3 erosion hot spots were found along the coast. (3) Sediment redistributed with cross-shore profile equilibration, it showed a pattern across beach profile as medium sand (0.4-0.5mm) in beach berm, smaller (0.3-0.4mm) in high and middle tidal zone, coarse sand(0.6-1mm) in beach slope transitional zone, fine sand(0.1-0.25mm) in low tidal zone. The sediment grain size change of foreshore was rapidly response to the passage of storm surge.

1. Introduction

Over the last two decades, beach nourishment has become a more and more popular solution to erosion problem and coastal restoration in China (Cai *et al.*, 2012), meanwhile, the debates over the effectiveness of this young technology continues, especially in some serious erosion beaches with high energy. Often, data produced from inadequately planned monitoring programs after nourishment is unable to address the pertinent issues, and crucial performance questions remain unanswered (NRC, 1995). It is noted that immediate high-resolution post-nourishment monitoring is seldom (Davis *et al.*, 2000; Gravens *et al.*, 2003). Nourished beaches are almost always constructed with sediment that differs from the native grain size of the natural beach, they also been constructed on considerably steeper slopes than natural profiles. Nourishment projects tend to be highly scrutinized by the public during construction and after project completion (i.e. Pilkey and Clayton, 1989). Understanding the project adjustment has an important implications (Elco & Wang, 2006), not only to coastal engineer but also to public. Nourished beach profile and planform adjustment must be explained to avoid misinterpretation of early stage adjustment as a permanent loss of sand or a misuse of public funds (NRC, 1995; Elko, 2005).

This study contributes to the understanding of processes governing profile equilibration and planform adjustment, particularly the appearance of erosion hot spot. The objective of this study was to understand the early stage of profile, planform, and sediment evolution responding to a high-energy and macro-tidal environment.

2. Study area

Pingtan Island is the fifth biggest island in China and locates in the west of the Taiwan Strait, Haitan Bay was formed in the right east of Pingtan Island (Figure 1). The study area Longfengtou Beach is located in the southeast of Haitan Bay. It is a high-energy, macro-tidal zone with an annual average $H_{1/10}$ (10% highest waves) of 1.56m, dominant wave direction is ENE with an frequency of 67.43%(Figure 2).The mean tidal range is 4.29 m with an average tidal cycle of 12.5 hours(Wang *et al.*, 2014). Affected by the monsoon, the seasonal wind direction shows an obvious variation in the study area, the annual average wind speed is approximately 8.4 m/s. The statistical data also indicates the predominance of northeast(NE) wind in the autumn, winter and spring seasons, whereas south(S) wind in the summer. Normally, this area is attacked by Typhoon from June to September every year.

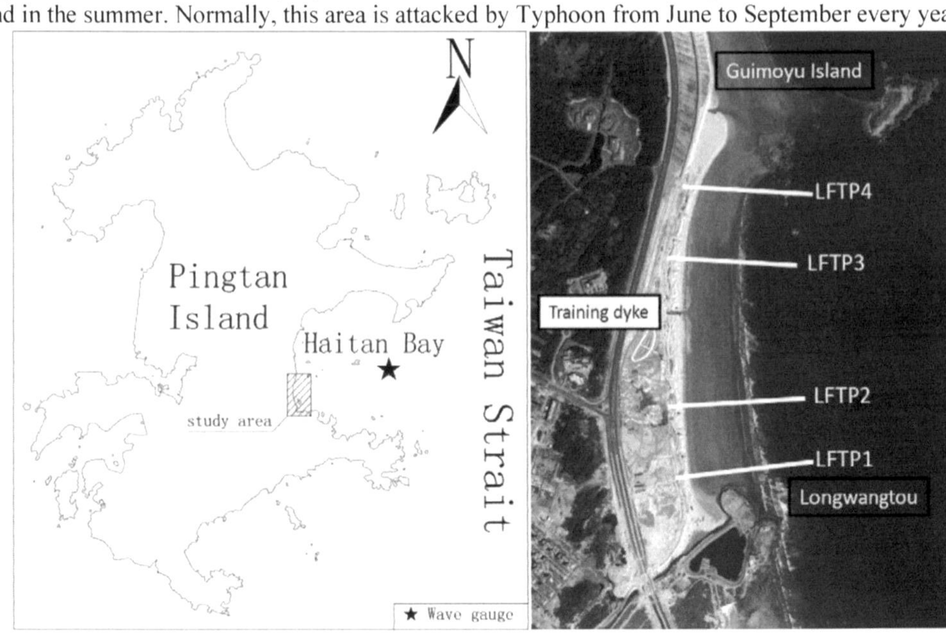

Figure 1. Location of study area of Longfengtou Beach and distribution of monitoring beach profiles.

Longfengtou beach used to be a famous leisure area which was destroyed by seawall building along the shoreface aiming to infrastructure construction in 1990s. To recover the beach view, nourishment project has been completed in August 2011, with a length of 1.3km. The wider north part is 120m in width and 200m long, transitional middle part is 130m long and the narrow south part with a width of 60m and length of 1000m.

It is worth to note that there are two specific geomorphologies which have significant impacts on the project. The first one is Longwangtou man-made offshore headland located about 100m away from the south end of the project (Figure1), it is a vertical elliptic construction with a long axis of 140m, a minor axis of 70m and a height of 6m from the beachface, and higher than the height of spring tides. The second one is Guimoyu Island which locates about 550m away from the north end of the project, the island is 200m long and 60m wide with a trend direction of NE. A training dyke was built in the middle of the project to drain away the wastewater and stormwater from urban city.

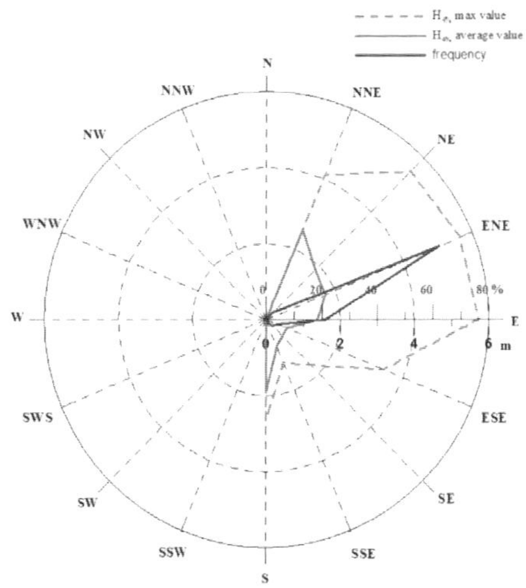

Figure 2. Wave rose of Haitan Bay measured from July 2011 to July 2012.

3. Methodology

Four beach profiles (Figure 1), shoreline position and sediment samples grain size were measured periodically from September to December 2011.

Based on the experience from previous monitoring efforts, traditional wading depth profiles were extended to approximately -2 m (85 Yellow Sea Height Datum) to capture nearshore changes and measure profile equilibration. These wading beach-profile surveys followed level-and-transit procedures using an electronic total station. Four surveys of the 4 profile lines were conducted during this field effort. Monthly beach profiles were surveyed 20 days after nourishment from 4 Sep.2011 to December 24, 2011.Beach shoreline position was identified as berm crest during field survey and mapped with the RTK-GPS (Real Time Kinematic-Global Positioning System) in an space interval of 0.5m.

Beach topography measurement was also completed by RTK-GPS, from the edge in front of seawall to Mean Low Water line across the shore with the scale of 1:100. Two times of topographic survey were carried out on the date of 20 Sep 2011 and 21 Dec 2011to compare the beach morphology change.

Sediment samples of top 5cm surface were collected synchronously with beach profile survey. The sampling locations were distributed along the profiles normally with an elevation interval of 30cm. Specifically, they would be more dense in some observed grain size transitional areas. The sediment grain size was determined using Mastersizer 2000 laser diffraction particle size analyzer. AWAC-Acoustic Wave and Current gauge was deployed about 5000 m offshore of the centre of Haitan Bay in depth approximately 15 m (Figure 1). Wave conditions were measured every 60 min and sampled at 2 Hz from July 23 2011 to July 21 2012. A large amount of detailed field data was collected over 4 months in the initial post nourishment phase of the project. The data allowed for analysis of the immediate post nourishment response, which is the focus of this paper.

4. Results

4.1. Shoreline change

After construction, morphologic changes were measured. Shoreline were surveyed at the first time on 4 Sep.2011 about 20 days after the project finished, and then 3 times on 20 Sep.2011,6 Nov.2011, and 20 Dec.2011, respectively. The comparison of shoreline position in different periods was displayed in Figure3. It is seen the shoreline has a steadily regression with an average distance of 11.1m and a maximum of 31.02m which was found in the right north of a training dyke. Shoreline retreat was characterized by 3 segments: (1)serious erosion happened in the south segment(from the south end to training dyke) with a mean erosion distance of 13.2m, shoreline retreated consistently and synchronously; (2) gentle erosion happened in the middle segment (from training dyke to 1100m from the south beginning) with a mean erosion distance of 7.1 m exceptional two erosion hot spots were found in both ends of this segment(see Figure 3); (3) beach shoreline remained basically stable in the north segment. It can be seen from Figure 4, shoreline retreated speed in last month during survey period was much less than earlier.

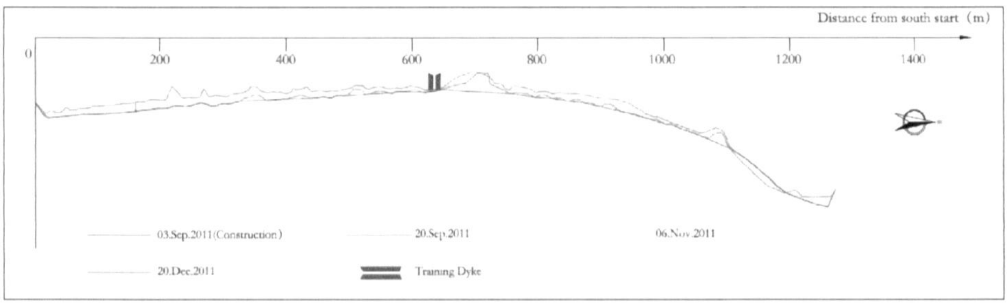

Figure 3. Shoreline position measured in 4 different periods.

Three erosion hot spots were found during survey(see Figure 4). The original definition of an Erosion Hot Spot (e.g., Bridges 1995; Dean, Liotta, and Simón 1999) was intended to cover erosional phenomena that were unanticipated and primarily local (e.g., a well-identified area located within a beach fill). Erosion hot spots are herein defined as local sections of beach—alongshore—that show obvious anomalous erosion or higher erosion rates compared to the adjacent segment. As statistics indicate (Table 1), the maximum retreat distance of each erosion hot spots are 31.02m, 30.3m and 30.01m from the south to the north respectively with different scales varied from 20m to 150m owing to the locations along the coast.

Table 1. Parameter of erosion hot spots occurred in Longfengtou nourished beach

Erosion hot spot	Location（from the south beginning m）	Scale （m）	Average retreat（m）	Maximum retreat（m）
1	210-360	150	15.85	31.02
2	665-730	55	25.84	30.3
3	1075-1095	20	21.13	30.01

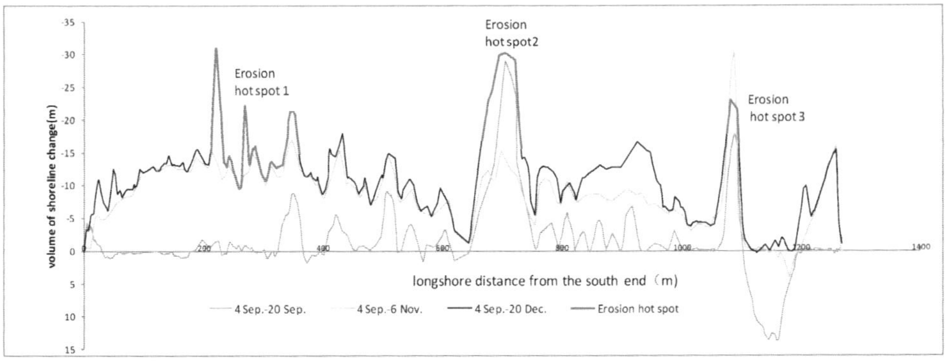

Figure 4. Value of shoreline change between 4 Sepetember and different periods and the locations of erosion hot spots.

4.2. Beach profile evolution

Longfengtou beach was a dissipative beach with a gentle original slope which is 1:35 for upper beach above MHW and 1:80 for shore face between MHW and MLW. As noted, beach nourishment projects are generally placed with profiles which are steeper than the natural profile for the coarser size of sediment that is used in the beach nourishment project. Thus over several tidal circles affected by the wave action beach profile will tend to equilibrate to a new shape. Figure 5 shows the evolution of different beach profiles from 4th Sep. 2011 to 20 Dec. 2011, it indicates that upper beach slope has become gentler especially for profile LFTP2 which is located in the area of erosion hot spot 1. There was an obvious transitional zone in each cross-profile, which is normally equal to the height of local mean high water, basically kept dynamic balance during the process of profile change. Sediment accumulated beneath the transitional zone while the beachface eroded above it, the slope between the beach berm and the landward edge of low tidal zone became gentler accompanied with seaward transportation of beach sediment. At the first month, beach slope changed tremendously, and then the velocity of beach slope adjustment in later stage was slower than earlier.

4.3. Sediment variation

It was found that choosing sampling sites along the profile based on hydrodynamic zonation on the beach (i.e. area of maximum run-up, mid-tide area and mean low tide area) gives the best representative picture of grain size distribution. The mean grain size of sediment sampled from original beach profile showed the distribution trend as 0.15-0.25mm in high tide zone, 0.2-0.4mm in middle tide zone and 0.1-0.2mm in low tide zone. Ideally, for any given beach nourishment project, the borrow material should be similar in grain size distributions to the native beach. Often the composition of fill material in available borrow areas is not the same. Fill material placed in Longfengtou beach nourishment project was borrowed from offshore in the north of Pingtan Island, the mean grain size is 0.47mm.

Fill was placed in the intertidal area and sediment redistribution occurred violently owing to the action of wave, tide and wind forces. 211 samples were collected during field survey synchronously with beach profile measurement, the change of sediment mean grain size distribution in each profile for 4 times is shown in Figure 6.

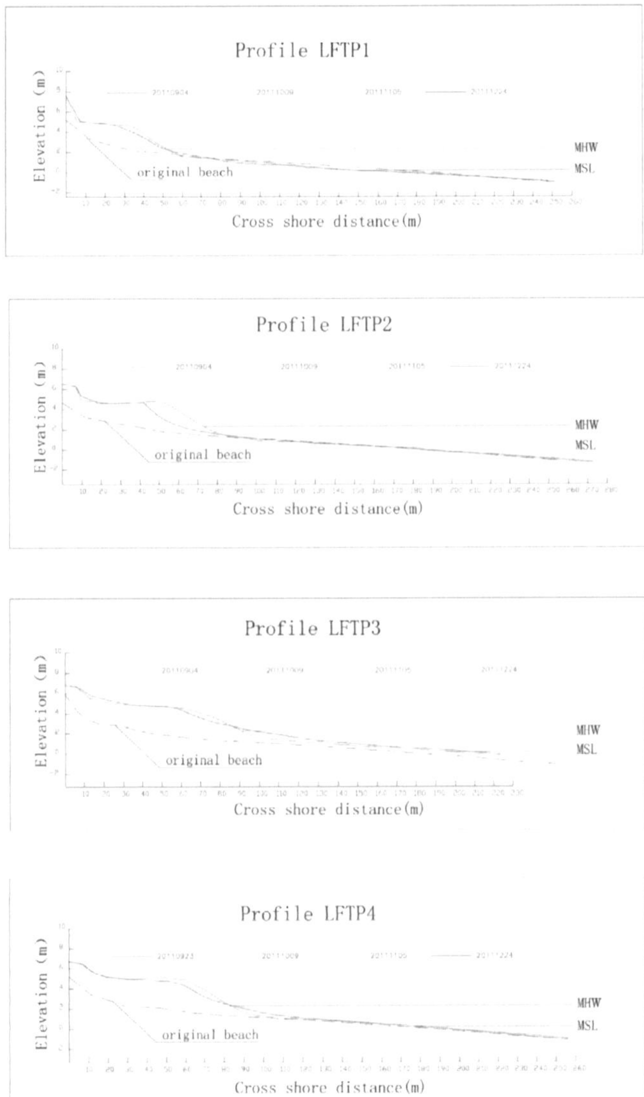

Figure 5. Beach profile changes during 4 months after nourishment project construction.

(a) The character of sediment grain size distribution across beach profile

For each profile at the upper beach higher than 3.3m(which is about the height of Mean High Water Spring), sediment shows a slight variation and the mean grain size is about 0.45-0.55mm generally. Beneath this height, sediment grain size varied significantly. For profile LFTP3 and LFTP4 sediment grain size varied quickly and enormously from the height 3.3m to 1.2m seaward, the coarsest mean grain size of about 0.8-1.2mm was found in the height between 1.7m-2.3m, and the grain size decreased to ~0.2mm where in the height of 1.2m, and beneath it the grain size showed only slight change. For profile LFTP1 and LFTP2, the span of sediment varying cross shore was not as large as the other profiles, showed a decreasing trend generally.

(b) Variation of beach sediment distribution in different periods

Four profiles distributed from the south to the north showed different performance of sediment change in each survey period. For profile LFTP1, there was a slight change of sediment distribution characteristics during each time mainly happened in low tide zone. For profile LFTP2, the change is significant with a wide grain size variation span from 0.2mm to 0.6mm in middle tide zone. For profile LFTP3 and LFTP4, as mentioned above the particle size distribution cross shore was roughly similar in each period, except on the date of 22 Sep. 2011, it was 3 days after the landfall of typhoon Nanmadol.

The sediment grain size change in foreshore was complicated, and rapidly responded to the passage of storm surge. Normally, sediment across beach profile showed a distribution regulation as medium sand (0.4-0.5mm) in beach berm and dry beach, smaller (0.3-0.4mm) in high and middle tide zone, coarse sand(0.6-1mm) in beach slope transitional belt, fine sand(0.1-0.25mm) in low tide zone.

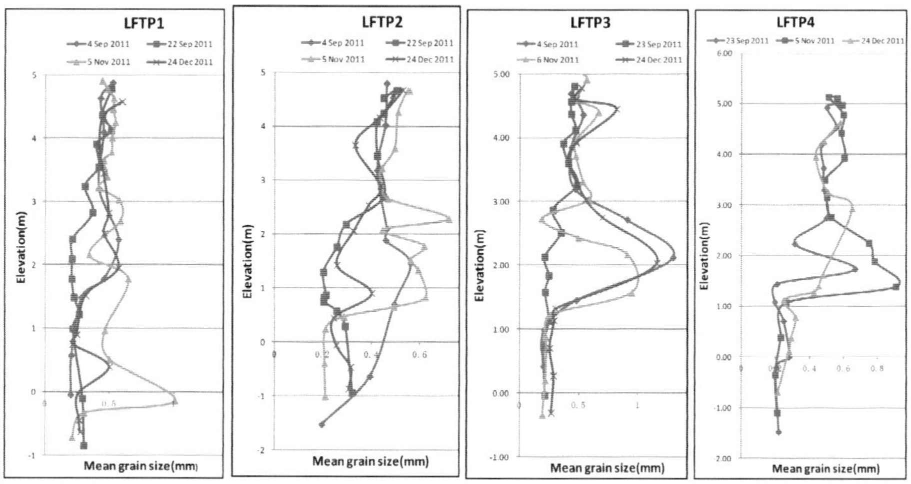

Figure 6. Variation trend of mean grain size of beachface sediment at different elevation along the beach profile.

5. Discussion

5.1. Early stage beach evolution

Study area is characterized by high-energy and macro-tidal environment, stronger nearshore hydrodynamics contributes to rapid nourished beach profile equilibration which also results in a violent adjustment of beach planform. During the process of profile equilibration, most of the volume of placed material remains within the project area landward of the closure depth, and is simply redistributed across the profile, the dry beach width is usually reduced during this process (Elco & Wang, 2006). Comparing the value of accumulation and erosion across beach profile during the first 2 months, net sediment deficit is small except in some erosion hot spot area. And in the later period after then, net sediment deficit was observed gradually.

A new equilibrium beach profile was created in high energy macro-tidal environment by fill material cross-shore sediment transport which was the dominant process causing profile re-equilibration in the early stage. Nourished beach planform adjustment in early stage was result from combined action of cross-shore and longshore sediment transport, and the former is the controlling process.

5.2. Erosion Hot Spot (EHS)

An EHS is an area that experiences sediment transport potential without having adequate sediment supply. EHSs erode more rapidly than the adjacent beaches or more rapidly than anticipated during design. Some coasts to be nourished are sediment deficient, obscuring the presence of latent EHSs that emerge once material is in place. Kraus & Galgano (2001) identified and classified all known types of EHSs and points to possible underlying causes as presently understood. In this study 3 EHSs observed along the Longfentou Beach can be classified as two types: 1)Structure-induced, and 2) Non-uniform offshore translation of beach. EHS2 was a structure-induced erosion because the training dyke was located in the south and blocked the sediment transport from the south to the north. The training dyke was made of geotextile with sand inside, the protruding part of the dyke could be easily destroyed by wave action. EHS1 and EHS3 can be classified as the type of Non-uniform offshore translation, the former was affected by Longwangtou man-made offshore headland and the latter might be induced by the complicated bathymetry comprised by Guimoyu Island and a series of submerged reefs arranged in the direction of ENE.

5.3. Cross-shore distribution of sediment

The cross-shore distribution of sediment grain size is not always uniform even in different parts of same bay beach. The distribution pattern of grain size across the surf zone is determined mainly by the rate of wave-energy dissipation (e.g., Dean 1977, 1991), local balance in the transport energetics (e.g., Inman and Bagnold 1963; Bowen 1980; Bailard and Inman 1981), and sediment supply. High wave energy and macro tidal hydrodynamic characteristic in study area determine the sediment distribution pattern of coarse sediment in middle zone and fine sediment in upper and lower zone. Owing to the complexity of nearshore bathymetry, attached morphology and man-made coastal constructions, hydrodynamic condition varied significantly from the north to the south part along Longfengtou beach, thus result in a regional difference of sediment distribution even developed from same fill material after beach nourishment construction.

6. Conclusion

Beach morphology evolution happened soon after beach nourishment project completed, the primary performance of beach change showed as beach profile equilibration, beach planform adjustment and sediment distribution tends to be in corresponding zones gradually. High wave energy macro-tidal environment is the main reason for quick beach change. Beach profile evolved to new equilibration adapting to macro-tide and fill material composition. Nourished beach shoreline change resulted from the combination of two processes as beach profile equilibration and longshore sediment transport, the complexity of nearshore bathymetry and attached man-made constructions induced hydrodynamic difference along study area caused the regional distribution of erosion hot spots. Accompanied with cross-shore profile evolution, sediment distribution was formed in a pattern as fine sediment in upper zone, coarse sediment in middle zone, and finer sediment in lower zone.

Acknowledgements

The study was supported by the public science and technology research funds projects of ocean (201405037) and National Natural Science Foundation of China (41406070).

References

[1] Bailard, J.A., and Inman, D.L., 1981. An energetics bedload model for a plane sloping beach: local transport, *Journal of Geophysical Research*, 86(C3), pp. 2035–2043.

[2] Binyu Wang, Cuiping Kuang, Yunchi Shan, 2014. Hydrodynamic Response to a Groin System at Longfengtou Beach, China, in *Proceedings of the Twenty-fourth (2014) International Ocean and Polar Engineering Conference*.

[3] Bowen, A.J., 1980. Simple models of nearshore sedimentation: Beach profiles and longshore bars, in McCann, S.B., ed., *The Coastline of Canada: Littoral Processes and Nearshore Bars: Geological Survey of Canada, Ottawa, Ontario*, pp. 1–11.

[4] Bridges, M. H., 1995. "Analysis of the processes creating erosional hot spots in beach nourishment projects," *Coastal & Oceanographic Eng. Program*, U. of Florida, Gainesville, FL., pp.135.

[5] Cai F, Dean R G, Liu J, 2011. Beach nourishment in China: status and prospects, in *Coastal Engineering Proceedings*.

[6] Davis, R.A., Wang, P., Silverman, B.R., 2000. Comparison of the performance of three adjacent and differently constructed beach nourishment projects on the Gulf peninsula of Florida, *Journal of Coastal Research*, 16(2), pp.396-408.

[7] Dean, R. G., Liotta, R., and Simón, G., 1999. Erosional hot spots, *Coastal &Oceanographic Eng. Program*, pp. 60.

[8] Dean, R.G.,1977. Equilibrium beach profiles: U.S. Atlantic and Gulf coasts, *Ocean Engineering, University of Delaware, Newark, Delaware*, pp.45.

[9] Dean, R.G.,1983. "Principles of beach nourishment," in: *Komar, P.D. (Ed.), CRC Handbook of Coastal Processes and Erosion*. Boca Raton, CRC Press.

[10] Dean, R.G.,1991. Equilibrium beach profiles: Characteristics and applications, *Journal of Coastal Research*, 7(1), pp.53–84.

[11] Dean, R.G., Campbell, T.J.,1999. Recommended beach nourishment guidelines for the state of Florida and unresolved and related issues, FDEP Workshop on Beach Nourishment. University of Florida, *Coastal & Oceanographic Engineering Dept. Tech*. 10 pp.

[12] Elko, N.A., 2005. Management of a beach nourishment project during the 2004 hurricane season, *Shore and Beach*, 73(2), pp.49-54.

[13] Gravens, M.B., Ebersole, B.A., Walton, T.L., Wise, R.A., 2003. Beach fill. In *Curtis, W. (Ed.), Coastal Engineering Manual, Part V, Coastal Project Planning and Design*,pp.2100.

[14] Inman, D.L., and Bagnold, R.A., Littoral processes, in *Hill, M.N., ed., The Sea*: New York, Wiley-Interscience, 1963, p. 529.

[15] National Research Council, Beach Nourishment and Protection, Washington, D.C. National Academy Press, 1995.

[16] Nicole A. Elko, Ping Wang.2007. Immediate profile and planform evolution of a beach nourishment project with hurricane influences, *Coastal Engineering*, 54(1), pp.49-66.

[17] Pilkey, O.H., Clayton, R.D., 1989. Summary of beach replenishment experience on U.S. east coast barrier islands, *Journal of Coastal Research*, 5(1), pp.147-159.

[18] Weggel, R.J., 1995. A primer on monitoring beach nourishment projects, *Shore & Beach*, 63(3), pp.20-24.

Storm surge observed by Chinese HY-2A satellite radar altimetry

J Yang[1], X Li[1], G Han[2], N Chen[1] and D Chen[1]

[1] State Key Laboratory of Satellite Ocean Environment Dynamics, Second Institute of Oceanography, State Oceanic Administration, Hangzhou, China.

[2]Biological and physical Oceanography Section, Fisheries and Oceans Canada, Northwest Atlantic Fisheries Centre, St. John's, Newfoundland, Canada.

Corresponding author's e-mail: jsyang@sio.org.cn

Abstract. HY-2A is the first Chinese ocean dynamic environment monitoring satellite, which was launched in August 2011. The satellite repeats its ground track every 14 days. It plays an important role in global monitoring of sea surface winds (especially extreme winds like typhoons and hurricanes), ocean waves, currents, eddies, and extreme events like storm surges by using its four major payloads, i.e. radar altimetry, microwave scatterometer, scanning microwave radiometer and calibration microwave radiometer. The HY-2A data are obtained from China's National Satellite Ocean Application Service (NSOAS). We use 1 s along-track data with a nominal spatial resolution of about 7 km. For example, a storm surge induced by tropical cyclone Funso in the Southwest Indian Ocean near Mozambique in January 2012 is observed by HY-2A satellite altimetry. The storm surge magnitude is estimated to be 0.49 m and the cross-shelf e-folding decay scale to be 92 km. The present study shows that the HY-2A satellite altimetry is a useful tool for monitoring storm surges and their impacts in the Indian Ocean.

1. Introduction

Storm surge induced by an intense tropical cyclone (or typhoon/hurricane) is often the greatest threat to life and property of coastal areas. Intense Tropical Cyclone Funso (Category 4) was a powerful tropical cyclone which produced flooding in Mozambique and Malawi in January 2012 (figure 1). It is the first storm to affect Mozambique since Jokwe in 2008.

Satellite altimetry is a useful tool for observing and studying storm surge features. Scharroo et al. [2005] reported for the first time the altimetry observation of a storm surge in the Gulf of Mexico associated with Hurricane Katrina. Han et al. [2012] demonstrated the utility of satellite altimetry in observing and studying storm surge features and understanding dynamic mechanisms off Newfoundland and in the Gulf of Mexico. Lillibridge et al. [2013] and Chen et al. [2014] analyzed the HY-2A captured storm surge (with a surge magnitude of about 1.83 m at the coast and a cross-shelf decaying scale of 68 km) which was induced by Hurricane Sandy off New York. They pointed out that the along-track surge profile was approximately linear over the continental shelf. One of the key components in the e-Surge initiative is to integrate satellite altimetry and other space observations into monitoring and forecasting of storm surges.

2. Data and Method

2.1. HY-2A Data

The HY-2A sea surface height data and sea surface wind field data are obtained from China's National Satellite Ocean Application Service (NSOAS). The satellite repeats its ground track within ±1 km every 14 days.

2.2. *Other Data*

We also use 3-hourly pressure fields from the North American Regional Reanalysis (NARR) project. The NARR is a high-resolution extension of the National Centers for Environmental Prediction (NCEP) Global Reanalysis over the North American Region.

3. Method

We use 1 s along-track data with a nominal spatial resolution of about 7 km. The altimetry data are corrected for wet tropospheric (based on the onboard calibration microwave radiometer) and ionospheric path delays, and for ocean, solid earth and pole tides [NSOAS, 2011]. The inverse barometric correction is not applied, unless specified otherwise. During Cycle 9, HY-2A flew over Quelimane, Mozambique at 15:32, 25 January 2012 UTC. Along-track data from 5 October 2011 to 26 December 2012 UTC are processed for track 127 from 18° to 30°S. There are a total of 32 cycles of good data.

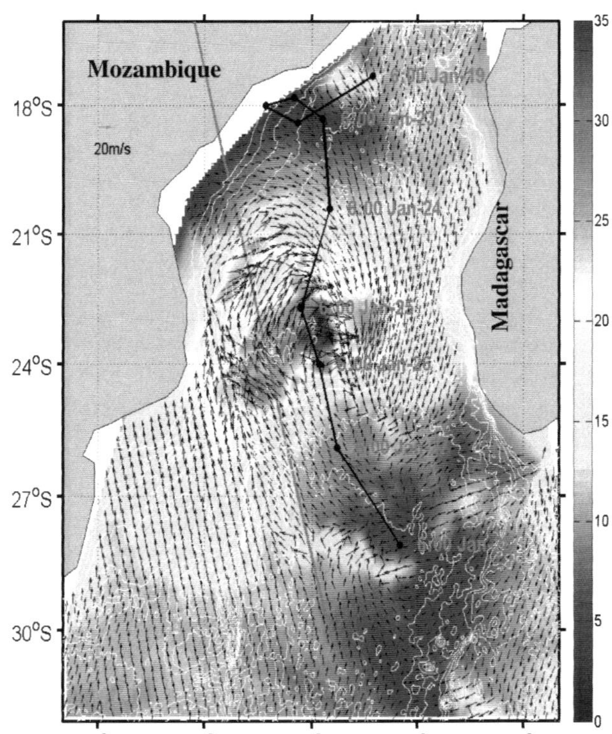

Figure 1. Map showing the study area between Mozambique and Madagascar, and sea surface wind fields retrieved from microwave scatterometer on board HY-2A at 15:32, 25 Jan 2012, UTC. Intense tropical cyclone Funso track and locations at specific time are shown as black lines and dots. Red line is the HY-2A satellite ground track.

Dry tropospheric delays are corrected based on the sea level pressure field described in section 2.2. Ocean wave effects are not corrected. To suppress range noise in the HY-2A altimetry data, we applied a 50 km Butterworth low-pass filter, following Lillibridge et al. [2013]. The data from the other 31 cycles are linearly interpolated onto the data locations (based on latitude) of Cycle 9. Mean sea level for the satellite data period from 5 October 2011 to 26 December 2012 UTC is then calculated by averaging sea levels of all cycles at each location. Finally sea level anomalies are obtained by subtracting the mean sea level from sea levels of each cycle.

Continental shelf waves, with its sea surface height decaying exponentially across the shelf, could be generated by passing hurricanes [LeBlond and Mysak, 1978]. Thus, we fit the altimetry along-track sea surface height anomalies corrected for the inverse barometric effect at Cycle 9 to an exponential function with the maximum value at the coast. It is important to remove the inverse barometric effect before the exponential fitting so that shelf wave properties can be retrieved better. The inverse barometric effect is determined based on the atmospheric pressure data. The coastal surge is then determined by adding the fitted value to the inverse barometric effect at the coast. Different offshore extents are considered.
The sea level anomalies corrected for the inverse barometric effect are expressed as

$$S(x) = S0 + S1*\exp(-x/L) \quad (1)$$

where S is the sea level anomalies, x is the coordinate in the cross-shelf direction (positive offshore) with the origin at the coast, and S0, S1, and L are constants to be determined. L is the cross-shelf e-folding scale associated with a continental shelf wave. The fitted coastal value is calculated as

$$S(0) = S0 + S1 \quad (2)$$

The phase speed c for the continental shelf wave can be obtained as

$$c = L*f \quad (3)$$

where f is the Coriolis frequency.

Finally the total surge magnitude can be determined by adding the inverse barometric effect to S(0).

4. Results and Conclusions

For the impacts of storm surge, we will focus on the satellite-derived sea surface height anomalies relative to the mean sea surface profile estimated using HY-2A data. The storm surge magnitude at the location closest to the coast (about 6.2 km away) is 0.49 m from satellite data (figure2).

Table 1. Cross-Shelf Exponential Length Scale of the Sea Level Variation and Extrapolated Coastal Surge, by Fitting the HY-2A Observations Corrected for the Inverse Barometric Effect over Different Segments Along Track 127.

Latitude of Offshore Extent (°S)	Length Scale L (km)	Coastal Surge (m)	RMS Difference (m)
20.077	67.5	0.51	0.00
20.020	85.8	0.51	0.00
19.963	122.8	0.50	0.00
Mean ±SD	92.0 ± 20.5	0.51 ± 0.00	0.00.± 0.00

The surge level anomalies increase slowly from the shelf edge at the 1000 m isobath toward the coast, resulting in a small cross-shelf slope. In contrast, the near-coast sea level anomalies at Cycle 8 and 10 are close to zero. By fitting the altimetry-derived surge height profile corrected for the inverse barometer effect (Cycle 9) over selected coastal segments to an exponential function, i.e., equation (1), the cross-shore decay scale is estimated to be 92 km on average, with a standard deviation of 20.5 km with respect to fitting segments (table 1). In comparison, the shelf width of Mozambique Channel is about 200 km. The phase speed based on the altimetry data is 4.34 m/s from equation (3). The extrapolated surge magnitude corrected for the inverse barometer effect is, on average, 0.51 m at the coast, with a standard deviation of 0.00 m with respect to fitting segments (table 1). Therefore, the total coastal surge magnitude with the inverse barometer effect included is 0.43 from HY-2A. This estimated surge magnitude is also subject to the uncertainty associated with the exponential fitting, which is calculated as the root-mean-square (RMS) differences between fitted values and altimetry data. The RMS difference is 0.00 for each segment (table 1). The storm surge magnitude is estimated to be 0.49 m and the cross-shelf e-folding decay scale to be 92 km. It shows that the HY-2A satellite altimetry is a useful tool for monitoring storm surges and their impacts in the Indian Ocean.

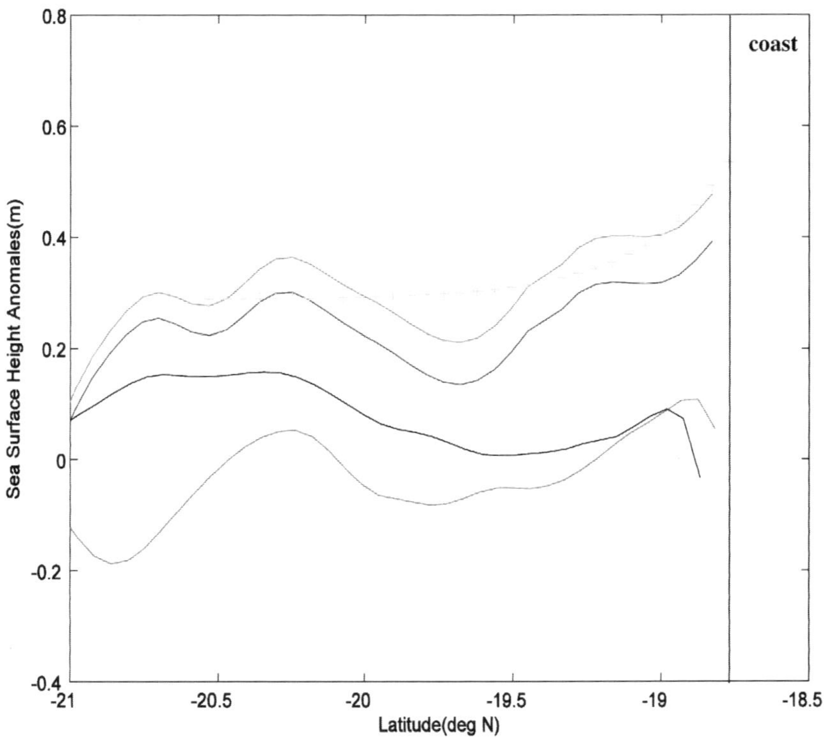

Figure 2. Altimetry-derived sea surface height anomalies along Track 127 across part of Mozambique Strait, observed by HY-2A. Blue: Cycle 9, shortly before the landfall of Tropical Cyclone Funso. Red: Cycle 8. Black: Cycle 10. The sea surface height anomalies corrected for the inverse barometric effect (magenta) are also shown for Cycle 9, along with fitted values (green crosses) by the exponential fitting to the anomalies over 18.77°S - 20.36°S.

5. Acknowledgements
The altimetry data are from NSOAS. This work is partially supported by the National 863 Program of China (2013AA09A505), the Global change and air sea interaction program of China (GASI-IPOVAI-04), the National Science Foundation (41321004, 41306191, 41306192, and 41406203) of China and the European Space Agency and Ministry of Science and Technology Dragon 4 Cooperation Programme (32249).

6. References

[1] Chen N, Han G, Yang J and Chen D 2014 Hurricane Sandy storm surges observed by HY-2A satellite altimetry and tide gauges *J. Geophys. Res.Oceans*. **119** doi:10.1002/2013JC009782.

[2] Han G, Ma Z, Chen D, deYoung B and Chen N 2012 Observing storm surges from space: Hurricane Igor off Newfoundland, *Sci. Rep.* **2** 1010 doi:10.1038/srep01010.

[3] LeBlond P and Mysak L 1978 Waves in the ocean *Elsevier Oceanography Series* vol **20** (Elsevier/North Holland) 602.

[4] Lillibridge J, Lin M, and Shum C K 2013 Hurricane Sandy storm surge measured by satellite altimetry *Oceanography*. **26** 8–9.

[5] National Satellite Ocean Application Service 2011 User Handbook: HY-2A Satellite Radar Altimetry Data [in Chinese] Beijing China.

[6] Scharroo R, Smith W H F and Lillibridge J L 2005 Satellite altimetry and the intensification of Hurricane Katrina, *EOS Trans. AGU*. **80** 366 doi:10.1029/2005EO400004.

Two-dimensional Phase Unwrapping Method Using Cost Function of L^0 Norm

J GAO [1,2], L LI[3] and L SHI[3]

[1] College of geographic and Biologic Information, Nanjing University of Posts and Telecommunications, Nanjing 210023, China
[2] Health Services Network Systems Engineering Research Centre of Education Ministry, Nanjing, 210023 China
[3] College of Telecommunications & Information Engineering, Nanjing University of Posts and Telecommunications, Nanjing, 210023, China

E-mail: gaoj@njupt.edu.cn

Abstract. Considering cost model and convergence speed of the minimum norm unwrapping, a highly efficient two-dimensional global phase unwrapping method optimized with L^0 norm is proposed. As analysing features of cost model in phase unwrapping with minimum norm, a cost function definition is provided in line with the L^0 norm, which impose a stronger constraint in the tangent direction of phase discontinuity boundary than that in normal direction, in order to preserve integrity of discontinuity during iterative unwrapping processing for continuous phase. For the sake of slow speed of low-frequency error convergence during linear solving, a data partitioning strategy is introduced into unwrapping processing. Due to independence of minimum norm method in blocks, linear solving only focus on high-frequency information and improve efficiency of iterative work, and the low-frequency processing part is transferred to offsetting-aligning between blocks. With experiments and analysis, reliability and efficiency of the novel phase unwrapping method are certified comparing to existing methods.

1. Introduction

Interferometric Synthetic Aperture Radar or Sonar (InSAR/InSAS) has arisen in topographic observing for decades. Wherein, two-dimensional phase unwrapping is a key step of DEM generating in interferometry. Existing unwrapping algorithms can be basically divided into two classes, path following and global optimization.

Unwrapped phase is obtained by integrating phase gradients along paths from reference point in following algorithms, auxiliary information from wrapped phase or external data are taken into account to avoid disturbance of phase discontinuity by posing barriers or setting priority levels to integrating paths. Such algorithms include branch-cuts method [1], the quality map guided method [2-4], Flynn minimal discontinuity method [5], network flow method [6-7], and these belong to local methods. Global optimization unwrapping is mostly an approaching towards the global extremal value of optimization objective function in order to obtain the optimal solution. The global methods usually take cost functions in form of norms, also known as minimum norm methods, including least-squares method, L^p Norm method [8].

IOP Conf. Series: Earth and Environmental Science 57 (2017) 012041 doi:10.1088/1755-1315/57/1/012041

Minimum norm unwrapping method take measure of difference between unwrapped and wrapped phases on gradient as cost function, and search for the optimal solution by minimizing the objective function globally. One classical approach is taking least squares method with L^2 norm, which translates unwrapping problem into a Poisson's equation, but it is unpractical for the sake of detail loss by reason of over-smoothing. On basis of the least square method, weighted least squares method [8] was presented to improved conditions. Ghiglia [8] introduced a general framework L^p norm method, as a generalization of minimum norm unwrapping methods. Recently a variety of mixed norm [9] was proposed in unwrapping program.

Minimum norm unwrapping research focus on selecting an appropriate cost model, another application difficulty is low convergence speed of iterative procedure. Based on existing achievement, a cost function similar to L^0 norm is provided here for two-dimensional phase unwrapping. A data partitioning scheme is introduced to overcome the low-frequency error convergence problem in linear differential equation solving, and improve the efficiency of unwrapping process.

2. Principle

2.1. Norms in phase unwrapping

The relation between wrapped phase ψ and true unwrapped ones φ can be described like this

$$\psi = W(\varphi)$$

Where W is the wrapping function modulo 2π. Phase unwrapping is a process of recovering φ from ψ, and when a result meets certain constraint from true values, that means the unwrapping process achieves a desired goal. The unwrapping resolution guidelines generally are about optimization issues of gradient differences, such as

$$\inf_{\phi} \int_{\Omega} g(|\nabla u|)\,\mathrm{d}x \tag{1}$$

Where $u = \phi - \varphi$, ϕ is the unwrapped phase, Ω refers to definition domain of phase, $g(\cdot)$ is so-called cost function making effects of derivative difference in the statistics. Such optimization constraint makes unwrapped result ϕ maintain some consistency with φ.

Assuming $g(\cdot)$ is a continuous function and first order derivable, the corresponding Euler-Lagrange equation of minimal problem in Eq. (1) is

$$\mathrm{div}\left(\frac{g'(|\nabla u|)}{|\nabla u|}|\nabla u|\right) = 0 \tag{2}$$

Where div is the divergence operator. In the condition of Nyquist theorem, gradient of u can be obtained by

$$\nabla u = \nabla\phi - \nabla\varphi = \nabla\phi - \nabla W(\psi) \tag{3}$$

A key of building Eq. (2) is definition of the cost function g, which plays a dominant role in unwrapping process. The primary difficulty of unwrapping just lies in local phase discontinuity caused by distinct features of object observed or environmental noises, that contradicts with given assumptions in Eq. (3). If phase discontinuities occur, cost function should preserve congruency of continuous data, and meanwhile suppress disturbance from phase discontinuity.

Different definitions of the cost function would bring about different results, for example, in form of L^0, L^1 and L^2 norms which are shown in (a) of Fig. 1 [10]. As one-dimensional example shown in (b), there are two fixed values at points A with zero phase residual u and B with nonzero residual. Among the three given connection paths, curve l_1 refers to an arbitrary monotone function, curve l_2 refers to a line with constant derivative value and curve l_3 refers to a connection with minimal residual influence. And curve l_2 is accordant minimal L^2 norm of ∇u, l_3 corresponds to L^0 norm and all the three curves to L^1 norm.

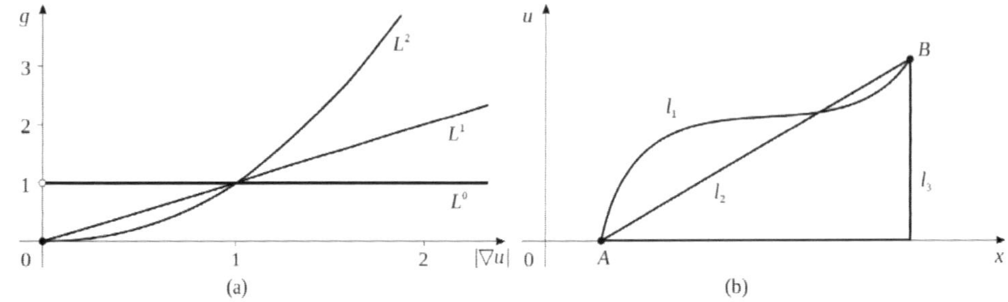

Figure 1. Cost function in form of L^0, L^1 and L^2 norms (a) and residual of different connection paths (b) between point *A* and *B*.

The best choice is L^0 norm cost function [10], while L^2 norm take strong smoothing effects, and L^1 norm is prone to staircase phenomenon. But classic L^0 norm function is not continuous or first order derivable, and it is a typically and untreatable problem (NP Hard) in network-flow optimization. Here a novel expression of L^0 norm cost function is proposed for applicable minimal norm unwrapping method.

2.2. *L^0 norm cost function*

In the two-dimensional case, the divergence in Eq. (2) can be decomposed into two second order derivate parts, $c_\xi u_{\xi\xi} + c_\eta u_{\eta\eta}$, where $\eta = \frac{\nabla u}{|\nabla u|}$ and ξ the normal vector to η, $u_{\xi\xi}$ and $u_{\eta\eta}$ refer to second order derivate in ξ and η directions. Namely,

$$\operatorname{div}\left(\frac{g'(|\nabla u|)}{|\nabla u|}|\nabla u|\right) = \nabla\left(\frac{g'(|\nabla u|)}{|\nabla u|}\right)\cdot\nabla u + \frac{g'(|\nabla u|)}{|\nabla u|}\Delta u = \frac{g'(|\nabla u|)}{|\nabla u|}u_{\xi\xi} + g''(|\nabla u|)u_{\eta\eta}$$

Where Δ is Laplacian operator and

$$u_{\xi\xi} = \tfrac{1}{|\nabla u|}(u_x^2 u_{yy} + u_y^2 u_{xx} - 2u_x u_y u_{xy}) \quad u_{\eta\eta} = \tfrac{1}{|\nabla u|}(u_x^2 u_{xx} + u_y^2 u_{yy} + 2u_x u_y u_{xy})$$

Equation (2) can be rewritten as [11]

$$\tfrac{g'(|\nabla u|)}{|\nabla u|}u_{\xi\xi} + g''(|\nabla u|)u_{\eta\eta} = 0 \tag{4}$$

Such a second order differential equation can be taken as a smoothing constraint, and coefficient values of derivate modulate the acting strength. Normally, smoothing in ξ and η directions both are expected to work in continuous phase region, whereas only in ξ is expected to run at discontinuity phase. That means

$$\lim_{s\to 0^+}\tfrac{g'(s)}{s} = \lim_{s\to 0^+} g''(s) = g''(0) > 0 \quad \lim_{s\to +\infty}\tfrac{g'(s)}{s} > 0 \quad \lim_{s\to +\infty} g''(s) = 0 \tag{5}$$

But the two conditions are not compatible, Equation (5) can be moderated as following

$$\lim_{s\to 0^+}\tfrac{g'(s)}{s} = \lim_{s\to 0^+} g''(s) = g''(0) > 0 \quad \lim_{s\to +\infty}\tfrac{g'(s)}{s} = 0 \quad \lim_{s\to +\infty}\tfrac{sg''(s)}{g'(s)} = 0 \tag{6}$$

Then gradient coefficients of the two components approach to zero in this case, and the tangent component more slowly than the normal component parts. In fact, domain of variable s is usually finite, namely, no more than a half cycle in wrapped phase data.

According to requirements from (6) for building cost function, Geman-Reynolds model [12] scheme is employed here for L^0 norm, namely

$$g(|\nabla u|) = \tfrac{|\nabla u|^2}{\kappa^2 + |\nabla u|^2} \tag{7}$$

Where κ is a positive scale coefficient. The function value at $|\nabla u| = 0$ is zero, and tends to a constant when $|\nabla u|$ get large, the cost function with different scale coefficient is shown in (a) of Fig. 2. And the corresponding dominating smoothing coefficient c_ξ is shown in (b). If value of κ is too large, for example $\kappa = 1$, the cost function approximate to L^1 norm, and if too little, effect of the coefficient c_ξ is limited in a small range, just like that with $\kappa = 0.01$ in (b) of Fig. 2. Empirical value of κ is about 0.1.

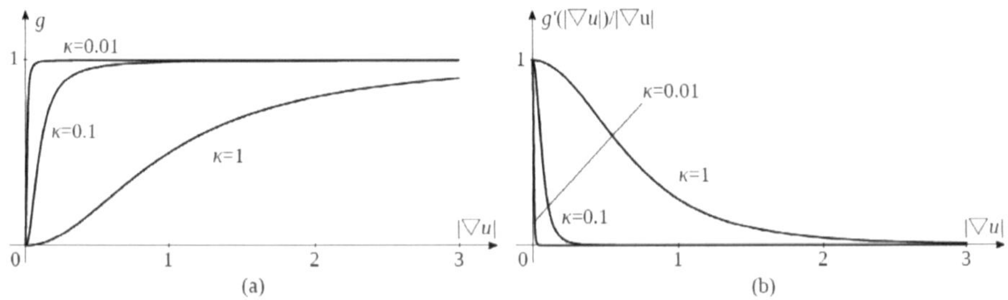

Figure 2. Cost function (a) and coefficient c_ξ(b) with different scale coefficients.

Substituting the cost function (7) to (2), a new unwrapping equation is obtained

$$\mathrm{div}\left(\frac{\nabla u}{(\kappa^2 + |\nabla u|^2)^2}\right) = 0 \tag{8}$$

Which describes constraint of L^0 norm unwrapping problem and is not easy to solve directly.

3. Numerical method

3.1. Solution of nonlinear equation

Dealing with numerical differential equation with discrete data, the first thing need to do is converting differential equations into difference ones, and then make them solved. An iterative scheme is employed here, each iteration runs on the basis of its predecessor during the cycle procedure, and the difference between results of the two-step process become an important part to focus on.

According to Eq. (3), we let

$$b = 1/(\kappa^2 + |\nabla\phi - \nabla W(\varphi)|^2)^2$$

Which can be taken as a constant, when solving

$$\mathrm{div}(b\nabla\phi) = \mathrm{div}(bW(\nabla\varphi)) \tag{9}$$

And its solution would be used to update b. Assuming unwrapping results of step n and step $n+1$ meet with

$$\mathrm{div}(b_n\nabla\phi_{n+1}) = \mathrm{div}(b_n W(\nabla\varphi))$$

By means of correction value, it can be rewritten as

$$\mathrm{div}(b_n\nabla\delta\phi_n) = \mathrm{div}(b_n(W(\nabla\varphi) - \nabla\phi_n))$$

Where $\delta\phi_n = \phi_{n+1} - \phi_n$. In this case the right-hand-side refers to original wrapped phase information exclude known unwrapped part, then solving for correction values $\delta\phi_n$ and a new unwrapped phase in loops for final result. Such a linear equation with sparse coefficient matrix, whose condition number is often large, and the solver generally take Krysolv subspace algorithms, just like Preconditioned Conjugate Gradient (PCG) method [13-14] employed here.

The cycle terminates in some conditions, such as inexistence of residual point in residual wrapped phase $\delta\psi$

$$\delta\psi_n = \psi - \phi_n$$

When there is no phase residue, it suggests no phase discontinuity in $\delta\psi_n$, whose unwrapped counterpart can be calculated following arbitrary integration paths. The whole process of unwrapping is the following.

1. let $n = 0$, initialize unwrapped phase by $\phi_n = 0$;
2. calculate coefficient matrices b_n;
3. calculate right-hand-sides of equations $\operatorname{div}\left(b_n(W(\nabla\varphi) - \nabla\phi_n)\right)$;
4. solve equations to obtain solutions $\delta\phi_n$;
5. update unwrapped phase $\phi_{n+1} = \phi_n + \delta\phi_n$;
6. calculate residual wrapped phase $\delta\psi_{n+1}$;
7. check existence of residue, if true, let n=n+1 and go to step 2;
8. unwrap residual unwrapped phase $\delta\psi_{n+1}$, finish unwrapping process.

3.2. *Data partitioning*

For the sake of low-frequency residual attenuation, if the size of wrapped phase is very large, inefficiency in convergence of linear equations will be more evident. Such a time-consuming method would not be practical in real interferometry processing.

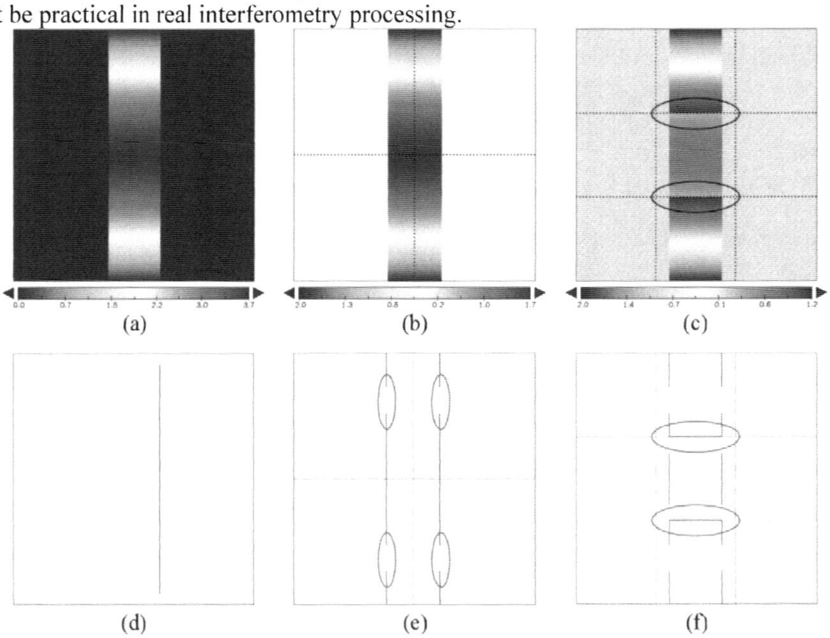

Figure 3. Unwrapping results and discontinuity edges of original, 2×2 blocks and 3×3 blocks.

Taking into account of data partitioning, big blocks are divided into small size ones, then unwrapping is carried out on small size data. After unwrapping, small blocks would be adjoined together for an integral result. Only with high frequency residual, solving on small block data would convergence faster. Parallel processing mode can be adopted in unwrapping of small blocks for their independence.

The only thing need to note here is that, partitioning should not divide discontinuity boundaries into little chips, when discontinuity of wide range appears, though that occur scarcely. For example, synthetic phase data is generated by adding a vertical half-cycle sine band to background plane, shown as (a) of Fig. 3 and the discontinuity boundaries are shown in (d).

The synthetic phase data, is divided into 2×2 and 3×3 blocks respectively. After unwrapping on each sub-block and then assembling for the unwrapping results. The results and corresponding phase discontinuity boundaries are shown in (b), (c), (e) and (f). As in the former scheme, discontinuity boundaries are broken, though that does not destroy the unwrapping result.

In the latter scheme, the discontinuity boundaries are divided into more little sections. The smaller blocks scheme prevents to recover the wrapped phase correctly, there is an error in the central location.

Therefore, when designing a data partitioning scheme, it needs attention to choose block size, avoid breaking discontinuity boundaries, although process run faster on smaller blocks. In general, the phase discontinuity is often due to signal noise or abnormal feature of target observed, and there is seldom a discontinuity boundary over wide range, so this problem is not very serious.

4. Experiments and analysis

The machine used for processing in experiments is equip a CPU with 8 logical cores at 2.7GHz, dual-channel 8GB DDR3 memories at 1600MHz and a 64-bit operating systems. Processes work in a shared memory multi-core parallel processing mode [15], using a linear equation solver of good stability, half coarsening multi-grid preconditioned conjugate gradient method.

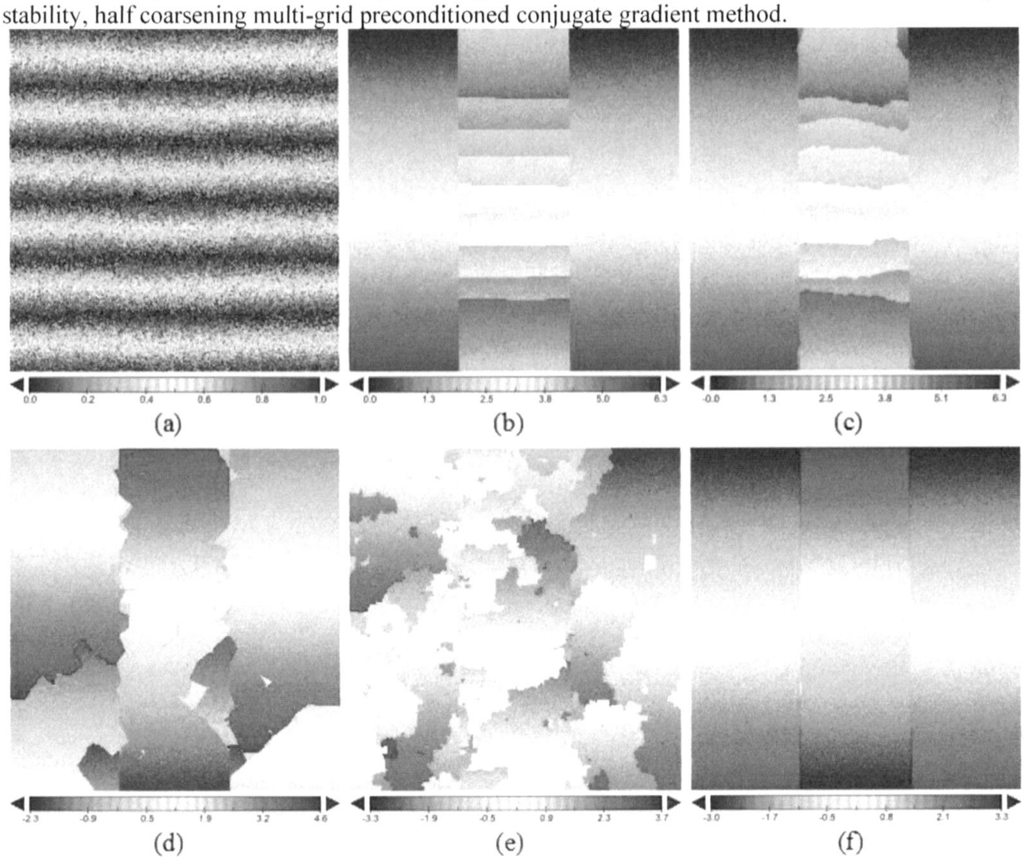

Figure 4. Wrapped noisy shear phase data (a) and unwrapped phase from Flynn's (b), MCF method(c), Goldstein's method (d), quality-guided method (e) and proposed method (f).

Experimental data include synthetic wrapped phase data and real InSAR data, the synthetic ones are generated by wrapping noisy shear data, 257×257 pixels, shown as (a) of Fig. 4. The synthetic data are unwrapped by Flynn's Minimal Discontinuity method, Minimum Cost Flow (MCF) method,

Goldstein's branch cut method, derivative variance based Quality Map Guided method and the proposed method. The results are shown in Fig. 4.

As shown, Flynn's and MCF and methods failed in recovering at central area for their L^1 norm cost schemes, Goldstein's branch cut method with L^0 norm get a better result but not eligible, for wrong connections of residues. The Quality Map Guided method did not well for fragment data and the proposed L^0 norm method works well.

Using real InSAR interferometric phase as the experimental data, whose size is 1398×622 pixels, contains a global gently fluctuation, and serious noise at several locations. Especially some noises are located in areas near boundaries, whose bad impact would bring obstacle to convergence for stretching from only one side, and cycle operation within a larger area may be more time-consuming.

Experimental data are divided into 7×3 and 14×6 blocks at 200-pixel and 100-pixel scales respectively and then unwrapping. As shown in Fig. 5, the original InSAR wrapped phase data (a), 14×6 partitioning diagram (b), partition unwrapping results (c), offsets of blocks (d), final results (e), the corresponding discontinuity edges (f), it suggests that unwrapping the result is able to maintain good integrity even partitioned in 100-pixel scale.

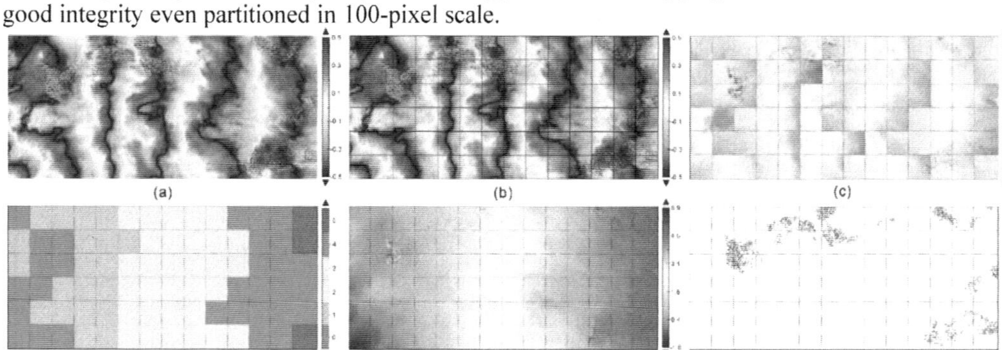

Figure 5. Unwrapping process of InSAR phase in 14×6 blocks.

In order to verify validity of proposed method, different unwrapping algorithms run on the InSAR phase data. The quality of unwrapping a result and the efficiency of unwrapping process are described by quantitative statistical indicators, including standard deviation of difference between re-wrapped results and original data, length of phase discontinuity edges, and the running time of process, statistical values are shown in Tab. 1.

It should be noted that the data are stored in 8-bit float type, and there are no more than 7 significance digits for a value in such case, so if the order of a precision index is higher than 10^{-6}, that is mainly caused by truncation errors. As shown in the table, the standard deviation of Goldstein branch cut method is more prominent, mainly because the method may bring about islands or hole in the noisy area of poor quality. Considering indicators of length of discontinuity boundaries, Flynn's is best, whose most important principle just is that, minimal discontinuity. The proposed method, in both partitioning schemes get good performance better than Goldstein's, Quality Map Guided and SNAPHU.

Table 1. Statistics Indicators of the Results from Different Unwrapping Methods

Algorithm	L^0 norm/px	L^1 norm/px	Time/sec
Goldstein's	17546	34553	0.156
Quality-Guided	62881	124229	3.927
Flynn's	11984	11989	6.938
MCF	12997	12997	15.265
SNAPHU	12989	12989	3.534
PUMA	13012	13012	152.507
Proposed(7×3)	13714	13730	17.979
Proposed(14×6)	13729	13748	7.564

Finally, taking into account of running time, the branch-cut is fastest due to its simplicity on operation, followed by the quality-guided and PUMA slowest method. The running time of proposed method in this paper, in both partitioning scheme, are all better than SNAPHU's and PUMA's, and especially in scheme of 14×6 blocks which is also better slightly than Flynn's. That proves the efficiency of new method.

Owing to the table, the proposed unwrapping method achieve a good performance in experiments, obtaining reliable result with small discontinuity and big efficiency. That are not found in classical minimal norm unwrapping methods.

5. Conclusion

This paper proposed a global phase unwrapping method using cost function of L0 norm, which makes different constraints intensity on normal and tangent directions at discontinuity edges, obtaining reliable unwrapped phase. The strategy of data partitioning gets rid of low-frequency residual in linear differential equation to improve processing efficiency. The experimental results and comparing with other popular methods verify its validity and efficiency.

Acknowledgements

This work was supported by the Scientific Research Foundation of Nanjing University of Posts and Telecommunications under Grant NY213056, NY215109, by the Natural Science Foundation of Jiangsu Province under Grant BK20130864, and by the National Natural Science Foundation of China under Grant 41401480, 41501378, 41501497.

References

[1] Huang Q., 2015. Parallel branch-cut algorithm based on simulated annealing for large-scale phase unwrapping. *IEEE Transactions on Geoscience and Remote Sensing*, 53(7), pp.3833–3846.

[2] Gao D., 2012. Mask cut optimization in two-dimensional phase unwrapping. *IEEE Geoscience and Remote Sensing Letters*, 9(3), pp.338–342.

[3] Zhong H., 2014. A quality-guided and local minimum discontinuity based phase unwrapping algorithm for InSAR/InSAS interferograms. *IEEE Geoscience and Remote Sensing Letters*, 11(1), pp.215–219.

[4] Liu G., 2014. A new quality map for 2-d phase unwrapping based on gray level co-occurrence matrix. *IEEE Geoscience and Remote Sensing Letters*, 11(2), pp.444–448.

[5] Flynn T., 1998. Phase unwrapping using discontinuity optimization. In: *Geoscience and Remote Sensing Symposium Proceedings*, Vol. I, pp.80–82.

[6] McKilliam R. G., 2014. Polynomial phase estimation by least squares phase unwrapping. *IEEE Transactions on Signal Processing*, 62(8), pp.1962–1975.

[7] Dong J., 2015. Simultaneous phase unwrapping and removal of chemical shift (spurs) using graph cuts: Application in quantitative susceptibility mapping. *IEEE Transactions on Medical Imaging*, 34(2), pp.531–540.

[8] Ghiglia D. C., 1994. Robust two-dimensional weighted and unweighted phase unwrapping that uses fast transforms and iterative methods. *Journal of the Optical Society of America*, A 11(1), pp.107-117.

[9] Liu H., 2015. A novel mixed-norm multi-baseline phase-unwrapping algorithm based on linear programming. *IEEE Geoscience and Remote Sensing Letters*, 12(5), pp.1086–1090.

[10] Chen C. W., 2001. Statistical-cost Network-flow Approaches to Two-dimensional Phase Unwrapping for Radar Interferometry. Ph.D. thesis, Stanford University.

[11] Kornprobst P., 1997. Nonlinear operators in image restoration. In: *IEEE Comput. Soc.*, pp.325-330.

[12] Geman D., 1992. Constrained restoration and the recovery of discontinuities. *IEEE Transactions on Pattern Analysis and Machine Intelligence*, 14(3), pp.367–383

[13] Trefethen L. N., 1997. Numerical Linear Algebra, Society for Industrial and Applied Mathematics.
[14] Demmel J. W., 1997. Applied Numerical Linear Algebra, SIAM.
[15] Barbara C., 2007. Using OpenMP: Portable Shared Memory Parallel Programming. The MIT Press.

IOP Conf. Series: Earth and Environmental Science 57(2017) 012042 doi:10.1088/1755-1315/57/1/0120

Extraction and Analysis of Mega Cities' Impervious Surface on Pixel-based and Object-oriented Support Vector Machine Classification Technology: A case of Bombay

S S Yu [1,2], Z C Sun [2,3], L Sun [1], M F Wu [1,4]

[1] Shandong University of Science and Technology, Qingdao Shandong 266590, China
[2] Key Laboratory of Digital Earth Science, Institute of Remote Sensing and Digital Earth, Chinese Academy of Sciences, Beijing 100094, China
[3] Hainan Key Laboratory Earth Observation, Sanya Hainan 572029, China
[4] Hunan Normal University, Changsha Hunan 410006, China

Email: yussta@163.com; sunzc@radi.ac.cn

ABSTRACT. The object of this paper is to study the impervious surface extraction method using remote sensing imagery and monitor the spatiotemporal changing patterns of mega cities. Megacity Bombay was selected as the interesting area. Firstly, the pixel-based and object-oriented support vector machine (SVM) classification methods were used to acquire the land use/land cover (LULC) products of Bombay in 2010. Consequently, the overall accuracy (OA) and overall Kappa (OK) of the pixel-based method were 94.97% and 0.96 with a running time of 78 minutes, the OA and OK of the object-oriented method were 93.72% and 0.94 with a running time of only 17s. Additionally, OA and OK of the object-oriented method after a post-classification were improved up to 95.8% and 0.94. Then, the dynamic impervious surfaces of Bombay in the period 1973-2015 were extracted and the urbanization pattern of Bombay was analysed. Results told that both the two SVM classification methods could accomplish the impervious surface extraction, but the object-oriented method should be a better choice. Urbanization of Bombay experienced a fast extending during the past 42 years, implying a dramatically urban sprawl of mega cities in the developing countries along the One Belt and One Road (OBOR).

1. Introduction

In today's world, urbanization has become a significant problem of environment in countries and regions on earth, especially in the developing countries. As a typical product of urbanization, mega cities, carrying more than 1,000,000 persons, play a more and more important role in economic development, infrastructure construction and ecological environment response [1]. Meanwhile, mega cities have to face more challenges on the environment protection, resources sustainable utilization as well as people's daily life [2]. There are only 2 mega cities around the world in 1950. According to the UN data in 2014, the number of mega cities had increased from 2 to 29 at an alarming speed over the past 60 years. And it is forecasted that there will be more than 40 megacities along the One Belt and One Road (OBOR) until

2030. Hence, monitoring the urban sprawl of mega cities so as to macroeconomic regulation and control the mega cities development allows of no delay.

Influence of urbanization on environment is mainly from impervious surface. Impervious surface is defined as the entire impermeable surfaces such as buildings, asphalt roads, parking lots, sidewalks and other infrastructural elements of urban areas, and has become a key indicator of urbanization assessment. Monitoring urbanization of mega cities via impervious surface extraction has become a hot topic nowadays. Due to the relatively low cost and is suitable for large area mapping, remote sensing imagery has been widely used in studying urban space distribution since 1970s [3]. And more researchers have been concentrated on urbanization and urban sprawl by impervious surface extraction during the past 20 years [4]. Sun et al. [5] estimated Beijing urban impervious surfaces in 2009 using multilayer perceptron neural network (NLPNN) and SVM classification method. Subsequently, Sun et al. [6] derived impervious surface of Beijing from 1992 to 2009 and investigated the long-term effects of LULC change on surface runoff. Im et al. [7] quantified the urban impervious surface using a synthesis of artificial immune networks, based on the LiDAR nDSM data and WorldView-2 multispectral imagery. Yang and Li [8] extracted Beijing impervious surface in 2003 from QuickBird imagery using linear spectral unmixing models. Nie and Xu [9] quantified the impervious surface of Shanghai, China in 2010 using ETM+ images with the linear spectral mixture analysis (LSMA). Song et al. [10] delineated Washington urban growth in the period 1984-2010 using post per-pixel classification method. Zhang et al. [11] mapped the urban impervious surface of the Pearl River Delta (PRD) using dual- polarimetric SAR data. Amir et al. [12] summarized urban expansion of Kandun Town in Cixi County, Zhejiang Province, China using time series of impervious surface fractions (ISFs). All studies mentioned-above have shown a satisfying result in impervious surface extraction and urban expansion patterns analysis. However, the SVM method was proved to be a relatively simple and typical approach to accomplish LULC classification and urbanization monitoring [13].

The object of this study is to monitor and analyze Megacity Bombay's urban extent and urbanization patterns. By comparing the efficiency of the pixel-based and object-oriented SVM classification methods operating on Landsat TM of Bombay in 2010, the better method was selected out and applied to extracting the impervious surface of Bombay in the period 1973-2015 from Landsat MSS/TM/ETM+/OLI imagery. This study aims to serve a useful approach for further applications of mega cities' urbanization planning and sustainable development.

2. Materials and method

2.1 Study area

The study area is Bombay, locating in the western India and covering about 603km^2 (Fig.1). The central latitude and longitude are separately 18°56'N and 72°49'E. Bombay is city of tropical monsoon climate, possessing plenty of moisture and heat, which is suitable for human survival. As the capital and the largest harbor city of India, Bombay has centralized the primary economic, political and cultural activities around the country and experienced rapid urbanization during the past 42 years. According to the UK (2014), population of Bombay was less than 7,000,000 in 1973. In 1985, its population exceeded 10,000,000. Then the number reached about 12,436,000 in 1990. And till 2015, the population of Bombay should be more than 21,000,000, which would be over 3 times of that in 1973. Population growth always results in urban sprawl, so Bombay could be a suitable demo to monitor mega cities' urbanization patterns.

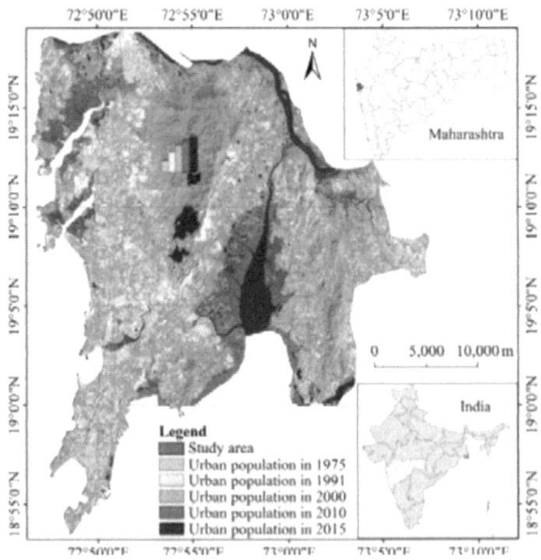

Figure 1. Study area

2.2 Dataset and data preprocessing

The Landsat MSS/TM/ETM+/OLI imagery, acquired in the period 1973-2015, was selected as the study data, due to their low cost, wide coverage and short cycle. Datasets used in the study are all cloudless so as to omit the atmospheric correction [14].

The visible bands, near-infrared (NIR) bands and short wavelength infrared (SWIR) bands were used in this study. During the data preprocessing, the Landsat imagery was resampled to 30m resolution and unified to the Universal Transverse Mercator (UTM) and WGS84 geodetic datum. Additionally, the Normalized Difference Vegetation Index (NDVI) [15], the Soil-adjusted Vegetation Index (SAVI) [16] and the Modified Normalized Difference Water Index (MNDWI) [17] were calculated to participant in the follow-up classification together with the visible, NIR and SWIR bands.

2.3 Methodology

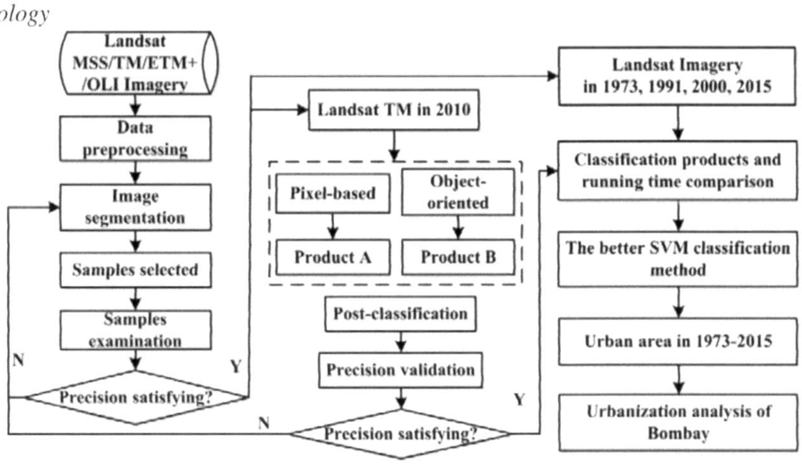

Figure 2. The detailed study framework

Fig.2 was the detailed study framework. Firstly, the data preprocessing was executed and LULC samples were selected. Secondly, the pixel-based and object-oriented SVM classification methods were compared using the Landsat TM imagery of Bombay in 2010. Thirdly, images in the period 1973-2015 were classified using the better method. At last, the urbanization pattern of Bombay was analyzed.

The concept of the SVM derived from a nonparametric machine learning methodology based on Vapnik's [18] structural risk minimization (SRM) principle, aiming to map data into a high-dimensional space and trying to find the optimal hyperplane of different classes. The theory of SVM has been extensively described in the literature by Burges [19] and Brown et al [20]. Therefore, the paper will only describe the basic concepts about image segmentation, the pixel-basic and the object-oriented method.

Table 1. Image segmentation parameters of Bombay

Parameters	MSS	TM	ETM+	OLI
Scale	5	5	10	35
Shape	2	1	2	1
Compactness	0.8	0.9	0.8	0.9

The multiresolution segmentation is a semi-automatic process before the SVM classification. Users defined segmental scale, shape and compactness parameters manually and the Landsat imagery was divided into different objects named "image samples". The scale parameter directly defines size of the image samples, it shouldn't be too large or small because a suitable image sample should only contain one LULC type. Usually, the scale of Landsat MSS/TM/ETM+ and HJ-1 CCD imagery is less than the Landsat OLI imagery, due to their diffident bit numbers. The shape parameter defines the weight the shape criterion should have. The higher its value, the lower the influence of color feather on image samples. And the compactness parameter defines the weight of the compactness criterion. The higher the value, the more compact image objects may be. According to plenty of experiments, the three parameters of Landsat imagery for Bombay are set in table 1.

During the SVM classification, the same training samples were used in the pixel-based and the object-oriented methods. The pixel-based classification implemented supervised classification, using maximum likelihood algorithm, according to the spectrum and geometrical characteristics of each pixel. Each pixel is an individual to participant in the classification. So each pixel was classified to the most similar class. However, the basic unit in the object-oriented classification is not pixels but image samples. Each image sample is composed of multiple pixels. Pixels of the same image sample share similar size, shape, spectrum, texture, context and geometrical characteristics. During the object-oriented classification, pixels of the same image sample are classified into the same class due to their common features.

3. Results and discussion

3.1 Comparison of the pixel-based and object-oriented SVM classification methods

The Landsat TM imagery of Bombay in 2010 was classified using the pixel-based and object-oriented methods in this study. In Fig. 3, the red represents impervious surface, the buff represents bare soil, the green represents vegetation and the blue represents water area. At a first glance, it is difficult to find out any differences between the two results. However, differences indeed existed. The lower-left corner gives the enlarged views near 18°9'11"N, 72°9'8"E. It proves that the pixel-based SVM classification method has a better ability in classification, especially for impervious surface and bare soil.

The Landsat TM imagery of Bombay in 2010 is classified into impervious surface, bare soil, vegetation and water in the classification product. In order to compare the two methods' precision, 199 random Points of each class were selected. By the cross-validating with google earth images as reference, the accuracy and running time of the two classification methods were calculated (Table 2).

Table 2. Accuracy of pixel-based and object-oriented SVM classification methods

		a. Pixel-based SVM Classification Method					
		Impervious surface	Bare Soil	Vegetation	Water	Total	UA(%)
Classified Data	ISA	187	7	4	1	199	93.97
	BS	8	187	2	2	199	93.97
	Veg	2	13	183	1	199	91.96
	Water	0	0	0	199	199	100.00
	Total	197	207	189	203	796	
	PA(%)	94.92	90.34	96.83	98.03		
	OA(%)	94.97					
	OK	0.96					
	Time	78minutes					
		b. Object-oriented SVM Classification Method					
		Impervious surface	Bare Soil	Vegetation	Water	Total	UA(%)
Classified Data	ISA	184	13	2	0	199	92.46
	BS	11	184	1	3	199	92.46
	Veg	5	13	179	2	199	89.95
	Water	0	0	0	199	199	100.00
	Total	200	210	182	204	796	
	PA(%)	92.00	87.62	98.35	97.55		
	OA(%)	93.72					
	OK	0.94					
	Time	25seconds					

ISA: the impervious surface area; BS: the bare soil; Veg: the vegetation; PA: the producer's accuracy; UA: the user's accuracy; OA: the overall accuracy; OK: the overall kappa coefficient

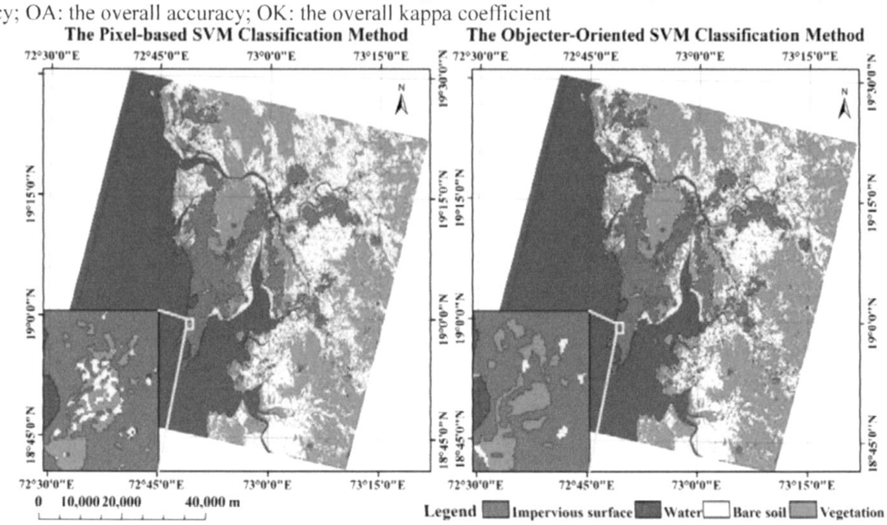

Figure 3. Results of pixel-based and object-oriented SVM classification methods

The overall accuracy (OA) and overall Kappa (OK) of the pixel-based SVM classification method were 94.97% and 0.96, the OA and OK of the object-oriented SVM classification method were 93.72% and 0.94. Both the two methods could acquire a satisfying result, but the precision of the pixel-based SVM classification method was a little higher. However, table 2 also gives the running time, the pixel-based SVM classification method needs 78 minutes but the object-oriented SVM classification method just needs 17s. In comparison, the object-oriented SVM classification method along with post-classification process should be a better choice.

3.2 Accuracy validation

The impervious surface of Megacity Bombay in the period 1973-2015 was extracted using the object-oriented SVM classification method. By the same validation techniques as table 2, the accuracy of the five classification products was estimated. Table 3 showed the user's accuracy (UA), producer's accuracy (PA), OA and OK of impervious surface in the period 1973-2015. According to Table 3, UA are all above 79.00%, OA are all above 88.91%, OK are all above 0.80, and PA in the period 1973-2010 are all above 89.24%. Year 2015 doesn't show a satisfying result due to none Landsat OLI images of good quality were found. The main problem is the impervious surface and bare soil share similar spectral signature in the used OLI image. However, the precision in 2015 was still better than that in 1973 because OLI images often possessed of more bands and better definition than MSS images. In short, the accuracy proved that the object-oriented SVM classification and post-classification method is able to acquire impervious surface of mega cities and analysis of Bombay urbanization was feasible.

Table 3. Accuracy of impervious surface in the period 1973-2015

Year	UA(%)	PA(%)	OA(%)	OK
1973	69.20	72.24	88.91	0.87
1991	88.00	97.24	93.51	0.91
2000	90.50	94.00	93.12	0.92
2010	93.10	97.42	95.80	0.94
2015	79.00	40.72	93.70	0.80

3.3 Analysis of urban expansion

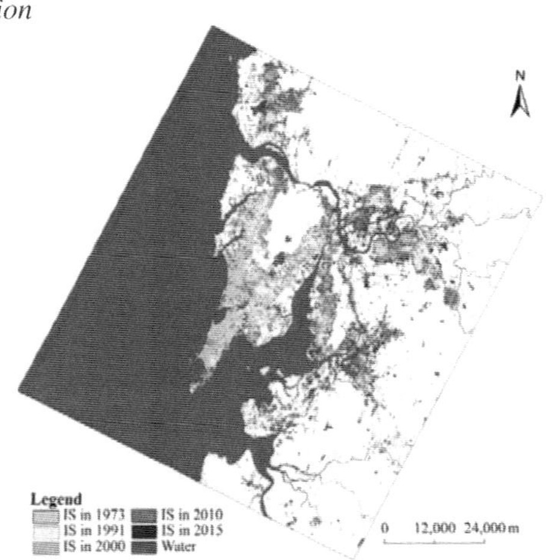

IS: the impervious surface

Figure 4. The urban sprawl of Bombay in the period 1973-2015

Fig.4 shows the impervious surface's dynamic changes of Bombay in the period 1973-2015. It is obvious that the impervious surface experienced tremendous changes during the past 42 years. The biggest change occurred in the period 1973-1991. There was only limited impervious surface in 1973, and it mostly located along the coast and the main river channel. However, quantities of small towns appeared in the period 1973-1999. They just used less than 3 decades to develop into large cities. Until 2015, areas of impervious surface had expanded about five times with 1973 as baseline. The basic expansion pattern was from coastal areas to inland. And impervious surface in the inland area showed a concentric expansion pattern.

According to the urbanization pattern of Bombay, it could be speculated that expansions of mega cities along the OBOR were analogous to Bombay. The urban sprawl direction is from the coastal areas to the inland. And it's mainly "concentric expansion pattern" in the inland area. Additionally, there are also a large number of satellite town generated.

4. Conclusions

Using the Landsat MSS/TM/ETM+/OLI imagery, this study accomplished two tasks.

1) Based on the Landsat TM imagery of Bombay in 2010, the pixel-based and object-oriented SVM classification methods were available to extract impervious surface and analyze urbanization of mega cities. According to the classification products and running time, the object-oriented SVM classification method was suggested to be more efficient.

2) Using the object-oriented SVM classification method, the impervious surface of Megacity Bombay in the period 1973-2015 was delineated out. And the urbanization process of Bombay was analyzed.

This study explored an effective method to monitor and analyze mega cities' urbanization patterns. Analysis about Bombay was a helpful reference for other mega cities in the developing countries along the OBOR. The study findings provided scientific theory basis for relevant departments to regulate new type of urbanization, preserve the ecological environment, develop the economy and ensure supplies in the future.

However, there are still several weaknesses in the study. 1) The study dataset are 30m moderate resolution remote sensing images, which inevitably lead to errors in the classification products. 2) Impervious surface and bare soil share the same spectral signature, bringing mix-classification problems.

Aiming to provide more and better references for mega cities' sustainable development, the high resolution remote sensing images should be used to the impervious surface extraction, and urbanization patterns of more mega cities will be monitored and analyzed in the next study.

Acknowledgements

The paper is supported by the National Natural Science Foundation of China (No.41201357). The authors wish to thank the anonymous reviewers and the editor for their helpful suggestions and constructive comments.

References

[1] Aguilar A G, Ward P and Smith C B 2003 Globalization, regional development, and mega-city expansion in Latin America: Analyzing Mexico City's peri-urban hinterland *Cities*. **20** 3-21

[2] Esch T, Taubenbock H, Felbier A, Heldens W, Wiesner M and Dech S 2012 Monioring of Global Urbanization-time series analyses for maga cities based on optical and SAR data *EORSA*. 21-25

[3] Zhang L and Weng Q H 2016 Annual dynamics of imperious surface in the Pearl River Delta, China from 1988 to 2013, using time series Landsat imagery *ISPRS*. **113** 86-96

[4] Guo H D, Yang H N, Sun Z C, Li X W and Wang C Z 2014 Synergistic use of optical and PolSAR imagery for urban impervious surface estimation *Photogramm. Eng. Remote Sens*. **80** 91-102

[5] Sun Z C, Guo H D, Li X W，Liu L L and Du X P 2011 Estimating urban ISAs from Landsat-5 TM images using multilayer perceptron neural network and support vector machine J *J. Appl. Remote Sens.* **5** 053501
[6] Sun Z C, Li X W, Fu W X, Li Y K and Tang D S 2014 Long-term effects of land use/land cover change on surface runoff in urban areas of Beijing, China *J. Appl. Remote Sens.* **8** 084596
[7] Im J, Lu Z Y and Rhee J 2012 Impervious surface quantification using a synthesis of artificial immune networks and decision/regression trees from multi-sensor data *Remote Sens. Environ.* **117** 102-113
[8] Yang J and Li P J 2015 Impervious surface extraction in urban areas from high spatial resolution imagery using linear spectral unmixing *Remote Sens. Appl. Soc. Environ.* **1** 61-71
[9] Nie Q and Xu J H 2015 Understanding the effects of the ISAs pattern on land surface temperature in an urban area *Front. Earth Sci.* **9** 276-285
[10] Song X P, Sexton J O, Huang C Q, Channan S and Townshend J R 2016 Characterizing the magnitude, timing and duration of urban growth from time series of Landsat-based estimates of impervious cover *Remote Sens. Environ.* **175** 1-13
[11] Zhang H S, Lin H, Li Y, Zhang Y Z and Fang C Y 2016 Mapping urban impervious surface with dual-polarimetric SAR data: An improved method *Landscape Urban Plan.* **151** 55-63
[12] Shahtahmassebi A R, Song J, Zheng Q, Blackburn G A, Wang K, Huang L Y, Pan Y, Moore N, Shahtahmassebi G, Haghighi R S and Deng J S 2016 Remote sensing of impervious surface growth: A framework for quantifying urban expansion and re-densification mechanisms *Int. J. Appl. Earth Obs.* **46** 94-112
[13] Gong P, Wang J, Yu L, Zhao Y, Zhao Y, Liang L, Niu Z, Huang Z, Fu H and Liu S 2013 Finer resolution observation and monitoring of global land cover: first mapping results with Landsat TM and ETM+ data *Int. J. Remote Sens.* **34** 2607-2654
[14] Taubenbock H, Esch T, Felbier A, Wiesner M, Roth A and Dech S 2012 Monitoring urbanization in mega cities from space *Remote Sens. Environ.* **117** 162-176
[15] Townshend J R G and Justice C O 1986 Analysis of the dynamics of African vegetation using the Normalized Difference Vegetation Index *INT. J. Remote Sens.* **7** 1435-1445
[16] Baret G 1991 Potentials and limits of vegetation indices for LAI and APAR assessment *Remote Sens. Environ.* **35** 161-173
[17] Xu H Q 2005 A study on information extraction of water body with the Modified Normalized Difference Water Index (MNDWI) *J. Remote Sens.* **9** 589-595
[18] Vapnik V N 1997 The nature of stature of statistical learning theory *Neural Networks.* **8** 1564-1564
[19] Burges C J C 1998 A tutorial on support vector machines for pattern recognition *Data Min. Knowl. Discov.* **2** 121-167
[20] Brown M, Lewis H G and Gunn S R 2000 Routear spectural mixture models and support vector machines for remote sensing *Geosci. Remote Sens.* **38** 2346-1360

Study on Mobile Object Positioning and Alarming System Based on the "Map World" in the Core Area of the Silk Road Economic Belt

Kai.Mu[1,2,3]

[1] Xinjiang of Academy Surveying and Mapping, Urumqi 830002, China;
[2] Engineering Research Center of Central Asia Geoinformation Development and Utilization, National Administration of Surveying, Mapping and Geoinformation, Urumqi 830002, China;
[3]NO.2 Institute of Surveying and Mapping Xinjiang Uygur Autonomous Region, Urumqi 830002, China;
author's e-mail 21650998@qq.com

Abstract. The established "Map World" on the National Geographic Information Public Service Platform offers free access to many geographic information in the Core Area of the Silk Road Economic Belt. Considering the special security situation and severe splittism and anti-splittism struggles in the Core Area of the Silk Road Economic Belt, a set of moving target positioning and alarming platform based on J2EE platform and B/S structure was designed and realized by combining the "Map World" data and global navigation satellite system. This platform solves various problems, such as effective combination of Global Navigation Satellite System (GNSS) and "Map World" resources, moving target alarming setting, inquiry of historical routes, system management, etc.

Key words: Positioning system; Alarming system; Ray casting algorithm; J2EE platform; B / S structure; GNSS; Silk Road Economic Belt.

1. Introduction

Locating in the northwest border of China, Xinjiang where covers an extensive territory and has special natural environment, climate conditions and frequent natural disasters is facing with very severe splittism and anti-splittism struggles, which is related with the complicated surrounding environment. This brings potential threats against stability. Therefore, the position-based geographic information service platform is crucial to autonomous region management.

The established system is to support service demands based on accurate positions in different industries and departments in the study area, promote the Map World service platform, make full use of the Map World data, fulfil the Map World safeguarding service based on the overall goal of social stability and long-term peace in Xinjiang, and protect life safety of stability protectors.

2. Materials and methods

2.1 System structure

The system employs the MVC framework, releases JSP (Java Server Pages) to server through Tomcat, accesses to the database by JavaBean, constructs the platform using JSP, JavaBean and Java Servlet of J2EE, and establishes the dynamic portal system[1][2][3].

Model layer: it realizes business logics of the system, that is, the JavaBean [4].

View layer: it is responsible for communicating with users. In other words, data objects are displayed on the interface to users, that is, html and JSP[5]..

Control layer: it serves as the communication bridge between the Model layer and View layer, and can assign user requests, choose appropriate views for display, explain user inputs and map them into executable operations of the Model layer. This is the responsibility of Servlet [6].

2.2 Map World API

As an application programmers interface based on new JavaScript upgrading, the Map World Web API (JavaScript) V2.1.2 provides developers a channel for fast call of online geographic information service of the Map World, including quick map creation, map access, POI search and adding covering on the map. Based on Web API of the Map World, developers can embed its abundant map functions into different application systems or websites as well as realize various value-added services and applications based on its data and function service resources [7].

The Map World Web API (JavaScript) V2.1.2 has following characteristics:

1). Quicker map loading and smoother display.

2). More stable map service and convenient use.

3). The latest map style and better visual effect.

4). Compatible with mainstream browsers.

5). Map browsing and switchover based on two projections: geographic coordinate system and spherical Mercator

2.3 Ray casting algorithm

Positional relationship between personnel and the "fence" was calculated through the underlaying programming and superposition of geographic information algorithm in the Ray casting algorithm (induce the ray through points, judge parity of times for the ray to pass through the polygon, and calculate hypernymy relation between the point and the polygon). Next, the logic relationship between personnel position and other "fences" was distinguished through the logic operation (in the administrator mode, it is also necessary to distinguish the logic relationship with other users). Finally, whether users are in the "fence" is determined and then alarming information is sent. It realizes communication with smart phone end through Web Service technology.

In this paper, the positional relationship between personnel and the "fence" was determined by encoding of the Ray casting algorithm. Firstly, personnel position information is the only one "fence", which can refer to one or several persons. If the personnel is in the "fence", it is "true"; otherwise, it is "false". The positional relationships between this personnel and all "fences" were calculated by this algorithm. If all positional relationships are "false", he sends the alarm information out of the "fence". If one positional relationship is "true", exit from the calculation and conclude that he is in the "fence". In the administrator mode, it is necessary to analyze the multi-users situation by the same method. Finally, the communication between the platform and smart phone end was realized and alarm was sent to users through the Web Service technology when the personnel is out of the "fence".

A concave polygon with 14 sides is shown in Fig.1. Whether the red point is in the polygon shall be determined.

The Y coordinate of the test point was compared with each point of the polygon, getting the list of intersection between the line where the test point lies on and the polygon. In this case, 8 sides intersected with the line and rest 6 sides didn't. If the point number at two sides of the test point is odd, the test point is in the polygon; otherwise, it is out of the polygon. In this case, the test point had five intersections in the left and three in the right, indicating that it is in the polygon[8].

(Note: This algorithm is applicable to both clockwise and counterclockwise polygons.)

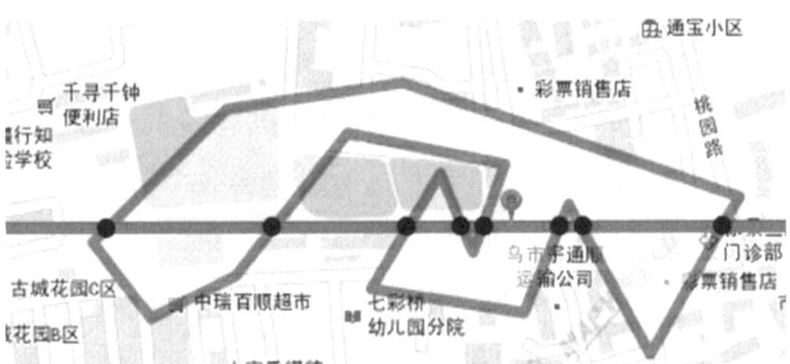

Figure 1.normal situation of ray casting algorithm.

However, there are some special situations. Self-intersecting of a polygon is shown in Fig.2. In this case, 10 sides of the polygon cross mutually, which is similar with "XOR" in assembly language. Overlaps in the polygon were deleted[9]. Therefore, the test point is out of the polygon, which was proved by the intersection number (2 intersections) at two sides of the test point .

In Fig.2, the polygon has no overlap, but two sides intersect. The algorithm is also applicable under this circumstance.

Figure 2. Self-intersecting polygon situation.

In Fig.3, the scanning line where the test point lies on just passes through one vertex of the polygon and intersects with side a and side b, forming two intersections. Similar phenomenon is found on the right of the test point. According to the proposed algorithm, the test point is out of the polygon. However, this is obviously wrong.

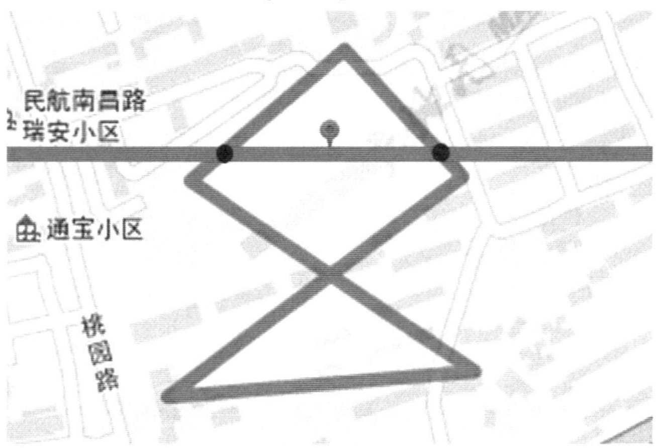

Figure 3.Two edges intersect but no overlap polygon.

Figure 4 shows that test point leads rays through exactly one vertex of the polygon. Therefore, there is a ray with a side of the intersection, and the side b is also a point of intersection, a total of two corners, right test point is the same situation. According to the algorithm that test point outside the polygon conclusion [11].

It is simple to solve this problem. Two end point at two sides of the scanning line is the key to determine whether the point on the side intersects with the scanning line. In this paper, end points on or above the scanning line is viewed same. In Fig.4, one end point of side a is below the scanning line and the other one is on or above the scanning line, indicating that side a intersects with the scanning line. Two ends of sides are on or above the scanning line, so it doesn't intersect with the side b.

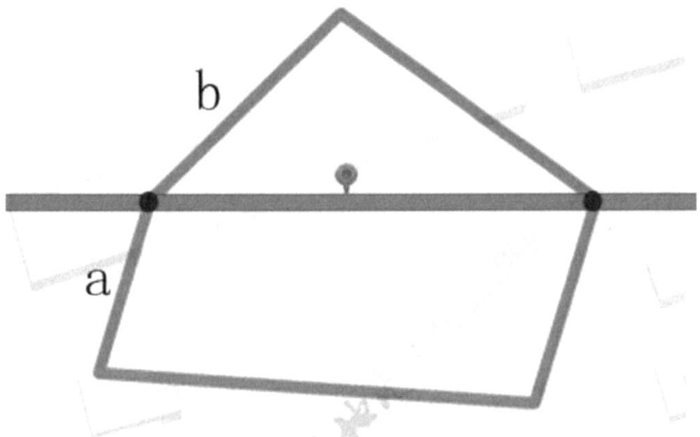

Figure 4.The ray through exactly one vertex of the polygon.

One side of the polygon exactly coincide with the ray of the case shown in Figure 5. According to one FIG. 4 specifically described an endpoint of the edge c below the rays and another endpoint on or above the ray, so there is the edge c and ray have one intersection point. Two endpoints of the edge d are on the ray, so no intersection point. Two endpoints of the edge e are on or above the ray, so there is no intersection point. Finally judgment test points inside the polygon because test point and c, d, e the edge have an intersection point [12].

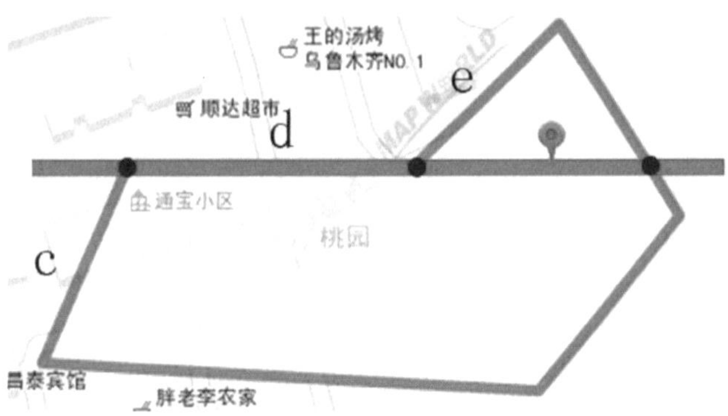

Figure 5. One side of the polygon exactly coincide with the ray.

3. Results and test

3.1 System Architecture

The system is mainly divided into 8 modules (positioning module, alarm module, query model, communication model, user module, map module, linking module and maintenance module) and 22 main functions. Modules are described as follows:

Linking module: it is responsible for connection of GPS, transmission data of BD satellite receiver and the database.

Positioning module: it is mainly to feedback data through the receiver and realize real-time personnel positioning, assigned "fence" positioning and historical track positioning on the Map World based on "fence" setting data.

Alarm module: it calculates positional relationships between the personnel and the corresponding "fence" by the ray method, realize "fence" alarm based on logics operation, and give power alarm and SOS alarm by judging feedback data of the receiver.

Query module: it is mainly to search with key words, give amplified display the interested personnel information in the middle of the map, browse complete information in the left column, search "fences" in list pattern, query historical records based on the structure tree, and search information under limited conditions, including unit, instrument mode, alarm type, etc.

User module: It is mainly in charge of user registration, account management, personnel management, permission assignment and login verification.

Map module: it is equipped with interested information selection, personnel display and "fence" display except for the basic map service functions of the Map World API.

Maintenance module: it offers database maintenance, including adding, deletion, modification and query of personnel information as well as visual adding and deletion of "fences".

Communication module: alarm information is sent to appointed cell phone as SMS.

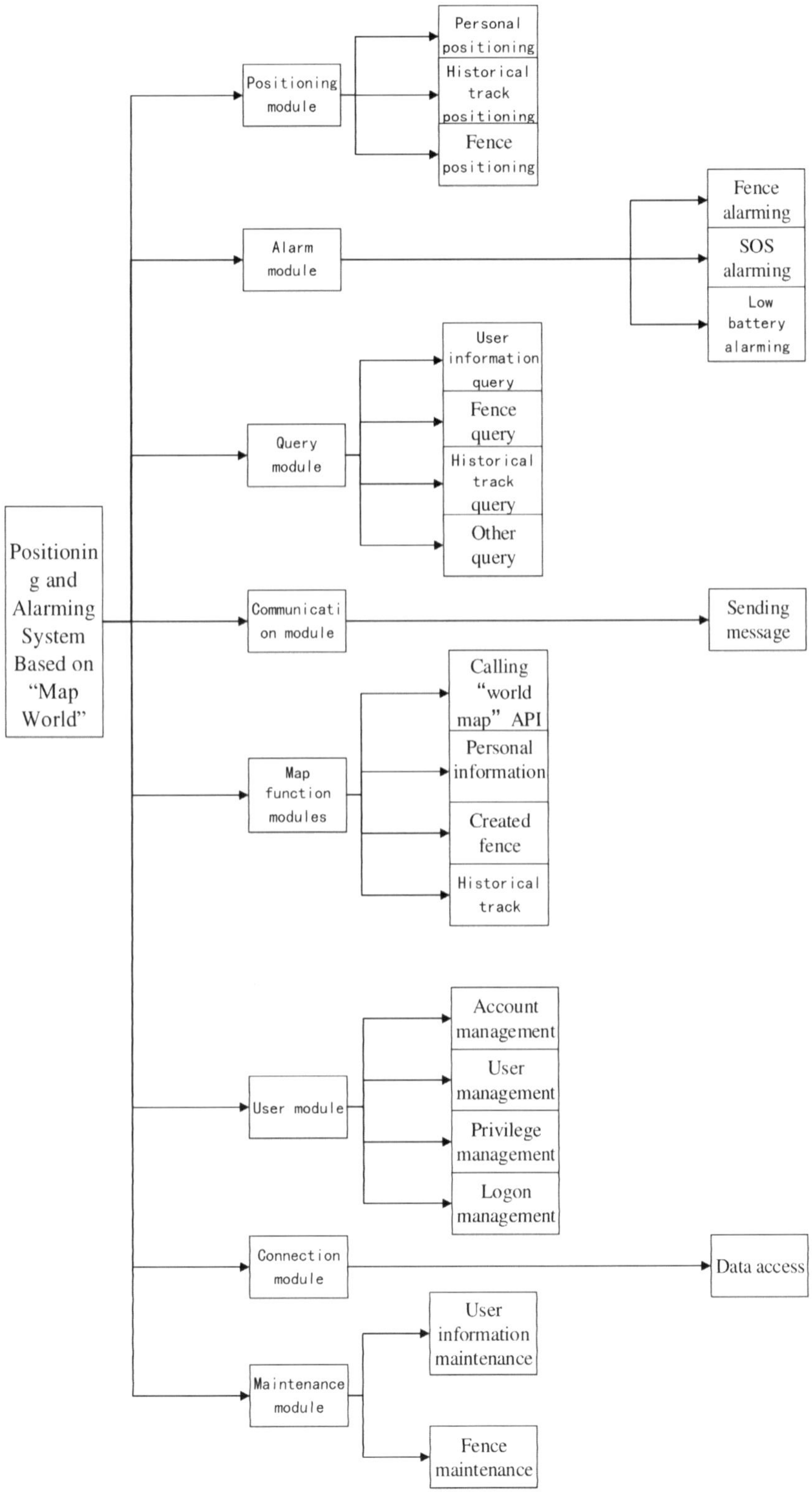

Figure 6. System structure drawing

3.2 System test

System was tested by STONEX S7-S receiver several times in August 2015 at Urumqi. STONEX S7-S is able to connect GPS single point positioning accuracy of 2.5m, post-processed differential precision 0.2m, real-time differential accuracy of less than 0.5m e.g. Figure7.

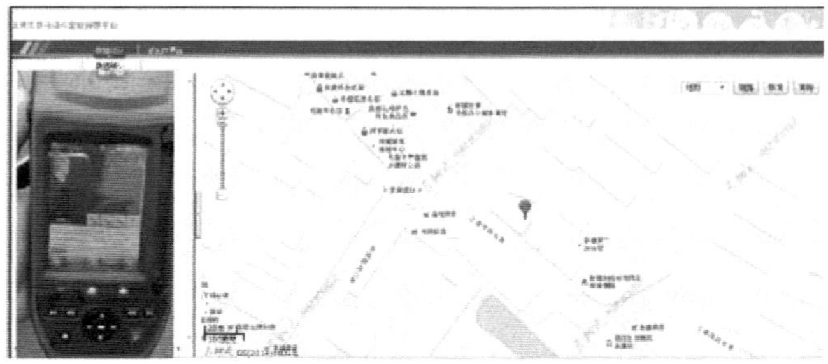

Figure 7. STONEX S7-S receiver

Tests to determine the real-time Positioning by receiving GPS receiver and set the time to send Positioning information to the server at intervals of four seconds. Test distance of about 200 meters. After the test found that the process of sending data from the receiver to the server and then from the server to appear on the system average delay of 1 second. Under built-up blocking more serious cases the position error is within 10 meters. In normal circumstances, the position error is about 2.5 meters. After several tests proven to target the mobile "fence" alarm and the alarm system is successful e.g. Figure 8. After several tests proven to the system is able to alarm and send text message to phone of user when target move out "fence".

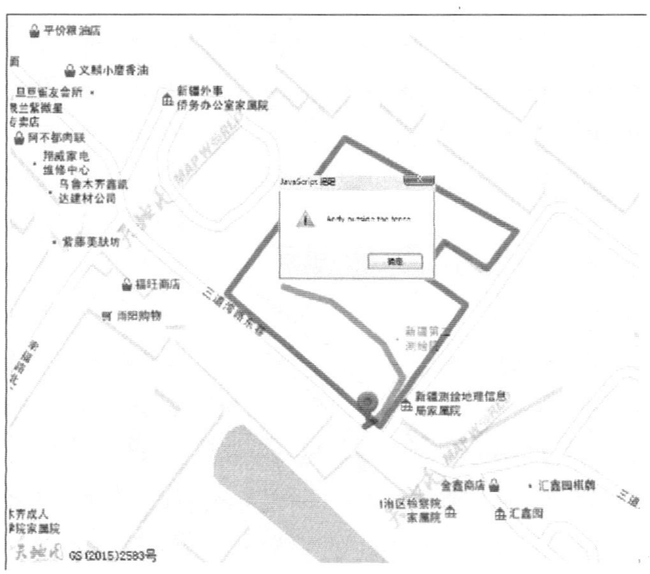

Figure 8. Alarm Demo

4. Conclusion

The "Map World" possesses abundant accurate basic surveying and mapping data and image data, which are valuable information resources for construction of the economic belt core of the Silk Road. The moving target positioning and alarm platform based on "Map World" is one application that integrates "Map World" data, GNNS data and GIS technology. Except for some GIS functions, this platform realizes the alarm function through the ray method. Its practicability is further proved by tests and use. In future, the platform will employ smaller receivers (e.g. necklace and watch) and achieve more accurate positioning through cooperation with CORS.

References

[1] S.A.Hameed, O.Khalifa, M.Ershad, F.Zahudi,B. Sheyaa, andW. Asender, Car monitoring, alerting and tracking model: enhancement with mobility and database facilities, in Proceeding of the International Conference on Computer and Communication Engineering (ICCCE'10), PP.1-5, May 2010.

[2] H.Tan, J.Feng, G.Feng, W.Wang, andY.J.Zhang, Traffic volume data outlier recovery via tensor model, Mathematical Problem in Engineering, vol.2013, ArticleID 164810, 8pages, 2013.

[3] Jun Zeng, Minbo Li, and Yuanfeng Cai.A Tracking System Supporting Large-Scale Users Based On Gps and G-Sensor, International Journals of Distributed Sensor Networks, vol 2015, Article ID 862184 ,13 pages,2015.

[4] C.Bo,C.Xin, Z.Zhongyi, Z.Chengwen, and C.Junliang, Web of tings-based remote monitoriing system for coal mine safty using wireless sensor network, International Journal of Distributed Sensor Networks ,vol.2014,Article ID 322127,14 pages 20014.

[5] S.Lee,G. Tewolde, and J. Kwon, Design and implementation of vehicle tracking system using GPS/GSM/GPRS technology and smartphone application, in Proceedings of the IEEE World Forum on Internet of Things(WF-IoT'14),pp. 353–358,IEEE, March 2014.

[6] P.Liang, H.Wang; C.H.Pengetal, Digital transmitter design for mobile devices, IEEE Communications Magazine ,vol. 51,no.10,2013.

[7] Wensheng Liu;Yu Guo, The Development of Web Service-Based Remote Control and Montoring System, Computational Intelligence and Software Engineering, 2009. CiSE 2009.

[8] Peng Liu; Guojun Dai; Tingting Fu, A Web Services Based Email Extension for Remote Monitoring of Embedded System, Eighth ACIS International Conference on Software Engineering, Artificial Intelligence, Networking, and Parallel/Distributed Computing (SNPD 2007),2007.

[9] Gupta, V.;Poursohi, A.;Udupi, P.Sensor. Network: Anupen data exchange for the web of things, Pervasive Computing and Communications Workshops (PERCOM Workshops), 2010 8th IEEE International Conference on, 2010.

[10] Xiang Feng Kong, Monitoring System Design for Ships Network, Advanced Materials Research, Vol.846-847,pp. 1542-1545, 2013.

[11]El Zawawi, Amr; Ibrahim, Ali, Using ZigBee to Build a Web- Based DCS System,2012 IEEE POWERAND ENERGY SOCIETY GENERAL MEETING,2012.

[12] P.Wei. Internet of things set off a new wave of information technology revolution, LOGISTICS&MATERIAL HANDLING, Vol14, No.11, 42—45, 2009.

Improved algorithm for hyperspectral data dimension determination

CHEN Jie[1,2], DU Lei[1], LI Jing[2], HAN Yachao[1], GAO Zihong[1]

[1] China Aero Geophysical Survey and Remote Sensing Centre for Land and Resources, Beijing 100083, China
[2] Institute of Remote Sensing and Digital Earth, Chinese Academy of Sciences, Beijing 100101, China

Corresponding author Chen Jie's e-mail address: 6592296@qq.com

Abstract. The correlation between adjacent bands of hyperspectral image data is relatively strong. However, signal coexists with noise and the HySime (hyperspectral signal identification by minimum error) algorithm which is based on the principle of least squares is designed to calculate the estimated noise value and the estimated signal correlation matrix value. The algorithm is effective with accurate noise value but ineffective with estimated noise value obtained from spectral dimension reduction and de-correlation process. This paper proposes an improved HySime algorithm based on noise whitening process. It carries out the noise whitening, instead of removing noise pixel by pixel, process on the original data first, obtains the noise covariance matrix estimated value accurately, and uses the HySime algorithm to calculate the signal correlation matrix value in order to improve the precision of results. With simulated as well as real data experiments in this paper, results show that: firstly, the improved HySime algorithm are more accurate and stable than the original HySime algorithm; secondly, the improved HySime algorithm results have better consistency under the different conditions compared with the classic noise subspace projection algorithm (NSP); finally, the improved HySime algorithm improves the adaptability of non-white image noise with noise whitening process.

1. Introduction

Adjacent bands of hyperspectral data acquired with less than 20nm spectral resolution generally have a strong correlation, and their DN value and visual images are often very similar. The hyperspectral image is recorded in the form of a matrix. Each column of the matrix represents the corresponding band spectral response data value. Through a series of matrix calculation, the front rows of the matrix band data can represent most information of the original image, and the back rows of the matrix represents the image noise. Low-dimensional data can be used to represent high-dimensional data with this method, reducing the amount of data processing workload and suppressing the impact of the noise. The quality of dimension reduction results will directly affect the next end-member extraction algorithms' final results. In the data dimension reduction process, how to select the appropriate dimension methods and how to accurately determine the required low-dimensional data are the most important two procedures.

After years of research, there are some relatively mature and widely used methods in hyperspectral data de-correlation transformation. De-correlation transformation is the removal of inter-band correlation of hyperspectral data transformation that converted data bands in different dimensions of space. The principal component analysis (PCA) algorithm and the minimum noise fraction (MNF) algorithm are the most two frequently used de-correlation transformation algorithms. The PCA algorithm is a linear algorithm based on K-L transformation. It sets the original image matrix variables into a set of uncorrelated random variables, arranges them in order of the covariance value to form a new image matrix. Typically, the first principal component matrix contains 80% variance information of all the bands. However, the PCA algorithm is sensitive to image noise. Principal components with large amount of information of all bands do not necessarily mean they have high signal to noise ratio (SNR). When an image noise variance of the principal component is greater than the variance of the signal, the image quality formed with the principal components is poorer. Therefore, Andrew A. Green et al (Andrew A. Green, 1988) proposed the MNF transformation. What is different with the PCA transformation is that the vectors are arranged by their SNR and the impact of noise on image quality is eliminated. Later on, James B. Lee, etc. proposed the noise-adjusted principal component transformation algorithm (NAPC).

However, the data dimension determination is still in the research stage. And for the PCA and MNF transformation algorithms, a key issue is to determine how many dimensions should be chosen for the low-dimensional data to represent high-dimensional data. In practical applications, researchers often select dimensions on their own experiences which make the dimensions determination process tend to be subjective or even blind. The correctness of dimension determination will directly affect the accuracy of the subsequent data processing precision and data analysis results. Therefore, choosing the appropriate rules and methods to determine the optimal number of dimensions plays a significant important role for the subsequent data processing and analysis procedures. Based on the above considerations, this paper analyzes and reviews the commonly used noise-whitening algorithms as well as the subspace dimension determination algorithm, proposes a hyperspectral data subspace dimension determination algorithm based on noise whitening processing, uses the algorithm to calculate the dimension of both the simulated and real hyperspectral data, and finally confirms its stability and accuracy by comparing with different algorithms.

2. Hyperspectral image Noise estimation and whitening

Under the influence of various factors, it is inevitable to bring in image noise while acquiring hyperspectral images. Currently, there are three main methods for remote sensing image noise estimation: laboratory method, dark current method and image method. It is hard for both the laboratory and the dark current method to apply in practical application because prior knowledge of the image is needed and the calculation requires a complex series of accurate measurement. The image method uses remote sensing images to directly analyze and estimate image noise, so this method is widely used to estimate the hyperspectral image noise (Zhang B, 2011). For hyperspectral images, the image noise estimation method mainly utilizes three features: the type and nature of hyperspectral images, pixel spatial correlation and hyperspectral band correlation.

2.1. Hyperspectral Image Noise Estimation

Since hyperspectral image between adjacent bands have a strong correlation, there are two commonly used algorithms to estimate hyperspectral image noise. One is the de-correlation algorithm with whole spectral dimensions proposed by Roger (Roger, 1996), and the other one is the spectral and spatial de-correlation method (SSDC) proposed by Arnold (Arnold, 1996). The advantage of SSDS algorithm is that it makes use of hyperspectral image spatial and spectral correlation between adjacent bands. And it is little affected by ground covering vegetation types, has a high degree of automotive calculation and does not need any human intervention. So, the SSDC algorithm is one of the stable algorithms to estimate hyperspectral image noise.

2.2. Hyperspectral Image Noise Whitening

Noise whitening is a process which first does the data de-correlation processing, and then does the noise covariance unitization processing. Researches confirm that, for a random vector, the transformation matrix which can do the whitening process is not unique. Meanwhile, when adopting different whitening matrix, the whitening results are also different. Generally, when doing the hyperspectral image processing, this paper assumes that the hyperspectal image information and the noise is irrelevant. This means the covariance matrix of the noise is a diagonal matrix.

3. The HySime algorithm

Based on the Principle of Least Squares, José M. Bioucas-Dias (José M. Bioucas-Dias, 2008) proposes the hyperspectral signal identification by minimum error algorithm (HySime). The HySime algorithm is used to estimate the dimension of hyperspectral subspace data. It estimates the signal as well as the noise correlation matrix first, and then best represents the feature vector subset of signal subspace in the form of minimum mean square error.

Suppose the observation value y consists of the signal x as well as the noise n, shown in the form of vector formula that y=x+n. The signal correlation matrix estimation value is referred as $\hat{R}_x$. Suppose the mean value of noise is 0, and $\hat{R}_x$ is the covariance. Then the signal correlation matrix can be separated as:

$$\hat{R}_x = E\Sigma E^T \tag{1}$$

The space of L formed from $E=\left[e_1,\cdots,e_L\right]$ can be separated into two subspace $\langle E_k\rangle$ and $\langle E_k\rangle^{\perp}$ which are mutually orthogonal. And the corresponding feature vectors $E_k=\left[e_{i_1},\cdots,e_{i_k}\right]$, $E_{\frac{\perp}{k}}=\left[e_{i_{k+1}},\cdots,e_{i_L}\right]$ and $\pi=\{i_1,i_2,\cdots,i_L\}$ are used to record the sequence of feature matrix.

Suppose the projection matrix of subspace $\langle E_k\rangle$ is referred as $U_k=E_kE_k^T$, and the projection of y on $\langle E_k\rangle$ is referred as $\hat{x}_k=U_ky$.Try to find the best feature vector sequence $\pi=\{i_1,i_2,\cdots,i_L\}$ and k to make $mse(k)$ minimized. The calculation formula is as follows:

$$(k,\pi)=\arg\min_{k,\pi}\left\{tr\left(U_k^{\perp}R_y\right)+2tr\left(U_k\hat{R}_n\right)\right\} \tag{2}$$

Since $U_k=E_kE_k^T$, $U_k^{\perp}=I-U_k$, formula 2 can be put as:

$$(k,\pi)=\arg\min_{k,\pi}\left\{c+\sum_{j=1}^{k}\left(\underbrace{-p_{i_j}+2\sigma_{i_j}^2}_{\delta_{i_j}}\right)\right\} \tag{3}$$

In the formula above, c is a constant, and p_{i_j} 、 $\sigma_{i_j}^2$ refers to the binomial observation signal and noise correlation matrix respectively.

$$\begin{aligned} p_{i_j} &= e_{i_j}^T R_y e_{i_j} \\ \sigma_{i_j}^2 &= e_{i_j}^T \hat{R}_n e_{i_j} \end{aligned} \tag{4}$$

To minimize the right of the equal sign of formula 3, it is necessary to find all the negative value of δ_i, and the corresponding (k,π) is the calculation result .

4. The Improved HySime Algorithm

In HySime algorithm, it is easy to known from formula 3 and 4 that the estimation of image noise as well as signal matrix are two important procedures. And the calculation of corresponding feature vectors affects the final results directly. Meanwhile, the noise spectrum estimation algorithm adopts the full image spectral de-correlation method to estimate the dimensions, calculating pixel by pixel. Then it uses the original observation data minus the noise estimation value to get near true signal value, and the signal related estimation matrix can be calculated this way.

The original HySime algorithm is applicable when image noise is accurately estimated. However, besides the huge calculation, it is very difficult to accurately estimate a whole image's noise value. So it is not applicable to estimate the image noise and it cannot be an ideal algorithm to estimate the image noise in order to have a good precision with the original HySime algorithm.

Suppose the observation image is white noise. The noise covariance matrix can be referred as $\hat{R}_n = \sigma^2 I$, and the feature value λ and its corresponding feature vector x of the signal related matrix fits into the formula 5:

$$\left(\hat{R}_x - \lambda I\right)x = \left(R_y - \hat{R}_n - \lambda I\right)x = \left(R_y - \left(\sigma^2 + \lambda\right)I\right)x = 0 \quad (5)$$

It can be known after solving formula 5 with the HySime algorithm that when the noise is white noise, no matter big or small, the calculated feature vectors keep the same. This means the subspace determined by the feature vectors is not affected by image noise. And when the observation image noise is white noise, the impact of the image noise can be eliminated following two steps: first do the whitening transformation to the image noise, and then estimate its image noise with HySime algorithm. This method can improve the adaptability of the algorithm to image noise.

For a certain original observation data y, suppose $\hat{R}_n$ is the calculated noise covariance matrix. Separate $\hat{R}_n$ with its feature value to get its feature value and feature vectors, referred as Λ_n and A. Use matrix $F = A\Lambda_n^{-1/2}$ to do the image noise whitening process and mark the calculated observation data as y_w. Put y_w as the original HySime algorithm input value. Then the noise is white noise and its covariance matrix is a unit matrix, that is $\hat{R}_{nw} = I$. Mark the observation related matrix as R_{yw} and the signal related matrix equals to the observation data subtracting the noise value, that is $\hat{R}_{xw} = R_{yw} - \hat{R}_{nw} = R_{yw} - I$. Then do the calculation with HySime algorithm. Mark the feature vector of $\hat{R}_{xw}$ as $E = \left[e_1, \cdots, e_L\right]$, and mark the sequence of $\pi = \{i_1, i_2, \cdots, i_L\}$. Then it is easy to know that $\sigma_{i_j}^2 = e_{i_j}^T \hat{R}_{nw} e_{i_j} = 1$. So formula 3 can be put as formula 6. The corresponding (k, π) value of all the negative value of $\left(2 - p_{i_j}\right)$ is the subspace dimensions needed to be calculated.

$$\left(k, \pi\right) = \arg\min_{k,\pi}\left\{c + \sum_{j=1}^{k}\left(2 - p_{i_j}\right)\right\} \quad (6)$$

5. Experiments and analysis

5.1. Simulated Data Experiment

The simulated data used in this paper is made by manually adding noise into the real hyperspectral data. Signal data can be calculated by the number of end-member spectra multiplied by their corresponding abundance. And end-member spectra are selected from USGS spectral libraries.

Resample the spectra for 224 bands following the AVIRIS response functions and center wavelengths. The abundance data is distributed randomly with the Dirichlet method. The experiment is carried out this way: Set the number of different end-members to 5, 10, 15 and 20 respectively. The abundance of end-members follows the Dirichlet distribution which forms the signal data. Add a certain amount of noise to these four signal data to make their signal to noise ratio to be 15dB, 25dB and 35dB. Added noise can be separated into two types which are the white noise and colored noise. Abundance threshold value is set to 1. Each of the above combination can produce 100 sets of simulated data. Use the original Hysime algorithm, the improved Hysime algorithm and the NSP algorithm to do the calculation test respectively. Signal with manually added noise is relatively simple, so noise spectrum estimation adopts the full image spectral de-correlation method to estimate the dimensions, calculating pixel by pixel. The NSP algorithms with false alarm probability value of 10e-3 and 10e-4 are referred as NSP_10e-3 and NSP_10e-4. Signal subspace estimation is carried out for each 100 simulated data sets, and the mean and standard deviation of calculation results are used to be the evaluation indicators.

Conclusions can be drawn from the results of comparison of different algorithms with different parameters:

1) When the noise is white noise, under different parameters, results from the improved HySime algorithm as well as the original HySime algorithm have some consistency. However, when the noise is colored noise, the accuracy and stability of improved algorithm on the end-member estimation is much better than the original HySime algorithm.
2) Under different parameters, end-member estimation results from the improved Hysime algorithm and the NSP algorithm are basically the same.
3) When white noise exists, once the SNR increases, signal subspace estimation accuracy of each algorithm rises. Because the higher the SNR is, the better signal to noise suppression capability it is, and the smaller impact noise on the algorithm it is. This also illustrates the importance of noise estimation and removal.

5.2. Real Data Experiment

The AVIRIS data was collected from the Cuprite mining district, Nevada, in June, 1997. The DN value represents the reflectance ratio. It contains 224 bands between 400 nm to 2500 nm, its radiometric resolution is 10 nm and its spatial resolution is 20 m. To start with, remove water vapor absorption or low SNR bands so as to 192 bands are usable. The Cuprite mining district has a detailed ground survey result and geological background information. Gregg Swayze et al. have used the AVIRIS data which acquired in 1990 to identify 18 kinds of mineral types. They have carried out an X-ray diffraction analysis with field samples to determine their mineral categories and verified the results of the analysis (Swayze, 1992). Thereafter, Gregg Swayze et al. used the AVIRIS data acquired in 1995, analyzed the distribution of 25 kinds of minerals and did the mineral mapping with 23 kinds of minerals (Swayze, 1997).

From previous studies it can be concluded that the dimension estimation value of Cuprite mining zone basically lies between 18 and 30 (Ciobanu C, 1999). This paper uses the same four algorithms with simulated data, including: HySime algorithm, improved HySime algorithm, false alarm probability value with NSP algorithm 10e-3 and 10e-4's. The results are as follows:

Table. 1. Hyperspectral dimension estimation in Cuprite mining area

Dimension Estimation / Noise Estimation	HySime algorithm	Improved HySime algorithm	NSP_10e-3	NSP_10e-4
With Spectral de-correlation algorithm	20	27	27	27
With SSDC algorithm	-	26	26	26

The dimension number estimation value is 26 and 27 in the table which has a good consistency with the estimation value of 18-30 by other researchers (Ciobanu C, 1999）. The MNF transformation is adopted to check the correctness of dimension estimation in Cuprite mining district hyperspectral images. Choose different band combinations to calculate the pure pixel index (PPI), set the projected number of iterations to 20,000, set the threshold value of the coefficient to 2 and count the pixel number when the PPI index is greater than 1 in results. The results are shown in Table 2:

Table. 2. PPI calculation results with different parameters

Bands Number	10	20	26	30	40	50
PPI	639	874	1003	1060	1176	1355

The greater PPI is, the greater possibility pure pixels exist and the greater possibility the spectral space can be separated. Data in table 2 shows that when the number of bands is 26 or 30, the number of pixels is close to each other. The main image information is mostly concentrated in the top 40 bands after the MNF transformation. With the increase of the number of bands, the noise it contains is also rising. In this case, select the first 26 or 30 bands to calculate the PPI, the results are close to each other. This shows that the top 26 and top 30 bands can represent the original hyperspectral images good enough and have a good consistency. On the contrary, when the band number is 10 or 20, they cannot represent the original hyperspectral image completely. To further verify the above analysis, select the intersection of the above six cases and then 475 pixels were obtained. Mark these 475 pixels as pure pixels and label the pixel set as Φ. These pixels are little influenced by image noise and represent the pixels which are most likely to be end-members. Analyze each pixel's independence, not including the 475 pixels of pixel set Φ for the above six cases. The greater the spectral angle is, the more independent each pixel is to pixel set Φ, and the more possibility they belong to pure pixels. For the above six cases, the minimum angle of all spectral angle maximum value is 2.756° and its band number is 10. Meanwhile, the maximum angle of all spectral angle maximum value is 4.118° and its band number is 50. And the spectral angles are basically the same of 4.114° which is significantly different with other spectral angles of different band numbers, when band number is 26 and 30. This suggests that the information subspace obtained by dimension estimation can represent spectral information very well. And data analysis is effective with the help of dimension estimation number.

6. Conclusions

1) The improved HySime dimension estimation algorithm does not need to be performed pixel by pixel to estimate the hyperspectral image noise. It improves the efficiency as a result of reducing the amount of computation. What's more, it improves the accuracy thanks to eliminating errors cause by the noise estimation inaccuracy.

2) The improved HySime algorithm does not depend on the assumption that the mean value of hyperspectral image noise is zero. This makes the original algorithm to be more applicable. For both white and colored noise data, it can estimate the dimension number which is close to the true value.
3) Noise estimation plays an important role in real hyperspectral data processing and analysis. No matter the noise is white or colored, to accurately estimate or eliminate the noise will significantly improve the accuracy of image interpretation.

7. Acknowledgement
This paper is sponsored by two scientific research projects from China Geology Survey: one is the technical research of hyperspectral geological surveying, and its project number is 1212031513012; the other one is the Tianshan-Beishan key mining belt surveying. Its project number is 121201003000150008.

References
[1] Bioucas Dias, J.M., J.M.P. Nascimento. Hyperspectral subspace identification [J]. IEEE Transactions on Geoscience and Remote Sensing, 2008, 46(4): 2435-2445.
[2] Cawse, K., A. Robin, M. Sears. The effect of noise whitening on methods for determining the intrinsic dimension of a hyperspectral image in 3rd Workshop on Hyperspectral Image and Signal Processing: Evolution in Remote Sensing, WHISPERS 2011, June 6, 2011 - June 9, 2011. 2011. Lisbon, Portugal: IEEE Computer Society.
[3] Chang, C. I. , Q. Du. Estimation of number of spectrally distinct signal sources in hyperspectral imagery [J]. IEEE Transactions on Geoscience and Remote Sensing, 2004, 42(3): 608-619.
[4] Eldar, Y.C., A.V. Oppenheim. MMSE whitening and subspace whitening [J]. IEEE Transactions on Information Theory, 2003, 49(7): 1846-1851.
[5] Green, A.A., M. Berman, P. Switzer, et al. A transformation for ordering multispectral data in terms of image quality with applications for noise removal [J]. IEEE Transactions on Geoscience and Remote Sensing, 1988, 26(1): 65-74.
[6] Lee, J.B., A.S. Woodyatt, M. Berman. Enhancement of high spectral resolution remote-sensing data by noise-adjusted principal components transform [J]. IEEE Transactions on Geoscience and Remote Sensing, 1990, 28(3): 295-304.
[7] Roger, R.E. Principal Components transform with simple, automatic noise adjustment [J]. International Journal of Remote Sensing, 1996, 17(14): 2719-2727.
[8] Roger, R.E., J.F. Arnold. Reliably estimating the noise in AVIRIS hyperspectral images [J]. International Journal of Remote Sensing, 1996, 17(10): 1951-1962.
[9] Swayze, G., R.N. Clark, F. Kruse, et al. Ground-Truthing AVIRIS mineral mapping at Cuprite, Nevada. Summaries of the third annual JPL airborne geosciences workshop. AVIRIS Workshop: JPL Publication 92-14, 1992: 47-49.

Spatiotemporal changes of built-up land and population distribution patterns in China during 1990-2010

Y H Ren, Y L Liu (corresponding author) and L J Yu

Institute of Remote Sensing and Digital Earth, Chinese Academy of Sciences,100101 Beijing, China.

Email: liuyl@radi.ac.cn

Abstract. With the population growth, built-up land is expanding constantly in China. The paper aims to analyse spatiotemporal pattern of built-up land and population, and their relationship in China mainland from 1990 to 2010. With population census data and built-up land data, it calculates population density of counties, built-up surface proportion and population density by built-up land, as measures of population and built-up land distribution. The result shows that, the total population and built-up area have increased by 17.5% and 33.9%, respectively, in China mainland from 1990 to 2010. Meanwhile, the population density of built-up land has decreased by 12.0%, indicating that built-up land growth has outpaced population growth overall. However, such changes are not evenly distributed in space and time. Change in the later decade is much more significant than that in the earlier decade. For built-up land, the increases don't show obvious characteristics of spatial aggregation. Correspondingly, most increases in population density of counties and population density of built-up area are in the northwest area, divided by Aihui-Tengchong line, while most decreases are in the southeast area. These analyses help to explain the overall impact of political-economic environment and the policy changes on urbanization processes for China mainland.

1. Introduction

China is the most populous country in the world, with nearly 19 percent of the world's population. The estimated population in China mainland was 1.368 billion in 2014, about 2.5 times of its population in 1949. Indeed, during 1949-1990, the population of China mainland increased at a rate of 1.8% annually and no less than 1.5% for almost every year. However, the rate dropped down to 0.5-1.5% a year during 1990-2014, and basically stable at 0.5% after 2005.

With the population growth, built-up land has experienced rapid extension in China mainland over the last three decades. From 2000 to 2010, the area of built-up land in urban and rural has increased 1.61×104 kilometers and the growth rate is 11.17% [1]. In particular, the built-up land in urban has extended significantly. A study based on remote sensing images indicates that the urban built-up land have tripled in China mainland from 1990 to 2010 [2]. This urbanization has inevitably caused a series of resource and environmental problems and protection issues throughout China [3]. In addition, urban growth has caused unprecedented changes in land use patterns, especially loss of arable land [4].

Satellite-based remote sensing observation with various spatial, temporal and spectral resolutions has been extensively applied in investigations of urban dynamics and land cover changes, because remote sensing can provide timely and spatially explicit information for land cover change processes. Zhang et al. used the Landsat images to study the annual dynamics of impervious surface in the Pearl River Delta, China, from 1988 to 2013 [5]. The result showed that the Pearl River Delta had experienced a dramatic urban expansion from 1988 to 2013 and impervious surfaces increased more than ten times.

The population problem is an important part of the relationship between human and land which is the eternal theme of geography study. Simultaneously, studying the spatial and temporal characteristics of built-up land growth in China has important significance for promoting urban sustainable development and maintaining ecological safety. Therefore, with the built-up land data from remote sensing images and census data, the aim of this paper is to analyse the spatiotemporal distribution patterns of population and built-up land quantitatively in China mainland during 1990-2010, and reveal the relationship between these two issues, utilizing GIS spatial analysis combined with statistical analysis method.

2. Materials

2.1. Remote sensing image and built-up land map

Since the beginning of the Landsat program in the early 1970s, it has become one of the most widely used remote sensing imagery. Thematic Mapper (TM) images, with 30 m ground resolution and seven spectral bands, have proven to be a very effective tool in regional-scale land use/cover and urban studies [6]. The later series of Landsat satellite with Enhanced Thematic Mapper (ETM+) and Operational Land Imager (OLI) improve in spectral resolution and number of band.

Landsat 5 TM images and Landsat 7 ETM+ images during later years of 1980s, 2000 and 2010 are collected and used to acquire the built-up land distribution in 1990, 2000 and 2010, respectively. All the images are classified based on automatic classification combining with visual interpretation method, and then the land cover maps in 30 m resolution are obtained. We define all man-made surfaces as built-up surfaces, representing all types of land uses for urban and rural development. Because of China's vast expanses, the land cover data are resampled to 1 km resolution. Then, the built-up land density maps are produced in different years for built-up land change analysis, in which the value for each pixel represents the area proportion of built-up land in the pixel.

2.2. Population data and county map

Population census data are the most reliable data for population distribution study in China. The population data used in this study include data from the 1990, 2000 and 2010 censuses conducted by the State Council of the People's Republic of China. The census data are collected at the smallest possible census unit-the county. There are permanent and registered population, male and female population, and some others information. Because of a large number of floating populations, the population of permanent residents more than census register population can reflect the actual distribution of the population. Here, we mainly use the permanent population for each county. Since there is no information for Taiwan province in population census data, this study take China mainland as study area.

County map is not only the basis of spatial distribution of statistical population, but also the basis of relation analysis between population and built-up land. However, we have only collected the 2008 county map of China mainland at 1:1,000,000 scale from the State Bureau of Surveying and Mapping (SBSM) of China. For the reason of administrative division readjustment, the number and boundaries of counties are not completely equivalent. Creation of the polygons for counties of China mainland for 1990, 2000 and 2010 is done with vectorization method by reference to historic cadastral maps.

3. Methods

3.1. Spatial association of population data and county map

In order to analyse spatial pattern of population, the distribution map of population should be produced firstly. Since there are counties with same names, county name can't be used as the index field. According to the county name and its location, the county code is added to the population record for each county. With the county code, the population census data are associated with vector county map. Then, the population distribution vector data are produced for the year 1990, 2000 and 2010.

3.2. Calculation of population density

Population density is distribution of population in relation to land resources, which is a widely used measure of population spatial distribution and urbanization intensity. Population density generally defines as population of unit area of land. The most common unit of population density is the permanent population per hectare or per square kilometre. In this paper, it assumes population equally distributed within each county unit, and the population density for each county is calculated with formula as follow:

$$D_i = P_i / S_i \tag{1}$$

where D_i is population density of county i (unit: person per square kilometre), P_i is total permanent population of county i (unit: person), and S_i is area of county i (unit: square kilometre).

With the population distribution vector data which contain the information of permanent population and area for each county, the population density map of counties in China mainland are produced.

3.3. *Space matching of population distribution data in different periods*

As mentioned, the number and boundaries of counties change over time. For comparative study of population distribution in different periods, it takes county map in 2010 as standard and adjusts the population data of counties in 1990 and 2000.Taking population distribution data in 2000 as an example, the procedure of space matching of population distribution data is shown as follows in detail.

(1) For the counties whose boundaries are completely equivalent in 2000 and 2010, assign the permanent population in 2000 to the counties.

(2) For the counties whose boundaries are not completely equivalent in 2000 and 2010, convert the population density maps of counties in 2000 from vector to raster in resolution 1 kilometre firstly. Next, perform zonal statistical analysis on the population density data in raster with the county map in 2010 to get adjusted value of population for each county.

3.4. *Calculation of built-up land proportion*

Built-up land proportion defines as area of built-up land per unit area, which is another measure of urbanization intensity. In some studies, proportion of built-up land is called density of built-up land. In this paper, the built-up land proportion is calculated with formula as follow:

$$p_i = \frac{SB_i}{S_i} \times 100\% \tag{2}$$

where p_i is built-up land proportion of county i, SB_i is built-up land area in county i (unit: square kilometre), and S_i is area of county i (unit: square kilometre).

Once appropriate county maps are constructed, the proportion of built-up surface, based on image classification, is summarized for each county during study period.

3.5. *Calculation of population density of built-up land*

Although population density by census units is the typical measure of population distribution, it may not fully reflect the population distribution pattern because different land uses may have significantly different distribution of population. To illustrate the dynamics in both population and built-up land growth, it calculates population density by built-up surface with the aid of satellite imagery in this project. This indicator is especially important for understanding population change in outskirt of cities, where census units still contain much undeveloped or agricultural land. The population density of built-up land is defined as population per unit area of built-up surface. It is a measure of intensity of urban expansion. The calculation formula is shown as follow:

$$DB_i = \frac{P_i}{SB_i} \tag{3}$$

where DB_i is population density of built-up land in county i (unit: person per square kilometre), P_i is total permanent population in county i (unit: person), and SB_i is built-up land area in county i (unit: square kilometre).

3.6. *Spatiotemporal change analysis of population and built-up land*

With all the data of population density, proportion of built-up land and population density of built-up land at different periods, it analyses the spatiotemporal distribution patterns of population and built-up land quantitatively in China mainland during 1990-2010 using geographic information system (GIS) spatial analysis combined with statistical analysis method. Especially, it compares population density by total area of the census unit and that by built-up area to reveal the relationship between spatiotemporal change of population and built-up.

4. Result

4.1. Spatiotemporal pattern of population distribution

Hu Huanyong pointed to a line from Aihui in Heilongjiang province to Tengchong in Yunnan province as population density boundary of China in 1935. This line is called Aihui-Tengchong Line or Hu Line. The population density, geographical structure, the human lifestyle and economic development all show significant differences on different sides of this line. Figure 1 shows the population density by counties of China mainland in 1990, 2000 and 2010. It indicates that the population distributes densely in southeast and sparsely in northwest of China mainland, divided by Aihui-Tengchong Line, as Hu Huanyong proposed in 1935 [7]. The population density is less than 50 persons per kilometre for most of the counties on the northwest side of Aihui-Tengchong Line. On the southeast side, the population density is mainly more than 50 persons per kilometre and there are several larger population centers, such as North China Plain, Yangtze River Delta, Pearl River Delta, Sichuan Basin and the Central Shaanxi Plain, where the population density is more than 400 persons per kilometre. Although the overall pattern of population distribution is unchanged, changes of population distribution still have taken place. For the entire study area, population density has increased by 17.5% from 120 to 141 persons per kilometre from 1990 to 2010.

Figure 2 shows the changes of population density of China mainland during 1990 to 2010. Section (a) shows 2000 population density as ratios of the 1990 population density. Section (b) shows 2010 population density as ratios of the 2000 population density and section (c) shows 2010 population density as ratios of the 1990 population density. There are 1401 counties, about half of all the 2869 counties, presenting significant increase in population density during 1990-2010. The 2010 population density for these counties, when expressed as a ratio with 1990, commonly has change values above 1.10. These counties mostly locate at the large cities such as Beijing, Shanghai, Guangzhou, as well as areas northwest of Aihui-Tengchong Line where populations were so few in 1990. Counties with increasing population density are perhaps the clearest indicator of the driving forces of centralization in the east regions and development of the west regions in China mainland.

However, even more remarkable, it finds decreasing population density for many counties, especially in the southeast area of China. There are 495 counties presenting significant decline in population density during 1990-2010. The 2010 population density for these counties, when expressed as a ratio with 1990, commonly has change values below 0.90. Compared with period of 1990-2000, the number of counties with decreasing population has increased from 311 to 371 during 2000-2010. Many of these counties locate across Yangtze River economic belt, south of Qinling-Huaihe line, outside of big cities. It means populations have gathered from suburb to central cities and from peripheral area to the core of urban agglomeration. This is a performance of urban-rural dual structure effect and core- periphery structure effect. There are also some other areas with significant population decline, such as Inner Mongolia surrounding Hulunbuir and the three northeast provinces of China.

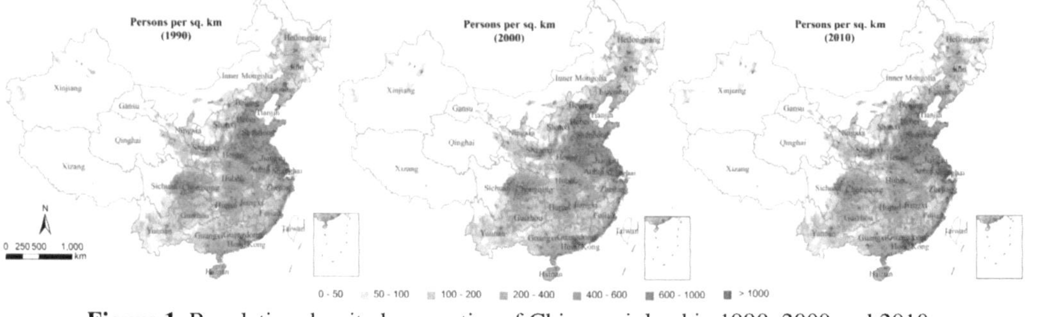

Figure 1. Population density by counties of China mainland in 1990, 2000 and 2010.

(b)

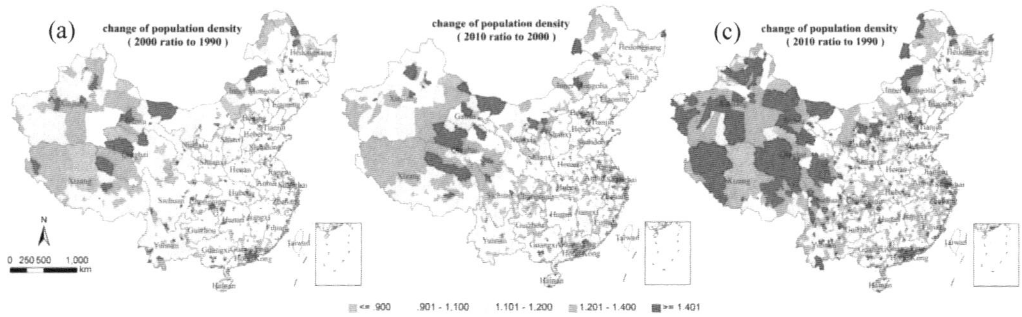

Figure 2. Changes in population density of China mainland during 1990-2010.

4.2. Spatiotemporal pattern of built-up land growth

Figure 3 shows the proportion of built-up land of counties in China mainland in 1990, 2000 and 2010. The overall distribution pattern of built-up proportion is similar to that of population density. The high values of built-up land proportion mainly distribute at the Northeast Plain, the North China Plain and the Middle-Lower Yangtze Plain, where are generally flat and suitable for built-up land development. For the areas on the northwest side of Aihui-Tengchong Line, the built-up proportion is generally low.

Figure 4 shows the change of built-up land proportion during the study period. Section (a) shows 2000 built-up land proportion as ratios of the 1990 built-up land proportion. Section (b) shows 2010 built-up land proportion as ratios of the 2000 built-up land proportion and section (c) shows 2010 built-up land proportion as ratios of the 1990 built-up land proportion. In the figure, the change value less than 0.85 means decrease of built-up land, while the value greater than 1.15 means increase of built-up land. The value between 0.85 and 1.15 means no significant change. The figure shows a significant growth of built-up land in China mainland during the study period and the growth during 2000-2010 have been even more pronounced than that during 1990-2000. The total area of built-up surface was 142,944 kilometres in 1990, 157,213 kilometres in 2000 and 191,385 kilometres in 2010, respectively. The area of the built-up surface has increased by 33.9% from 1990 to 2010. Of the total built-up area in 2000, 14,369 km2 or 10.0% were converted from other land cover from 1990 to 2000, while 34,172 km2 or 21.7% were converted from 2000 to 2010.

776 counties, about a quarter of all the counties, has experienced growth of built-up land proportion during 1990-2000, scattered the central core of big cities like Beijing, Shanghai, Chongqing, Guangzhou, Shenzhen, and so on. Correspondingly, 1630 counties, more than a half of all the counties, show obvious expansion of built-up land during the later decade, distributed nationwide. There are some counties showing very significant expansion of built-up land, with the change value greater than 1.5, 185 and 650 counties for 2000/1990 map and 2010/2000 map, respectively. While the central core of the city possessed high proportions of built-up surface, the greatest changes are seen along the peripheral areas.

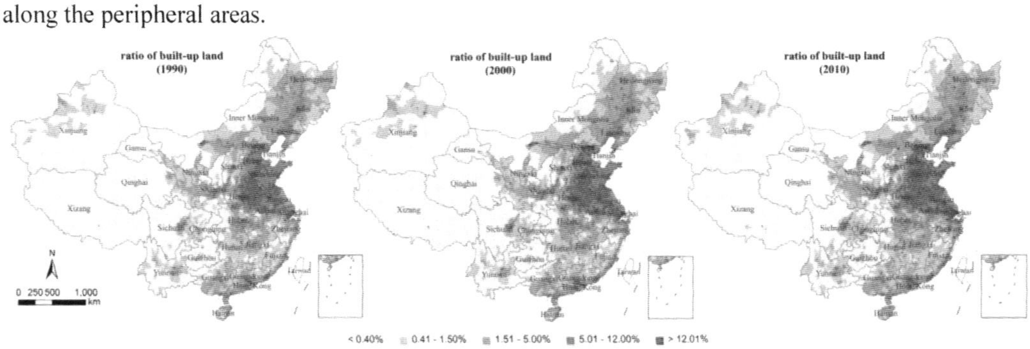

Figure 3. Proportion of built-up land images of China mainland in 1990, 2000 and 2010.

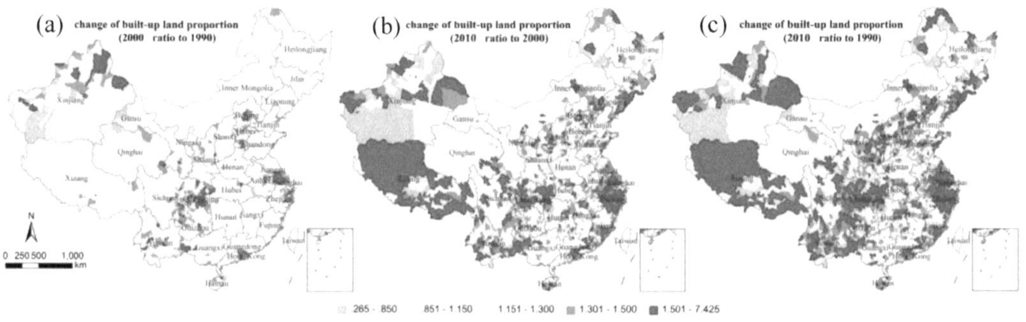

Figure 4. Changes of proportion of built-up land of China mainland during 1990-2010.

4.3. Analysis of population density of built-up area

Figure 5 shows the population density of built-up surface in 1990, 2000 and 2010. There are 36 counties that can't be calculated the population density of built-up land, since the built-up land area of these counties is equal to zero. It assigns 'nodata' to these counties. Unlike the spatial pattern of population density by counties, the population density by built-up land presents high in the south and low in the north of China mainland, mainly divided by Qinling-Huaihe-Qilian Mountains-Altun Mountain line. The high values generally locate in Qinghai-Tibet Plateau, Loess Plateau, Sichuan Basin, Yunnan-Guizhou Plateau and Jiangnan hilly land. In these areas, the surface condition is very complicated, landform undulates terribly and it is not suitable for built-up land development.

Figure 6 shows the changes of population density of built-up surface during the study period. Section (a) shows 2000 population density of built-up land as ratios of the value in 1990. Section (b) shows 2010 population density of built-up land as ratios of the value in 2000 and section (c) shows 2010 population density of built-up land as ratios of the value in 2000. In the figure, the value less than 0.90 means decrease of population density of built-up land, while the value greater than 1.10 means increase. The value between 0.90 and 1.10 means no significant change. Quantitative analysis shows that, population per square kilometres of built-up surface has decreased from 7913 in 1990 to 7901 in 2000 and 6964 in 2010, taking China mainland as a whole. There are 805 counties undergone decrease and 644 counties undergone increase of population density of built-up land during1990-2000, while during 2000-2010 the number of county changes to 1592 and 411for decrease and increase of population density of built-up respectively. Such changes reflect that, built-up land expands at a much faster rate than population growth for most of the time, and the level of intensive land utilization has reduced in China mainland.

The spatial change pattern of population density of built-up area is similar to that of population density of counties, but there exists more significant differences between northwest and southeast of China mainland. On the northwest side of Aihui-Tengchong line, the value of population density of built-up area is mostly increased, while on the southeast side, the value is mostly decreased.

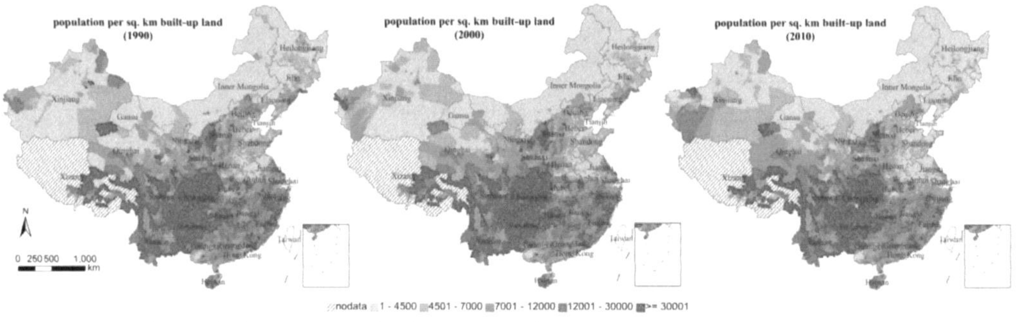

Figure 5. Population density of built-up land of China mainland in 1990, 2000 and 2010.

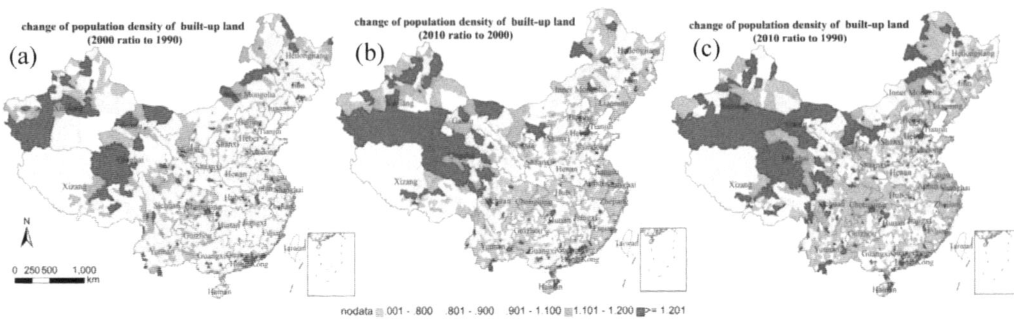

Figure 6. Changes of population density of built-up land in China mainland during 1990-2010.

According to causes, the changes of population density of built-up area may be roughly grouped to the following four types.

(1) Population density of built-up area declines, owning to decrease of population and no significant change of built-up land. There are some counties at the juncture of Hebei and Inner Mongolia provinces, falling into this type. Because these counties are generally the poverty county of national level, population move out of these areas.

(2) Population density of built-up area declines, owning to significant growth of built-up land and no significant change of population. These areas mainly locate around the central cities. In these areas, the speed of land urbanization is very rapid, but there is not enough attraction to population because of the lack of public service, poor transportation and some other reasons.

(3) Population density of built-up area declines, owning to decrease of population and growth of built-up land. As expected, counties located across Yangtze River economic belt south of Qinling-Huaihe line—which exhibit a decline in population density by census unit—also exhibit a decline in population density by built-up area. However, the area undergone a decline of population density by built-up area extends significantly comparing to that of population density by census unit.

(4) Population density of built-up area increases, owning to more significant increase of population than that of built-up land. These areas mainly locate in the northwest area and the core of big cities.

5. Conclusion and discussion

By integrating remote sensing and GIS technologies, the relationships between built-up surface and population are quantified. During 1990-2010, the rate of built-up surface growth in China mainland has outpaced the rate of population growth rate. This project results in a more detailed understanding of the relationship between change of land use and population. The analysis of population density patterns, by two different measures, census unit and built-up area, allow a better understanding of reducing of intensive land utilization in China mainland. The population density of built-up area offers a different measure from the traditional calculation of population density and provides a measure of living environment by reflecting the dynamics of both population and urban land use growth.

For the entire study area, population density has increased by 17.5% from 120 to 141 persons per kilometre during 1990–2010. In the meantime, the total built-up area has increased by 33.9% from 142,944 to 191,385 kilometres, while population per kilometre of built-up land has decreased by 12.0% from 7913 to 6964 persons per kilometre. This indicates that built-up land growth may have outpaced population growth in China mainland overall.

However, such changes are not evenly distributed in space and time. The rate of increase or decrease is much higher during the later decade than that during the earlier decade. For the change of built-up land, the increases in built-up surface don't have the characteristics of spatial aggregation. Correspondingly, most increases in population density are found in the northwest area of China mainland, divided by Aihui-Tengchong line, while most decreases are found in the southwest area, especially across Yangtze River economic belt south of Qinling-Huaihe line. The spatial change pattern of population density of built-up area is similar to that of population density of counties, but there exists more significant difference between northwest and southeast of China mainland. On the northwest side of Aihui-Tengchong line, the population density of built-up area is mostly increased, while on the southeast side, it is mostly decreased. Such changes reflects the nature of housing demand and the mechanisms that governed the urbanization process in China mainland.

Such analyses shed light on the overall impact of political-economic environment and policy changes on urbanization processes for China mainland. The information of land cover conditions and changes over time obtained through this study can be used in formulating management policies of the national and municipal governments.

Since the change pattern of population and built-up land are not the same in urban and in rural, we seek to assess changes in the spatial extent of built-up land and changes in population distribution patterns in urban and in rural respectively in the future work.

Acknowledgments
This study is supported by the major projects of high resolution earth observation system (03-Y20A04-9001-15/16), the "135" plan project of Institute of Remote Sensing and Digital Earth, Chinese Academy of Sciences (Y4SG0300CX) and National Natural Science Foundation of China (41601387). The authors would like to thank professor Zhao Xiaoli from the Institute of Remote Sensing and Digital Earth, Chinese Academy of Sciences, for providing built-up density maps.

References

[1] Chen J, Chen L, Li R, Liao A, Peng S, Lu N and Zhang Y 2015 Spatial distribution and ten years change of global built-up areas derived from Globeland30 *Acta Geodaetica et Cartographica Sinica* **44** 1181-8
[2] Wang L, Li C, Ying Q, Cheng X, Wang X, Li X, Hu L, Liang L, Yu L, Huang H and Gong P 2012 China's urban expansion from 1990 to 2010 determined with satellite remote sensing *Chin. Sci. Bull.***57** 2802-12
[3] Fang C 2009 *China's Urbanization Process and Resources Environment Protection Report* (Beijing: Science Press) p 30
[4] Wang J and Fang C 2011 Growth of urban construction land: Progress and prospect *Progress in Geography* **30** 1440–8
[5] Zhang L and Weng Q 2016 Annual dynamics of impervious surface in the Pearl River Delta, China, from 1988 to 2013, using time series Landsat imagery *ISPRS Journal of Photogrammetry and Remote Sensing* **113** 86-96
[6] Lillesand T M and Kiefer R W 1999 *Remote Sensing and Image Interpretation* (New York: John Wiley & Sons) p 724
[7] Hu L, Liu Y, Ren Y, Yu L and Qu C 2015 Spatial change of population density boundary in mainland China in recent 80 years *Journal of Remote Sensing* **19** 928-34

Effect of different forage species on the nitrogen uptake in Hulunbeir

Li- Jun Xu[1], Xing-Liang Xu[1], Xue-Juan Tang[1], Gui-Xia Yang[1], Zhao Zhang[1], Xiao-Ping Xin[1]*

[1] Institute of agriculture resources and regional planning, Chinese academy of agriculture sciences, No.12 Zhongguancun South Street, Haidian district, P.R.China.

Email: xinxiaoping@caas.cn

Abstract: Knowledge of determining factors for nitrogen uptake preferences and how they are modified in changing environments are critical to understand ecosystem nitrogen cycling and to predict plant responses to future environmental changes. However, it remains unclear in this aspect for the main managed grassland (Medicago sativa, Bromus inermis, Leymus chinensis) and crop (Brassica campestris) under field condition in Hulunbeir area of Inner Mongolia of China. Two 15N tracer experiments utilizing a unique differential labelled nitrogen source were employed in both managed grassland (M. sativa, B. inermis and L. chinensis) and crop (B. campestris) in Hulunbeir area of Inner Mongolia of China. Tracing both labels in the above- and below ground plant biomass, soil NH_4^+ -N or NO_3^- -N and NH_4^+ -N or NO_3^- -N uptake by plants. There were differences between soil NO_3^-N concentration and NH4+-N concentration, and NO_3^-N concentration was higher than NH_4^+-N concentration. NO_3^-N concentration was led by different grass species. The NH_4^+-N concentration in August were higher than in July on the whole, the highest value for B. campestris and the lowest for B. inermis. The plant N concentration in B. inermis, L. chinensis and B. campestris showed decreasing trend, its mean value decreased by 20.1, 47.9 and 26.7%, respectively, and M. sativa increased by 13.7%. Among the four species, the individuals exhibited a preference for $^{15}NO_3^-$, indicated by higher ^{15}N signatures in $^{15}NO_3$-treatment than in $^{15}NH_4^+$ treatment.

1.Introduction

Nitrogen (N) is an important limiting nutrient for net primary productivity in most terrestrial ecosystems (LeBauer & Treseder, 2008). Nitrogen is made available to plants through decomposition of soil organic matter (SOM), but microbes also require N for growth. Plantemicrobe competition for N can be intense when both the demand for N by plants and microbes to grow is high (Dunn et al., 2006; Månsson et al., 2009; Inselsbacher et al.,2010;Xu et al.,2011) affecting plant community composition, ecosystem N retention and carbon (C) sequestration (Zak et al.,1990; van der Heijden et al., 2008; Bloor et al., 2009; Averill et al., 2014). Ammonium (NH_4^+) and nitrate (NO_3^-) being the two major inorganic N forms available for plants (Mahmood & Kaiser 2003; Wang et al, 2011). Recent studies reinforce the idea that plants clearly prefer inorganic N forms over organic forms in most systems (Harrison, Bol & Bardgett 2007; Ashton et al. 2008), and the optimization of the plant preference of different forms of inorganic N presumably enhances the survivorship and the fitness of plants (Wang & Macko, 2011). Recent years, some scholars studied the N uptake characterizes mainly including forest species (Zhang et al., 2014), crops (Sauer et al. 2015; Cicek, 2015; Chakwizira, 2015), and alpine ecosystems (Xu et al., 2006; Miller & Bowman, 2002) and so on. Whereas there are only a few reports on grasses (Ramirez-Garcia1 et al., 2015; Schmitt et al., 2013), especially perennial grasses (*Medicago sativa, Bromus inermis, Leymus chinensis*) (Zebarth et al., 2015; Lv et al., 2013; Mitra & Mandal, 2012; Chen et al., 2003).

Grasslands, one of the most dominant vegetation types over the world, occupy about 40% of the land area in China (Chen, 2000). In recent years, the natural grasslands in China are experiencing degradation in different degrees. A sharp decline in the grassland productivity often occurs during the degradation, and further affects the animal husbandry. To chieve high quality forage supply, managed grasslands with cultivation have been developed in the severely degraded grasslands in China. Cultivation has merged to collectively affect the productivity of the Chinese grassland ecosystems. Till to 2013, the total area of managed grasslands with cultivation was increasing to 1.2×10^7 hm^2 (National animal husbandry, 2014). It was more and more important to select the suitable grasses species.

Inner Mongolia possesses a large area of temperate grasslands and some of them have been intensive managed. Hulunbeir located in northeast Inner Mongolia, China. The natural grassland in Hulunbeir City, with 88000 km^2 in area which covers 33.5% of the total area. Hulunbeir was a typical area, natural grassland degradation and managed grassland planted. At present, *Medicago sativa, Bromus inermis, Leymus chinensis* was mainly planted in these areas. The planted areas are enlarging year by year.

However, few studies have been carried out to investigate N uptake characteristics of soil and plants in different grass species and crop. Species differ in their ability to take up different forms of N, such partitioning may enable plants to efficiently utilize available soil N (Chapin et al., 1993; Leadley et al., 1997; Näsholm et al., 1998; Lipson et al., 2001). This paper reports how we met the gap through a field experiment in Inner Mongolia, China. Specifically, we assessed managed grassland & crop nitrogen uptake using stable nitrogen isotope in Hulunbeir Inner Mongolia, China. Experiments focusing on N dynamics was conducted at selected grasslands in Inner Mongolia in 2014. The hypotheses backing the experiments were: (1) plants prefer NO_3^- in the four species because it is a dominant inorganic N form in the temperate grasslands; (2) The preference for N form is affected by the soil NH_4^+/NO_3^- ratios.

2.Materials and methods

2.1. Research site

This study was conducted at Hulunbeir Grassland Ecosystem Observation and Research Station (N 49°19'35", E 119°56'52"), in the north-eastern Region of Inner Mongolia, China. The frost-free period is approximately 110 days in this area. The average precipitation is ranged from 350 mm to 400 mm from 2000 to 2010, about 80% of which falls between July and September. The annual mean air temperature ranged from -5 to -2°C, with the maximum monthly mean of 36.2°C in July and the minimum of -48.5 °C in January. The soil is chestnut soil or chernozem (Table 1). In this study, we selected three managed grasslands (Medicago sativa, Bromus inermis and Leymus chinensis) and one type of crop (Brassica campestris) as a reference. The plots were arranged in a split-plot design as the main treatment. Each plot consisted of four subplots (3×5m). The seedling amounts of the M. sativa, B. inermis, L. chinensis grasslands and B. campestris were 15.0, 22.5, 37.5,75.0 kg hm^{-2}, respectively. The managed grasslands were perennial grasses, and planted in 2010, B. campestris was annual crop and planted in 2014. No fertilizer was applied for all the experimental grasslands.

2.2.Experimental design

The experiment was done July and August in 2014, respectively. Each selected experimental samples was carefully injected as a water suspension into the soil of four plots at a depth of 5 cm using the method described by Schimel and Chapin (1996) for injecting amino acid solution. Four plots were used as control, by injecting the same amount of water. $NaNO_3$ (10%) and $(NH_4)_2SO_4$ (10%) inputs corresponded to 1.82 and 1.41g m^{-2}, firstly, the drug was dissolved in the distilled water, 45 ml once, and sprayed to soil, and another 45 ml distilled water was used to clean and then sprayed to soil again. The beginning time was calculated and 16 h later, the samples were taken back to laboratory, including the aboveground, root and soil samples.

Roots were carefully separated from soil cores. Plant material was separated from the aboveground and the root, and heated 90 s by microwave oven, and then was dried at 75℃ for 48 h, weighed for total dry biomass and ground for measurement of total N content, $^{15}N/^{14}N$ ratios. Soil samples were sieved to 2 mm to remove coarse fragments for measuring pH, $^{15}NH_4^+$ and $^{15}NO_3^-$.Total nitrogen (N) content was measured by dry combustion on a LECO CNS-1000 elemental analyzer (LECO Corporation, St.Joseph, MI) (Du & Gao, 2006).

2.3.Calculations and statistics

Plant ^{15}N uptake was calculated as the following:

$$\text{Plant } ^{15}\text{N uptake } [\mu g\ N\ g^{-1}\ \text{root}] = N\% \times \text{Biomass} \div 100 \times \text{APE} \div 100 \times 1000000 \tag{1}$$

$$\text{Plant uptake } NO_3^-\ [\mu g\ N\ g^{-1}\text{root}] = \text{Plant } ^{15}\text{N uptake} \div [NO_3^-]\text{soil} \times {}^{15}\text{Nadded} \tag{2}$$

$$\text{Plant uptake } NH_4^+\ [\mu g\ N\ g^{-1}\ \text{root}] = \text{Plant } ^{15}\text{N uptake} \div [NH_4^+]\text{soil} \times {}^{15}\text{Nadded} \tag{3}$$

$$\text{Plant } ^{15}\text{N uptake rate}[\mu g\ N\ g^{-1}\text{root } h^{-1}] = NO_3^- \text{ was absorbed by plants} \div \text{Biomass(BG)} \div t \tag{4}$$

And t means time from the beginning time of 15Nadded to the end of the samples taken away.

BG means belowground

The significance of differences in ^{15}N uptake as well as N contents between plant species was examined using one-way analysis of variance (ANOVA). Critical LSD values for 5% error probability were calculated. The standard errors of means are presented on the figures as a variability parameter. One-way ANOVA was used to test the difference in mean biomass, N content, ^{15}N uptake by plants between the treatments. Significance of all tests was accepted at $P < 0.05$.

3.Results

3.1.Biomass

Among the four species, plant biomass (aboveground + belowground) at the time of sampling was remarkably higher in *B. inermis* compared to the three other species, while plant biomass of *L. chinensis* secondary and *M. sativa* was lowest in July, and there was significant among four species (Fig. 1). *L. chinensis* is highest in August, but there is not significantly ($P>0.05$) (Fig.1). There were significant different in aboveground or belowground among four species both July and August ($P<0.05$). The R:S ratio(ratios of root to sheet) showed the similar trend, that is decreased after July on the whole. *B. inermis* (3.74 and 0.97) was higher than any other species. *L. chinensis* (1.57 and 0.52) secondary, *B. campestris* is the lowest, only 0.21(in July) and 0.07(in August).

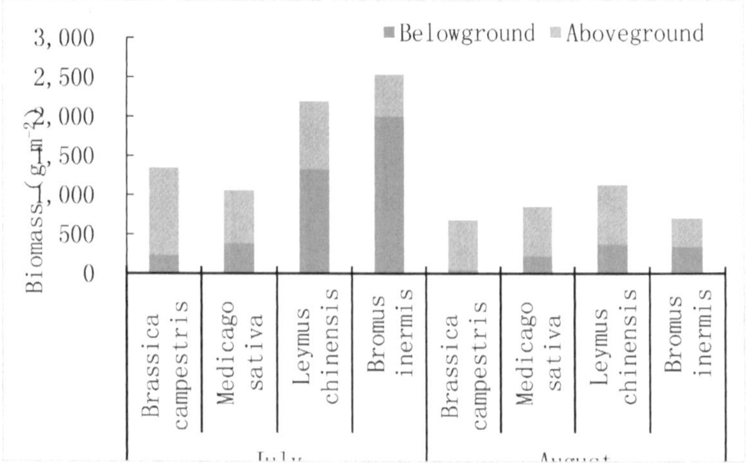

Figure1. The Biomass change in different treatments. The figure indicated the biomass change in July and August, biomass including aboveground and belowground. ANOVA method was used to analyze the difference of aboveground and belowground among four species, Values are means (±SE) of four replicates, corrected by those in the control plots. Letters indicate significant differences at 0.05 error probability level, small letters stand for the differences of grass species in same months, and capitals stands for the differences of months as to the same species.

3.2.Change of soil N concentration

There were differences between soil NO_3^--N concentration and NH_4^+-N concentration, and NO_3^--N concentration was higher than NH_4^+-N concentration. NO_3^--N concentration was led by different grass species. In July, NO_3^--N concentration in *M. sativa* was higher than any other grass species, which were 4.4, 1.8 and 2.5 times larger than in *B. inermis*, *L. chinensis* and *B. campestris*, respectively, and there was significant positive difference among four grass species ($P<0.05$) (Fig.2-a). In August, NO_3^--N concentration in *L. chinensis* was the highest, compared to in July, *B. inermis* and *B. campestris* have the tendency of increase, and *M. sativa* decreased, and significant among three species ($P<0.05$). The same specie was not significant between July and August ($P>0.05$).

The NH_4^+-N concentration in August were higher than in July on the whole, the highest value for *B. campestris* and the lowest for *B. inermis* in July, and the highest value for *B. campestris* were 1.6, 1.4 and 1.3 times larger than *M. sativa*, *B. inermis* and *L. chinensis*, and there was not significant difference between *B. campestris* and three other species, and there was no significant difference among these three species (*M. sativa*, *B. inermis*, and *L. chinensis*) (Fig. 2-b).

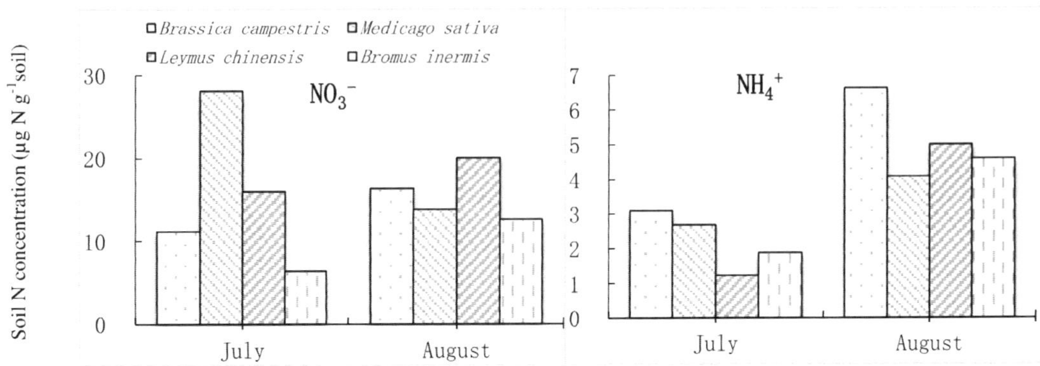

Figure2. Soil N concentration of different species in different months, including NO_3^- -N concentration and NH_4^+ -N concentration. Values are means (±SE) of four replicates, corrected by those in the control plots. Letters indicate significant differences at 0.05 error probability level, small letters stand for the differences of grass species in same months, and capitals stands for the differences of months as to the same species.

3.3.Change of plant N concentration

The plant N concentration in *B. inermis*, *L. chinensis* and *B. campestris* showed decreasing trend, its mean value decreased by 20.1, 47.9 and 26.7%, respectively, and *M. sativa* showed different trend compared to three other species, the plant N concentration showed increasing trend from July to August, its value increased by 13.7%. Among the four species, the plant N concentration was higher in *M. sativa* compared to the three other species, while the plant N concentration of *B. inermis* and *L. chinensis* was similar, and the lowest. *B. campestris* was secondary. There was significant different among four species in July and August ($P<0.05$). *M. sativa* and *B. inermis* were not different in different months, but *L. chinensis* and *B. campestris* not.

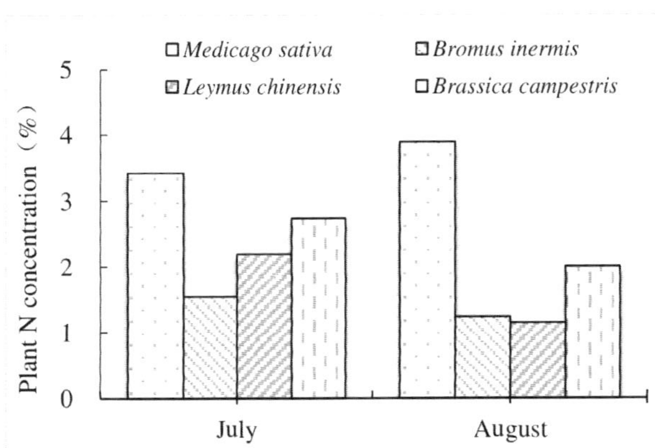

Figure3. Change of plant N concentration. Values are means (±SE) of four replicates, corrected by those in the control plots. Letters indicate significant differences at 0.05 error probability level, small letters stand for the differences of grass species in same months, and capitals stands for the differences of months as to the same species.

3.4.Change of plant N uptake rate

Among the four species, the individuals exhibited a preference for NO_3^-(Fig.3), indicated by higher ^{15}N signatures in $^{15}NO_3^-$treatment than in $^{15}NH_4^+$ treatment. On the whole, the plant N uptake rate to NO_3^- in July was higher than in August. And there was significantly different in *M. sativa* from July to August, decreased by 64.4% ($P<0.05$), but three other species uptake preferences did not obviously change ($P>0.05$).

Whereas four species had also a preference for NH_4^+, three species (*B. inermis*, *L. chinensis* and *B. campestris*) generally showed the same trends, which the plant N uptake rate to NH_4^+ in August was higher than in July, and there were significantly different in different months and species($P<0.05$), increased by 5.8, 5.2 and 8.8 times. But *M. sativa* showed different trend, decreased by 29.4%.

4.Discussion

4.1.Species preferred to NO_3^-

N mineralization and nitrification should be different for organic matter derived from different sources, i.e. soil, litter or added algae. Whatever the source material is, ammonium and nitrate produced are liberated into the soil solution from which they are taken up by the plants in an identical way (Xu et al.,2006). We studied the change of nitrogen content and soil ammonium and nitrate concentration of the four species, the results indicated that nitrogen content in *M. sativa* was higher than any others (Fig.3). M. sativa is legume, N fixed by nodule, which can improve available nitrogen content. From the flowering period to the fruit period, the nitrogen content of M. sativa is increasing.

Different plants showed different preferences for N (NH_4^+ and NO_3^-), a multitude of studies have shown that most plants prefer NO_3^-–N to NH_4^+–N (Britto and Kronzucker,2002; Marschner,1995). Azarmi and Esmaeilpour (2010) observed that cucumber grown in a 25:75 NH_4^+: NO_3^- solution had the highest total fruit yield. Strawberry has also been found to produce the highest vegetative growth and fruit yield even at 50:50 NH_4^+: NO_3^- (Tabatabaei et al., 2008). These studies imply that different plant species may have different optimal NH_4^+: NO_3^- ratios. We studied the change of nitrogen content of the four species, the results indicated that *M. sativa* was higher than any others. M. sativa is legume, N fixed by nodule, which can improve available nitrogen content.

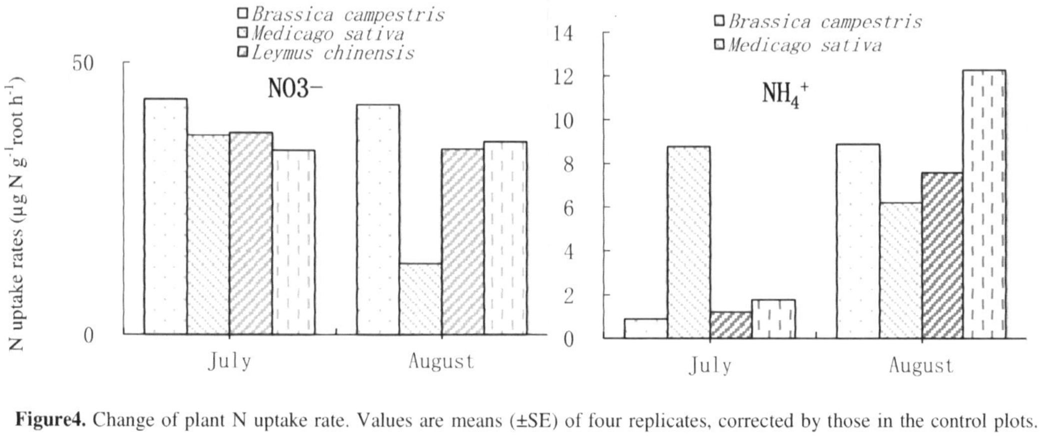

Figure4. Change of plant N uptake rate. Values are means (±SE) of four replicates, corrected by those in the control plots. Letters indicate significant differences at 0.05 error probability level, small letters stand for the differences of grass species in same months, and capitals stands for the differences of months as to the same species.

Acknowledgement

The authors thank Qing Ge-le for assistance in sample measurements. This research was supported by "973" Program (2015CB150800),public sector projects in the Ministry of Agriculture (201303060), China agriculture research system **"**China agriculture research system" (CARS-35).

References

[1] Ramirez-Garcia J Martens H J Quemada M and Kristian Thorup-Kristensen K 2015 *Intercropping effect on root growth and nitrogen uptake at different nitrogen levels* vol 8（Journal of Plant Ecology）p 380–389

[2] Andreas S A Pausch J and Kuzyakov Y 2013 *C and N allocation in soil under ryegrass and alfalfa estimated by ^{13}C and ^{15}N labeling* vol 368 (Plant Soil) p 581–590 DOI 10.1007/s11104-012-1536-5

[3] Mahmood T and Kaiser W M 2003 *Growth and solute composition of the salt-tolerant kallar grass [Leptochloa fusca (L.) Kunth] as affected by nitrogen source* vol 252 (Plant and Soil) p 359–366

[4] LeBauer D S and Treseder K K 2008 *Nitrogen limitation of net primary productivity in terrestrial ecosystems is globally distributed* vol 89 (Ecology) p 371–379

[5] Xu X L 2003 *The fate of ^{15}N-labelled ammonium and nitrate, nitrogen deposition and carbon sequestration, and uptake of organic nitrogen by plants in alpine meadows in the Qinghai-Tibet Plateau* (Institute of Geographic Sciences and Natural Resources Research)

[6] Xu X L Ouyang H Kuzyakov Y Richter A and Wanek W 2006 *Significance of organic nitrogen acquisition for dominant plant species in an alpine meadow on the Tibet plateau, China* vol 285 (Plant Soil) p 221–231

[7] Wang L X and Macko S A 2011 *Constrained preferences in nitrogen uptake across plant species and environments* vol 34 (Plant, Cell and Environment) p 525–534

[8] Wang L X and Macko S A 2011 *Constrained preferences in nitrogen uptake across plant species and environments* vol 34 (Plant, Cell and Environment) p 525–534 doi: 10.1111/j. 1365-3040. 2010. 02260.x

[9] Ashton I W Miller A E Bowman W D and Suding K N 2008 *Nitrogen preferences and plant-soil feedbacks as influenced by neighbors in the alpine tundra* vol 156 (Oecologia) p 625–636

[10] Harrison K A Bol R and Bardgett R D 2007 *Preferences for different nitrogen forms by coexisting plant species and soil microbes* vol 88 (Ecology) p 989–999

[11] Sauer T J Coblentz W K Thomas A L Brye K R Brauer D K Skinner J V Brahana J V DeFauw S L Hays P D Moffitt D C Robinson J L James T A and Hickie K A 2015 *Nutrient cycling in an agroforestry alley cropping system receiving poultry litter or nitrogen fertilizer* vol 101 (Nutr. Cycl. Agroecosyst) p 167–179 DOI 10.1007/s10705-014-9667-0

[12] Cicek H Thiessen Martens J R Bamford K C and Entz M H 2015 *Late-season catch crops reduce nitrate leaching risk after grazed green manures but release N slower than wheat demand* vol 202 (Agriculture, Ecosystems & Environment) p 31–41

[13] Chakwizira E de Ruiter J M and Maley S 2015 *Effects of nitrogen fertiliser application rate on nitrogen partitioning, nitrogen use efficiency and nutritive value of forage kale* vol 58 (New Zealand Journal of Agricultural Research) p 259–270 http://dx.doi.org/10.1080/ 00288233. 2015.1016538

[14] Zhang J J Li Y F Chang S C Jiang P K Zhou G M Juan Liu J Wu J S and Shen Z M 2014 *Understory vegetation management affected greenhouse gas emissions and labile organic carbon pools in an intensively managed Chinese chestnut plantation* vol 376 (Plant Soil) p 363–375 DOI 10.1007/s11104-013-1996-2

[15] Miller A E and Bowman W D 2002 *Variation in nitrogen-15 natural abundance and nitrogen uptake traits among co-occurring alpine species: do species partition by nitrogen form?* Vol 130 (Oecologia) p 609–616 DOI 10.1007/s00442-001-0838-8

[16] Zebarth B J McDougall R Neilsen G and Neilsen D 2015 *Availability of nitrogen from municipal biosolids for dryland forage grass* (Can. J. Plant Sci.) 575–583

[17] Lv X T Reed S S Yu Q He N P Wang Z W and Han X G 2013 *Convergent responses of nitrogen and phosphorus resorption to nitrogen inputs in a semiarid grassland* vol 19 (Global Change Biology) p 2775–2784 DOI: 10.1111/gcb.12235

[18] Chen S K Edwards C A and Subler S 2003 *The influence of two agricultural biostimulants on nitrogen transformations, microbial activity, and plant growth in soil microcosms* vol 35(Soil Biology & Biochemistry) p 9–19

[19] Mitra B and Mandal B 2012 *Effect of nutrient management and straw mulching on crop yield, uptake and soil fertility in rapeseed (Brassica campestris)–greengram (Vigna radiata)–rice (Oryza sativa) cropping system under Gangetic plains of India* vol 58 (Archives of Agronomy and Soil Science) p 213–222

The Cluster of City Buildings Based on the SOM Neural Network

Z. Wang [1], Q. Liu [1]

1. School of Resources and Environment, University of Electronic Science and Technology of China, 611731 Chengdu, China

E-mail: moniquewang@foxmail.com

Abstract. The clustering of buildings on the city map is an important field of cartographic and the main step to resolve issues related to the scale of the city map transformation process. The development of computer technology promotes the studies of cartographic and modelling algorithms. In order to achieve automatic clustering of map elements, the domestic and foreign scholars have proposed numerous automatic clustering algorithms. Although each algorithm works for specific problems, all algorithms are based on the graphical analysis transformation of building polygons on the map. By reading and analyzing current development of domestic and foreign researches, we find out that map building clustering needs more inspiring interactive features and automatic integrated decision-making simulation functions. It needs to improve the visualization software and integrated automatic intelligence, increase the proportion of the batch to support implementation of automated cartographic generalization to a higher level.

1. Introduction

Recently, with the growth of geographic information industry, the social demand of digital map is also increasing rapidly. Therefore, it is facing a higher challenge for the manual cartography. Map generalization is an important study area of cartography. It refers to the clustering of various elements, the operations such as mergers and simplification. The map generalization is essential to make a suitable map that can satisfy the requirements of the appropriate scale and actual needs when the map scale changes. The realization of the automatic map building clustering is the main step to provide conditions for the consolidation and simplification of cartography. It is not only conducive to streamline manual drawing operations and increase the productivity [1], but also it provides a technical support for scientific research in the field of cartography.

2. Methodology

2.1. SOM neural network

SOM refers to Self-Organizing Mapping, and SOM neural network algorithm is an unsupervised type of neural network algorithm [2]. It has been proposed by Teuvo Kohonen in 1981. The SOM algorithm only needs a few sample data and it implements automatic data clustering through clustering algorithm parameter settings [3]. Therefore, the algorithm has a high degree of intelligence. Typically, we use the competitive learning and self-organizing clustering feature of the SOM neural network to find the inherent similarity of the target experimental data. Without the requirement of human priori knowledge constraints or parameters, the algorithm can achieve automatic clustering. Clustering results are completely in accordance with the internal data similarity, and it is not subject to external constraints. Moreover, it can reflect the real characteristics of the data very well. The two-dimensional structure of SOM neural networks [4] is shown in Figure 1.

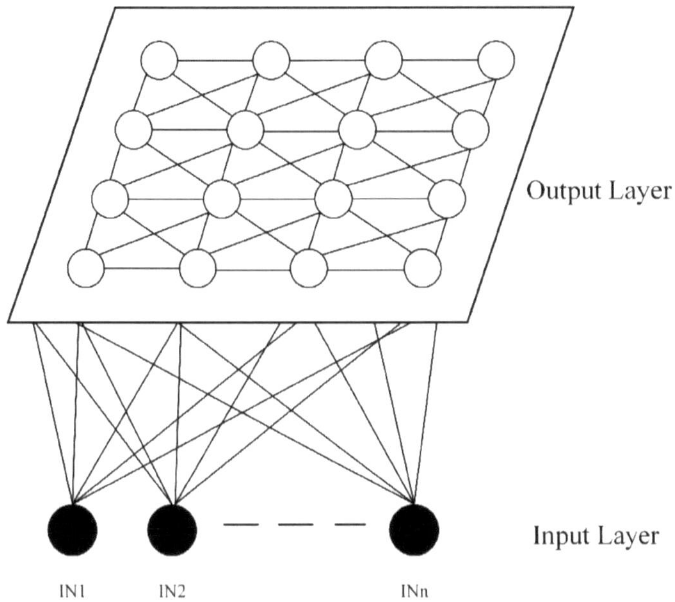

Figure 1. Structure of SOM neural networks

The method of using SOM network algorithm to achieve the building clustering is proposed by Boyan Chen [5]. This method uses the coordinates of the gravity centres of the buildings to characterize the feature of building positions. By using the property of SOM network, buildings with close coordinates of gravity centres can be clustered as one group.

2.2. Introduction of building index

According to the understanding and recognition of building polygon features, this paper sets up 3 building indexes to extract characteristics of the buildings on the vector map [6]. After expressing different factors to characterize different features of the buildings, the effects of various factors on the course of realization of SOM clustering have been analysed and discussed.

2.2.1. Building area index. The building area index is defined and it refers to a single building polygon area. We directly use the ESRI's packed area calculation application programming interface to call the calculations. The building area index is added during the clustering process, as community buildings in the standardized town are always of similar building areas. Thus, standardized buildings will be clustered together in the artificial drawing process. The clustering process based on the building area index is shown in Figure 2.

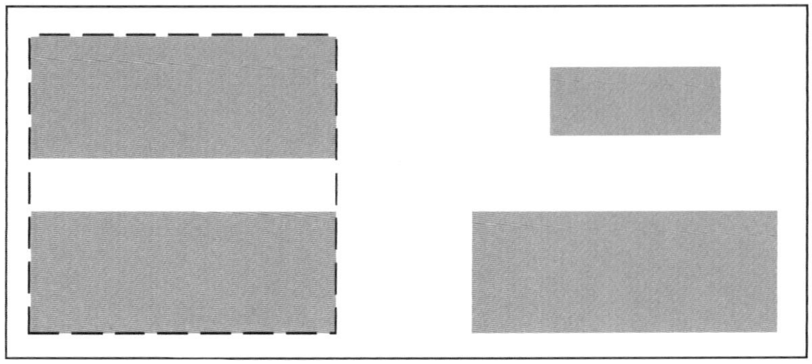

Figure 2. The building area index

2.2.2. Building density index. We make the current building's center of gravity as the center of a circle and calculate the total area of the buildings within a radius R of the circular. The definition of building density index is the area ratio, and the expression is shown in the formula 1. In the formula 1, the *Dens* refers to the density index, the *SAT* refers to the total area of the internal buildings in the circle, and *R* represents the radius of the circle. In addition, the definition of radius *R* is set in accordance with the scope of the map sheet in this paper. With the specific dimensions related to the expression shown in the formula, the density calculating method is shown in the formula 1.

$$Dens=\frac{S_{AT}}{\pi\times R^2} \tag{1}$$

The radius formula is shown in formula 2, where the *MapS* represents the map sheet range. The calculation results are obtained by the ESRI GIS .axMapControl1.MapScale interface.

$$R=\frac{5.0\times MapS}{1000.0} \tag{2}$$

2.2.3. Building shape index. Building shape index refers to a flat rate of building polygons. The expression of building polygon shape index is shown in formula 3, where *l* and S are the perimeter and the area of the building polygon shape, respectively. The more flat of the polygon is, the smaller of the shape index of the building will be. There are three types of building polygon shapes, including compact type, circle and expanded form. Circle is the most special example of polygons, and its shape index is 1. If the shape factor is less than 1, we call the building is compact. If the shape factor is more than 1, the building is refer to the expanded form. Even if two polygons have the same building area, their shape factor could be very different. Introducing the shape index aims at improving the limitations of the area index. If the sizes are similar between the buildings and the differences between the shape factors are large, we need to consider whether to implement clustering though the building shape index. Besides, the influence of the shape index is shown in Figure 3.

$$Shp=\frac{l}{2(\pi\times S)^{1/2}} \tag{3}$$

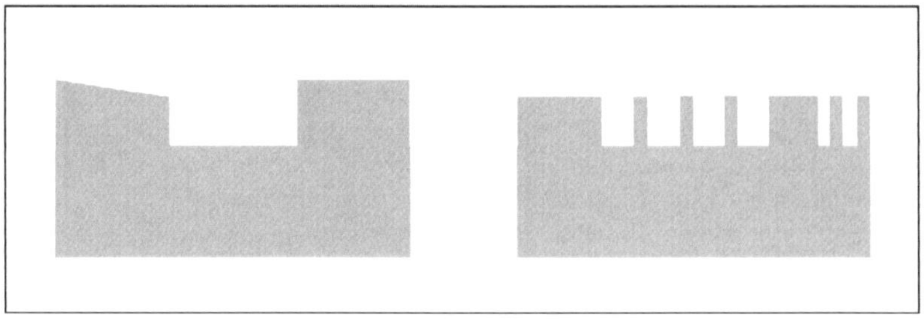

Figure 3. The building shape index

2.2.4. The improved algorithm of building clustering. The experiment of building clustering using building index is divided into two steps. A preliminary clustering is implemented in accordance with the gravity centers of buildings. , Then the secondary fine clustering is implemented based on the preliminary clustering result by considering other three index parameters. Before proceeding fine clustering, we need to normalize all three indexes [7]. This section carries on separate experiments with different indexes. According to the results of the experiments, the suitability of the indexes for building SOM clustering implementation is determined. The experimental flow chart is shown in Figure 4.

The improved algorithm is implemented by the secondary development of GIS based on C#. First of all, the improved algorithm needs to load the attribute information before the official building clustering. The attribute information includes various information of buildings and the details of the roads [8]. Second, the information is written into the attribute table of the map's shp files. After creating the initial clustering attribute and a secondary attribute clustering result, the grouped information is written among the attributes of each building. Based on the demand of the experiment, five new properties, including other indexes information besides the coordinate information are created for the building attribute table. After exporting the property sheet shp file, a simple txt file with only three new indexes is created. Secondary clustering algorithm reads the data from the generated txt file which can be considered as input of the SOM neural network.

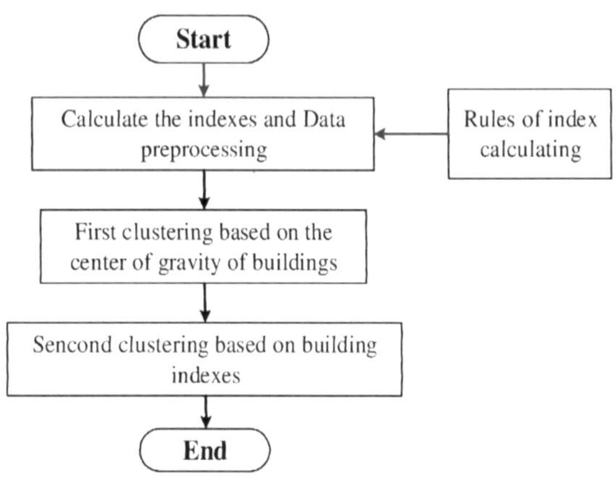

Figure 4. The flow chart of clustering

3. Results

We choose areas in downtown Boston with intensive-distribution buildings for experiment (Source: http://gis.dallascityhall.com/homepage/shapezip.htm). The size of cell of competitive layer in SOM neural network is set to 40 * 40, depending on the amount of buildings, and the total iteration is 10000. The clustering results of the build-up area is much better than the normal areas, benefitting from the suitable property of centre coordinate factor to describe feature distribution, especially under the road's constraints. However, from the results shown in Figure 5(a), we can see that the buildings of the eastern and northern part of the map sheets are clustered together due to the inconspicuous effects on gravity factor from the existing road constraints.

In the following experiment, we add area of the building for re-clustering. Clustering is to divide finely based on the original groups. The number of iterations and the size of competitive layer's cell remains unchanged. Results are shown in Figure 5(b). Compared with the building area index in Figure 5(a), the buildings have been divided well and excessive clustering results have been avoided. However, mixed clustering results have been generated in the southeast of the building map. The reason for this is the areas of these buildings are relative close. During the regrouping, that random assignment of values would have a certain impact on the original results.

As to the results of density factor, the fine clustering results are still based on the performance of the centre of buildings' gravity coordinates. We remain the parameter settings unchanged. The results are shown in Figure 5(c). As we can see, the building density is indeed performing better in built-up areas. Especially in the eastern and northern part of the map area, blocks with the similar buildings densities are clustered together. The blocks next to them with different densities are automatically grouped. The clustering itself maintains the initial clustering effect that use the centre coordinates of the buildings during the overall clustering process.

To avoid the limitation of area index, shape index is put forward. Shape index is the flat-ratio of building polygon in essence, and its clustering result is shown in Fig 5(d). In this map, we get local fine clustering for the building polygon with shape index.

(a) (b)

(c) (d)

Figure 5. The result of clustering

4. Conclusion

Clustering is achieved by utilizing the internal similarity of data [9]. This paper uses the morphological features of building polygon to get fine building clustering result. In order to show the different features of building polygon adequately, we define some indexes by descripting building polygon status, including area index, density index and shape index in detail. Then the 3 indexes are used to get a fine building clustering in vector map by using self-organizing neural network algorithm respectively. The results show that in the context of a wide range of initial clustering, fine clustering is feasible and can meet the needs of manual cartography [10].

5. References

[1] Liu Q and Cheng B 2006 *J. Geo-spatial Information Science* **9** 206
[2] Vesanto J, Alhoniemi E 2000 *J. IEEE Transactions on Neural Networks* **11** 586
[3] Hagenbuchner M, Sperduti A and Ah Chung Tsoi 2003 *J. IEEE Transactions on Neural Networks* **14** 491
[4] Bação F, Lobo V and Painho M 2005 *J. Computers & Geosciences* **31** 155
[5] Cheng B, Liu Q and Li X 2013 *J. Communications in Computer and Information Science* **399** 296
[6] Hequn Y, Yinming Y and Shengmao Z *International Conference on Geoinformatics (IEEE)* pp 1-5
[7] Argyrou A 2009 *Advances in Self-Organizing Maps (Springer Berlin Heidelberg)* pp 19–27
[8] Zhang LJ, Zhang F 2012 *J. Advanced Materials Research* **479-481** 640
[9] Shekhar S and Xiong H 2008 *Encyclopedia of GIS (Springer Nature)* pp 68–71
[10] Sester M 2008 Self-Organizing Maps for Density-Preserving Reduction of Objects in Cartographic Generalization *Self-Organising Maps (Wiley-Blackwell)* pp 107–120

Cloud removing method for daily snow mapping over Central Asia and Xinjiang, China

Xiaoqi Yu[1,2], Yubao Qiu[1], Huadong Guo[1], Lijuan Chen[1,3]

1 Institute of Remote Sensing and Digital Earth, Laboratory of Digital Earth Science, Dengzhuang South Road, Haidian District, Beijing, China, 100094, qiuyb@radi.ac.cn

2 College of Geographical Information System, Liaoning Technical University, Fuxin, China,123000

3 College of Geographical Science, Fujian Normal University, Fuzhou, ChinaType the author addresses here

Abstract. Central Asia and Xinjiang, China are conjunct areas, located in the hinterland of the Eurasian continent, where the snowfall is an important water resource supplement form. The induced seasonal snow cover is vita factors to the regional energy and water balance, remote sensing plays a key role in the snow mapping filed, while the daily remote sensing products are normally contaminated by the occurrence of cloud, that obviously obstacles the utility of snow cover parameters. In this paper, based on the daily snow product from Moderate Resolution Imaging Spectroradiometer (MODIS A1), a cloud removing method was developed by considering the regional snow distribution characteristics with latitude and altitude dependence respectively. In the end, the daily cloud free products was compared with the same period of eight days MODIS standard product, revealing that the cloud free snow products are reasonable, while could provide higher temporal resolution, and more details over Center Asia and Xinjiang Province.

1. Introduction

Central Asia and Xinjiang is one of the main snow area in the middle and low latitude. Snow and snow water resources plays an important role in the feedback regulation and supply of climate and water cycle in Central Asia and Xinjiang Province.

At present,optical remote sensing monitoring of snow cover interferenced by the weather, the cloud has the greatest impact on snow monitoring in the study area[1]. A lot of research based on MODIS Snow Cover products to the cloud removing has done, integrated Central Asia with unique geographical features, it's necessary to combined with a variety of algorithms for cloud removal. Firstly, the temporal and spatial distribution characteristics of snow cover are analyzed, then using the mountain-plain partitioned algorithm, formed a set of cloud algorithm processing of daily snow cover products suitable for Central Asia and China's Xinjiang region.The elevation and study area conditions are showed by the following figure.

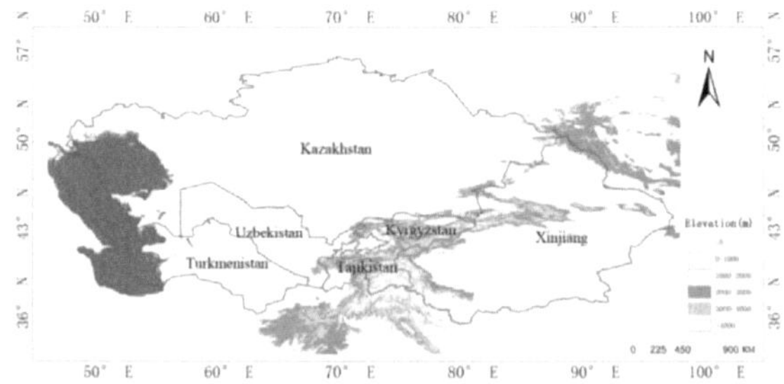

Figure 1. Distribution of Central Asia and Xinjiang region.

2. Data and characteristic analysis

MODIS snow product used in the study is based on the NDSI threshold method to get the daily global snow cover products. The V005 version of Terra MODIS and Aqua MODIS daily snow classification products MOD10A1 and MYD10A1 were download by NSIDC website. Digital elevation model DEM (Elevation Model) is selected by the USGSEROS data center where responsible for data archiving and reporting SRTM_DEM elevation data to the public.

Currently there are many cloud algorithms, Andreas et al[2] carrying out the study on the applicability of the different algorithms, including: MODIS daily combination, adjacent temporal deduction, near four pixels method, near eight pixels method, elevation filtering, long time series method in order to deal with the cloud. In this paper, based on the near eight pixels method to analyze characteristics. And try to improve the snowline method.

2.1. Variation of the snowline characteristics with altitude dependency

In Central Asia, there was less snow distributing in the valleys and basins, the article will divide the elevation into three segments, respectively, less than 0,0-2000meters, and 2000-8429meter.

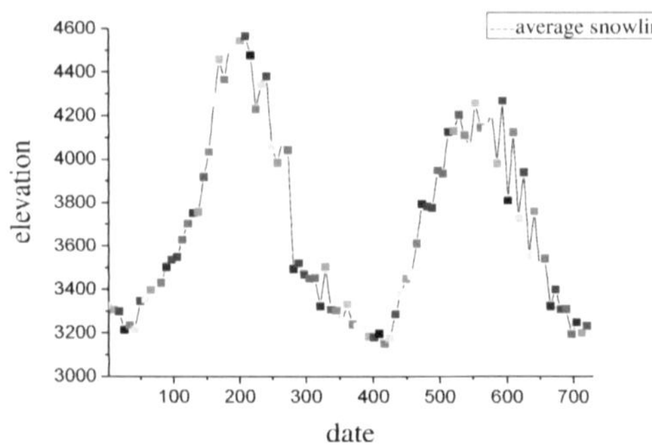

Figure 2. The average figure over mountain area.

Form eight days maximum snow products in 2010,2011 years,seeking average snowline of 2000-8429 meters, getting the change rule of mountain snowline for two years. (The horizontal coordinates of 100 shows 100th days in 2010, 400 shows 35th days in 2011).In 200-250, 580-650 days, elevation level fluctuation, which indicates that there is a phenomenon of sporadic snowfall and melting in the period of melting snow, but the snow accumulation season is a continuous snowfall, there is no melting phenomenon.

2.2. Variation of the snowline characteristics with latitude dependency

According to eight days maximum snow cover products in 2010, in Central Asia and China's Xinjiang region, when statistical longitude is 45 to 96 degrees and there are most land pixel value below snow pixels, statistics the latitude of variation range of snow pixel values, the results as shown in figure 3, the date of 2010 first to 365th days, straight line represents no snow.

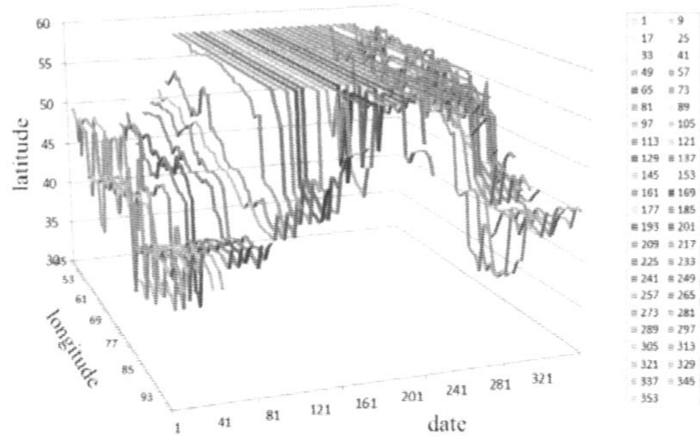

Figure 3. The snowline latitude curve.

Every 8 days to get a snowline of latitude, the latitude snowline range is approximately 30 to 58 degree, the range of 1-105 days and 297-365 days (i.e. integrable snowy season), longitude roughly in the range of 0 to 74, most of the terrain is plain areas, the latitude is linear, that is, there is no snow in the lower elevation areas in the summer, longitude in the range of 75 to 96, latitude curve fluctuation, contains mountains and high altitude areas where snow does not melt completely, so snowline of latitude lower than the plain areas.

3. Daily cloud-free algorithm

According to the snowline changes of latitude and elevation mentioned above, we draw a feasible scheme as figure 4 shows.

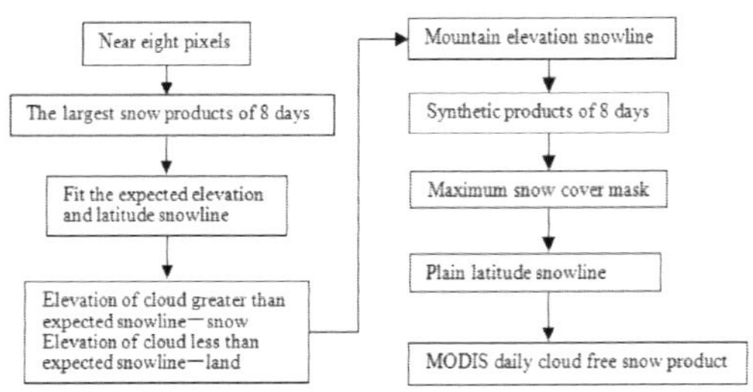

Figure 4. Cloud removal algorithm flow chart.

3.1. Snowline algorithm based on elevation over mountain area

Extraction of 2000 meters above the area, to deal with the cloud by the average snowline. Snowl method based on: 1) the same scene images,the cloud height is greater than or equal to all the snow pixel average elevation, cloud pixels are classified as snow; 2) the cloud pixel height less than or equal to all land pixel average elevation, cloud pixels are classified into land; 3) the cloud height between all the snow pixel average elevation and land elevation, cloud pixels are classified into a piece of snow. We using elevation snowline method to remove some cloud based on the near eight pixels data,and figure 5 is the comparison chart of near eight pixels and snowline processing.

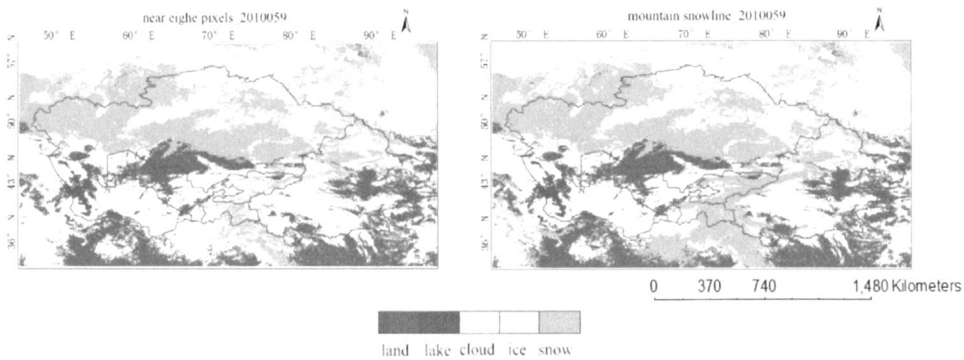

Figure 5. Comparison between near eight pixels and elevation snowline method.

3.2. Snowline algorithm based on latitude over plain area

Based on the products which are obtained by the mountain elevation snowline algorithm, 8 days products synthesized of maximum snow and land, get 8 days maximum difference of snow and land. According to the known latitude snowline on plain area,dealing with the maximum snow cover film products by latitude snowline method. Judge: above the latitude snowline and altitude is greater than 0, the cloud pixel assigned to snow. Under the latitude snowline, altitude area greater than 0, the cloud pixel assigned to land. Then the cloud is completely removed, and getting MODIS daily cloud-free snow products. Figure 6 is the comparison chart of maximum snow mask and latitude snowline method.

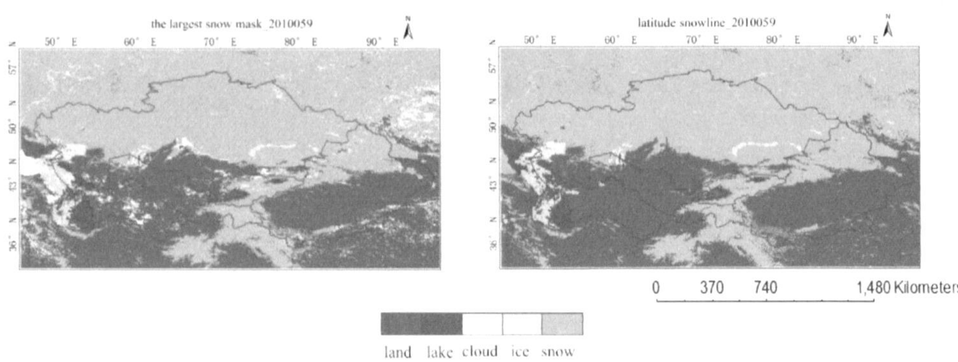

Figure 6. Comparison between the largest snow mask and latitude snowline.

4. Comparative analysis and verification

In order to assess and validate the snow classification accuracy, the snow cover area precision indexes are used to evaluate the snow classification for each step , indicators include the total classification accuracy (Accuracy), Recall, snow classification accuracy (Precision) and F value.The total accuracy represent the probability that a pixel is classified correctly, but when the image has large tracts of no snow regions, this index may be misleading, if the image is only 10% snow pixels, then the total accuracy will reach 90% because most pixels are divided into land . Recall is the probability of detection of snow pixels, i.e. a ratio between correct classification of snow pixel number and the actual number of all snow sample pixels. Precision is the ratio between correct classification of snow pixel number and classification of all the snow pixel. Recall and Precision are correlative interplay, high recall rates show that the correct classification of snow pixel number has high accuracy compared with the ground stations, and the high precision indicate that correct classification of snow pixels has high accuracy relative to the total snow pixels number of MODIS. Recall is high, does not mean that the Precision is higher, so using F value , perhaps the most useful indicators, balances Recall and Precision.

There are 363 temporal images from October 1,2012 to February 28,2014 in Central Asian region, and daily snow depth data provided by 103 ground stations, through the confusion matrix (Table 1)and formula (1), (2), (3), (4) to evaluate the snow classification image, analysis the precision of MODIS (MOD10A1, MYD10A1) data under the clear weather and the synthetic products in the process of cloud removing, considering the following samples: 1) the sample number a shows snow both in MODIS and ground station records (snow depth >0) ; 2) b shows that snow in ground station records but not in MODIS; 3) c shows that snow in MODIS but not in ground station records ; 4) d shows no snow in either the MODIS or ground station records.

Table 1. Confusion matrix of accuracy verification.

MODIS / stations	snow	land
snow	a	b
land	c	d

The index formula of snow accuracy is as follows:

$$Accuracy = \frac{a+d}{a+b+c+d} \times 100\% \tag{1}$$

$$\mathrm{Re}call = \frac{a}{a+b} \times 100\% \tag{2}$$

$$\mathrm{Pr}ecision = \frac{a}{a+c} \times 100\% \tag{3}$$

$$f = 2\frac{\mathrm{Pr}ecision \times \mathrm{Re}call}{\mathrm{Pr}ecision + \mathrm{Re}call} \times 100\% = \frac{2a}{2a+b+c} \times 100\% \tag{4}$$

The number of samples of raw data MOD10A1 ,MYD10A1 and each process product in the confusion matrix is shown in table 2.

Table 2. Number of samples in raw data and each process.

No.	product	a	b	c	d	a3	b3	Accuracy(%)		Recall(%)		Precision(%)		F value(%)	
								All station	Snow depth>3cm	All station	Snow depth>3cm	All station	Snow depth>3cm	All station	Snow depth>3cm
1	MOD10A1	1525	286	1404	14307	1364	125	90.35	91.11	84.21	91.61	79.72	77.85	81.90	84.17
2	MYD10A1	1237	193	1276	13761	1171	67	91.08	91.75	86.5	94.59	79.09	78.17	82.63	85.60
3	MOYD10A1	2065	340	1921	16549	1874	161	89.17	89.85	85.86	92.09	81.17	79.64	83.45	85.41
4	adjacent temporal deduction	2495	399	2253	19092	2253	186	89.06	89.75	86.21	92.37	82.13	80.58	84.12	86.07
5	near four pixels method	2560	404	2285	19229	2305	186	89.01	89.71	86.37	92.53	82.26	80.68	84.27	86.20
6	near eight pixels method	2975	404	2587	19229	2656	186	88.13	88.75	88.04	93.46	83.57	81.95	85.75	87.33
7	elevation snowline	3219	416	3053	19468	2768	186	86.74	87.29	88.56	93.7	83.35	81.15	85.87	86.98
8	maximum snow and land mask	5414	819	4784	24276	4708	385	84.12	84.87	86.86	92.44	81.04	78.79	83.85	85.07
9	latitude snowline	6091	895	5560	24615	5201	409	82.63	83.32	87.19	92.71	81.18	78.65	84.07	85.10

The cloud removal algorithm finally removed all cloud pixels by gradually controlling the cloud cover amount and the loss of precision, considering the factors of terrain features in the area and the correlation of snow in time and space, and satisfactory results were obtained. The ground stations observation data is used to validate the accuracy of MODIS daily cloudless snow products. Among them, Accuracy was 82.63%, Precision was 81.18%, Recall rate was 87.19%, F value was 84.07%. When the snow depth is small, MODIS Snow classification accuracy is not high, if only retain snow samples that snow depth greater than 3cm , the classification accuracy is greatly improved and can be used for the dynamic monitoring of snow cover area in Central Asia.

There are large area and complex terrain over Central Asia and Xinjiang,China, and the plateau snow distribution has a certain randomness and uncertainty. The snowline algorithm based on elevation in the mountain and latitude in the plain used to cloud removing, its uncertainty may be affected by the terrain, temperature, precipitation and other factors. Gafurov[3] considered the slope factors when using snowline method, there are more sunlight in sunny slope, because of the differences of local water sources and climate in different regions, the high altitude pixel located in sunny slope is land, adjacent low altitude pixel located in shady slope is snow. These conditions without considering may cause the cloud removal algorithm in the study area have error. 8 days maximum snow or land area mask discrimination method using snow cover information correlation in short time, could remove most of the cloud and the precision is higher, but still can not accurately return to the original situation of the day's snow cover, and a little miscarriage of justice arose. Therefore, developing more perfect cloud removal algorithm and the method to verify the precision is of great significance for the real-time dynamic monitoring of the snow and analyzing the effects of snow cover on climate.

5. Comparative analysis and verification

This work is supported in part by National Natural Science Foundation of China (Grant No. 41371351 and 41120114001) and Chinese Special Fund for Meteorology-Scientific Research in the Public (GYHY201206040).

References

[1] Xiaodong H and Xiaohua H 2012 *Cloud removal methodology from MODIS daily snow product* vol 34(Beijing : Journal of glaciology and geocryology) No.5

[2] Andreas J D and Claudia K 2012 *Remote sensing of snow –a review of available methods* ed U Gessner and S Dech (International Journal of Remote Sensing) p 4094-4134

[3] Gafurov A and Bardoyy A 2009 *Cloud removal methodology from MODIS snow cover product* (Hydrology and Earth System Sciences) p 1361-1373

[4] Parajka J and Pepe M 2010 *A regional snow-line method for estimating snow cover from MODIS during cloud cover* (Journal of Hydrology) p 203-212

Precipitation Retrievals in typhoon domain combining of FY3C MWHTS Observations and WRF Predicted Models

HE Jieying[1], ZHANG Shengwei[1], LI Na[1,2]

[1].Key Laboratory of Microwave Remote Sensing, National Space Science Center, Chinese Academy of Sciences, Beijing 100190, China
[2] University of Chinese Academy of Sciences, Beijing 100190, China

hejieying@mirslab.cn

Abstract. A passive sub-millimeter precipitation retrievals algorithm is provided based on Microwave Humidity and Temperature Sounder (MWHTS) onboard the Chinese Feng Yun 3C (FY-3C) satellite. Using the validated global reference physical model NCEP/WRF/ VDISORT), NCEP data per 6 hours are downloaded to run the Weather Research and Forecast model WRF, and derive the typical precipitation data from the whole world. The precipitation retrieval algorithm can operate either on land or on seawater for global. To simply the calculation procedure and save the training time, principle component analysis (PCA) was adapted to filter out the redundancy caused by scanning angle and surface effects, as well as system noise. According to the comparison and validation combing with other precipitation sources, it is demonstrated that the retrievals are reliable for surface precipitation rate higher than 0.1 mm/h at 15km resolution.
Key words: FY3C, MWHTS, 118.75GHz, WRF, precipitation

1. Introduction

Global monitoring of precipitation is important because of its significant human consequences. However, the multiplicity of hydrometeor types and their small-scale and large-scale spatial inhomogeneities make accurate measurements difficult. For example, rain gauge measurements are significantly impaired by wind, poor global coverage, and the non-uniformity of rain. Both ground-based radars and passive microwave satellite sensors sense precipitation aloft and are generally unable to discern how much of that precipitation evaporates before impact. Both are also sensitive to unknown local hydrometeor size, form, and vertical velocity distributions, as are simple single frequency radars on satellites. The resulting lack of adequate ground truth seriously complicates development and validation of global precipitation sensing methods.

This paper addresses the following: 1) the consistency between histograms of millimeter-wave atmospheric radiances observed by satellite instruments FY-3C MWTHS [1-2] and those predicted by a mesoscale numerical weather prediction (NWP) model WRF (http://www2.mmm.ucar.edu/wrf/users/) in combination with a four-stream radiative transfer model (VDISORT). WRF was initialized by National Center for Environmental Prediction NCEP [3] analyses on a 1° grid approximately 6 h prior to each FY-3C MWHTS transit and employed the

Goddard explicit cloud physics model. The scattering behavior of icy hydrometeors, including snow and graupel was considered. Retrievals of precipitation can have a coefficient of 0.61 now, and with the more and more observing data and validation data from radar and other techniques, this work is continuing and will have a better result.

2. Instrument description

MWHTS onboard FY-3C satellite is a fifteen-channel total power microwave radiometer in the range of 89GHz to 191GHz and has been launched in the September, 2013. Combining 13 horizontal polarized channels near 118.75GHz (which is the first operated to detect atmosphere on current operational polar orbiting satellite) and 183GHz, also window channels at 89GHz and 150GHz, precipitation content and path can be predicted.

Data assimilation of MWHS is demonstrated to improve the analysis of numerical weather prediction (NWP) and play an important role in the ECMWF forecast mode [4]. Based on sounding channels operated between 89 to 191 GHz (horizontal and vertical polarization), the MWHTS operated in cross-track scanning mode and will be used for deriving atmospheric vertical distribution of temperature and humidity, rainfall, etc. The chosen observation channels are at 89 GHz (vertical polarization), 118.75 GHz (horizontal polarization, eight-channel), 150 GHz (vertical polarization), 183.31 GHz (horizontal polarization, five-channel) [5].

Table1 Channel characteristics of MWHTS receivers

No.	Center frequency (GHz)	polarization	Bandwidth (MHz)	LO precision (MHz)	Dynamic range (K)	NEΔT (K)	Calibration accuracy (K)	3dB beam width
1	89.0	V	1500	50	3−340	1.0	1.3	2.0°
2	118.75±0.08	H	20	30	3−340	3.6	2.0	2.0°
3	118.75±0.2	H	100	30	3−340	2.0	2.0	2.0°
4	118.75±0.3	H	165	30	3−340	1.6	2.0	2.0°
5	118.75±0.8	H	200	30	3−340	1.6	2.0	2.0°
6	118.75±1.1	H	200	30	3−340	1.6	2.0	2.0°
7	118.75±2.5	H	200	30	3−340	1.6	2.0	2.0°
8	118.75±3.0	H	1000	30	3−340	1.0	2.0	2.0°
9	118.75±5.0	H	2000	30	3−340	1.0	2.0	2.0°
10	150.0	V	1500	50	3−340	1.0	1.3	1.1°
11	183.31±1	H	500	30	3−340	1.0	1.3	1.1°
12	183.31±1.8	H	700	30	3−340	1.0	1.3	1.1°
13	183.31±3	H	1000	30	3−340	1.0	1.3	1.1°
14	183.31±4.5	H	2000	30	3−340	1.0	1.3	1.1°
15	183.31±7	H	2000	30	3−340	1.0	1.3	1.1°

[a]Note: In column 3, V/H means quasi V/H polarization

3. Data collecting and processing

3.1. Data collecting

NCEP datasets(per 6 hours) from January, 1, 2014 to Oct, 31, 2014are chosen to run the Weather Research and Forecast model WRF, and derive the typical precipitation data from the whole world, especially for the extreme weather, such as typhoon, cyclone and rainfall. Bring out the incomplete cases and latitude values smaller than 0 degree and larger than 30 degree. The whole global datasets observed by FY-3C MWHTS from January, 1, 2014 to Oct, 31, 2014 are used to evaluated and analysis the validation of retrieval algorithm.

IOP Conf. Series: Earth and Environmental Scien(2057) 012049 doi:10.1088/1755-1315/57/1/

Radiance data in Level-1B format from FY-3C MWHTS are employed for this study [6-7]. Both FY-3C has afternoon-configureured orbit. MWHTS is a self-calibrating total-power radiometer, which is a cross-track step-scan instrument and executes one complete revolution every 8/3 s period. The MWHTS antenna systems have a nominal field of view (FOV) of 1.1 degree at the half-power points (covering a 16-km diameter footprint at nadir) and execute a cross-track scan with 98 Earth FOVs (within ±53.35degreefrom nadir), one cold space FOV, and one warm blackbody FOV per 8/3-s scan period. FOVs 1 and 98 (at ±53.35degree from the nadir, respectively) are the outermost scan positions of the Earth views. The FOVs 49 and 50 (at ±1.1◦ from nadir) straddle the nadir. The space view center is at 107.1◦. The space views and internal warm target views are used to complete on-orbit two-point calibration for MWHTS.

For FY-3C MWHTS, channel 9 and channel 15 are sensitive to precipitation. Combined with window channels distributed at 89GHz and 150GHz, people can retrieve precipitation rate using neural network estimators.

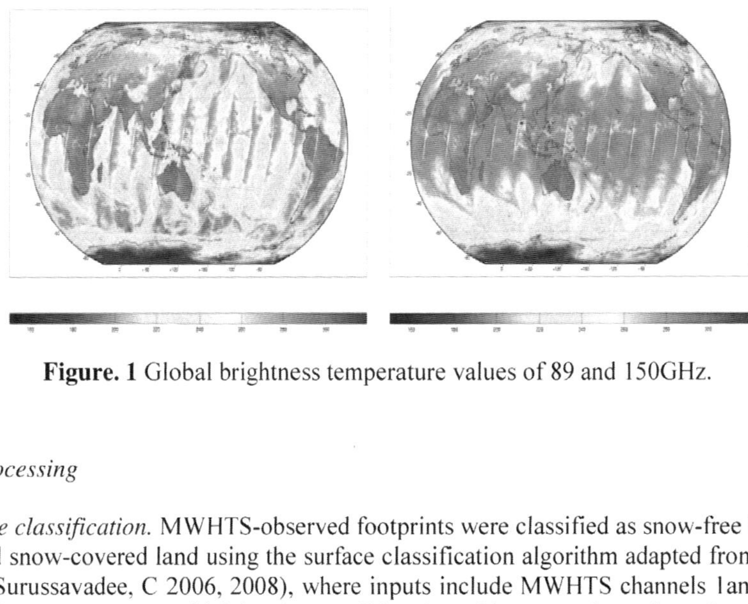

Figure. 1 Global brightness temperature values of 89 and 150GHz.

3.2. *Data processing*

3.2.1. Surface classification. MWHTS-observed footprints were classified as snow-free land, seawater, or sea ice and snow-covered land using the surface classification algorithm adapted from Chinnawat's paper [8][9](Surussavadee, C 2006, 2008), where inputs include MWHTS channels 1and 10, land/sea flag, and surface temperature, which has been validated. In this paper, the authors focus on MWHTS footprints classified as snow-free land or seawater for global.

3.2.2. Matching. Microwave humidity and temperature sounder (MWHTS) are onboard the Chinese FY-3Csatellite is a cross-track scanning radiometer with a swath width of 2650km. MWHTS has a nadir footprint size of 15 km, and 98 fields of view per scan line. The absolute visiting period is 5.5 days. The limitation threshold of matching domain is 0.05degree for both altitude and longitude with 20 mins.

The atmospheric and surface conditions are used for producing global simulations of brightness temperatures that are measured by MWHTS. The vertical profiles of temperature, specific humidity, and pressure, the surface parameters of surface skin temperature, 2-m wind speed, and wind direction from the National Center for Environmental Prediction (NCEP) 6-h reanalysis data are used as input to WRF. The NCEP/WRF data field shave a horizontal resolution 0.05◦×0.05◦ and 29 vertical levels, with the highest vertical level located near 0.1hPa.

3.2.3. Correction for small brightness temperature biases between NCEP/WRF/VDISORT and MWHTS. Considering the calibration accuracy and model error, bias correction of brightness temperature are in needed. According to the bias calculations and error corrections about cold and warm target, variable target, nonlinearity error and system random noise, the primary calibration results can be derived, which are shown in Figure. 2 and 3, which meet the requirements of design and development of MWHTS onboard FY-C satellite. To demonstrate that system requirements are appropriate and can be met, the paper provides the analysis of brightness temperatures through calibration process and profile retrievals using observing data of MWHTS.

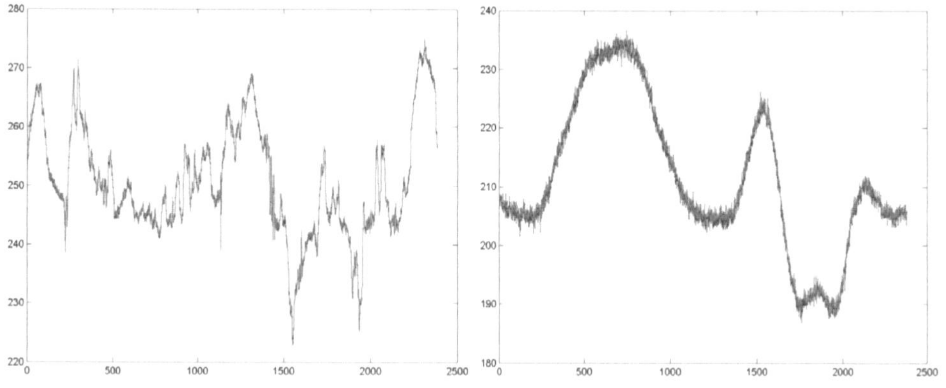

Figure.2 Comparisons of brightness temperature between observation and simulation.

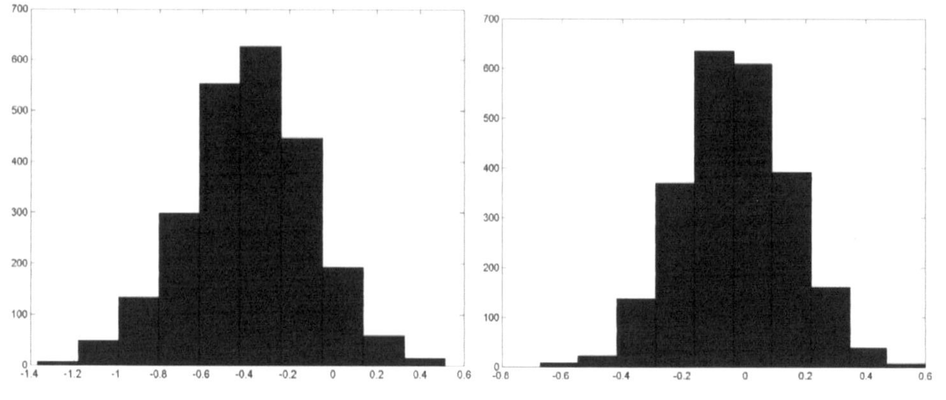

Figure.3 Bias distribution of brightness temperature between observation and simulation.

3.2.4. Omission of extreme values. The following values are omitted: 1) any brightness temperature for that footprint is less than 50 K or greater than 400 K, which is invalid, 2) the surface altitude is above 2 km for |latitude| < 60o, or above 1.5 km for 60o≤|latitude| < 70o, or above 0.5 km elsewhere, which could be covered by snow and are sensed more strongly and is called too-high, or 3) MWHTS channel 9 is less than 190 K, which implies that the atmosphere is so cold and potentially dry that precipitation is unlikely and that even the most opaque channel, i.e., 183±1 GHz, may sense the surface and yield false detections of precipitation, and is called too-cold.

3.2.5. Principal component analysis (PCA). The method of principal component analysis is to extract only good signals that provide useful information about precipitation but are insensitive to most angle and surface effects, and other noises. Therefore, this method will filter out the redundancy caused by scanning angle and surface effects, as well as system noise. Also, the dimensions of data matrix are reduced, which is helpful to save time and make the retrievals easier.

4. Retrievals and analysis

ANN is essentially a nonlinear statistical regression between a set of predictors (in this case the observation vectors X) and a set of predictands (in this case profiles of atmospheric temperature Z). The ANN model consists 3 layers 1, 2, and 3, which represent the input layer, the hidden layer, and the output layer, respectively, as shown in Figure.4.

The neurons of the input layer are represented by vector Xi (X1, X2X3, ..., XL), where L is the number of the input neurons. The neurons of the middle layer are represented by vector Yi (Y1, Y2, Y3, ...YM), where M is the number of the hidden neurons. The neurons of the output layer are represented by vector Zi (Z1, Z2, Z3), where Z is the rain detection and precipitation rate for different surface types and situations [10-12].

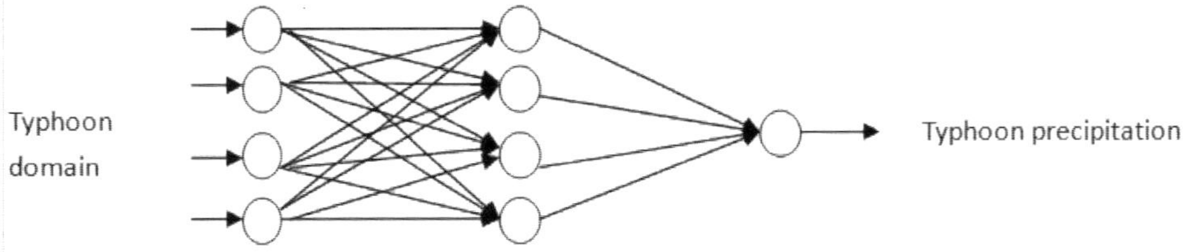

Figure. 4 The schematic of artificial neural network.

Estimates over land are usually slightly less accurate than those over sea, as expected, particularly at the lower rain rates where the surface remains visible. Using the artificial neural network retrieval algorithm of rain detection and precipitation presented above, and considering difference distributions of brightness temperatures as model inputs, as well as observing angles, the global map of rain detection can be retrieved as Figure 5 and Figure 6 shows.

In Table2, for most cases, the root mean square error (rms) errors are among the rain rate range, while for rain rate larger than 30mm/h, the rms error is less than the lower bound, which suggests good retrieval performance. While in typhoon region, the retrievals of rain rate are slightly worse and need more training data to improve the retrieving model.

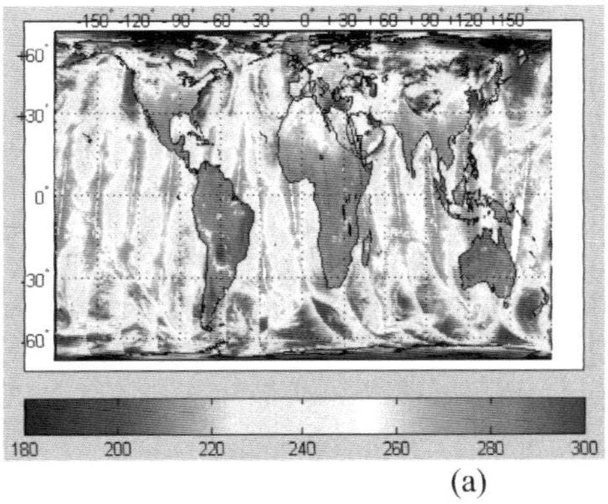

(a)

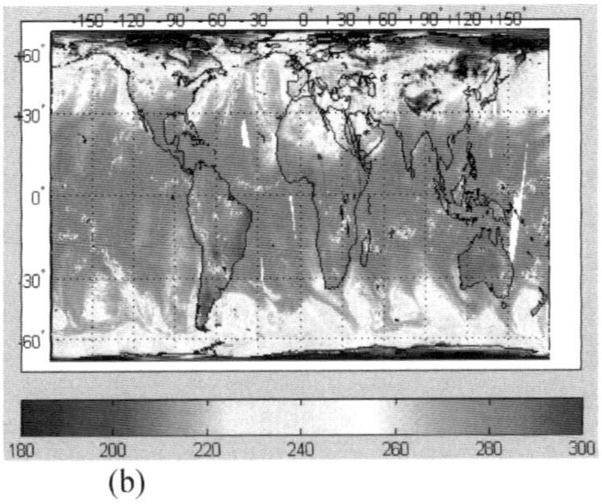

(b)

Figure.5 Global distributions of brightness temperatures on Jan20, 2015for (a) channel 1(b) channel 10 for FY-3C MWHTS

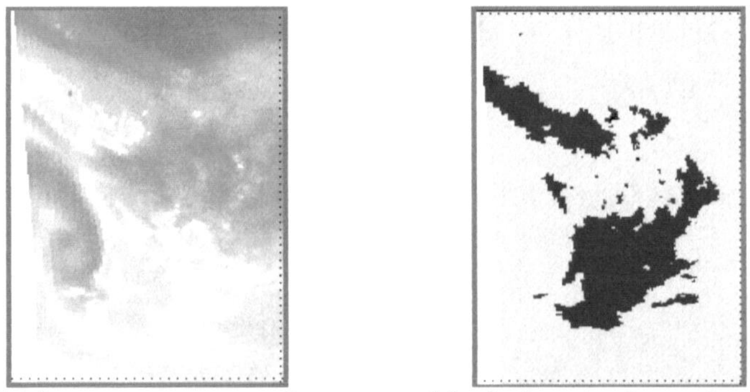

(a) Brightness temperatures of channel 10 in certain region (b) rain detection in certain region

Figure.6 Brightness temperatures and rain detection results based on MWHTS on time 22:44Jan, 20, 2015 to 00:30 Jan, 20, 2015.

Table2 Rain detection of typhoon domain by FY-3C MWHTS between Jan,1 to Oct,31, 2014

Location	rain Rate(mm/h)	Land rms/K	sea rms/K
Typhoon domain	0.1-1	0.88	0.84
	1-5	3.95	3.54
	5-10	7.11	6.78
	10-30	22.1	17.21
	30-50	24.21	21.23
	50-65	34.2	32.4

The correlation coefficients between rain rate detection and retrievals were found to be approximately 0.79 and 0.81 for land and sea, respectively.

5. Conclusion

For the recent work, the surface is classified the land mark as land and sea, and using the observing data to test and validate the accuracy of the rain rate. The work is not enough, and the authors are doing further improvement, like considering the surface covered by snow, ice and rainforest. Also, according to convection and stratiform, and classifying different rain rate from 0.1 to 50 as detail as possible to retrieve the rain rate. The radar data are also needed to validate the accuracy of rain detection. All of above will be described in future paper.

References

[1] S W. Zhang, J. Li, J. Jiang, Y. Zhang, Z. Wang and X. Dong.2006. Microwave Humidity Sounder (MWHS) of Chinese Meteorological Satellite FY-3. Proceedings of the Microwave Technology and Techniques Workshop -Enabling Future Space Systems, ESA/ESTEC, Noordwijk, The Netherlands, May 15-16, 2006, ESA SP-632.

[2] S W. Zhang, J. Li and Z Z. Wang 2012. Design of the second generation microwave humidity sounder (MWHS-II) for Chinese meteorological satellite FY-3. *IEEE International Geosciences and Remote Sensing Symposium (IGARSS), 2012,* 4672 –75, Munich, 22-27

July 2012.

[3] NCEP Global Tropospheric Analyses, 1 x 1 Daily Sep. 15, 1999–Present.[Online]. Available: http://dss.ucar.edu/datasets/ds083.2/

[4] http://www.ecmwf.int/en/about/media-centre/news/2014/ecmwf-starts-using-chinese-satellite-data

[5] J Y. He, S W. Zhang and Z Z. WANG 2015 Advanced Microwave Atmospheric Sounder (AMAS) Channel Specifications and T/V Calibration Results on FY-3C satellite. *IEEE Trans. Geosci. Remote Sens.* **1** 1481-93.

[6] X H. Tang and Z H. Cheng 2011 The data processing method of FY-3 meteorological satellite, Conference on Communications and Networking in China (CHINACOM). 2011 6th International ICST, 17-19 Aug. DOI: 10.1109/ChinaCom.2011.6158286.

[7] Z D. Yang, N M. Lu and Shi Jinming 2013 Overview of FY-3 Payload and Ground Application System. *IEEE Trans. Geosci. Remote Sens.* **12** 4846–53.

[8] Surussavadee, C and D. H. Staelin 2006 Comparison of AMSU Millimeter-Wave Satellite Observations, MM5/TBSCAT Predicted Radiances, and Electromagnetic Models for Hydrometeors. *IEEE Trans. Geosci. Remote Sens* **10** 2667-78.

[9] Surussavadee, C and D. H. Staelin 2008 Global millimeter-wave precipitation retrievals trained with a cloud-resolving numerical weather predictionmodel, Part I: Retrieval design," *IEEE Trans. Geosci. Remote Sens* **1** 99–108.

[10] J Y. He, S W. Zhang and Z Z. WANG 2012 The retrievals and analysis of water vapor density in arctic regions using FY-3A satellite MWHS, *Radio Science*. **2** 301-11.

[11] J Y. He and S W. Zhang 2012 Humidity retrievals in mid-latitude and tropical regions using FY-3 MWHS. *Journal of remote sensing*. **3** 581-97.

[12] Surussavadee, C. Blackwell, W.J.; Entekhabi, D. Leslie and R.V 2012 A global Precipitation retrieval algorithm for Suomi NPP ATMS, *International Geoscience and Remote Sensing Symposium (IGARSS)*, 22-27 July 2012. DOI: 10.1109/IGARSS.2012.6351128.

Assessment of Water Use in Pan-Eurasian and African Continents by ETMonitor with Multi-Source Satellite Data

Chaolei Zheng [1], Li Jia [1]*, Guangcheng Hu [1], Massimo Menenti [1,2], Jing Lu [1], Jie Zhou [1], Kun Wang [1], and Zhansheng Li [1]

[1] State Key Laboratory of Remote Sensing Science, Institute of Remote Sensing and Digital Earth, Chinese Academy of Sciences, 100101, Beijing, China
[2] Delft University of Technology, Delft, The Netherlands

Corresponding author: Li Jia (jiali@radi.ac.cn)

Abstract. The Pan-Eurasian and African Continents are characterized by large ranges of climates varying from humid, semi-humid, semi-arid and arid regions, and great challenges exist in water allocation for different sectors that related to water resource and food security, which depends strongly on the water use information. Quantitative information on water use is also important to understand the effectiveness of water allocation and further to prevent from water stress resulted by drought in water-scarce regions. Explosive development of satellite remote sensing observations provide great chance to provide useful spatiotemporal information for quantifying the water use at regional to global scales. In this paper, a process-based model ETMonitor was used in combination with biophysical and hydrological parameters retrieved from earth observations to estimate the actual evapotranspiration, i.e. the agricultural and ecological water use. The total water use is also partitioned into beneficial part, e.g. plant transpiration, and non-beneficial part, e.g. soil evaporation and canopy rainfall interception, according to the water accounting framework. The estimated water use show good agreements with the ground observation, indicating the ability of ETMonitor for global and continental scale water use estimation. The spatial and temporal patterns of the water use in the Pan-Eurasian and African Continents were further analysed, while large spatial variation of water use was convinced. Current study also highlights the great capability of satellite observations in studying the regional water resource and continental water cycle.

1. Introduction

Water is an important resource required for the daily sustenance and survival of human beings, and it is crucial to facilitate livelihoods and economic growth. As a result of climate change and population development pressures, water is becoming an increasingly scarce resource worldwide [1]. Coping with water scarcity and growing competition for water among different sectors requires proper water management strategies and decision processes. A prerequisite is crucial to understand the basin hydrological processes, manageable and unmanageable water flows, the interaction with land use and opportunities to mitigate the negative effects and increase the benefits of water depletion on society [2-4].

At global and continental scales, precipitation is the largest components of the terrestrial water budget which presents the water availability, while evapotranspiration (ET) is the second largest component presenting the water use [5]. Remote sensing has great potential but remains underutilized by practicing water resource managers. Accurately estimating consumptive water use using remotely sensed data helps water managers in planning, allocation, and management of water resources. Thanks to the continuous advanced hardware and data processing technologies, the open-access precipitation datasets based on satellite earth observation have a long heritage and are rapidly increasing, e.g. Tropical Rainfall Measuring Mission (TRMM), CPC MORPHing technique (CMORPH) product, which could be utilized for land surface modelling and global water budget study [6-7]. However, the generation of similar datasets for actual ET is in its infancy due to the complex water and energy transfer process. ET is limited by soil moisture supply and atmospheric moisture demand [8-9]. The former is largely linked to precipitation, while the latter relates to net radiation and advection which are impacted by surface and atmospheric temperature. A quantitative knowledge of the loss of water by actual ET (latent heat flux) is crucial in hydrological studies and water resource management, because it serves as a link between the land surface and the atmosphere [10-11]. Therefore, it is crucial to develop methods or tools to quantify water availability and water use (i.e. ET) over large spatial scales in order to inform decision makers on sustainable utilization and management of water resource.

The most popular remote sensing algorithms for continental or global scale ET estimation include surface energy balance models based on visible and infrared observations, and the micro-meteorological models based on the optical and microwave remote sensing data [12-18]. The first type algorithm was developed with the thermal infrared remote sensing, and was recognized to be able to generate accurate ET. However, it was restricted in cloud free conditions and difficult to provide continuous ET information at high temporal resolution since it relied heavily on the land surface temperature retrieved from thermal infrared bands. The land surface temperature was either assumed to be the aerodynamic temperature for surface sensible heat estimation, or believed to somehow reflect the surface moisture condition under the assumption that high temperature occurred over dry surface with low ET, while low temperature occurred over wet surface with high ET [19-20]. These assumptions were convinced in most cases, however failed in some cases, e.g. over the heterogeneous and non-isothermal surfaces, thermal infrared surface temperature cannot be directly used as the aerodynamic temperature in estimating surface fluxes [21]. Thus, the second one has grown up to the most attractive approach with the increasing earth observation dataset available [11, 16-17, 22-24].

The Pan-Eurasian and African Continents are characterized by large ranges of climates varying from humid, semi-humid, semi-arid and arid regions. Practical problems related to water resource and food security depends strongly on the water availability and water use in the Pan-Eurasian and African Continents. In many agricultural areas, crop water consumption often accounts for major elements of regional water use, which competes with many other purposes of water uses (e.g. cities and industries) in particular in water-scarce regions. Therefore, current study aims to analyze the quantitatively water availability and water use in Pan-Eurasian and African Continents to understand the effectiveness of water use and further to mitigate from water stress resulted by drought in water-scarce regions.

2. Study area and material

2.1 Study area

The Pan-Eurasian and African Continents characterized by large ranges of climates varying from humid, semi-humid, semi-arid and arid regions, account over half of the global land area. It is the area with the largest population density with its population account for about 85% of the world. This area is also characterized by uneven development, including the most developed region, e.g. west Europe, underdeveloped regions, e.g. Africa, and fast developing regions, e.g. China and India. The large amount of population and development of economy in recent year have great pressure on its water resources management. Practical problems related to water resource and food security depends strongly on the water availability and water use in the Pan-Eurasian and African Continents.

2.2 Methodology and Dataset

2.2.1 Water use estimation. The water use was estimated as the total actual evapotranspiration from the land surface to the atmosphere. A process-based ET estimation model named ETMonitor based on remotely sensed data was developed to estimate the daily actual evapotranspiration from 2008 to 2013 with spatial resolution of 1km. The ETMonitor model combined different ET parameterizations for the following land cover types: (1) water body; (2) snow/ice surface; and (3) soil–vegetation canopy. The details about ETMonitor were given in previous work [11, 25].

In current study, the biophysical variables derived from the Global LAnd Surface Satellite (GLASS) products including albedo and LAI were adopted [26]. The standard MODIS land cover product (MCD12) was used to infer information on different land cover classes, and 17 land cover types were identified based on the IGBP classification. The Advanced Scatterometer (ASCAT) soil moisture products estimated from the ERS scatterometer and soil data [27-28] was used. The near-surface meteorological forcing data, including air temperature, air pressure, dew point temperature, wind speed, precipitation, downward short-wave and long-wave radiation fluxes are essential to drive the ETMonitor, which were derived from the European Centre for Medium-Range Weather Forecasts (ECMWF) Interim Re-Analysis (ERA-Interim) meteorological product (http://www.ecmwf.int).

2.2.2 Statistical analysis for validation. The water use estimated by ETMonitor was validated according to the *in situ* observation from 10 observation sites (Table 1), with land cover types of forest, shrubland, grassland, and cropland, after careful quality evaluation. We do not intend to validate the precipitation data since it has been well validated in other studies [7, 29].

Table 1. Information of selected sites for validation

Site	Latitude	Longitude	Land cover	period
BE-Bra	51.30	4.52	Mixed Forest	2008-2011
CH-Oe2	47.28	7.7343	Croplands	2008-2011
IT-Ren	46.58	11.43	Evergreen Needleleaf forest	2009-2011
RU-Cok	70.82	147.49	Open Shrublands	2008-2009
Yingke	38.85	100.42	Croplands	2009-2011
Arou	38.05	100.45	Grasslands	2009-2011
Guantan	38.53	100.25	Evergreen Needleleaf Forest	2009-2011
Daxing	39.62	116.42	Croplands	2009-2011
Guantao	36.51	115.12	Croplands	2009-2011
Miyun	40.63	117.32	Mixed Forest	2009-2011

The statistical indices, including coefficient of determination (R^2) and root mean square error (RMSE), were calculated in this study to illustrate the difference between model outputs and ground measurement.

3. Results and discussion

3.1 Validation of water use estimated by ETMonitor

The result of water use validation is listed in Table 2. ETMonitor performs the best at RU-Cok site with very high R^2 of 0.93 and low bias and RMSE of 0.03 mm d^{-1} and 0.19 mm d^{-1}, respectively. The water use estimated by ETMonitor differ from observation mostly occurs in Guantao site, however its bias is still less than 1 mm d^{-1} and RMSE is less than 1.5 mm d^{-1}, which is acceptable in current large scale water flux estimation model. Overall, the ETMonitor based on mainly remote sensing datasets shows good accuracy in estimating water use.

Table 2. Statistics of estimated water use validation results.

Site	Bias (mm d^{-1})	R^2	RMSE(mm d^{-1})
BE-Bra	0.11	0.86	0.34
CH-Oe2	-0.74	0.61	1.14
IT-Ren	-0.50	0.65	0.86
RU-Cok	0.03	0.93	0.19
Yingke	-0.07	0.86	0.90
Arou	-0.30	0.86	0.51
Guantan	0.29	0.68	0.79
Daxing	0.09	0.36	1.34
Guantao	0.88	0.80	1.40
Miyun	0.72	0.78	1.20

3.2 Spatial variation of water availability and water use

CMORPH precipitation data was collected to present the water availability in current study, which showed clear spatial variations (Figure 1A). Extreme large amount precipitation was found near the equator area in Africa and Asia, with maximum value reaches 5000mm/yr. Extreme small amount precipitation was found near the desert area in Africa and Asia (e.g. the Sahara desert), with the minimum value less than 100 mm/yr. The water use, estimated by ETMonitor, also showed clear spatial variation (Figure 1B). Extreme large amount ET was found near the equator area in Africa and Asia, with the maximum value reaches 1500mm/yr. Extreme small amount ET was found near the desert area in Africa and Asia, with the minimum value less than 100 mm/yr. Regions with high precipitation are usually accompanied with high water use, most due to the limit of water available for water use is low in these regions, which is especially true in the equator region.

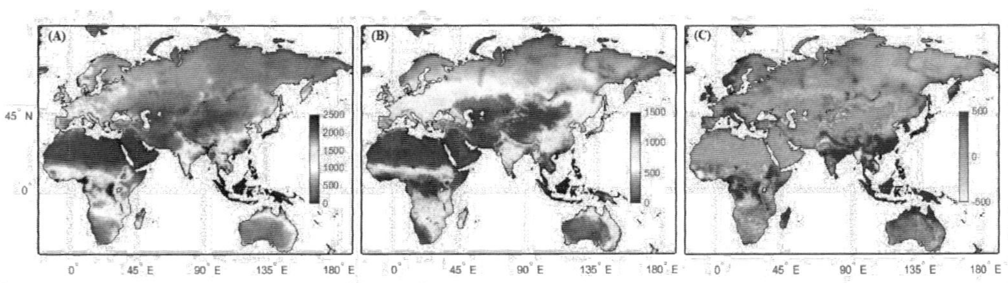

Figure 1. Spatial variation of water availability (A), water use (B), and the precipitation – evapotranspiration deficit (C). Annual mean value from 2008 to 2013 with the unit of mm yr^{-1}.

3.3 Spatial variation of precipitation – evapotranspiration deficit

The deficit between precipitation and evapotranspiration (P-ET) represents the degree of water availability to meet the water use, and its annual spatial variation is shown in Figure 1C. Regions with high precipitation were found to have high positive value of P-ET, including the forest area around the equator in Africa and Asia, west Europe, and monsoon impact area in east and southeast Asia, indicating the water availability could meet the water use and there exist water storage gaining potentially in these regions. These regions will generate surface runoff, interflow, drainage, groundwater recharge, seepage and base flow. Agro-ecosystems where P > ET are referred to as net producers of water and are typically present in the forested upstream end of river basins. Such excess water moves downgradient in a given tributary to be used by other agro-ecosystems [30]. For the regions with limited precipitation amount, negative values of P-ET were found, mainly located in the agricultural area with irrigation, e.g. the north China plain, the Indus basin, and the Nile basin, which could be found in previous studies based on different datasets [31]. Agro-ecosystems that are net water consumers have an incremental ET that cannot be attributed to precipitation only, but also to other water sources with a natural origin, such as groundwater seepage, shallow water tables, interflow or inundations during annual wet seasons with high river flow levels.

The P-ET value also showed large differences among different land cover types (Figure 2). Evergreen broadleaf forest showed the highest P-ET value, since it is mainly located in the tropical rainforest regions with large amount of precipitation. Cropland show low value of P-ET, but larger than grassland, mostly due to some cropland locate in the tropical area with high precipitation, while grassland mainly locate in the high latitude area with less precipitation. We notice that the averaged P-ET value for each land cover types shows a positive value, which is mainly caused by the averaged effect that large precipitation contribute too much to P-ET and blends the negative P-ET.

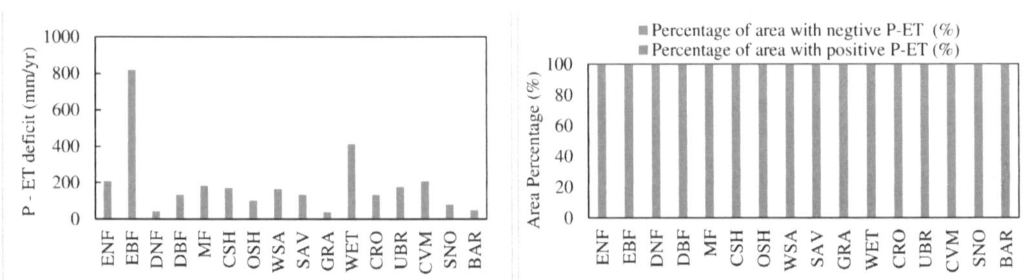

Figure 2. Variation of P-ET among different land cover types. (left panel: averaged P-ET in each land cover type; right panel: percentage of water gain and loss area for each land cover type)
ENF: Evergreen Needleleaf forest; EBF: Evergreen Broadleaf forest; DNF: Deciduous Needleleaf forest; DBF: Deciduous Broadleaf forest; MF: Mixed forest; CSH: Closed shrublands; OSH: Open shrublands; WSA: Woody savannas; SAV: Savannas; GRA: Grasslands; WET: Permanent wetlands; CRO: Croplands; URB: Urban and built-up; CVM: Cropland/Natural vegetation mosaic; SNO: Snow and ice; BAR: Barren or sparsely vegetated

In Figure 2, we also show the percentage of area with positive P-ET and negative P-ET for each land cover type. Evergreen broadleaf forest showed the highest percentage of positive P-ET since it mainly located in the tropical area with high precipitation. Deciduous forest, woody savannah, grassland, cropland, are among the land cover types showed the highest percentage of negative P-ET, about 30%.

Combing with the uneven spatial distribution as illustrated in Figure 2, it could be found that the cropland in the tropical area gain much water from precipitation than water loss by ET and thus irrigation is unnecessary. However, for the cropland in the temperate or arid area, many regions are found to be with negative P-ET since the ET is higher than precipitation, indicating the potential large amount of irrigation in these areas [31-32]. It also highlights the ability of groundwater management, since about 43% of irrigated water is pumped from groundwater globally [33]. Even large efforts have been done, the water-table in many areas, like the north China plain, are keeping declining, and potential solutions are urgently to be in action, including reducing the irrigated area, reintroducing fallow periods, and shifting water from agriculture to other less consumptive uses [34].

3.4 Spatial variation of water use components
The different water use components are also estimated in ETMonitor, including plant transpiration (Tr), soil evaporation (Es), canopy rainfall interception (EI), open water body evaporation, and snow

sublimation (Figure 3). Plant transpiration is the largest water use components, accounting for 54.83% of total water use, while soil evaporation is the second largest accounting for 33.33%. The canopy rainfall interception is much less than the plant transpiration and soil evaporation, only accounting for 9.01% of the total water use. The rest of the water use, about 2.83%, is lost through open water body evaporation, and snow sublimation.

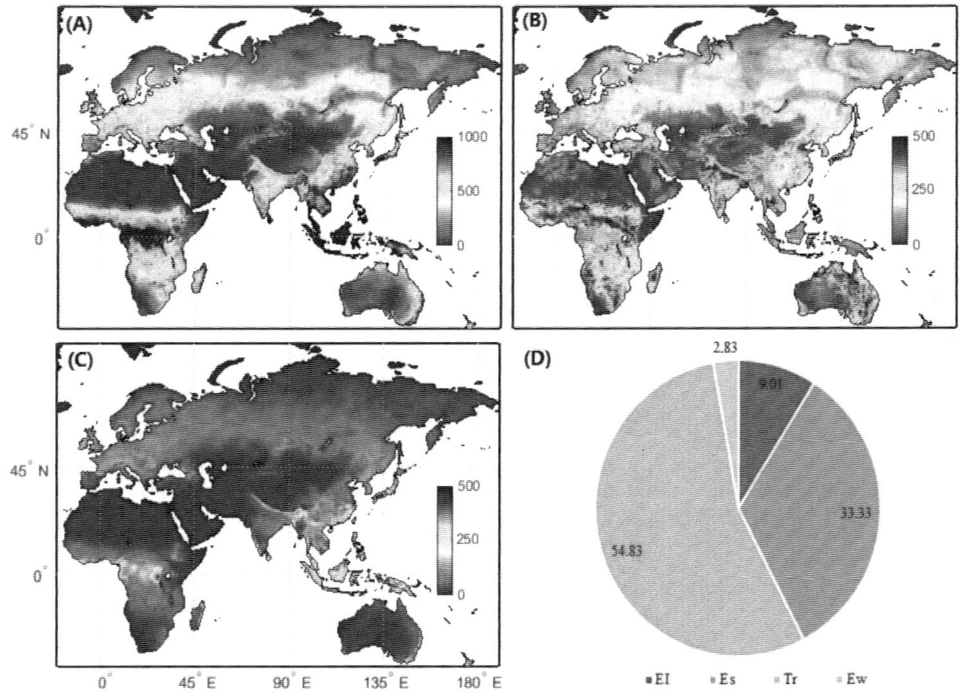

Figure 3. Spatial variation of plant transpiration (A), soil evaporation (B), canopy rainfall interception (C), and their percentage in % to total water use (D). Ew in (D) represent the water body evaporation and snow sublimation.

4. Conclusion

Agriculture is generally assumed to be the largest consumer of water in the Pan-Eurasian and African Continents, future increases in food production will be critical to ensure human wellbeing in both these regions and globally. The estimation of consumptive water use in the Pan-Eurasian and African Continents in current study is important for assessing and managing limited water resources. The estimated water use shows good agreements with the ground observations, indicating the ability of ETMonitor for global and continental scale evapotranspiration estimation. Large spatial variation of water availability and water use are convinced in the study area, and over a half of the total water use is by plant transpiration. The difference between precipitation and evapotranspiration (P-ET), is the difference between water availability and water use, also showed large spatial variations among different regions with different climate zones and different land cover types. The cropland is among the regions with low P-ET, highlighting its severe water scarce, especially in the temperate climate zones. Large amount of groundwater is used for irrigation in these areas, which threats its water security and leads to be severe problems in water management.

Acknowledgments

This work is supported by the National Key Basic Research Program of China (Grant no.

2015CB953702) and the National Natural Science Foundation of China (NSFC) (Grant no.91425303). We thank the Global LAnd Surface Satellite (GLASS) product generation system (http://glass-product.bnu.edu.cn/) for providing the input LAI and albedo datasets. We also thank HiWATER and EuroFlux providing the ground observation data for validation.

The authors declare no conflict of interest.

References

[1] Oki T and Kanae S 2006 *Science* 313(5790) 1068-1072.
[2] Allen R G, Pereira L S, Raes D and Smith M 1998 Crop evapotranspiration-Guidelines for computing crop water requirements- FAO Irrigation and drainage paper 56. 300(9) D05109.
[3] Huntington T G 2006 *Journal of Hydrology* 319(1) 83-95.
[4] Karimi P, Bastiaanssen W G M and Molden D 2013 *Hydrology and Earth System Sciences* 17 (7) 2459-2472.
[5] Mueller B, Hirschi M, Jimenez C and et al. 2013 *Hydrology and Earth System Sciences* 17 3707-3720.
[6] Joyce R J, Janowiak J E, Arkin P A and Xie P 2004 *Journal of Hydrometeorology* 5(3) 487-503.
[7] Xie P and Xiong A Y 2011 *Journal of Geophysical Research: Atmospheres* 116 D21106.
[8] Jung M, Reichstein M, Ciais P, et al. 2010 *Nature* 467(7318) 951-954.
[9] Zheng C and Wang Q 2014 *Hydrological Processes* 28(25) 6124-6134.
[10] Yao Y, Liang S, Cheng J and et al. 2013 *Agricultural and Forest Meteorology*, 171, 187-202.
[11] Hu G and Jia L 2015 *Remote Sensing* 7(3) 3056-3087.
[12] Jia L, Su Z, van den Hurk B, Menenti M and et al. 2003 *Physics and Chemistry of the Earth, Parts A/B/C* 28(1) 75-88.
[13] Mu Q, Heinsch F A, Zhao M and Running S W 2007 *Remote Sensing of Environment* 111(4) 519-536.
[14] Li Z L, Tang R, Wan Z and et al. 2009 *Sensors* 9(5) 3801-3853.
[15] Long D, Gao Y, and Singh V P 2010 *Journal of Hydrology* 388(3) 217-233.
[16] Michel D, Jiménez C, Miralles D G and et al. 2016 *Hydrology and Earth System Sciences* 20(2) 803-822.
[17] Miralles D, Jiménez C, Jung M and et al. 2016 *Hydrology and Earth System Sciences* 20(2) 823-842.
[18] Zheng C, Wang Q and Li P 2016 *Hydrological Sciences Journal*, 61(8), 1535-1547.
[19] Bastiaanssen W G M, Menenti M, Feddes R A and Holtslag A A M 1998 *Journal of hydrology* 212 198-212.
[20] Norman J M, Kustas W P and Humes K S 1995 *Agricultural and Forest Meteorology* 77(3) 263-293.
[21] Li Z L, Tang B H, Wu H and et al. 2013 *Remote Sensing of Environment* 131 14-37.
[22] Mu Q, Zhao M and Running S W 2011 *Remote Sensing of Environment* 115(8) 1781-1800.
[23] Bastiaanssen W G M, Cheema M J M, Immerzeel W W and et al. 2012 *Water Resources Research* 48(11) W11512.
[24] Hu G, Jia L and Menenti M 2015 *Remote Sensing of Environment* 156 510-526.
[25] Zheng C, Jia L, Hu G and et al. 2016 *IEEE Geoscience and Remote Sensing Symposium* (IGARSS).
[26] Liang S, Zhao X, Liu S and et al. 2013 *International Journal of Digital Earth* 6(sup1) 5-33.
[27] Wagner W, Hahn S, Kidd R and et al. 2013 *Meteorologische Zeitschrift* 22(1) 5-33.
[28] Wagner W, Lemoine G and Rott H 1999 *Remote Sensing of Environment* 70(2) 191-207.
[29] Wang S, Liu S, Mo X and et al. 2015 *Journal of Hydrometeorology* 16(6) 2577-2594.
[30] Bastiaanssen W G, Karimi P, Rebelo L M and et al. 2014 *Remote Sensing* 6(11) 10306-10334.
[31] Mo X, Liu S, Lin Z and Qiu J 2011 *Acta Geographica Sinica* 5 005.
[32] Sun H Y, Liu C M, Zhang X Y and et al. 2006 *Agricultural Water Management* 85(1) 211-218.
[33] Siebert S, Burke J, Faures J M and et al. 2010 *Hydrology and Earth System Sciences* 14(10)

1863-1880.
[34] Kendy E, Zhang Y, Liu C, Wang J and et al. 2004 *Hydrological Processes* 18(12) 2289-2302.

Improved Algorithm of SCS-CN Model Parameters in Typical Inland River Basin in Central Asia

Jin J Wang [1, 2, 3,4],Jian L Ding [1,3,4*], Zhe Zhang [1,3,4], Wen Q Chen [1,3,4]

[1] Ecological postdoctoral research station, Xinjiang University, 831401 Urumqi, China
[2] Xinjiang Vocational and Technical College of Communications , 831401 Urumqi, China
[3] College of Resources and Environment Science , Xinjiang University , 830046 Urumqi , China
[4] Key Laboratory of Oasis Ecology of Ministry of Education , Xinjiang University , 830046 Urumqi , China

Corresponding author: Ding Jian Li. E-mail: watarid@xju.edu.cn.

ABSTRACT. Rainfall-runoff relationship is the most important factor for hydrological structures, social and economic development on the background of global warmer, especially in arid regions. The aim of this paper is find the suitable method to simulate the runoff in arid area. The Soil Conservation Service Curve Number (SCS-CN) is the most popular and widely applied model for direct runoff estimation. In this paper, we will focus on Wen-quan Basin in source regions of Boertala River. It is a typical valley of inland in Central Asia. First time to use the 16m resolution remote sensing image about high-definition earth observation satellite "Gaofen-1" to provide a high degree accuracy data for land use classification determine the curve number. Use surface temperature/vegetation index (T_S/VI) construct 2D scatter plot combine with the soil moisture absorption balance principle calculate the moisture-holding capacity of soil. Using original and parameter algorithm improved SCS-CN model respectively to simulation the runoff. The simulation results show that the improved model is better than original model. Both of them in calibration and validation periods Nash-Sutcliffe efficiency were 0.79, 0.71 and 0.66,038. And relative error were3%, 12% and 17%, 27%. It shows that the simulation accuracy should be further improved and using remote sensing information technology to improve the basic geographic data for the hydrological model has the following advantages: 1) Remote sensing data having a planar characteristic, comprehensive and representative. 2) To get around the bottleneck about lack of data, provide reference to simulation the runoff in similar basin conditions and data-lacking regions.

1. Introduction

The direct runoff simulation provides total water resource for hydraulic structure design and the calculation of peak flow discharges. The Soil Conservation Service (SCS) which is known as curve number (CN) model developed by the U.S. Department of National Resources Conversion Service (NRCS) is an extensively used method for estimating and predicting runoff. Because of simplicity and stability, the model has fruitful application case all over the world.

The original Curve Number method can be traced back to infiltrometer tests carried out by the SCS. The purpose was to establish basic data to evaluate the effects of watershed treatments and soil conservation measures on the rainfall-runoff process. The curve number (CN) is derived from the tables given in the National Engineering Handbook for catchment characteristics, such as soil texture, land use, hydrologic condition, and initial soil moisture condition.

Another important parameter of SCS model is initial abstraction coefficient (λ), it plays an important role in calculating the direct runoff, the hydrograph of peak flow, and the runoff distribution time(Baltas et al., 2007a).The original SCS-CN model calculated the initial abstraction (Ia) as a constant 20% of the maximum potential retention (S), based on observed rainfall and runoff data in North America valleys in 1954.Since then, the initial abstraction ratio, defined as Ia /S, has been applied to estimate direct runoff in many different countries. Although λ is set to 0.2 in the original SCS-CN model adjusted from experimental data in North America, the validity and applicability of this result is ambiguous in other regions of the world (Ponce and Hawkins, 1996a; Shi et al., 2009a). For example, for 50% of rainfall events used in 0.2.(Tedela et al. 2012a), λ ranged is 0.095~0.38, whereas for all rainfall events λ ranged is set 0.013 to 2.20. λ is described as a regional parameter (Mishra and Singh, 2003a), because its selection requires refinement for regional watersheds.

The SCS-CN model supplies sufficient room for variation due to the rainfall difference on spatial and temporal. For example, the measured rainfall-runoff data, and the different grades of antecedent soil moisture content (AMC) (Ponce and Hawkins, 1996a; Mishra et al., 2006a). The AMC is a function of the total rainfall of 5 days ago and the infiltration capacity of watershed soil. it is divided into three classes, AMC-I(dry), AMC-II (normal), and AMC-III (wet), The ARC-II grade is considered as the reference condition for CN values choose from National Engineering Handbook tables (SCS, 1971a).But the evaporation is variably in different catchments caused the soil water retention quality is difference in the same rainfall situation.

In summary, we need to explore new ways to improved model parameter calculation method. The purpose is to reduce the model parameters, simplified model structure, decreasing model error due to different geographical environment of the river basin.

2. Material

2.1. The Study Area

The study area of this paper is located in Xinjiang of China in central Asia. It lies in the plain of upstream Boertala River in Xinjiang（44°50′~45°58′N，80°07′~81°05′E）, named as Wen-quan basin. The total area of Wen-quan basin is 999 km2, the annual average precipitation is 228 mm, the average DEM is 3500 m. This watershed surrounded by mountains and January to February more than half of the total mountain area is covered by snow, a depth of more than 20 cm, as depicted in Fig.1. (Dou Yan et al., 2010a). The basin accumulated snow from November to march every year. Snow melting flood started to happen with the temperature rising from April, and the snowmelt became the major supply for the basin in spring. Precipitation is increasing from June to September, and the rainfall became the major supply for the basin in summer. The direct runoff of Wen-quan basin is consists of two parts: rainfall and snowmelt, and with more flood disaster in April to September in the year, especially in Tuha farm, Zhalemute, chagange, angelige, and kundelun farm. The snowmelt flooding affects the continuous development of husbandry and peoples life-property safety in these regions. There have a hydrologic station and a weather station in Wen-quan basin.

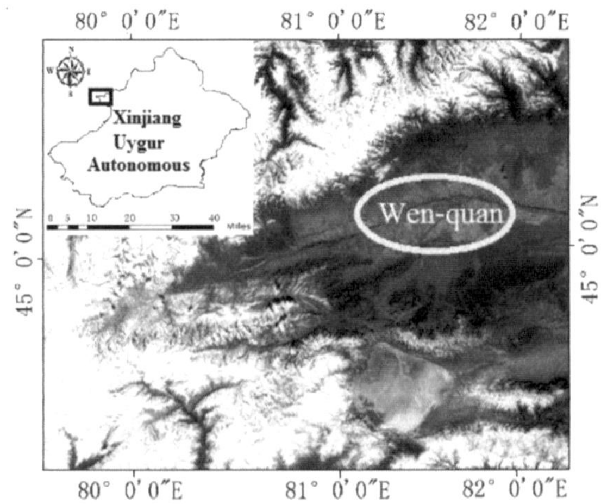

Figure 1. Watershed location in the study area of China.

This paper used many different sources of data which mainly include the meteorological, the hydrologic, the soil type, the land use types, the high-definition earth observation satellite "Gaofen-1" image, and the landsat8 OLI image. Depicted the details is shown in Table1.

Table 1. Geographical data for study area.

Data type	Data description
Meteorological data	Daily precipitation and temperature observations data from 3-4th in 2013 in study area
Hydrologic data	Daily runoff observation data of Boertala River hydrological Station from 3-4th in2013
Soil type and soil moisture	Filed observations data include land use and 36 points soil moisture
Soil texture	Extract from harmonized world soil database ;the resolution is 1km， It uses FAO90 soil classification system.
Remote-sensing image	① "GF-1"satellite image serial number：229827; DATE ACQUIRED：2014-05-19；② Landsat8 OLI image WRS_PATH:147; WRS_ROW：29; DATE ACQUIRED：2013-06-05；③SRTM90 m resolution DEM data.

3. Method

3.1. Original SCS-CN Model

The wildly used and popular method for direct runoff simulation is the SCS-CN model, it is based on water balance equation and two assumptions. The water balance equation is:

$$P = Q + F + I_a \quad (1)$$

Where, P is the total rainfall depth (mm), Q is the direct surface runoff (mm), F is actual retention after runoff begins, and Ia is the initial abstraction (mm).

We can write the two hypothesis as follows,

$$\frac{Q}{P - I_a} = \frac{F}{S} \quad (2)$$

$$I_a = \lambda S \quad (3)$$

Where, S is the potential maximum retention (mm) after runoff begins; λ is the initial abstraction coefficient. Depending on above equations, we can express the direct surface runoff simulation equation as follows:

$$Q = \frac{(P - \lambda S)^2}{P + (1 - \lambda)S} \quad (4)$$

S can be transformed to a dimensionless CN, varying in a logical range (0, 100) and can be expressed as in Eq. (5).

$$S = \frac{25400 - 254CN}{CN} \quad (5)$$

The CN values can be calculated from the valley characteristics, for example: the land use/cover, hydrologic condition, hydrologic soil group, and antecedent moisture condition) (SCS, 1971).

3.2. New Proposed Model

This paper through the following methods to improve the SCS model parameter algorithm:

3.2.1. Rainfall modified. Primary model of SCS is a pure study on rainfall and runoff. In order to satisfy the character of mix supplied runoff, the precipitation revised to the sum of rainfall and snowmelt. The snowmelt calculated by degree-day model. The equation written as follow:

$$P = P_r + P_s \quad (6)$$

Where, P is the daily total precipitation (mm), Pr is observed daily rainfall (mm), Ps is the daily snow-melting (mm). because of the simple structure and less parameters involved, the degree-day model used in calculate the Ps in Eq.(7).

$$P_s = \mathrm{DDF} \cdot \mathrm{PDD} \quad (7)$$

Where, DDF is degree-day factors (mm/（d·℃）), PDD is called the positive degree-day factor. Based on result of Zhang Yong(Zhang Yong, 2006), we used the method of Kriging spatial interpolation calculate PDD of Wen-quan basin is 2.7 mm/（d·℃）. we can express the PDD as follow:

$$\mathrm{PDD} = \sum_{t=1}^{n} H_t * T_t \quad (8)$$

Where, Tt is the average temperature for a period. Ht is logical variable. Tt≥0℃, Ht=1.0. Tt≤0℃, Ht=0.0.

3.2.2. Potential maximum retention (S) modified. Potential maximum retention (S) of original SCS-CN model is calculated by curve number. This paper use remote sensing estimation the S, in order to reduce the parameters of SCS model. The remote sensing data has polygon characteristics, through remote sensing obtain the value of S also has polygon characteristics. It has global and representative in large basin. We use the relationship of surface temperature and vegetation index (TS/VI) combined with technology of remote sensing information extract inversion the soil moisture of study area. And then combined with water absorption balance of soil to calculate the value of S. improve methods introduced described as below.

Potential maximum retention is determined by the soil texture, the soil infiltration, the surface vegetation coverage, the land use/cover, hydrologic condition, hydrologic soil group, and antecedent moisture condition in original SCS-CN model. The process of estimate to S is very complicated. Based on water absorption balance of soil to calculate the potential maximum retention, we can explain as follow:

$$S=C\left(W_{max}-W_{soil}\right) \quad (9)$$

Where, C is adjustment coefficient, W_{max} is saturated soil moisture content, obtained by field measuring the soil porosity and soil bulk density. In this paper, we take the 100% to calculate due to lack of data. W_{soil} is antecedent soil moisture, use the method of remote sensing and TS/VI to calculate. Model parameters unit unified mm. Moran results show that the changes relationship between the surface temperature and soil moisture is very close. Take the vegetation index (VI) as the x-axis, the surface temperature (TS) as y-axis can be constructed corresponding temperature - vegetation drought index (TVDI) and fitting equation to calculate the soil moisture content effectively. Where, the normalized difference vegetation index (NDVI) is calculated by the satellite of landsat8 OLI image combining Equation 5 as follow:

$$NDVI = (NIR-R) / (NIR+R) \quad (10)$$

Where, R is red band belong to band 4 of OLI; NIR is near infrared band belong to band 5 of OLI. The surface temperature is calculated by the Landsat8 TIRS based on atmospheric correction method.

4. RESULTS

Use improved method of SCS model and original SCS model respectively to simulate the 18 rainfall events runoff of Wen-quan basin, in 2013.

4.1. Calculation the Parameters of Original SCS-CN Model

Original SCS model simulation requires different parameters, for example: the curve number, the potential maximum retention (S), and the initial abstraction coefficient (λ). The cn value is the first parameter to calculate that synthesize the surface vegetation coverage, the land use/cover and classification, hydrologic condition, hydrologic soil group (HSG), and antecedent moisture condition (AMC) to determine. Depending on the soil infiltration rate and minimum soil texture, the hydrologic soil group is divided into A, B, C, D 4 categories, the soil infiltration rate reduce in turn from A to D. the HSG criteria can be find in Table 2. And then extract the soil texture of Wen-quan basin with GIS from the harmonized world soil database. The result show that the soil texture in the study area composed with sandy loam and loamy sandy. The soil hydrology group classified of Wen-quan basin is A. The curve number is affected by soil moisture, so the SCS model divided the soil moisture into three levels, AMCI is dry, AMCII is average, AMCIII is moist. Statistics the precipitation events of study area in 2013, we can identify the antecedent soil moisture condition of study area is AMCII. Use the 16m resolution remote sensing image about high-definition earth observation satellite "Gaofen-1" to provide a high degree accuracy data of the land use classification for SCS-CN to determine the curve number. The classification accuracy is 86.43% and kappa coefficient is 0.837, detail is shown in Figure 2. The classification area of land use and its AMCII curve number in study area shown in Table 3. The Wen-quan basin curve number is 73.33.

Table 2. Hydrologic soil group classification of SCS-CN model.

Soil classification	Minimum of infiltration rate/($mm \cdot h^{-1}$)	Soil texture
A	>7.26	sand; loamy sand; sandy loam;
B	>3.81～7.26	Loam; silt loam
C	>1.27～3.81	sandy clay loam
D	0～1.27	clay loam; silt clay; clay; sand clay

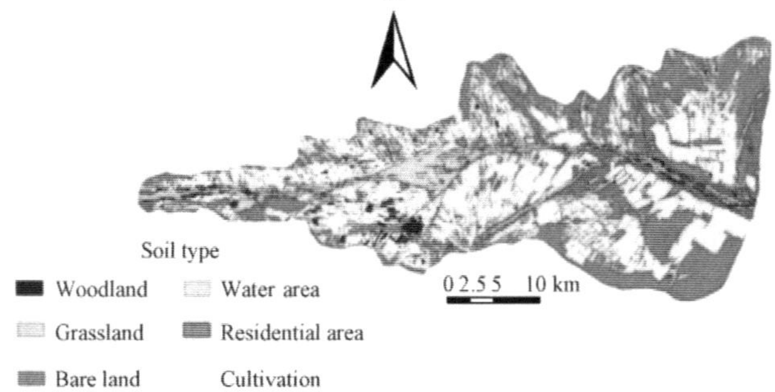

Figure 2. Land use classified of Wen-quan watershed based on GF-1 image.

Table 3. Curve number value and areas of different land use under medium soil moisture.

Soil type	Land use area/km^2	Total area percent/%	Curve number of hydrologic soil group A and AMCII
Residential area	21.1	2.1	1.06
Cultivation	449	44.9	31.91
Woodland	40	4.0	1.80
Bare land	375.8	37.6	28.97
Grassland	53.9	5.4	3.67
Water area	59.2	5.9	5.92
Total area	999	CN	73.33

Take the curve number to Eq. (5) and estimate the potential maximum retention in study area is 92.37. Statistics 101 rainfall events of Wen-quan basin in 2013, modified the precipitation to the sum of rainfall and snowmelt and use the degree-day model calculate the snowmelt.

Take the 101 rainfall events observed runoff, the modified precipitation, and the potential maximum retention into Eq. (4) calculate the initial abstraction coefficient (λ). Screen 18 rainfall events to simulate the runoff in study area by boundary condition of precipitation greater than infiltration, and use 12 events for parameter calibration, 6 events for validation. Selected median, mean, and mode of 12 events initial abstraction coefficient to simulation the runoff respectively and find that accuracy is the highest when the initial abstraction coefficient (λ) is 0.45 in study area.

4.2. Calculation the Parameters of Improved Method

After the improved method of SCS model to simulate the runoff need two parameters, potential maximum retention (S) and initial abstraction coefficient (λ). S is determined by saturated soil moisture content and soil moisture content. Saturated soil moisture content equals 1. The soil moisture content calculate as follow:(1) use of landsat8 remote sensing image extract the information of NDVI and LST to build the 2D scatter plots and TS / VI fitting equation to calculate the TVDI in Wen-quan basin, shown as Figure 3.

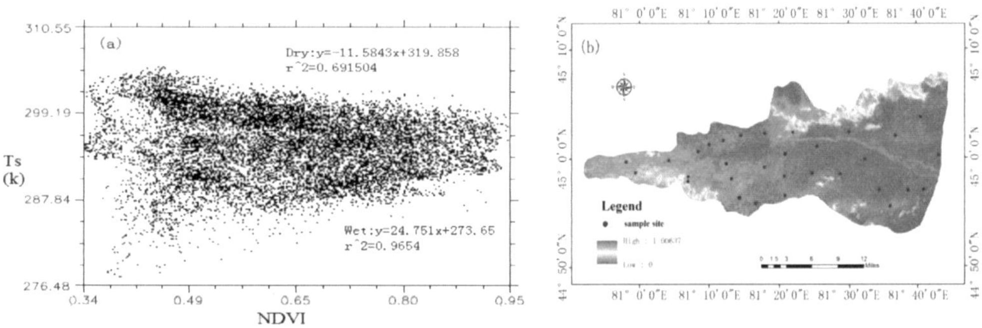

a. Ts/VI characteristic spaces Dimensional scatter plot of Wen-quan watershed.
b. Distribution of TVDI in Wen-quan watershed.

Figure 3. Characteristic spaces and TVDI distribution in Wen-quan watershed.

Use of space distribution of TVDI to calculate the surface soil moisture content space distribution of study area in June 5, 2013, described the map as Figure 4a. Spatial distribution of soil moisture consist by a number of pixels, each pixel corresponds to a soil moisture digital number representatively. Combined with formula (9) calculated potential maximum retention (S) is 85.61 mm, the initial abstraction coefficient (λ) is 0.499 in study area. Take the flied data verification the accuracy of soil moisture which inversion by remote sensing and TS/VI method. The result show that both calculated and measured soil moisture correlation coefficient is 0.6071 and relative error coefficient is -0.2%, shown as Fig.7b. It proved that accuracy of inversed soil moisture is reliability.

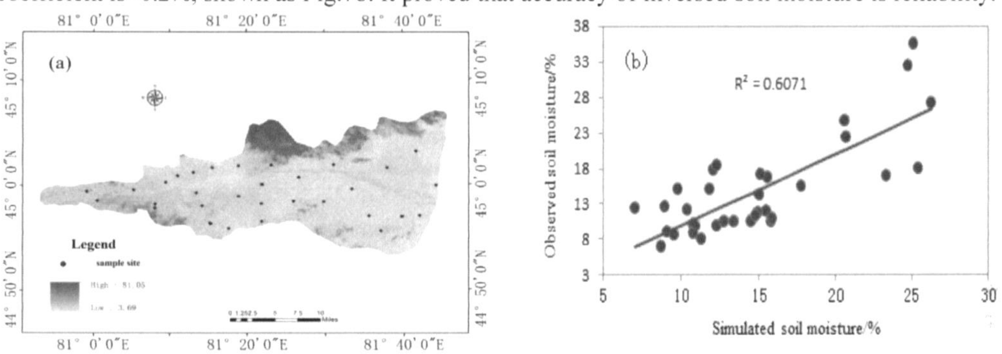

a. SRWC of the Wen-quan watershed

b. Correlation analysis of the soil moisture between measured and simulated

Figure 4. Soil moisture distribution and verification in Wen-quan watershed.

4.3. Comparing Estimated Runoff Results

The original and improved SCS model runoff simulation results of 18 rainfall events shown in Figure 5.Two models results compared to observed runoff had a relatively consistent trend. The simulation and observed runoff has a positive correlation with modified precipitation. Both original and improved SCS model Nash-Sutcliffe efficiency and relative error coefficients shown as Table 4.

5. DISSCUSSION

Compared with previous studies, this paper based on structure of original model improved the parameters calculation method. Use data observed by landsat8 and GF-1 satellite to calculate the potential maximum retention of study area. It has following advantages: 1) The calculated potential maximum retention(S) just like remote sensing data has planar characteristics. It has a good representation at large scale. 2) Because of the updated remote sensing data, the potential maximum retention(S) with better timeliness. 3) This method need fewer parameters and the data easy to obtain, it can addresses the lack of hydrological data on arid and semi-arid region.

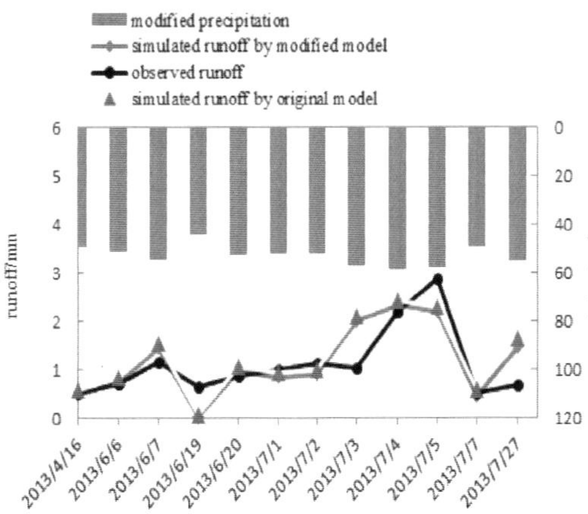

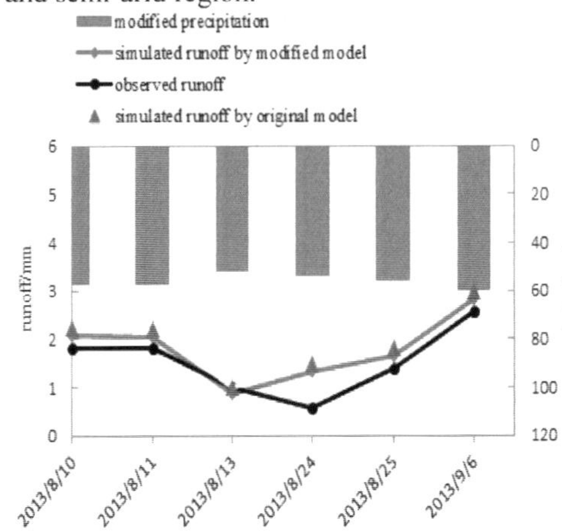

a. Compared of observed and simulated runoff by calibration periods. b. Compared of observed and simulated runoff by validation periods.

Figure 5. Compared of simulated and observed runoff between original and modified model in Wen-quan watershed.

Table 4. NSE (Nash-Sutcliffe efficiency) and RE (relative error) for simulation performance assessments.

Assessment indexes	Calibration periods		Validation periods	
	Original SCS model	Modified SCS model	Original SCS model	Modified SCS model
NSE Nash-Sutcliffe efficiency	0.71	0.79	0.38	0.66
Relative error/%	12	3	27	17

6. CONCLUSIONS

• This paper use landsat8 OLI remote sensing image and combine with TS/VI inversion the soil moisture of Wen-quan basin, compared with the measured results,soil moisture correlation coefficient is 0.6071 and relative error coefficient is -0.2%. The results indicated that the accuracy of inversed soil moisture is available to apply in the SCS model.

• Soil moisture which inversed by remote sensing composed of many pixels has a planar characteristic. It represents an average soil moisture value of study area. By using inversed data to calculate the potential maximum retention have already realized indirect the point data change to the planner. Results showed that the improved model simulation value is better than original model. Both of them in the validation periods correlation coefficient were 0.66,038. And relative error were17%, 27%. It is further explained that we can use remote sensing to inversion the land surface parameters for SCS-CN model and modified algorithm of SCS-CN model parameters can improve the accuracy of the simulation results. This method has the advantage of requiring less computation with few parameters and a simple structure. It can avoid being influenced by lack of data and have feasibility and practicability. The method can be used for similar conditions of study area, especially helpful the lack of data basin to simulate the runoff. With the large-scales and complex underlying surface conditions, the simulation precision should be further improved. Radar image is an effective way to improve the accuracy.

7. Acknowledgements (optional)

Acknowledgements of support for the professor Jian li Ding are welcome.

8. References

[1] Baltas E A, Dervos N A and Mimikou M A 2007.Technical note: determination of SCS initial abstraction ratio in an experimental watershed in Greece. *Hydrol. Earth Syst. Sci* **11** pp 1825-1829
[2] Dou Yan, Chen Xi and Bao Anming 2010. Study of the temporal and spatial distribute of the snow cover in the Tianshan mountains. *China. Journal of Glaciology and Geocryology* **32(1)** pp 28-34
[3] Ponce V M, Hawkins R H 1996. Runoff curve number: has it reached maturity? *J. Hydrol. Eng.* **1** pp 11-19
[4] Shi Z H, Chen L D, Fang N F, Qin D F and Cai C F 2009. Research on the SCS-CN initial abstraction ratio using rainfall-runoff event analysis in the three gorges area *China. Catena* **77** pp 1-7
[5] Mishra S K, Sahu R K, Eldho T I, Jain M K 2006. An improved I_a–S relation incorporating antecedent moisture in SCS-CN methodology *J.Water Resour. Manage* **20** pp 643- 660

Analysis of Spatial Pattern of Urban System along the Overland Silk Road Economic Belt Using DMSP-OLS Nighttime Light Data

Jianzhong Feng[a], **Linyan Bai**[b,*], **Kui wang**[b], **Xuefu Zhang**[a], **Nengfu Xie**[a], **Qiyun Ran**[b], **Mingqiu Guo**[b], **Lijun Xu**[c]
[a] Key Lab of Agri-information Service Technology, Ministry of Agriculture of P. R. China, Agricultural Information Institute, Chinese Academy of Agricultural Sciences, Beijing 100081, China;
[b] Institute of Remote Sensing and Digital Earth, Chinese Academy of Sciences, Beijing 100094, China;
[c] Institute of Agricultural Resources and Regional Planning, Chinese Academy of Agricultural Sciences, Beijing 100081, China.
Email: baily@radi.ac.cn

Abstract. As China promotes the Belt and Road (BAR) initiative, the overland SREB development is widely concerned. The cities (including towns), population centers, of urban system are the cores of the economy along the SREB. Therefore, it is necessary to monitoring the urbanization of the belt so that the new growing points of urban development and the valid coupling mechanism between human and nature will be explored to promote the regional socio-economic sustainable development and effectively implement the BAR initiative. Using the DMSP-OLS stable nighttime lights (NTL) data in 1992, 2003, and 2014, in this paper we studied the urbanized spatial patterns of and the urbanized characteristics and trends of the main city system along the SREB in the view of the whole regionalized economic zone and typical cities and settlements (towns). The results showed that in general the NTL intensities in the SREB's city system had the obvious geographical differentiation characteristics where there was maximum brightness of NTLs over the western European countries as well as being gradually decreasing from west to east. There were obvious increases of the NTL digital number (DN) values and NTL covering areas in 2003 and 2013 comparatively with that of 1992, which indicates the great urbanization development during this period. As for the four types of urban development process, there was an apparent consistency in a certain local area but a large heterogeneity among different areas. By analyzing the 273 pivot cities and the most pivot 26 cities, we found the number of the relatively small cities being decreasing but that of the large and medium-sized cities increasing. This study would provide the scientific support for the related researches and decision making of urbanization and urban economic development to promote the socio-economic comprehensive development of the overland SREB.

1. Introduction

As China promotes the Belt and Road (BAR), i.e., Silk Road Economic Belt (SREB) and 21st-Century Maritime Silk Road (21st-CMSR), initiative, the overland SREB development, which is a significant measure for mutually beneficial cooperation and common development among China and countries of central Asia and Europe, is widely concerned. It is also very important to expand and deepen the national strategy of China's Opening-up to the Outside World and brings to the corresponding provinces (such as Shaanxi, Gansu, and Xinjiang) along it opportunities of speeding up their social and economic development and promoting their deepening of reform (http://www.mofcom.gov.cn/article/resume/n/201504/20150400929655.shtml). The cities (including towns), population centers, of urban system are the cores of the economy along the SREB, which crucially support and drive its socio-economic development. Therefore, it is necessary to monitoring the urbanization of the belt (such as urban development situation, urbanization mode and trend so that the new growing points of urban development and the valid coupling mechanism between human and nature will be explored to promote the regional socio-economic sustainable development and effectively implement the BAR initiative.

The Operational Linescan System (OLS), which can detect light emission from the earth surface at night, is carried on a series of satellites of the US Air Force (USAF) Defense Meteorological Satellite Program (DMSP). The DMSP-OLS has a ground swath of about 3000 km. It has two broad spectral bands, one covering the visible-near infrared region (0.4 - 1 μm) and the other is in the thermal infrared region (10 - 13 μm). The OLS data are acquired in two spatial resolution modes: 'fine' and 'smoothed'. The full resolution fine data have a nominal spatial resolution of 0.56 km. On board averaging of five by five blocks of fine data produces "smoothed" data with a nominal spatial resolution of 2.7 km. The DMSP-OLS has a ground swath of about 3000 km. It has two broad spectral bands, one covering the visible-near infrared region (0.5 - 0.9 μm) and the other is in the thermal infrared region around 10 μm. The OLS data are acquired in two spatial resolution modes: 'fine' and 'smoothed'. The full resolution fine data have a nominal spatial resolution of 0.56 km. On board averaging of five by five blocks of fine data produces "smoothed" data with a nominal spatial resolution of 2.7 km. Most of the data received by NOAA-NGDC is in the smoothed spatial resolution mode (http://www.crisp.nus. edu.sg/~research/tutorial /dmsp.htm) (Liao et al., 2012). The DMSP-OLS has a unique capability to observe faint sources of visible-near infrared emissions present at night on the Earth's surface, including cities, towns, villages, gas flares, heavily lit fishing boats and fires (Chen et al., 2003; & Wang et al., 2012). By analyzing a time series of DMDP-OLS images, it is possible to define a reference set of "stable" lights, which are present in the same location on a consistent basis, and easily explore the significant human activities such as characterization of urban extent and spatial change. Specifically, there are caveats inherent to the fact that the area and intensity of illumination (detected by the DMDP-OLS) are known to vary significantly with energy availability, economic development and density of settlement (Elvidge et al., 1997; Sutton et al, 1997; Small et al., 2005; Elvidge et al., 1999; Hao et al., 2014). At present, the digital data from DMDP-OLS (since 1992) can provide three products of stable, radiance calibrated, and non-radiance calibrated nighttime lights (Hao et al., 2014; Ziskin et al., 2010; Elvidge et al., 2001).

Using the DMDP-OLS stable nighttime lights data in 1992, 2003, and 2014, in this paper we studied the urbanized spatial patterns of and the urbanized characteristics and trends of the main urban system along the SREB.

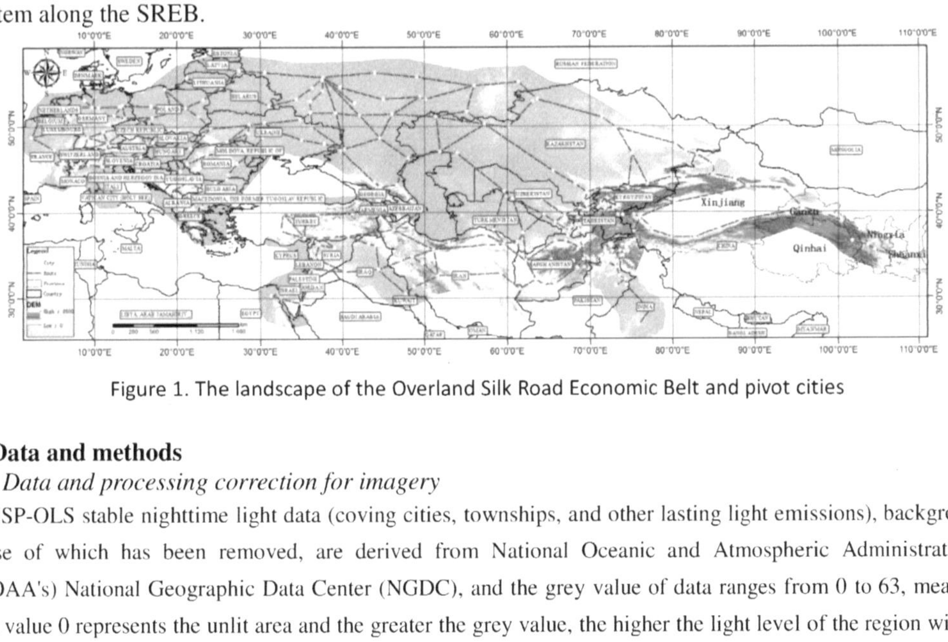

Figure 1. The landscape of the Overland Silk Road Economic Belt and pivot cities

2. Data and methods

2.1 Data and processing correction for imagery

DMSP-OLS stable nighttime light data (coving cities, townships, and other lasting light emissions), background noise of which has been removed, are derived from National Oceanic and Atmospheric Administration's (NOAA's) National Geographic Data Center (NGDC), and the grey value of data ranges from 0 to 63, meaning that value 0 represents the unlit area and the greater the grey value, the higher the light level of the region will be (Wu et al, 2013). Now the products of 34 satellite years (Version 4) derived from six satellites (F10, F12, F14, F15, F16, and F18) from 1992 to 2013 are available for downloading from NGDC (http://www.ngdc.noaa.gov /dmsp/download.html). In this study we used the stable light data, which have the same reference ellipsoid as the World Geodetic System 1984 (WGS-84), derived from the satellites, i.e., F10 (1992), F14 and F15 (2003), and F18 (2013), where nighttime passes of the DMSP typically occurred between 20:30 and 21:30 local time (Elvidge et al, 2001; Yang et al., 2011). The spatial resolution of data is 30 arc-seconds, about one kilometer from the equator, spanning -180° to 180° longitude and -65° to 75° latitude, which represents the basic global coverage of human activity areas, and then the data were transformed into the Lambert azimuthal equal-area projection so as to accurately represent area in all regions of the sphere and define the extent of interest areas. Given the fact that the continuity and comparability of images derived from the DMSP different satellites are often lacking, e.g., the amounts and geographical locations of lots of pixels in the same annual light images of different sensors are inconsistent with each other and/or the pixels' digital number (DN) values are very different (Zou et al., 2014), we employed a linear regression model to empirically fix the stable light data for 1992-2003 and 2003-2013 respectively, as shows:

$$DN_{fixing}(year) = \alpha(year \,\|\, average\_year) \cdot DN_{original}(year) + \varepsilon_{year} \quad (1)$$

where DN_{fixing} is a correct value of each pixel, and α is a correction coefficient, and *year* is 1992 and 2013, and *average_year* is a benchmark year of 2003, and $DN_{original}$ is a original value of corresponding pixel, and ε is a constant term.

Owing to obviously stable nighttime lights in different annual images, their corresponding pixels can easily be probed through valid spatial features. Based on the DN average values of each pixel of two chose stable nighttime light images (from F14 and F15) in 2003, we used a moving window with 8×8 pixels to detect the matching spatial textures of the light data for 1992 and 2013, respectively, in order to obtain the available pixels for training the above correction formula (1). There are two achieved correction coefficients of $\alpha(1992\|2003)$ = 0.980 with ε_{1992} = 0.255 (R^2 = 0.953, n = 6358) and $\alpha(2013\|2003)$ = 1.008 with ε_{2013} = 0.737 (R^2 = 0.981, n = 8121) (see Table 2), which means their own fine fitting relationships greatly valid for correction of the light data, respectively.

The high-resolution dataset of country administrative areas, derived from the Database of Global Administrative Areas (downloaded from URL of http://www.gadm.org on March-30-2016), with a few export formats of shapefiles used in most common Geographic Information System (GIS) applications, were utilized to overlap on the corrected nighttime light data, and the main cities and towns along the Silk Road Economic Belt (SREB) were selected from high-definition (see Figure 1), high-resolution imagery (e.g., Google Earth imagery with Version V7.1.2.2041, 2013) and located to the light data. Based on the broken linking line of the chose cities and towns, a set-on-demand buffer with different diameters was produced in terms of its local actual situation by using the ArcGIS software, and the study zone (region) of the SREB was thus obtained.

Additionally, The land cover data was derived from the annual GlobCover 2009 land cover product with the full resolution of 300m, produced by the European Space Agency (ESA) in collaboration with its partners (i.e., EEA, FAO, GOFC-GOLD, IGBP, JRC and UNEP) (downloaded from the website of http://www.esa.int/due/ionia/globcover). The global land cover product was produced from an automatic and regionally-tuned classification of a time series of global Medium Resolution Imaging Spectrometer Instrument (MERIS on board the ENVISAT satellite mission) Fine Resolution (FR) for the year 2009, which counts 22 land cover classes defined with the United Nations (UN) Land Cover Classification System (LCCS), i.e., croplands, forests, shrubland, grassland, artificial surfaces and associated areas (including urban areas), bare areas, water bodies, and permanent snow and ice, etc. The Digital Elevation Model (DEM) dataset was derived from the NASA/USGS Shuttle Radar Topography Mission (SRTM) 90m resolution digital elevation database with the geographic coordinate system of WGS84 (Version 4.1) (downloaded from the website of http://srtm.csi.cgiar.org /SELECTION/inputCoord.asp or http://srtm.datamirror. csdb.cn/search.jsp, i.e., http://www.gscloud.cn/), offered by the Consortium for Spatial Information (CGIAR-CSI) of the Consultative Group for International Agricultural Research (CGIAR). The GlobCover 2009 land cover and SRTM 90m DEM data are used to analyze the study area's urban spatial dynamics and development trend of urbanization synthetically associated with the previous DMSP nighttime light data (Li et al., 2007).

2.2 Extracting urban spatial scopes

Urban spatial area size, shape, and distribution are the basic elements to study urban types, spatial pattern, and population, etc.. It was necessary to extract the spatial scopes of urban cities and large settlements along the SREB, and we chose key 322 cities and towns (settlements) in the economic zone, where being a principle to choose traffic hubs, i.e., indicative population, supplies and goods distribution centers, and typically examined them to further understand the spatial pattern and urbanization of this city system. Based on the DMSP-OLS stable lights data, a spatial contrast method was used to determine the optimal group of a minimum marginal digital number value (which initially defined a urban space) and dynamical match a DN threshold so as to exactly detect urban local areas, which was iterated to extract the spatial scopes of urban cores, and it has a significantly easy, effective advantage with a high accuracy (Mi et al., 2013). Then, in this study region the spatial pattern of urban system were analyzed and the urban expansion mode, velocity, and trend were discussed with the urbanization process of the SREB by using urban land-cover spatial indices such as land cover diversity, area shape, fragmentation, and compactness parameters (Wang et al., 2006; Ma & He, 2008; Li & Yang, 2012).

2.3 Urbanized spatial intensity and pattern transition exploration for urban system

Given that urban land is a base of urban social and economic development, the reasonably intensive land development is known as a basic tactics to save land resource. Thus, urban central areas often have a high level of spatial accumulation, which indicates their intensity and concentration, and corresponding land uses while peripheral areas have lower levels of accumulation. Urbanized (land use) spatial intensity (USI) is commonly defined with a intensive degree of land use and/or land cover per unit urban area, which is measured by population density, invested capital density of urban land, or energy consumption per unit area, etc. (Yue, 2002).The urban stable night-light intensity can generally well represent utilization intensity of urbanized (land use) space because of their strong correlation (Elvidge et al. 2001; Amaral et al. 2006). Using the DMSP-OLS stable night-light data as proxy data of urbanized spatial intensity (USI), we are able to readily probe the general utilization efficiencies of urban land-use/cover spaces, spatial pattern and variation characteristics, and their spatially collaborative coupling relationships through the urban night-light intensity monitoring and analyses. These are very significant to understand and predict the changes of urbanized spatial pattern (e.g., conversion of parcel land use and extension of urban areas) and urbanization process laws, etc. and promote the urban ecological security and urban socio-economic sustainable development along the SREB.

Obviously, spatial pattern of urbanized (land use) spatial intensities, as shown at the city level or at the regional city group level, is essentially the comprehensive mapping of regional economic, social, and ecological structures, that is, the orderly combination conformation of the quantity structure and spatial forms of urbanized spatial intensity index within a scope of regionalized urban space, and the change of the fore can often inevitably lead to the correspondence of the latter and vice versa; meanwhile, spatial changes of urban land use make the quantity structure and spatial forms of urbanized spatial intensities present specific spatial and temporal evolution characteristics and laws in regional areas of a city or city and town system.

We chose the DMSP-OLS stable nighttime light data in 1992, 2003, and 2013 as proxy data sources of urbanized spatial intensity to explore the spatial-temporal development characteristics of urban land use/cover for the SREB's urban system in nearly 20 years. In order to efficiently reveal the gradually gradient development tendency of spatial pattern of urbanized spatial intensity (or, urban land cover) in this study area, a false color composite (FCC) image was created based upon three bands of Red (R), green (G), and blue (B) orderly assigned to the time series NTL data, respectively, where being stretched for color composition, as shows:

$$MONO\_DN_{\mathrm{color}(i)} = \alpha/\beta \times DN(i) \qquad (2)$$

where $MONO\_DN_{\mathrm{color}(i)}$ is a color component of each pixel of the false color image at color(i) = R, G or B band, and $\alpha+1 = 256$ and $\beta+1 = 64$ are the color grey-scale number and rank number of NTL data, respectively, and

DN represents the DN value of each pixel of NTL images in i = 1992, 2003, and 2014.

There is an evident merit as to each urban pixel (or potential urban pixel) of the false color image that apparently displays the dynamic situation of its urbanized spatial intensity and corresponding spatial combination (or accumulation) characteristics of pixels, i.e., urbanized spatial pattern gradient changes in nearly 20 years for the SREB's city system, so it effectively transforms the contents of static night-light panel data of different periods into a visual process array with intuitive spatial segregation information. The subtle changes at the pixel level are thus easily identified as well as it avoids that error accumulation caused by urbanized spatial intensity classification results in reducing detection accuracy of land use/cover changes. As shown in table 1, there are 8 types of dynamic modes of pixels for urbanized spatial intensity along the SREB during nearly two decade period, and we can lightly identify the urbanized spatial intensity changes, urban spatial development types and trends: for example, the urbanized spatial intensities of urban land use and cover in type 1-6 areas were increasing, which indicate that the areas are emerging urban plats, but the type 7 and 8 areas are recessing urban places because their corresponding urbanized spatial intensities were decreasing.

ID	Condition	Express	Meaning
1	A≥0, b≥0 & c≥1	E-E-A	Expansion during 1992-2003, and a accelerated expansion during 2003-2013
2	A≥0, b≥0 & c<1	E-E-S	Expansion during 1992-2003, and a slow expansion during 2003-2013
3	A≥0, b<0 & c≥1	E-S-A	Shrinkage during 1992-2003, and a accelerated expansion during 2003-2013
4	A≥0, b<0 & c<1	E-S-S	Shrinkage during 1992-2003, and a slow expansion during 2003-2013
5	A<0, b≥0 & c≥1	S-E-A	Expansion during 1992-2003, and a accelerated shrinkage during 2003-2013
6	A<0, b≥0 & c<1	S-E-S	Expansion during 1992-2003, and a slow shrinkage during 2003-2013
7	A<0, b<0 & c≥1	S-S-A	Shrinkage during 1992-2003, and a accelerated shrinkage during 2003-2013
8	A<0, b<0 & c<1	S-S-S	Shrinkage during 1992-2003, and a slow shrinkage during 2003-2013

*Note: a = DN_{2013} - DN_{2003}; b = DN_{2003} - DN_{1992}; $c = \left|\frac{a}{b}\right|$, and Expanding: E; Shrinking: S; A: Accelerating.

Table 1. Gradient development modes of urbanized spatial pattern

Year	Function construction			Validation		
	Pixels	Function	R^2	Pixels	r^2, P	F, t Test
1992	6358	$DN'_{1992}=0.255+0.980\times DN_{1992}$	0.953	1281	R^2=0.93, P<0.001	F=18132.08, T=7.50
2013	8121	$DN'_{2013}=0.737+1.008\times DN_{2013}$	0.981	1497	R^2=0.92, P<0.001	F=17717.53, T=7.85

Note: where R^2 and r^2 are the fitting, testing coefficients of determination, respectively, and P for the significant levels.

Table 2. Construction and validation of the Modify function

3. Results and discussion

3.1 Overall existing situation and trends of the SREB's urbanization development

We employed the 4 nighttime light (NTL) images from 4 satellites, F10 (1992), F14 and F15 (2003), and F18, and then, using the corresponding linear regression models that were relied on a stable light detection algorithm with a texture detector of a moving window with 8 × 8 pixels (see Section 2), the bi-temporal images (1992 and 2013) were corrected to reduce the system errors of NTL data, which were caused by the different performances of the satellite sensors, based upon the average value of two images in 2003 as a benchmark. As shown in Table 2, there are good consistencies between the correct and benchmark data.

The three temporal revised NTL data (in 1992, 2003, and 2013) were utilized to extract the urban spatial areas of the cities and towns along the Silk Road Economic Belt (SREB) by using a spatial contrast method at the pixel level (Mi et al., 2013) (see Figure 2) and farther explore the urbanized expansion. Meanwhile, the NTL intensities in the three images served as the proxy of urbanized (land use/cover) spatial intensities to detect their spatial characteristics and dynamic trends. In general, the NTL intensities in the SREB's city system had the obvious geographical differentiation characteristics where there was maximum brightness of NTLs over the western European countries as well as being gradually decreasing from west to east, but the NTLs over a small area of the southeast (namely, in parts of India) always had relatively high brightness during this period of years. Contrasted on that of 1992, there is an apparent increase of pixels with lower non-zero DN values in the image of 2003, suggesting that a large number of small towns (settlements) may be emerging until the time, whereas the NTL intensities remain about the same or increase slightly over the large and medium cities of Central Asia, which means a low speed urbanized expansion over the cities. However, the number of light pixels and their light brightness in the image of 2013 all have obviously increased, comparing with that of 2003, meaning there was a rapid urbanized expansion in this study region. We aimed at the 273 pivot cities from the chose 322 cities and towns, and summed up their urban spatial areas and DN values of NTL intensities (namely, urbanized spatial intensities) in 1992, 2003, and 2013, respectively, in order to further analyze the urbanized development (urban sprawl or urban decay) of the SREB's city (and towns) system. Although the total urban spatial area and total urbanized spatial intensity value in 2003 increase about 15.87% and 16.33% respectively in the case of that of 1992, there are obviously greater corresponding increases (about 66.13% and 75.28%, respectively) about that of 2013 and 2003 (see Table 3), which indicates faster urbanized expansion and regionalized economic development speeds during 1992-2003 that of 2003-2013, that is to say, we reached the same conclusion with what's mentioned above.

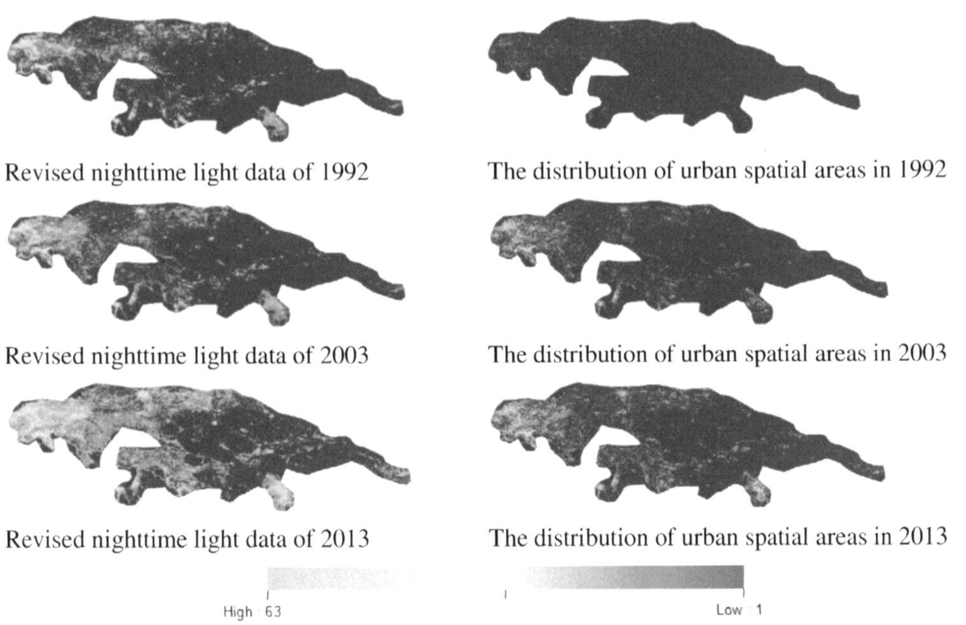

Figure 2. The three temporal revised nighttime light data and the distribution of urban spatial areas of the cities and towns in 1992, 2003, and 2013 along the Silk Road Economic Belt

			1992 to 2003		2003 to 2013	
1992	2003	2013	increased value	Growth ratio	increased value	Growth ratio

DN	37491 13	43614 32	76448 06	61231 9	0.1633	328337 4	0.7528
Areas (km2)	39195	45416	75450	6221	0.1587	30034	0.6613

Table 3. The statistic characteristics of area and sum of DN with the cities in 1992, 2003 and 2013

3.2 Urbanized spatial pattern and gradient characteristics along the SREB

In order to efficiently reveal the gradually gradient development tendency of spatial pattern of urbanized spatial intensity (or, urban land cover) in this study area, a false color composite (FCC) image was created based upon three bands of Red (R), green (G), and blue (B) orderly assigned to the time series NTL data (i.e., in 1992, 2003, and, 2013), respectively, which were used as proxy data sources of urbanized spatial intensities. Due to the close correlation between the urban NTL intensity and the associated urbanized spatial intensity and level of urban economic development, the trend of the NTL intensity change represents the corresponding urban development process. As shown in Figure 3(a), the red areas, which display scattered distribution mainly in the past Soviet Union (now parts of Russia and Ukraine, etc.), indicate that NTL maximum brightness values appeared in 1992, meaning their corresponding urbanized spatial intensities (or economic development situations) may present continuous falter and underperform after 1992. The green areas appeared in 2003, located mainly in the urban areas of Central and Eastern Europe, urbanized spatial intensities of which could increase first and then fall off. The blue areas, being scarce and mainly in small parts of Russia and India, appeared in 2013, which presented fast growth during 2003-2013, and the white indicates developed areas that are huge metropolises (such as Moscow and Delhi) and the vast Western Europe. The yellow areas presented more bright lights in 1992 and 2003, but the light intensities somewhat decreased until 2013, which means their urbanized spatial intensities increasing during 1992-2003 and to a certain declining during 2003-2013. The purple areas showed more bright lights in 1992 and 2013, but being weaker in 2003, economies of which represented being recession first and then growing. The blue-green were due to more bright lights in 2003 and 2013 but being weaker in 1992, urbanized spatial intensities of which had continued growth first with a high speed and then somewhat were slowing sales growth.

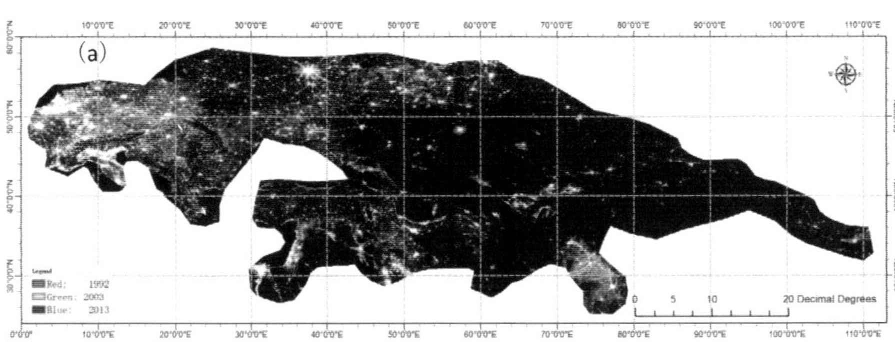

IOP Conf. Series: Earth and Environmental Science **57** (2017) 012052 doi:10.1088/1755-1315/57/1/012052

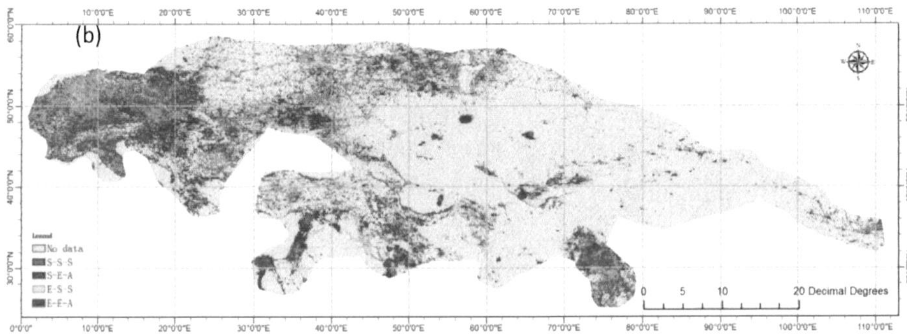

Figure 3. Nighttime light composites. (a) is represented by a primary colour (R:1992/G:2003/B:2013). (b) is Urbanized gradient mode distribution along the Silk Road Economic Belt

Hereon, according to the multiple gradient development modes of urbanized spatial pattern (see Table 1), we used the NTL data in 1992, 2003, 2013 to produce an urbanized spatial gradient mode image (see Figure 3(b)) so as to further identify the changes of NTL intensities, analyze and understand the urbanized spatial patterns and regional urban economic development and trends in this study area. There are four types of S-S-S, S-E-A, E-S-S, and E-E-A where S, E, and A correspondingly represent shrinking, expanding, and accelerating, respectively. It is obvious that the blue areas are the largest scope which indicates their on-going economic growth in faster speeds, and the green areas are located in the Moscow region of Russia and the northwest of Europe, etc., which suggest recession during 1992-2003 but slow growth after 2003. The small red and purple areas lie mainly in the urban areas of Eastern Europe and parts of the Middle East, and the Southern Ukraine, respectively.

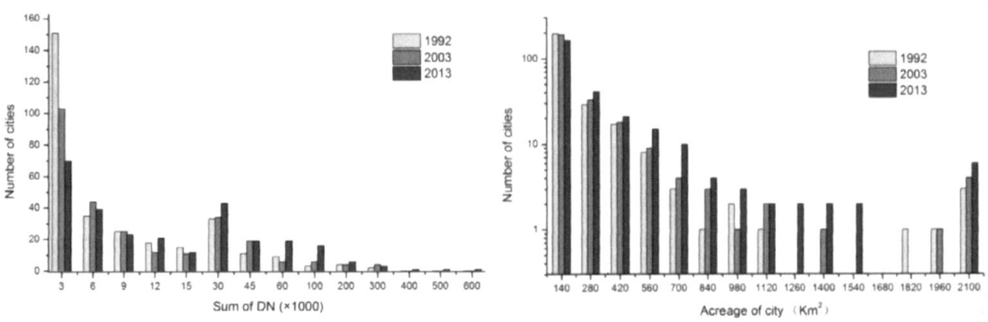

Figure 4. The frequency of the sum of DN value and acreage of 273 pivot cities

Name	Shape index			Fragmentation			Compactness		
	1992	2003	2013	1992	2003	2013	1992	2003	2013
Almaty	0.90	0.67	0.46	0.08	0.02	0.01	0.33	0.38	0.42
Amsterdam	0.64	0.48	0.35	0.03	0.01	0.00	0.25	0.30	0.37
Astana	1.27	0.91	0.48	0.10	0.02	0.02	0.37	0.39	0.39
Paris	0.27	0.24	0.21	0.01	0.01	0.00	0.28	0.32	0.34
Berlin	0.31	0.34	0.30	0.01	0.01	0.00	0.46	0.45	0.46
Bishkek	0.92	0.58	0.45	0.07	0.02	0.01	0.46	0.76	0.67
Brussel	0.49	0.43	0.31	0.02	0.01	0.00	0.32	0.37	0.44
Tehran	0.58	0.52	0.44	0.04	0.02	0.01	0.15	0.15	0.14
Dushanbe	0.51	0.69	0.54	0.01	0.02	0.01	0.46	0.71	0.68
Warszawa	0.79	0.73	0.42	0.04	0.04	0.01	0.21	0.22	0.31
Kyiv	0.45	0.46	0.27	0.01	0.01	0.00	0.47	0.51	0.64
Kabul	0.90	1.54	0.49	0.07	0.13	0.01	0.59	0.47	0.55
Kashi	2.06	1.22	0.81	0.35	0.08	0.02	0.58	0.48	0.47

Lanzhou	1.06	0.72	0.51	0.03	0.02	0.01	0.34	0.39	0.43
Lyon	0.72	0.52	0.41	0.04	0.01	0.00	0.23	0.28	0.32
Rotterdam	0.36	0.34	0.23	0.01	0.00	0.00	0.32	0.34	0.45
Mashhad	0.77	0.71	0.72	0.05	0.03	0.03	0.25	0.25	0.19
Milano	0.57	0.32	0.17	0.02	0.01	0.00	0.14	0.22	0.38
Moskva	0.52	0.46	0.17	0.02	0.01	0.00	0.15	0.18	0.33
Samarqand	0.58	0.63	0.60	0.02	0.02	0.02	0.76	0.77	0.77
Venezia	1.05	0.56	0.20	0.07	0.01	0.00	0.16	0.23	0.48
Urumchi	1.39	0.64	0.29	0.11	0.02	0.00	0.25	0.41	0.54
Xi'an	1.13	0.84	0.42	0.09	0.06	0.02	0.18	0.19	0.22
Islamabad	0.83	0.63	0.41	0.06	0.04	0.01	0.34	0.39	0.41
Yining	0.00	0.00	0.48	0.00	0.00	0.01	0.00	0.00	0.81
Duisburg	0.00	0.46	0.39	0.00	0.01	0.01	0.00	0.63	0.67

Table 4. Shape index, fragmentation, compactness of special pivot cities in 1992, 2003, 2013

In order to deeply understand the urbanization process of the SREB's city system in nearly 20 years, we chose the specially pivot cities and analyzed their landscape indices, i.e., Shape Index, Fragmentation, and Compactness. From table 4, we could find that in general the big cities often have their own low Shape Indices and Compactness degrees, compared with that of small and medium-sized cities, but there are no significant differences in terms of Fragmentation degrees. These may result from the fact that with their urbanization development, the core areas of a big city were gradually joining with the surrounding satellite cities, which leads to their Shape Indices and Fragmentation degrees declining but their Compactness degrees increasing.

4. Conclusion

We employed the 4 NTL images from 4 satellites, F10 (1992), F14 and F15 (2003), and F18, and then, using the corresponding linear regression models that were relied on a stable light detection algorithm with a texture detector of a moving window, the bi-temporal images (1992 and 2013) were corrected to reduce the system errors of NTL data, based upon the average value of two images in 2003 as a benchmark. The revised NTL data were utilized to extract the urban spatial areas of the cities and towns along the Silk Road Economic Belt (SREB) by using a spatial contrast method at the pixel level, and Meanwhile the NTL intensities in the three images served as the proxy of urbanized (land use/cover) spatial intensities, in order to detect their spatial characteristics and dynamic trends. In order to efficiently reveal the gradually gradient development tendency of spatial pattern of urbanized spatial intensity in this study area, a false color composite (FCC) image was created based upon three bands of Red (R), green (G), and blue (B) orderly assigned to the time series NTL data respectively, which were used as proxy data sources of urbanized spatial intensities. According to the multiple gradient development modes of urbanized spatial pattern, we used the NTL data in 1992, 2003, 2013 to produce an urbanized spatial gradient mode image and further identify the changes of NTL intensities, analyze and understand the urbanized spatial patterns and regional urban economic development and trends in this study area.

In this paper, we studied the urbanized spatial patterns of and the urbanized characteristics and trends of the main city system along the SREB in the view of the whole regionalized economic zone and typical cities and settlements (towns). The results showed that in general the NTL intensities in the SREB's city system had the obvious geographical differentiation characteristics where there was maximum brightness of NTLs over the western European countries as well as being gradually decreasing from west to east. There were obvious increases of the NTL DN values and NTL covering areas in 2003 and 2013 comparatively with that of 1992, which indicates the great urbanization development during this period. As for the four types of urban development process, there was an apparent consistency in a certain local area but a large heterogeneity among different areas. By analyzing the 273 pivot cities and the most pivot 26 cities, we found the number of the relatively small cities being decreasing but that of the large and medium-sized cities increasing. This study would provide the scientific support for the related researches and decision making of urbanization and urban economic development to promote the socio-economic comprehensive development of the overland SREB.

Acknowledgements

The work was supported by the Natural Science Foundation of China (NSFC) under Grant Nos. 41171340 and 41101390. This study was also supported by the Major International Cooperation and Exchange Project of National NSFC (Grant No. 41120114001), the Western Light Talent Culture Project (YBXM-2014-05), and the Science and Technology Innovation Program (for Agricultural Information Institute) of Chinese Academy of Agricultural Sciences (2016).

References

Amaral, S., A. M. V. Monteiro, A. M. V., Camara, G. & Quintanilha, J. A., 2006. DMSP/OLS night-time light imagery for urban population estimates in the Brazilian Amazon. International Journal of Remote Sensing, 27, pp. 855–870.

Chen, J., Zhuo, L. & Shi, P. J., et al., 2003. The study on urbanization process in china based on DMSP/OLS data: development of a light index for urbanization level estimation. Journal of Remote Sensing, 7, pp. 168-175.

Elvidge, C. D., Baugh K. E. & Dietz J. B., et al., 1999. Radiance calibration of DMSP-OLS low-light imaging data of human settlements. Remote Sensing of Environment, 68, pp. 77-88

Elvidge, C. D., Baugh, K. E., Kihn, E. A., Kroehl, H. W., Davis, E. R., & Davis, C. W., 1997. Relation between satellite observed visible–near infrared emissions, population, economic activity and electric power consumption. International Journal of Remote Sensing, 18, pp. 1373-1379.

Elvidge, C. D., Imhoff, M. L. & Baugh K. E., et al., 2001. Night-time lights of the world: 1994-1995. ISPRS Journal of Photogrammetry and Remote Sensing, 56, pp. 81-99.

Hao, R. F., Yu, D. Y., Liu, Y. P. & Sun, Y., 2014. Use of DMSP/OLS nighttime light data in urbanization monitor. Journal of Beijing Normal University (Natural Science), 50, pp. 407-413.

Li, J. G., He, C. Y., Shi, P. J. & Chen, J., et al., 2007. The Use of multisource satellite and geospatial data to study the ecological effects of urbanization: a case of the urban agglomerations in Bohai Rim. Journal of Remote Sensing, 11, pp. 115-126.

Li., C. Y & Yang, M., 2012. Quantity and spatial structure analysis of land resources in Xishan area of taiyuan city. Journal of Anhui Agri. Sci., 40, pp. 11088-11089, 11107.

Liao, B., Wei, K. X. & Song, W. W., 2012. Assessment and application of DMSP/OLS nighttime light data in the spatial structure of urban system: a case of Jiangxi province in nearly 16 years. Resources and Environment in the Yangtze Basin, 21, pp. 1295-1300.

Ma, Y. X. & He., B., 2008. Quantitative and spatial structure analyse for land use in Taiyuan city. J. Shanxi Agric. Univ. (Natural Science Edition), 28, pp. 121-124.

Mi, X. N., Bai, L. Y., Tan, X. H. & Lu., N., et al., 2013. A new method of extracing areas of center city regions based on DMSP/OLS data. Journal of Geo-information Science, 15, pp. 255-261, 279.

Small, C.; Pozzi, F. & Elvidge, C. D., 2005. Spatial analysis of global urban extent from DMSP-OLS night lights. Remote Sensing of Environment, 96, pp. 277-291.

Sutton, P., Roberts, C., Elvidge, C., & Meij, H., 1997. A comparison of nighttime satellite imagery and population density for the continental united states. Photogrammetric Engineering and Remote Sensing, 63, pp. 1303-1313.

Wang, H., Zheng, X. Q. & Yuan, T., 2012. Overview of researches based on DMSP/OLS nighttime light data. Progress in Geography, 31, pp. 11-18.

Wang., Z. L., Yang., Q. Y., Lu., C. Y. & Zhang, Z., D., 2007. Analysis on spatial structure of land utilization in Shapingba district of Chongqing city. Resources & Industries, 9, pp. 90-93.

Wu, J. S., He, S. B., Peng, J. & Li, W. F., et al., 2013. Intercalibration of DMSP-OLS night-time light data by the invariant region method. International Journal of Remote Sensing, 34, pp. 7356–7368.

Yang, Y., He, C. Y., Zhao, Y. Y. & Li, T., et al., 2011. Research on the layered threshold method for extracting urban land using the DMSP/OLS stable nighttime light data. Journal of Image and Graphics, 16, pp. 666-673.

Yue, J. F., 2002. Real estate law dictionary. The Post & Telecom Press, in Beijing.

Ziskin, D., Baugh, K. & Hsu, F. C., et al., 2010. Methods used for the 2006 radiance lights. Proceedings of the 30th Asia-Pacific Advanced Network Meeting, pp. 131-142.

Zou, J. G., Chen, Y. H., Tian, J. & Wang T., 2014. Construction of the calibration model for DMSP/OLS nighttime light Images based on ArcGIS. Journal of Geomatics, 39, pp. 33-37.

Investigating the origin of the Chinese name for alfalfa

Q Z Sun[1], L J Xu[2*], X J Tang[2], J T Ma[3], D Wang[2], D Li [2], Q Liu [4], Y Tao [1] and F Li [1]

[1] Grassland Research Institute of Chinese Academy of Agricultural Sciences, Hohhot 010010, China
[2] Institute of Agriculture Resources and Planning, Chinese Academy of Agriculture Science, 12 South Street of Zhongguancun, Beijing 100081, China
[3] Institute of Animal Science, Chinese Academy of Agricultural Sciences, Beijing 100193, China
[4] Animal Husbandry Institute of Liangshan Prefecture, 62 Xichang Road, Sichuan 615042, China
* Corresponding author. Email：xulijun_nmg@163.com

Abstract: It is assured that alfalfa (*Medicago sativa* L.) was introduced in Han dynasty. There are cognitive differences on whether Zhang Qian introduced alfalfa.. Based on the previous studies, research inductive method was used. The relationship between Zhang Qian and alfalfa introduction was analyzed from the motivation, experience and influence of Zhang Qian to the Western Regions and the image generation of Zhang Qian brought back alfalfa's seeds. Till to now, there are four opinions about Zhang Qian introducing Alfalfa seeds, including : (1) Zhang Qian introduced alfalfa seeds;(2) Zhang Qian did not introduce alfalfa seeds;(3) the information of Zhang Qian transferring alfalfa;(4)for commemoration Zhang Qian to the Western Regions. Although there are not direct historical materials to support Zhang Qian brought alfalfa seed to Han dynasty, it believes and confirms that the introducing of alfalfa is inextricably interwoven with Zhang Qian's western travel. Zhangqian brought relative information from western regions during the introduction, which was the basis of non-native theory, and after that, Chinese began to plant alfalfa in Han dynasty., According to historical literatures, it is clear that the Chinese diplomat brought alfalfa seeds back to China. Alfalfa, as the favorite forage to Ferghana horse, have been already planted in Dawan in Han dynasty. Despite the debate, Zhangqian played an important pioneering role in introducing alfalfa.

The Silk Road was a network of trade routes, formally established during the Han Dynasty of China, which linked the regions of the ancient world in commerce. As the Silk Road was not a single thoroughfare from east to west, the term 'Silk Routes' has become increasingly favored by historians, though 'Silk Road' is the more common and recognized name. Both terms for this network of roads were coined by the German geographer and traveler, Ferdinand von Richthofen, in 1877 CE, who designated them 'Seidenstrasse' (silk road) or 'Seidenstrassen' (silk routes). The network was used regularly from 130 BCE, when the Han officially opened trade with the west, to 1453 CE, when the Ottoman Empire boycotted trade with the west and closed the routes (Figure1).

According to historical records, the land and maritime silk roads of ancient China has had huge influence on promoting eastern and western cultural and economic exchange. Meanwhile, crops originated from Europe, West Asia, America and other areas started to be introduced to China. More than 20 species of crops were imported through the land silk road including grape and alfalfa (forage) as two of the firsts. In the Song and Yuan Dynasties, introduced a lot of foreign crops are mainly through the Silk Road on the sea. The introduction of overseas crops has a profound impact on China's agriculture.

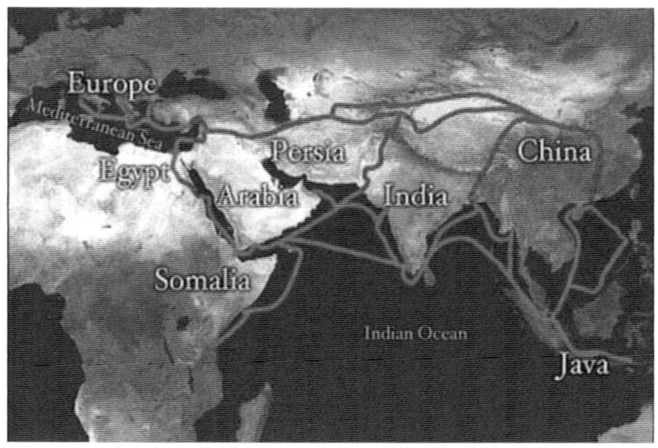

Figure 1. Main routes of the Silk Road

Alfalfa is not only an excellent forage worldwide, but also is a symbol of the cultural exchange between China and western regions, being regarded as a dazzling bright pearl on the Silk Road. Although alfalfa has more than 2000 years of cultivation history in China, scholars in pratacultural science, agricultural history, and China's history has different opinions on when alfalfa had been spread into China. China has a long history of alfalfa (Mu Xu in Chinese) cultivation. Because of the vast territories and the diverse ethnicities and dialects of the people in China, the Chinese names of alfalfa demonstrate diversity and hybridity, whether in people's interactions in the modern world, or in the vast literature of ancient classics, with the phenomenon of the long-term coexistence of official and alternative names. According to the Zhang Pingzhen's research [1], alfalfa has more than 30 official names, alternative names, and nicknames in Chinese. In this paper, Xia Weiying [2] and Ding Guangqi's [3] plant name authentication method would be used to explore the origin of the names of alfalfa in China. The formation of synonyms with the same sounds but different characters, or with different sounds, was also discussed, in order to understand the history of alfalfa names clearly and provide a theoretical basis of alfalfa history and culture in China.

1. Etymology of the Chinese word for alfalfa

Under the Records of the *Records of the Grand Historian·Ranked Biographies of Dayuan*, alfalfa was brought from Dayuan (Fergana) by an envoy in the Han Dynasty. The *Book of Han·Traditions of the Western Regions* documented it as Mu Xu. Dayuan is the current Fergana basin in central Asia. In 1919, the famous American Sinologist Laufer [4] thought that the Dayuan language was an Iranian dialect, and that Mu Xu was a borrowed word from Iranian language. In 1961, the Polish Sinologist Chmielewki [5] doubted Laufer's finding. According to Chmielewki, the word Mu Xu could not be found in the Iranian language and needed another etymology. He believed that the alfalfa found in the Han dynasty was likely from Kashmir. Kashmir also produced alfalfa, and the Han envoy had also seen alfalfa there. At that time, the Kashmir language was an Indian dialect related to the Sanskrit language. Although the early corresponding word of Mu Xu could not be found in the Kashmir language, alfalfa was a honey-producing plant and the word "honey" in the Sanskrit language was translated as "māksika", which was likely used as the name of honey-producing alfalfa. Thus Mu Xu was likely the translation of māksika in the Kashmir dialect. Chen [6] suggested that Mu Xu and its other names that had the same sounds but different characters were associated with the Sanskrit and Persian languages.

Based on the *Chinese Dictionary of Foreign-origin Words*, Mu Xu originated from * buksuk, * buxsux, and * buxsuk of the Eland or Dayuan languages [7]. Sun [8] (2005) pointed out that Mu Xu was derived from * buksuk, or * buxsux (Old Persian or Tocharian (= Dayuan)). Laufer [5] suggested that Mu Xu might be associated with "buksuk" or "buxsux" in the old Eland language [9]. Feng [10] (2004) also thought that Mu Xu was the translation of "buksuk" in the ancient Dayuan language. Zhang [1] (2006) pointed out that Mu Xu was derived from the geographical introduction of the ancient Dayuan language. Because of the frequent exchanges between Central and Western Asia, and common national languages, Dayuan and Persian languages were very close to each another, thus Mu Xu was the transcribed name of "buksuk" or "buxsux" in the ancient Islamic and Dayuan languages. The *Chinese Dictionary* and *Cihai (Agriculture Volume)* also stated that Mu Xu was a translation of "buksuk" in the ancient Dayuan language [11-12]. Xu [13] (2002) believed that the different ways of writing Mu Xu in Chinese, as well as other names, such as Xu, Mao Xu, and Mu Su, were all different translations of the same foreign word [9,14,15]. Yu [15] (1958) also found that the Chinese term Mu Su corresponded to the pronunciation of Musu in Iran. Because of the etymology of alfalfa and the contrasting views concerning its origins, Xu [16] (2007) said the Chinese interpretation of "alfalfa" in the dictionary should include different opinions (Table 1).

2. Zhang Qian and alfalfa

2.1 Theory that Zhang Qian brought alfalfa seeds to China

In the *Compendium of Materia Medica*, Li [17] said, "Miscellanies recorded that alfalfa was from Dayuan and the Han envoy Zhang Qian brought it to China." Li Shizhen was a great botanist, and the *Compendium of Materia Medica* is recognized worldwide for its high status in botanical history. Thus, its broad and profound impact led to the world belief that "alfalfa was from Dayuan and the Han envoy Zhang Qian brought it to China." In 1881, De Candolle [18] clearly stated in the *Origin of Cultivated Plants* that *Compendium of Materia Medica* recorded that Zhang Qian returned with items not present in China, including cucumbers and alfalfa. In 1907, Japanese scholar Matsuda Sadahisa [19] stated, in the *Records of the Grand Historian·Ranked Biographies of Dayuan*, in the sentence that "grapes were fermented into wine and horses were fed with alfalfa; the Han envoy brought them back so the emperor began to grow alfalfa and grapes," the emperor was Emperor Wu of Han and the Han envoy was Zhang Qian. The authors of *Ranked Biographies of Dayuan* were officers in the Han Dynasty, so this story is true. In the late Qing dynasty, Huang [20] also suggested that Zhang Qian brought alfalfa into China. In "Alfalfa planting," a chapter in *Qimin Yaoshu Translation and Annotation*, Mou [21] said "Alfalfa is a plant brought back by the Han envoy Zhang Qian from the Western Regions." In the *Complete History of China*, Shi Zhongwen and Hu [22] also pointed out that the authors of the *Book of Han·Traditions of the Western Regions* wrote that the Kashmir area had alfalfa, the Han envoy Zhang Qian retrieved it, and that the emperor treated it as a rare plant species, planting it in the garden of an annexed palace for appreciation.

Table 1. Alfalfa name evolution

Author	Dynasty	Literature	Name	Description

Ban Gu[23]	Han	Hanshu·Xiyu Zhuan	Mu su	Kasmira horizon, mild, with musu. The customs like to drink, horse like to eat musu. In the Han Dynasty, the officials brought back the seeds of the host.
Xu Shen[24]	Han	Shuowen Jiezi	Mu su	Yun is a kind of grass, it looks like musu.
Cui Shi[25]	Han	Siminyueling	Mu su	A lunar January, spring wheat, or musu. Pasture seed and miscellaneous garlic, can also be planted.
Guo Pu[26]	Jin	Erya Zhushu	Mu xu	Yellow flower. Guo Pu note, Niu Yun grass is Huang hua. Leaves like muxu.
Du You [27]	Tang	Tong Dian Bian Fang Dian	Alfalfa	Kasmira horizon, mild, with alfalfa.
Luo Yuan[28]	Song	Eryayi	Mu su	(Alfalfa) autumn fruit, seeds look like Ji, and named Mu Su, and used for rice.
Li Shi-zhen[17]	Ming	Bencao Gangmu	Mu su	Li Shizhen said that alfalfa was called Musu by Guo Pu, which grew naturally and was feed to cattle or horses.

2.2 Theory that Zhang Qian did not bring alfalfa to China

It is indisputable that alfalfa was introduced to China in the Han Dynasty. However, there are no official records that support that Zhang Qian brought alfalfa to China from the Western Regions. According to *Er Ya Yi*, which was written by Luo Yuan in the Song dynasty [28], alfalfa was produced in the Western Regions and was introduced to China at the beginning of the Han Dynasty, but names of envoys that possibly brought alfalfa into China were not mentioned. Luo Yuan also said, "*Naturalis Historia* described that Zhang Qian made other envoys bring back grapes, Hu onions, and alfalfa seeds in the Western Regions, and because Qian was the most famous one, he was the one mainly mentioned." This is the earliest historical record that disputes the theory that Zhang Qian introduced alfalfa seeds to China. Shi Shenghan believed that Jin Zhanghua and Lu Ji had most likely repeated Wang Yi's statement or the legend based on Wang Yi, and that Ren Fang's record was wrong. Zhang [29] believed that Wang Yi lived in the period of the Emperor Shun of Han, and it was speculated that Zhang Qian introduced garlic, grapes, and alfalfa to China. *Records of the Grand Historian* and *Book of Han* detailed the earliest historical record of alfalfa, but only described two diplomatic events involving Zhang Qian and his exploration of the Western Regions, and did not mention Zhang Qian bringing alfalfa to China. Moreover, the *Tongjian Chronicle* [30] and *Zizhi TongJian* [31] did not mention Zhang Qian bringing alfalfa to China.
However, all of these books recorded that Han envoys brought alfalfa seeds to China.

2.3 Theory of Zhang Qian's delivery of information about alfalfa information

Li [31] believed that Zhang Qian was the first person in the Han Dynasty to discover Fergana horses and alfalfa in the Western Regions. He said when Zhang Qian first returned from the Western Regions, he not only had information on a special horse in Western Dayuan, but also knew about a favorite forage grass for horses, which was alfalfa. In *Records of the Grand Historian·Ranked Biographies of Dayuan*, the authors stated, "Dayuan is southwest of Xiongnu and straight west of the Han dynasty, thousands of miles from Han. The indigenous people live by farming, grow rice and wheat, and have grape wine and lots of good horses, including the Ferghana horse, a heavenly horse." The authors also stated, "Around Dayuan, grapes are fermented into wine, and the rich store lots of wines, which do not deteriorate for a long time. people are addicted to wine, while horses to alfalfa " This is what Zhang Qian saw in Dayuan and shared with Emperor Wu of Han [32]. This was also the first record of Zhang Qian introducing agricultural production or Dayuan products into China [33-35]. Li [36] pointed out that the theory of Zhang Qian introducing many Western products to China, such as grapes, alfalfa, and pomegranates, etc., might not be entirely true, but that it was certainly true that Qian introduced the information about alfalfa, Ferghana horses, and other Dayuan products to China.

2.4 Memorial theory of Zhang Qian's opening of western regions

In *Records of the Grand Historian·Ranked Biographies of Dayuan*, the authors stated, "Around Dayuan, grapes are fermented into wine, and the rich store tons of wines, which do not rot for a long time. People are addicted to wine and horses are addicted to alfalfa. The Han envoy brought them back so the emperor began to grow alfalfa and grapes. Because of the many heavenly horses and foreign envoys that visited, large areas of grapes and alfalfa could be viewed from outside of the palace." Li [37] (2011) believed that this was events that happened after the death of Zhang Qian, and Sima Qian did not state who was the Chinese envoy in the *Records of the Grand Historian*. Because of the habit of the Chinese to give the credit to famous people, and that Zhang Qian's exploration of the Western Regions had made him famous, the credit of bringing alfalfa to China was given to him. Many studies stated that Zhang Qian did not really bring back seeds of alfalfa and other cultivated plants to China, but that he often talked about good products in the Western Regions, such as alfalfa, grapes, and Ferghana horses. *Records of the Grand Historian* and *Book of Han* both documented that he was also introduced to Emperor Wu of Han. At the same time, Zhang Qian's visit to the Western Regions provided a precondition for the Han Dynasty to seek grapes, alfalfa, Ferghana horses, and other products from the Western Regions. This provided a good reason for why he was praised [38-40]. Zhang Qian was thought to be responsible for all the Chinese and Foreign exchanges during this period, and, with the development and prosperity of economic and cultural exchanges between China and the Western Regions, his impact has grown. According to the legend, Zhang Qian brought back grapes, alfalfa, pomegranates, walnuts, and flax seeds, etc. from the Western Regions, but this might not be true. However, Zhang Qian made outstanding contributions to the Silk Road, opening up the Western Regions to China, which is still a universally praised achievement [36].

3. Summary

Alfalfa is an exotic plant from the Western Regions. It was initially planted in China during the Han Dynasty and from then the word Mu Xu appeared in the Chinese language. So far, there is still a big difference between Chinese and other foreign scholars in understanding of the Chinese "alfalfa" etymology. It needs our more exploration and investigation in future to confirm whether Zhangqian introduced alfalfa to Han dynasty.Although the belief that "alfalfa was originally produced in Dayuan and brought to China by Han envoy Zhang Qian" has been widely accepted, it lacks direct historical evidence. Referring to *Ranked Biographies of Dayuan, Zhang Qian and Li Guangli, Traditions of the Xiongnu, Traditions of the Western Regions,* and *Traditions of the Yi of the Southwest* from *Records of the Grand Historian* and *Book of Han*, only two of Zhang Qian's diplomatic events and exploration have been documented, but the plants that Qian might have brought back are not mentioned. The information that Zhangqian brought back helped the people learn how to produce alfalfa and raise Ferghana horses in DawWan. These, probably, finally facilitate the introduction.Thus, Zhang Qian played an important role in the process of introducing alfalfa to China.

As with other crops, researches of alfalfa have developed rapidly in modern China, although it is close to our life, we know litter about them because of a lack of relative knowledge. Lots of literature were consulted and a classification and summary was made. The results showed that lots of researches on alfalfa have been carried out, including original, species, utilization, etc. They also collected and identified the samples, such as *Medicago sativa*, *M. Lupulina*, *M. Minima*, and studied alfalfa nodule bacteria in northwest of China, moreover, the habitat, adaptation and the requirement of moisture and fertilizer were studied. The western science and technology and local test results have been rapid and widespread dissemination, the results and scientific technological knowledge have been spread in the relevant journals, newspapers and other media. All of these have laid the foundation for the research and the construction of the discipline framework of alfalfa in modern China.

Acknowledgments

The authors thank Qing Ge-le for assistance in references correction. This research was supported by the "973" Program (2015CB150800), the China Agriculture Research System (CARS-35), Public Sector Projects in the Ministry of Agriculture (201303060).

References

[1] Zhang P Z 2006 *In China Vegetables* (Beijing: The Peking University Publishing House)
[2] Xia W Y 1990 *Botanical Name Release Notes* (Beijing: China Agriculture Pres)
[3] Ding G Q 1957 *Plant Name Release* (Beijing: Science Publishing House)
[4] Chmielewski J 1961 Two Early Loan-words in Chinese *Rocznik Orientalistyczny* **24** 69–83
[5] Lao F E (Translated by Lin X Y) 1964 *Sino-Lranica* (Beijing: The Commercial Press)
[6] Chen Z T 1957 *Economic and Cultural Exchanges Between the Han Dynasty and the Western Regions and Other Places* (Shanghai: Shanghai People's Publishing House)
[7] Liu Z Y 1984 *Chinese Dictionary of Foreign Word* (Shanghai: Shanghai Lexicographical Publishing House)
[8] Sun J T 2005 *Linguistics* (Beijing: The Commercial Press) pp 48–71
[9] Feng F R 2009 *Analysis and Research on the Phenomenon of Language Evolution in the Five Four Period of the New Youth* (Taiwan: Tamkang University)
[10] Feng T Y 2004 *The Origin of New Words* (Beijing: Chinese Publishing House)
[11] Luo Z F 1986-1993 *Chinese Dictionary* (Shanghai: Chinese Dictionary Press)
[12] Editorial Board of the Dictionary 1978 *Ci Hai Section on Agriculture* (Shanghai: Shanghai Lexicographical Publishing House)
[13] Xu W H 1992 *Chinese Lexicology Introduction* (Beijing: The Commercial Press)
[14] Sun F X 2008 Analysis of "Bencaogangmu" in Foreign Words Error Training *Literature Education (on)* **11** 148–50
[15] Yu J R 1952 Sweat Blood Horse and Clover *Continental magazine* **5** 24–5
[16] Xu W K 2007 A brief discussion on the origin of Chinese loan words and the compilation of dictionaries *The 2nd Session of the Int. Academic Symp. on Traditional Chinese Research (Shanghai)* vol 2 pp148–56
[17] Li S Z 1982 *Ben Cao Gang Mu* (Beijing: People's Health Publishing House)
[18] Dole D (Translated by Yu D J and Cai X T) 1936 *Originated in Agronomic Plants* (Shanghai: The Commercial Press)
[19] Matsuda J 1907 Study on Alfalfa's Names and Species of Medicago L in China *Journal of Botany* **21** 1–6
[20] Huang Y R 1911 Textual Research on Alfalfa *Eastward Magazine* **8** 26–31
[21] Miao Q Y 2009 *The Notes of Qiminyaoshu* (Shanghai: Shanghai Ancient Books Publishing House)
[22] Shi Z W and HU X L 1994 *The Whole History of China* (Beijing: People's Publishing House)
[23] Ban G 2007 *Han Shu* (Beijing: Chinese Publishing House)
[24] Xu S (Revised by Xu X) 2013 *Shuo Wen Jie Zi* (Beijing: Chinese Publishing House)
[25] Cui S 1965 *Si Min Yue Ling Jiaozhu* (Beijing: Chinese Publishing House)

[26] Guo P and Xing B 2010 *Erya zhushu* (Shanghai: Shanghai Classics Publishing House)
[27] Du Y 1982 *Tong Dian* (Beijing: Zhonghua Shuju)
[28] Luo Y 1991 *Eryayi* (Hefei: Mount Huangshan Publishing House)
[29] Zhang B 1989 *The History of Agriculture and Animal Husbandry in Northwest China*(Xi'an: Shaanxi Science and Technology Press)
[30] Yuan S 1964 *Tong Jian Chronicle Beginning-End* (Beijing: Chinese Publishing House)
[31] Si M G 1956 *Zi Zhi Tong Jian* (Beijing: Chinese Publishing House)
[32] Li J X and Yu T S 2009 "*Tong Dian*" *Document from Western Regions to Note* (Shanghai: Shanghai people's Publishing House)
[33] Chinese Academy of Agricultural Sciences Chinese Agricultural Heritage Research Laboratory of Nanjing Agricultural College 1959 *Chinese Agricultural History* (Beijing: Science Press)
[34] Feng H M 1979 *Zhang Qian to the Western Regions* (Beijing: Chinese Publishing House)
[35] Chen Y 2004 *The Outline Of Chinese History* (Shanghai: Shanghai University Press)
[36] Li D F 2007 *Continually Talk about the History of Qin and Han* (Shanghai: Shanghai people's Publishing House)
[37] Li C D 2011 The Fate of the Grape in China *Examination weekly* **51** 232–4
[38] Shi S H 1963 Discussion on the relationship between China introducing plant from Western Regions and Zhang Qian *Chinese Journal of History of Science* 16–33
[39] Hou P X 2006 *Research on Ferghana Horse* (Lanzhou: Gansu Culture Press)
[40] Li R H 2014 Zhang Qian Image in Wei, Jin, Southern and Northern Dynasties *Chinese Culture Forum* **94** 100–3

The Crucial Records Number to Retrieve Offshore Directional Wind Distribution

X Zhu[1, 2], Z Li[1] and X Yang[1]

[1] State Key Laboratory of Remote Sensing Science, Institute of Remote Sensing and Digital Earth, Chinese Academy of Sciences, 100101 Beijing, China

[2] University of Chinese Academy of Sciences 100040 Beijing, China

zhuxh@radi.ac.cn

Abstract. The wind energy production estimates are very important to a wind power project. And, the remote sensing technique has been widely used to obtain the offshore wind speed and direction which could be used to calculate the wind energy of potential wind farm. However, the directional wind energy distributions are rarely studied, which also play important roles in analysis of wind farms' potential power. In this article, the minimum number of records to obtain offshore directional wind distribution is stated by simulation experiment on In-situ dataset. The NDBC buoy dataset is randomly and multiply sampled to build new dataset under different numbers of observation records, which vary from 21 to 800. The resample under the same number of observation is repeated for 100 times to build dataset group. The directional wind distribution of new dataset is compared with the one of original buoy dataset, and errors made by dataset with fewer records are calculated. Besides, the 10th largest error in the sampled dataset group, which have the same number of observation records, is regarded as the error bound for those dataset. The change rule of the error bound is shown by fitted curves. Based on the fitted curves, minimum number of records is calculated. By this simulation experiment, the minimum number of records to represent wind direction frequency is 350, and 800 for annual direction distributions of wind energy density. To reduce the number of records needed in retrieval, some methods are discussed and tested.

1. Introduction

1.1 *Research Background*

Wind energy brings about ecological and social benefits and contributes to sustainable energy development. Using electricity from wind turbine instead of electricity from coal power plant could save an average of about 820-910 tonnes of CO_2 for every GWh [1]. The wind is usually stronger and steadier in offshore areas, making offshore wind energy a major trend in wind power industry [2]. And, offshore directional wind distribution plays a valuable role in this trend. Using wind direction distributions information to analysis potential wind power of a specific site plays an important role when selecting the location of new wind energy facilities, and has impacts on both, price of generated energy and return of investment [3]. Some researches about how to use the multiple satellite data to assess offshore wind resources have been done [4]. The mean wind speed and wind power density are the focus of attention in those studies. The parameters of directional wind distribution, however, have not been sufficiently studied.

1.2 *Parameters*
According to the Chinese national standard GB/T 18710-2002, the direction distributions used to investigate wind farm site are annual wind direction frequency (DF) and annual direction distributions of wind energy density (DDED). Both parameters could be shown by rose diagram. (Showed in Figure 1) Those two distributions both have 16 directions.

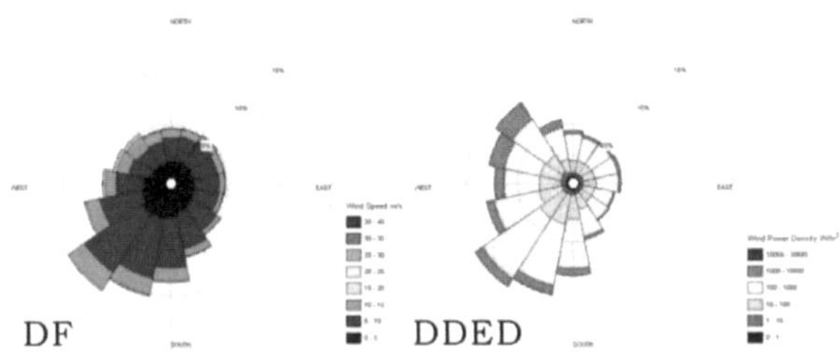

Figure 1. Rose diagram of two wind direction distributions

The equation of annual wind direction frequency:

$$DF_j = \frac{n_j}{\sum_{j=1}^{16} n_j} \quad (1)$$

where DF_j = the frequency of direction j(When j=1, j means north. When j=5, j means east, etc.)
n_j = the number of wind records whose directions are j

The equation of annual direction distributions of wind energy density:

$$DDED_j = \frac{D_j}{D} \quad (2)$$

$$D = \sum_{j=1}^{16} D_j \quad (3)$$

$$D_j = \frac{1}{2}\sum_{i=1}^{m}(\rho)(v_{ji}^3)t_{ji} \quad (4)$$

where $DDED_j$ = the wind energy density distribution of direction j
D_j = the wind energy density of direction j
D = the total wind energy density
ρ = the air density
v_{ji} = the wind speed of record NO.i in direction j
t_{ji} = the duration represented by the record
m= the records number of direction j

1.3 *Research Purposes*
In this paper, how many records are needed to retrieve reliable directional wind distributions is discussed. Moreover, some approaches to reduce the record number in retrieval are presented. And, the results of those approaches are also showed.

2. Data Description

2.1 *In-situ buoy measurements*
The In-situ wind measurements at 3~10 m above the ocean surface, with mean values at time intervals of 1 h are provided by National Data Buoy Centre (NOAA/NDBC). The detail of station information is described in Table 1. Those four buoys include diverse offshore wind condition, giving this study

IOP Conf. Series: Earth and Environmental Science **57** (2017) 012054 doi:10.1088/1755-1315/57/1/012054

enough representative records. All the in-situ data used in this study were checked for quality control to delete inaccurate and unrealistic records [5].

The criteria to discard unreliable records:

$$|\vec{u}| < 0.4ms^{-1}; |\vec{u}| > 75ms^{-1} \tag{5}$$

$$\text{When } |\vec{u_i}| > 15ms^{-1}, |\vec{u_i}| > |\vec{u_{i-1}}| \times 1.5 \text{ and } |\vec{u_i}| > |\vec{u_{i+1}}| \times 1.5 \tag{6}$$

$$0^{\circ} > \theta > 360^{\circ} \tag{7}$$

where, $\vec{u}$ = the wind vector

θ = the angle of wind direction

Table 1. The description of In-situ wind measurements

Station	41048	42040	46001	46086
Location	31.87°N 69.57°W Atlantic	29.21°N 88.21°W Gulf of Mexico	56.30°N 147.92°W Gulf of Alaska	32.49°N 118.04°W Pacific
Sea water depth	5340 m	164.6 m	4206 m	1828.8 m
Anemometer height	5 m	10 m	5 m	5 m
Buoy	3-meter discus	10-meter discus	6-meter NOMAD	3-meter discus
Payload	ARES	ARES	ARES	ARES
Data period	2008-2014	2005-2014	2005-2013	2005-2014

2.2 *ASCAT wind observations*

The ASCAT L2 12.5-km swath near real-time wind products are provided by the NASA/JPL Physical Oceanography Distributed Active Archive Centre [6].The wind products contain quality control information, according to which unreliable records will be deleted. Because of the orbit characteristics of the satellite, MetOp, which carries the ASCAT, one specific sits' wind records only include several specific hours' data. In this study, ASCAT wind products in 2010 are used to analyse the effectiveness of approaches to reduce the minimum number of records.

2.3 *Data analysis*

To get the basic understanding of those data, the directional wind distribution parameters of those buoy observation dataset groups are calculated. And, circular statistics of those distributions are also analysed, which include average, standard deviation, skewness and kurtosis. These analysis results are showed in Figure 2 and Table 2. Moreover, the time series characteristic of ASCAT data is analysed in Figure 3, which prepares for the research work of over-sample.

Table 2. The circular statistics of In-situ wind datasets

Station	41048	42040	46001	46086
		Circle statistics of DF		
Mean	212.57°	100.42°	226.00°	293.30°
Std dev	72.39°	72.88°	68.87°	46.20°
Skewness	0	0.06	0.12	0.06
Kurtosis	0.12	0.04	0.02	0.46
		Circle statistics of DDEF		
Mean	245.18°	61.35°	226.28°	292.94°
Std dev	70.55°	71.76°	69.86°	34.44°
Skewness	-0.06	0.02	0.20	0.11
Kurtosis	0.04	-0.15	-0.07	0.71

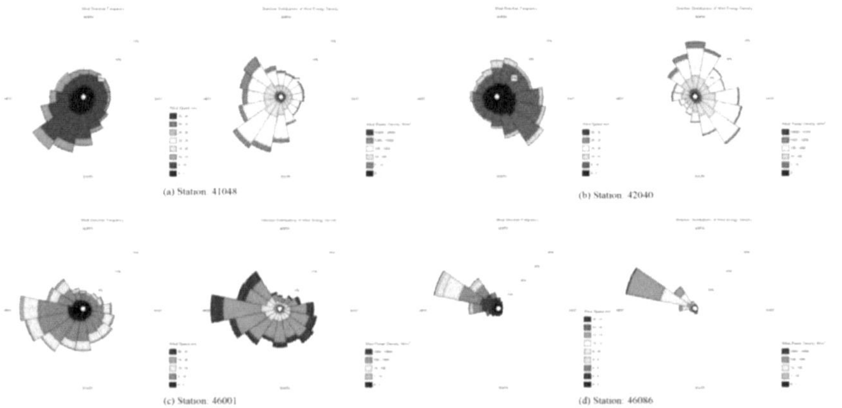

Figure 2. The DF and DDED of those stations

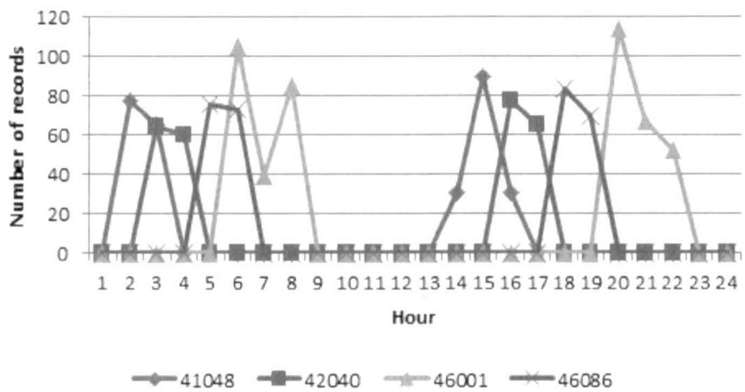

Figure 3. The time series of ASCAT data

3. Methodology

3.1 *Fitting of errors change and finding the crucial records number*

To examine the influence of the record number on retrieval accuracy, the dataset from In-situ buoy measurements was randomly and multiply sampled for a range of number of observations from n=21 (assumed to be the lower bound on the dataset likely to be obtained using remote sensing) to n=0.1 of the actual number of observations available from buoy for the annual data collection period [7]. The resampling was undertaken 100 times for each n to build 100 data sets for every n, number of records. Those new-built data sets' offshore directional wind distributions, DF and DDED, are calculated and compared with the one of the original dataset, which is regarded as real distribution. In this compare, the total difference of distribution's 16 directions is summed and normalized to demonstrate the error caused by fewer observations. Obviously, the error reduces swiftly with the number of records' increase. To avoid the extreme condition, the 10th largest error in the group of 100 new-built data sets of the same n is regarded as this records number's error bound at a confidence level of 90%. One sample is presented in Figure 4(a) and (c).

Different types of fitted curves are tested to give a good fit to the change rule of error bound. In these fitted curves, the 2-terms exponential function gives superior and robust fit. The fitted curve of one sample is presented in Figure 4(b) and (d). With this function, the minimum number of records at which the error bound decline to below 10% could be calculated. 2-terms exponential function:

$$f(x) = a \times e^{bx} + c \times e^{dx} \tag{8}$$

where, a, b, c and d are parameters in fitted function

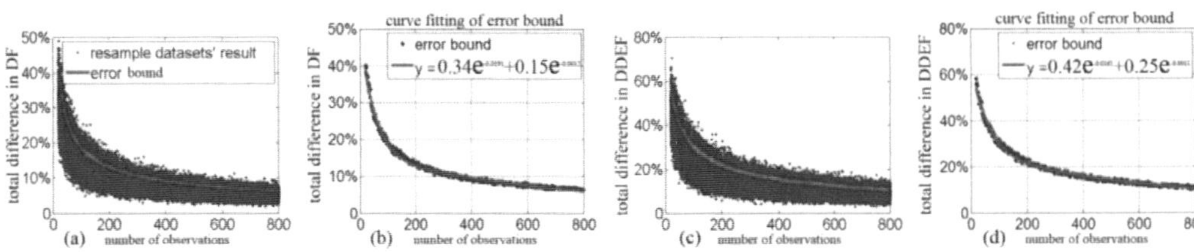

Figure 4. The error bound of sampled dataset and fitted curve

3.2 *Over-sample*

The different revisiting period of satellites makes them observe the same site different times in one year, and, the polar orbit satellite usually revisiting a specific place at fixed local time. As a result of this, when multiple satellite data is used to represent directional wind distribution, the number of records of different hours might be imbalanced. Because wind observations at some offshore area exhibit significant diurnal cycles due largely to advective effects [8], this imbalance data might distort the retrieval results. To handle with imbalanced data, the sub-dataset of hour with fewer records is over-sampled. Moreover, over-sampling also adds new records to dataset which might reduce the number of records of original dataset needed to obtain reliable distribution.

The over-sample technique used in this research work is improved SMOTE over-sample algorithm [9]. In the improved SMOTE method, new synthetic samples created by randomly interpolating pairs of selected nearest neighbours are added into minority sub-datasets. The number of new synthetic samples and the way to find the nearest record are changed in test to find the best over-sample method which could be used to reduce the minimum number of records. In the test, the imbalanced experiment dataset is built by conditionally sampling from in-situ records, which imitates the time series of ASCAT data.

The new synthetic sample $S_j(j = 1, 2, 3 \ldots\ldots m)$:

$$S_j = x_i + \text{rand}(0,1) \times (x_{ij} - x_i) \tag{9}$$

where, m = the oversample multiple

x_i = the record in minority sub-dataset

x_{ij} = the No. j nearest record in the same sub-dataset

$\text{rand}(0,1)$ = a number randomly chosen from (0,1)

3.3 *Weighting*

By analysing the In-situ data, the seasonal cycle and diurnal cycle of offshore wind in those areas are discovered, which are illustrated in Figure 5. The diversity of wind in difference months and hours means the records in some sub-dataset are more representative than others. Using the more representative sub-dataset to represent directional wind distribution will get more reliable result. In this condition, putting difference weight on records according to their representativeness could be a method to get better retrieval result with fewer records.

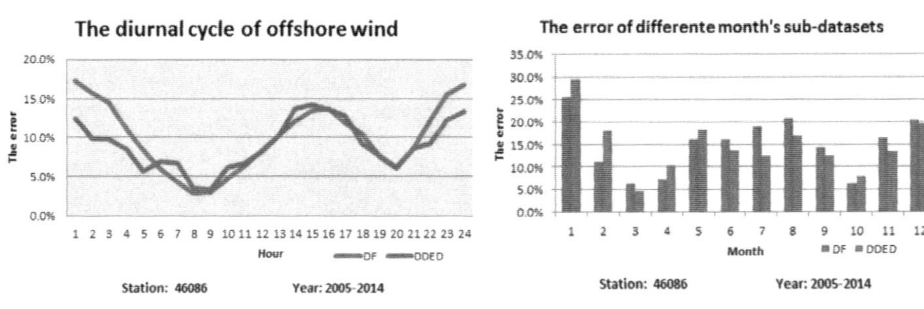

Figure 5. The seasonal cycle and diurnal cycles of offshore wind

The new equation for DF with weighting:

$$n_j = \sum_{i=1}^{k} \varphi\ (t)\ ;\ DF_j = \frac{n_j}{\sum_{j=1}^{m} n_j} \quad (10)$$

where DF_j = the frequency of direction j(When j=1, j means north. When j=5,j means east, and so on.)
n_j = the number of wind records whose direction
φ (t) = the function to calculate the weight of one record according to its time

The new equation for DDED with weighting:

$$DDED_j = \frac{D_j}{D};\ D = \sum_{j=1}^{16} D_j \quad (11)$$

$$D_j = \frac{1}{2}\sum_{i=1}^{m}(\rho)(v_{ji}^3)\,\psi\ (t) \quad (12)$$

where $DDED_j$ = the wind energy density distribution of direction j
D_j = the wind energy density of direction j
D = the total wind energy density
ρ = the air density
v_{ji} = the wind speed of record i in direction j
t_{ji} = the duration represented by the record
m= the records number of direction j
ψ (t) = the function to calculate the weight of one record according to its time

The weighting function φ (t) or ψ (t) :

$$\varphi\ (t)\ \text{or}\ (t) = 1 + \frac{weight(t) - minimumweight}{maximalweight - minimumweight}\ ;\ weight(t) = \frac{1}{error_{month}(t)} \times \frac{1}{error_{hour}(t)} \quad (13)$$

where $error_{month}(t)$ = the DF or DDED error of the month t sub-dataset
$error_{hour}(t)$ = the DF or DDED error of the hour t sub-dataset
weight(t) = the weight of the record whose time is t
minimumweight = the minimum weight in every record's weight
maximalweight = the maximal weight in every record's weight

4. Results

4.1 *Minimum number of records*

The results of the minimum number of records needed to represent DF and DDED realistically are showed in Table 3. The results are different for different offshore areas because of these areas' characteristic. Generally, when n >350, the error for DF declines to below 10% at an acceptable confidence level. And, to make the error under 10% for DDED, the n need to be more than 800.

Table 3. The minimum number of records to get reliable retrieval DF and DDEF

	Station 41048		Station 42040		Station 46001		Station 46086	
	DF	DDED	DF	DDED	DF	DDED	DF	DDED
2005	/	/	363.85	1102.10	344.55	637.38	263.42	569.54
2006	/	/	370.50	843.84	344.21	689.07	266.41	518.73
2007	/	/	365.03	728.29	308.93	605.94	259.12	528.17
2008	367.99	808.52	368.21	903.55	353.66	780.59	264.49	567.19
2009	366.15	785.23	330.37	586.52	367.28	777.51	255.03	541.13
2010	361.72	752.46	360.63	747.23	357.81	722.62	267.12	645.36
2011	371.71	1041.09	374.51	845.73	325.42	615.48	267.02	575.45
2012	365.69	860.04	375.46	NULL [a]	368.85	790.12	272.58	562.45
2013	372.71	785.60	366.07	NULL [a]	334.02	585.23	285.39	495.39
2014	369.68	806.60	368.09	800.10	/	/	265.62	533.95
Average	367.95	834.22	364.272	819.67	344.97	689.32	266.62	553.736
	Average for DF		335.95		Average for DDEF		724.23	

[a] The crucial number could not be calculated because of the interference of hurricane records

4.2 *The influences of over-sample*

Because of the representativeness diversity of different hour sub-dataset, the imbalanced dataset's influence is complex. When the majority sub-datasets are more representative than minority sub-datasets, the retrieval result's error will be smaller. So, the over-sample method could reduce the minimum number of records only in condition that the imbalanced data's minority sub-datasets are much more representative than majority counterparts. One example is showed in Figure 6.

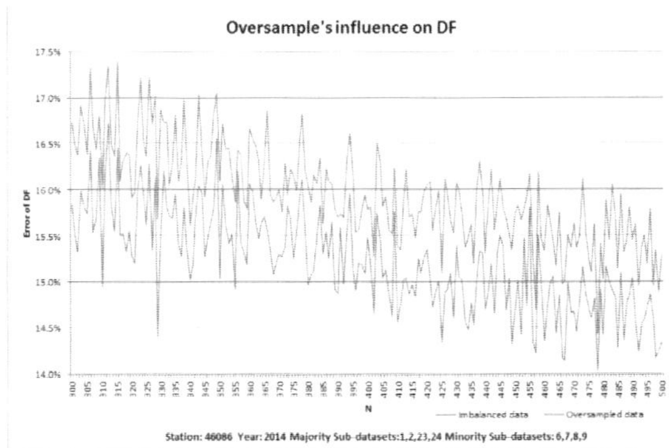

Figure 6. The over-sample's influence

4.3 *The influences of weighting*

The weighting's influence on the minimum number of records largely depends on the weighting function φ(t) or ψ(t). The weighting function use in this study could reduce the minimum number of records for retrieval of DF or DDED for the buoy data tested in research. Result of this is demonstrated in Table 4. The weighting method reduces 15 records for DF retrieval and 33 for DDED in this test.

Table 4. The weighting's influence on crucial records number

		2005	2006	2007	2008	2009	2010	2011	2012	2013
DF	Original	363.85	370.5	365.03	368.21	330.37	360.63	374.51	375.46	366.07
Station:	Weighting	346.83	358.78	350.27	349.24	329.63	347.52	355.75	353.63	345.82
	Reduced	17.02	11.72	14.76	18.97	0.74	13.11	18.76	21.83	20.25
42040	Average	15.19								
DDED	Original	569.54	518.73	528.17	567.19	541.13	645.36	575.45	562.45	495.39
Station:	Weighting	514.33	493.60	492.10	569.23	532.68	593.33	531.89	557.75	443.06
	Reduced	55.21	25.13	36.07	-2.04	8.45	52.03	43.56	4.70	52.33
46086	Average	33.28								

5. Discussion

With the changing of methods of over-sample or function of weighting, the retrieval results and the minimum number of records also changed. To get better retrieval results, much more weighting functions and over-sample methods should be tested and compared in further study.

Those methods used in this work rely on basic understanding of specific area's wind. So, when use those methods to obtain directional wind distribution from satellite data, historical In-situ observations are needed to build weighting function or decided whether over-sample is helpful in retrieval. Moreover, In-Situ observations usually only demonstrate the wind condition of a specific location.

Whether the information from this specific location could be used to build weighting function for near area is needed to answer in further research.

6.Conclusions

By this simulation experiment, the minimum number of records for annual wind direction frequency is 350, and 800 for annual direction distributions of wind energy density. And, the over-sample for minority sub-datasets of imbalanced data and using weighting function could have positive influences on retrieval of directional wind distribution and reduce the records needed in retrieval.

References

[1] Nguyen K Q 2007 Wind energy in Vietnam: Resource assessment, development status and future implications *Energy Policy* **35** pp. 1405-1413

[2] Zhao S, Jiang B, Xu H., Li S. and Ding J 2010 Exploration and Application of Ocean Wind Energy Resources in Coastal Sea of China *Ocean Technology* **29** pp. 117-121.

[3] Heckenbergerova J, Musilek P and Krömer P 2015 Optimization of Wind Direction Distribution Parameters Using Particle Swarm Optimization *Afro-European Conf. for Industrial Advancement* vol 334, ed A Abraham, P Krömer and V Snasel (Gewerbestrasse: Springer International Publishing) pp. 15-26.

[4] Chang R, Zhu R, Badger M, Hasager C B, Xing X and Jiang Y 2015 Offshore Wind Resources Assessment from Multiple Satellite Data and WRF Modeling over South China Sea *Remote Sens.* **7** pp. 467-487

[5] Doubrawa P, Barthelmie R J, Pryor S C, Hasager C B, Badger M and Karagali I 2015 Satellite winds as a tool for offshore wind resource assessment: The Great Lakes Wind Atlas *Remote Sens. Environ.* **168** pp. 349-359

[6] Yang X., Li X, Pichel W G and Li Z 2011 Comparison of Ocean Surface Winds From ENVISAT ASAR, MetOp ASCAT Scatterometer, Buoy Measurements, and NOGAPS Model *Ieee T. Geosci. Remote.* **49** pp. 4743-4750

[7] Barthelmie R J and Pryor S C 2003 Can satellite sampling of offshore wind speeds realistically represent wind speed distributions *J. Appl. Meteorol.* **42** pp. 83-94

[8] Barthelmie R J, Grisogono B and Pryor S C 1996 Observations and simulations of diurnal cycles of near-surface wind speeds over land and sea *J. Geophys. Res.* **101** pp. 21327-21337

[9] Gu Q 2009 Research of Machine Learning in Imbalanced Data Set and its Application in Geosciences Data Processing, China University of Geosciences

Acknowledgements

This work was supported by the National Natural Science Foundation of China under Grant 41371355.

Temporal and Spatial Variation Characteristics of Natural Woodland in the Upper Reaches of the Tarim River in Recent 25 Years

Q Y Ran[1], L Y Bai[1], J Z Feng,[2] Y Zhao[1], M Q Guo[1], K Wang[1]

[1]Institute of Remote Sensing and Digital Earth, Chinese Academy of Sciences, Beijing 100094, China;
[2]Key Lab of Agri-information Service Technology, Ministry of Agriculture of P. R. China, Agricultural Information Institute, Chinese Academy of Agricultural Sciences, Beijing 100081, China

Abstract. Natural woodland is highly prized in the Tarim River Basin for its great ecological, economic and social benefits, and it occupies an important position in the "The Belt and Road" strategy, but influenced by the natural environment changes and human activities, the natural woodland has changed in some way. So we implemented the dynamic monitoring of land use and natural woodland in the upper reach of the Tarim River and analyzed their changes, based upon the Landsat remote sensing images of 1990-2015. The results showed that: (1) From 1990 to 2015, the land use degree in the study area showed a significantly increasing trend, yet the land use degree and development intensity stayed in a relatively low level between two counterparts of the natural and agricultural land use. (2) In the past 25 years, the natural woodland area reached higher level in the middle terms (1995-2005), while remained lower at both ends. Additionally, the dynamic changes had a high increase speed during 1990-1995 as well as had a high decrease speed during 2005-2010. (3) The transfer features of natural woodland from 1990 to 2015 was: natural woodland were mainly changed into farmland, and the newly increased natural woodland mainly came from grassland and saline alkali land.

1. Introduction

March 27, 2015, China National Development and Reform Commission, the Ministry of Foreign Affairs and the Ministry of Commerce jointly issued the 《Vision and Actions on Jointly Building Silk Road Economic Belt and 21st-Century Maritime Silk Road》 in Hainan, the Boao forum for Asia, marking the "The Belt and Road" strategy had entered the comprehensive construction phase.

Xinjiang autonomous region, as a core area of the Silk Road Economic Zone in western China, has an important port to center and southern Asian countries (such as Pakistan, Afghanistan, and Tajikistan). The Tarim River Basin (TRB), as one part of Xinjiang, with large population plays a significant role in regional economic activities and ecological function and environment protection[1]. However, owing to the influence of natural environment changes and human activities, ecological environment problems such as the shortage of water resources, desert riparian forest degradation and land desertification have presented increasingly serious situation in the region during recent years. The TRB thus faces the challenges on the deterioration of ecological environment. The natural woodland, as the main body of riparian vegetation along the TRB, is very significant for the stability and healthy development of ecological environment in the basin.

In recent years, many researches on the woodland of TRB have been down, i.e., the relationship between woodland soil properties and vegetation[2-4], Populus euphratica's individual and population characteristics[5-6], groundwater and vegetation changes under multiple ecological water transfer in the lower reach of the Tarim River[7-9], and the simulation of water consumption and transpiration of woodland based on relevant models[10-12].

With the advantages of high timeliness, large range and low cost, remote sensing technology has been widely used. Many scholars at home and abroad thus have carried out a series of research on LUCC (Land-Use and Land-Cover Change) in TRB by using this technology. By using the Landsat8 OLI remote sensing images and measured LAI (Leaf Area Index) data, Zhu Xuchao et al. established a LAI estimation model to estimate the LAI of riparian forest along the lower reach of the Tarim River[13]. With the help of satellite remote sensing and GIS (Geographic Information System) technologies, Bai Yuan et al. adopted the buffer gradient method to analyze the horizontal distribution structure of desert riparian forest along the Tarim River[14]. Based on nearly 40 years of Landsat remote sensing images and field survey data, Chen Haiyan and Chen Yaning analyzed the temporal and spatial variation of the vegetation zone with which the main body was desert riparian forest along the Tarim River[15]. The results showed that the vegetation coverage degree along the Tarim River presented a tendency of rise, being faster in the upstream and slower in the midstream, while the downstream presented a downward trend; In terms of the different levels of the vegetation coverage, the area of the low vegetation coverage zone reflected the desert riparian forest showed a decreasing trend, and the area of the high vegetation coverage zone reflected the farmland and artificial vegetation area showed an increasing trend.

There were a lot of researches about woodland in the TRB. However, influenced by the zero flow on the lower reach of the Tarim River, more researches focused on the lower reach, and the UTR as the core area of the TRB drew less attention. Therefore, in this study, we selected the Upper reach of the Tarim River (UTR) as a study area, and implemented the dynamic monitoring of land use and natural woodland along the UTR and analyzed their changes, based upon the Landsat TM/ETM+/OLI remote sensing images of 1990-2015. In this way, we hope that our study can provide help for scientific research and ecological environment conservation and government decision making in this region.

2. Data and methodology

2.1 Study area

The TRB is the largest inland river basin in China, which is located in southern Xinjiang (between 73°10'E-94°05'E longitude and 34°55'N-43°48'N latitude) with an area of about 102km^2[16]. The Tarim River is consisted of four tributaries and one mainstream, and has a length of 1321km divided into three sections (upper, middle and lower reach). The upper reach is from Alaer to Yingbaza, and the river length is 495km; the middle reach is from Yingbaza to Kala with the river length of 398km; the lower reach is from Kala to Taitema Lake with the river length of 428km, but 321km of the lower reach currently has been drying up[17]. The Tarim River was surrounded by the Taklimakan Desert in northeast, and the river basin is a continental warm temperate extreme arid climate with the annual average temperature in 10.7℃ and the average precipitation in 17.4-42.8mm[14]. The zonal vegetation in the area affected by the arid desert climate is temperate and semi shrub, populous is the main part of the desert riparian forest that mainly distributed in the floodplain and on both sides of the lower terraces.

The study area contains the upper reach region of the Tarim River with a buffer at width of 12.6×2km. It crossed the Akesu and Bayingolin Mongol Autonomous prefectures, being across six county-level administrative units, i.e., Akesu City, Xinhe County, Shaya County, Kuche County, Luntai county, and Weli County, with an area of 7747 km^2. The main land use types in this area are woodland, grassland, farmland, and sand.

2.2 Data source

The land use classification data of the study area during 1990-2015 were derived from the historical Landsat TM/ETM+/OLI remote sensing images (Table 2), which were acquired from the website of Geospatial Data Cloud (http://www.gscloud.cn/) and Institute of Remote Sensing and Digital Earth, Chinese Academy of Sciences (http://www.radi.ac.cn/). The data of each study year includes five images, and the time of image mainly are concentrated in August with high vegetation coverage, where the available and cloud-free image will be selected.

2.3 Data processing

Firstly, this paper pre-processed the remote sensing images by the software of ENVI5.1, including layer stacking, clipping, and band mathing, then obtained the synthetic images that contained the data of band and spectral indices such as NDVI (Normalized Difference Vegetation Index)、NDBI (Normalized Difference Building Index) and MNDWI (Modified Normalized Difference Water Index). Secondly, the region of interest was selected by using the Google Earth software (version 7.1). Thirdly, the decision tree classifier that is a distributed model based on cart algorithm was used to classify the synthetic images, and get the land use classification data which had a total of nine categories included farmland (including artificial woodland), natural woodland, grassland, waters, beach land, residential land, sand, saline alkali land and swamps. Finally, verified classification data by GF-1 PMS (a spatial resolution of 2m), the results demonstrated that the accuracies of land use classification data (2015) with the overall accuracy of 87.74% and kappa coefficient of 0.84. Eventually there are six periods of land use classification data used in the analysis of this study.

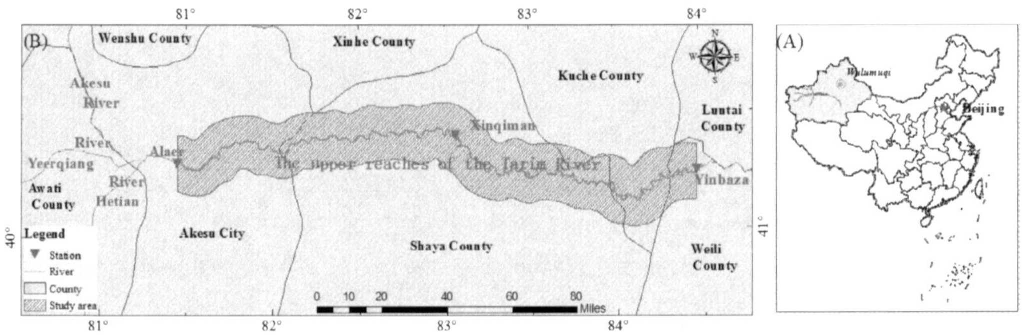

Figure 1. Location (A) and scope (B) of the study area

Table 2. Attributes of the Landsat TM/ETM+/OLI images used in the study

Year of study	Satellite	Sensor	Data imaging time					Spatial resolution /m	Time resolution /day
			144/31	144/32	145/31	145/32	146/32		
1990	Landsat5	TM	1992.09.26	1990.10.23	1992.04.10	1989.08.08	1990.08.02	30	16
1995	Landsat5	TM	1996.09.21	1995.11.06	1995.05.05	1996.08.27	1994.08.13	--	--
2000	Landsat7	ETM+	2000.08.07	2000.08.07	2000.05.10	2001.09.02	2000.08.05	--	--
2005	Landsat7	ETM+	2005.08.21	2005.08.21	2005.05.08	2005.09.13	2005.07.18	--	--
2010	Landsat5	TM	2010.08.11	2010.08.11	2010.08.02	2010.08.02	2010.08.06	--	--
2015	Landsat8	OLI	2015.07.24	2015.07.24		2015.07.15	2015.07.22	--	--

2.4 Models and methods

In order to research the land use situation, especially the temporal and spatial change of natural woodland, we analyzed the land use degree, the area variation and transfer features of natural woodland using the following models (Table 3). The land use degree index calculation formula is used to calculate the land use degree index, which is worked out by Liu Jiyuan et al[18-19], The land use degree index can be used to quantitatively reveal the comprehensive land use level, and has an important role in reflecting the change of land use degree. Besides, for the purpose of showing the intensity of land use change, this paper respectively used the single land use dynamic degree and the comprehensive land use dynamic degree index to analyze the dynamic change of natural woodland and all land use types[20]. Finally, the transfer matrix was used to analyze the transfer features of natural woodland and other land use types for the sake of making clear the natural woodland's transfer source and destination[20].

3. Results and discussion

3.1 Change of land use degree

The area proportion of each land use type were extracted from the land use classification data of 1990-2015, then calculated the land use degree index by the land use degree index calculation formula, as shown in Figure 4

Table 3. Formula of calculating index

Calculation index	Calculating formula	Variable introduction
Land use degree index	$K = 100 \times \sum_{i=1}^{n} (A_i \times C_i)$	where: K = land use degree index; A_i = land use degree grading index of i type; C_i = area proportion of i type; n = number of land use type.
Single land use dynamic degree	$L = \frac{(U_b - U_a)}{U_a} \times \frac{1}{T} \times 100\%$	where: L = single land use dynamic degree; U_b = area of land use type at the beginning; U_a = area of land use type at the end; T = length of study period;
Comprehensive land use dynamic degree	$LC = \left[\frac{\sum_{i=1}^{n} \Delta LU_{i-j}}{2\sum_{i=1}^{n} LU_i} \right] \times \frac{1}{T} \times 100\%$	where: LC = comprehensive land use dynamic degree; LU_i = area of i type in the early stage of the study; ΔLU_{i-j} = absolute value of the area of i type to non i type (j type).

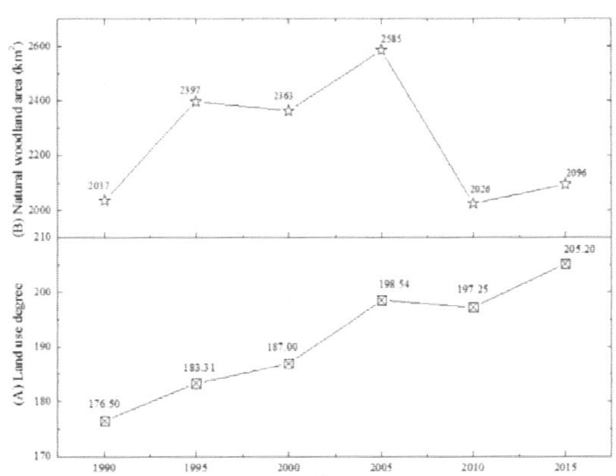

Figure 4. Variation tendency of the natural woodland area (B) and land use degree index (A) in the study area from 1990 to 2015.

From Figure 4, the land use degree index showed a gradually increasing trend from 176.05 in 1990 to 205.20 in 2015, in particular, the period of 2005-2010 decreased with a small amplitude. The fastest growing period of land use degree index was 2000-2005 with the value increased from 187 to 198.54. In the other periods, the change value from high to low was 7.95 of 2010-2015, 6.81 of 1990-1995, and 3.69 of 1995-2000. During the past 25 years, the most intense change period of land use degree was 2000-2015 with a high amplitude and unfixed change direction.

On the other hand, the maximum land use degree index in all years was 205.2, i.e., just half size of the limit value (namely, 400)[18-19], and the average land use degree index was 191.3 in a low level. It showed that the land use degree and development intensity stayed in a relatively low level between two counterparts of the natural and agricultural land use.

By comparing it with the change of the natural woodland, the land use degree index increased slowly (1995-2000) or decreased (2005-2010) comparatively with the decrease of the natural woodland in the corresponding periods, which indicated that the change of land use degree was to a certain extent affected by the change of natural woodland.

3.2 Dynamic change of natural woodland

The area and conversion area of each land use type were extracted from the land use classification data of 1990-2015, and the single land use dynamic degree of natural woodland and comprehensive land use dynamic degree were calculated by the single and comprehensive land use dynamic degree formula, as shown in Figure 5.

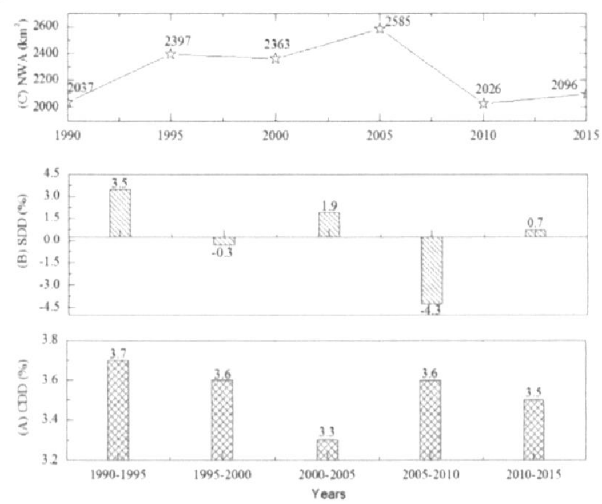

Figure 5. Variation tendency of the natural woodland area (C) and land use dynamic degree (A and B) in the study area between 1990 and 2015. (Note: CDD is the comprehensive dynamic degree, SDD is the single dynamic degree, NWA is the natural woodland area).

Figure 5 shows the natural woodland area of study area in the last 25 years, and the maximum and minimum value was respectively 2585km^2 in 2005 and 2026km^2 in 2010. Besides, the area of natural woodland in 1995 and 2000 is at the same high level as in 2005, and in 1990 and 2015 at the same low level as in 2010. From 1990 to 2015, the overall trend of natural woodland area in the study area increased firstly and then decreased with a total increase value of 59km^2.

We can discover the dynamic change of natural woodland from Figure 5, the results shows three increase periods and two reduction periods, and the increase periods was 1990-1995、2000-2005 and 2010-2015, as well as the reduction periods was 1995-2000 and 2005-2010. Among them, the fastest increasing period was 1990-1995 with a positive dynamic degree of 3.5% and the area increased by 360km2, while the period of 2005-2010 was turned out to be the fastest decreasing period with a negative dynamic degree of -4.3% and the area decreased by 559km^2. Variation range of the remaining three periods were small, and their single dynamic degree of natural woodland respectively were 1.9% during 2000-2005, 0.7% during 2010-2015, and 0.3% during1995-2000.

By comparing the single land use dynamic degree of natural woodland with the comprehensive land use dynamic degree, they both had significant values in the first and fourth periods, which indicated that the overall land use change was somehow influenced by the change of natural woodland.

3.3 Transfer features of natural woodland

The transfer matrix was constructed by the land use classification data of 1990 and 2015, as described in Figure 6 and Figure 7. Figure 6 shows that there were vast areas of natural woodland on both sides of the Tarim River, being no change during the last 25 years, but large areas of natural woodland changed into farmland not far from the river, and moreover, there were a lot of newly increased natural woodland at the borders between oasis and desert.

The transfer data of natural woodland with other land use types are mentioned in the Figure 7. From 1990 to 2015, a total of 846km^2 natural woodland changed into the non-natural woodland, of which about 44.8% transferred into farmland with an area of 379km^2. Furthermore, the grassland and waters were turned out to be the second and third, respectively transferred out of 337km^2 and 71km^2, accounted for 48.22% of all transfer area. Moreover, the transfer area of the rest types did not exceed 20km^2, and the roll out percentage did not exceed 2%.

Besides, there were 904km^2 newly increased natural woodland derived from the other land use types that are mainly grassland with an area of 502km^2 being more than half of the total transfer area. The saline alkali land, sand and waters in the following three, and the transfer area were 114km^2, 103km^2 and 83km^2. accounted for 33.18% of the total transfer area. The remaining four categories accounted for only 11.39%, and no changes had occurred on residential land.

It can be known that there were no obvious changes on the natural woodland area with only increased 59km^2 in the past 25 years. However, the transfers between natural woodland and other land use types were remarkable, mainly were the natural woodland changed into farmland and the newly increased natural woodland came from grassland, saline alkali land and sand.

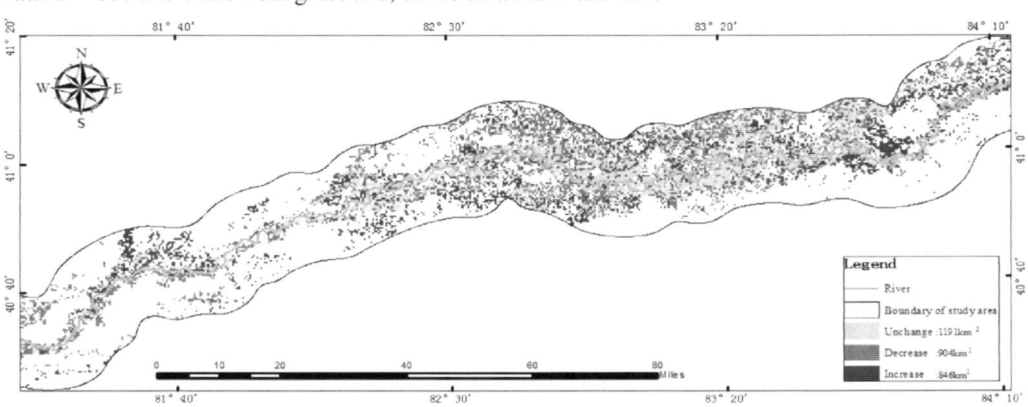

Figure 6. The spatial distribution of different change of natural woodland in the study area from 1990 to 2015.

3.4 Driving forces of changes

From 1990 to 2015, the land use degree of the study area showed a significantly increasing trend, which was mainly caused by the expansion of farmland and residential land with a high land use degree grading index (Figure 8), while the influence by the other land use types was not remarkable. In addition, the overall land use degree was low between the natural and agricultural land use level, and the reason was that there were a high ratio of the land use types with a low land use degree grading index, such as natural woodland, grassland, sand and saline alkali land (Figure 9).

In the past 25 years, the natural woodland area in the study area reached higher level in the middle terms (1995-2005), while remained lower at both ends, and the overall natural woodland area increased 59km2. It was influenced by natural factors in early stages, and later by the impact of human activities. The increasing farmland and residential land squeezed the living space of the natural woodland (Figure 8), which made the natural woodland area of 2005-2015 tended to decrease. On the other hand, the dynamic changes of natural woodland in 1990-1995 and 2005-2010 were the most dramatic, which increased significantly in 1990-1995, and 2005-2010 for a substantial reduction.

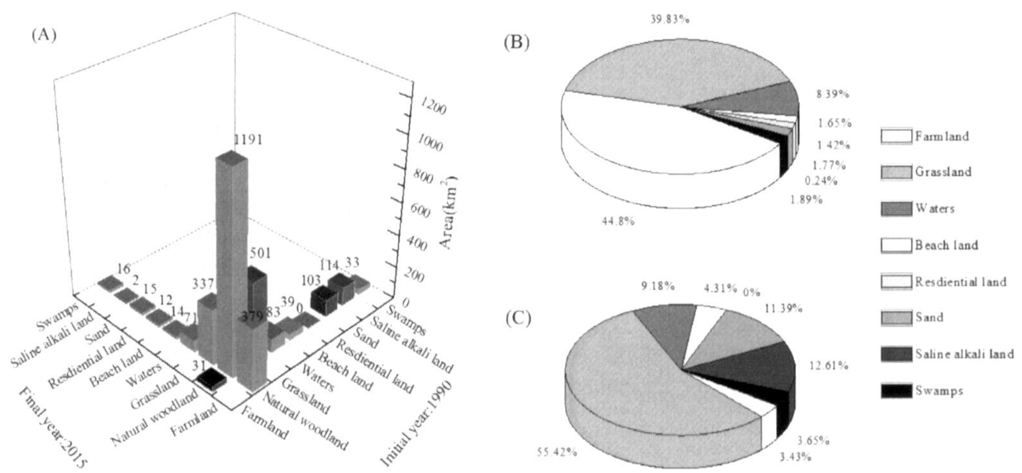

Figure 7. The transition area (A) and percentage (B and C) of natural woodland in the study area from 1990 to 2015.

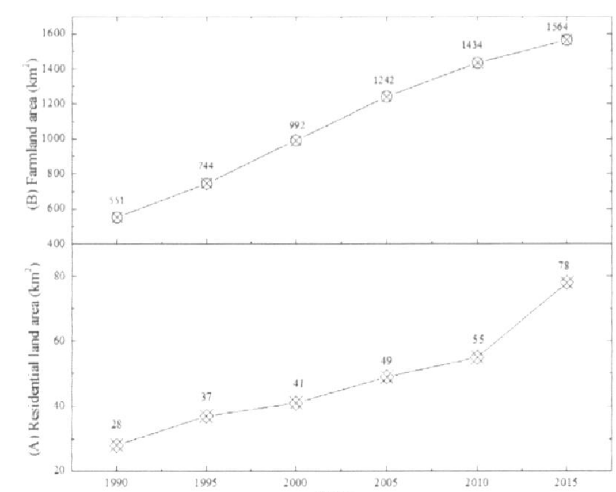

Figure 8. Variation tendency of the residential land area (A) and farmland area (B) in the study area from 1990 to 2015.

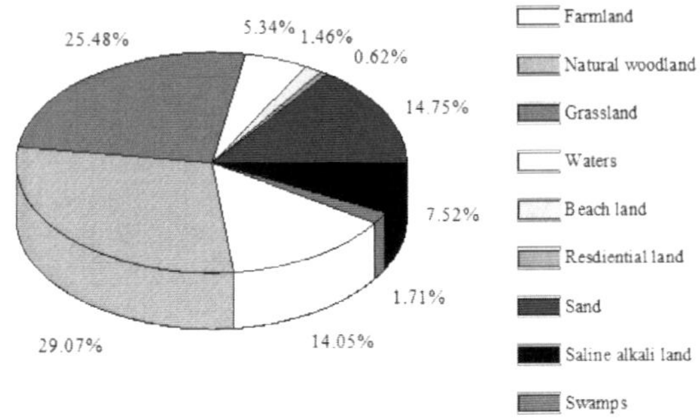

Figure 9. The proportion of each land use type in the study area

The transfer features of natural woodland mainly were natural woodland turned to farmland while grassland and saline alkali land turned to natural woodland during the last 25 years. Influenced by the agricultural and economic development, a large number of original natural woodland had been developed into farmland and residential land, therefore natural woodland tended to develop outward under the artificial squeeze. So there was a large part of the grassland and saline alkali land (which is located at the border between oasis and desert) transferred to natural woodland. However, there were no changes of vast areas of natural woodland on both sides of the Tarim River in the past 25 years, because it is difficult to develop in the region along the river, and the natural woodland along the river has very important value of ecology, economy and sociology so as to be protected by the government.

4. Uncertainty analyses

Data is the basis of the research, and the long-term temporal LUCC analysis usually requires the utilization of multi-temporal and multi-sensor images, which may incur problems such as unmatched image parameters. In order to minimize the impacts of inconsistency among images, we chose the Landsat TM/ETM+/OLI series data as the data source and ensured its imaging time mainly concentrated in August with high vegetation coverage. But influenced by the seasonal rivers, missing of some data and the uncertainty of interpretation, the consistency of the data was not high enough, so there was some difference between the research results and the actual level.

5. Conclusions

In this study, we quantitatively analyzed the dynamic change of the natural woodland over UTR in recently 25 years. This study indicated that the natural woodland had an intense change, especially the transition changes as to grassland and farmland. The changes were significantly influenced by human activities, particularly on the agriculture activity. The results of this study can offer an useful insight on reasonable land use, and provide available information to the policy makers so as to make scientific decision, and promote the implementation of the "The Belt and Road" strategy and regional ecological environment conservation and socio-economic sustainable development.

Acknowledge

The Western Light Talent Culture Project (YBXM-2014-05), the Natural Science Foundation of China (NSFC; Grant Nos. 41171340 and 41101390), and the Science and Technology Innovation

Program (for the Agricultural Information Institute) of the Chinese Academy of Agricultural Sciences supported this work.

Reference

[1] Wu L, Hou X Y and Xu X L 2013 *Transactions of the Chinese Society of Agricultural Engineering* **29** 207-16
[2] Zhang H T, Liang J Y, Zhou Z L and Lv R H 2016 *Research of Soil and Water Conservation* **23** 6-12
[3] Han L, Wang H Z and Yu J 2013 *Journal of Soil and Water Conservation* **27** 124-29
[4] Li L, Zhou Z L, Lü R H and Liang J Y 2015 *Journal of Northeast Forestry University* **43** 75-78
[5] Han L 2014 Beijing: China Agriculture University
[6] Wang J S, Halik U, Bernd C, Osman K and Peng S H 2008 *Chinese Bulletin of Botany* **25** 728-33
[7] Halik U, Chai Z, Habibulla A, Bernd C and Christian O 2011 *Bulletin of Soil & Water Conservation* **31** 18-22
[8] Maimaiti G, Halik U, Aishan T and Wang J S 2010 *Xinjiang Agricultural Sciences* **47** 241-45
[9] Keyimu M, Halik U, Aishan T, Abliz T and Baidourela A 2014 *Journal of Northeast Forestry University* **42** 15-18
[10] Yuan G F, Luo Y, Shao M A, Zhang P and Zhu X C 2015 *Science China Earth Science* **58** 1032-42
[11] Su L T, Li Y S, Aibibula A and Zhao T Y 2015 *Journal of Desert Research* **35** 380-84
[12] Wang H Z, Han L, Xu Y L, Niu J L and Yu J 2016 *Scientia Silvae Sinicae* **52** 136-42
[13] Zhu X C, Yuan G F, Yi X B and Du T 2014 *Arid Land Geography* **37** 1248-56
[14] Bai Y, Xu H L, Liu X H and Ling H B 2013 *Journal of Natural Resources* **28** 776-85
[15] Chen H Y and Chen Y N 2015 *Journal of Ecology* **34** 3166-73
[16] Deng X Y, Yang Z F and Long A H 2013 *Journal of Glaciology & Geocryology* **35** 1600-09
[17] Xu Y F 2013 *Zhengzhou: Zhengzhou University*
[18] Liu, J. Y., 1992. Land use in Tibet autonomous region. *Beijing: Science Press*
[19] Gao Z Q, Liu J Y and Zhuang D F 1999 *Journal of Remote Sensing* **3** 134-38

IOP Conf. Series: Earth and Environmental Science **57** (2017) 012056 doi:10.1088/1755-1315/57/1/012056

Urban Change Detection of Pingtan City based on Bi-temporal Remote Sensing Images

JIANG Degang, XU Jinyan*, GAO Yikang

Island Research Center, State Oceanic Administration, No.1 Tianmei Road, 350400 Pingtan, Fujian, China

Emai: xujy_liesmars@whu.edu.cn

Abstract. In this paper, a pair of SPOT 5-6 images with the resolution of 0.5m is selected. An object-oriented classification method is used to the two images and five classes of ground features were identified as man-made objects, farmland, forest, waterbody and unutilized land. An auxiliary ASTER GDEM was used to improve the classification accuracy. And the change detection based on the classification results was performed. Accuracy assessment was carried out finally. Consequently, satisfactory results were obtained. The results show that great changes of the Pingtan city have been detected as the expansion of the city area and the intensity increase of man-made buildings, roads and other infrastructures with the establishment of Pingtan comprehensive experimental zone. Wide range of open sea area along the island coast zones has been reclaimed for port and CBDs construction.

1. INTRODUCTION

Currently, the development of marine economy and population has caused great pressure to coast areas as the immigration from inland to the developed coastal cities, and push the human beings to find new living space. Island coastal zones, which are specific ecosystems along east China, make great opportunity to the Chinese people. With the proposed strategy of "Marine Ecological Civilization (MEC)" and "One Belt, One Road" [1], more and more efforts have been made to exploit the islands. It is quite essential to monitor the dynamic spatio-temporal changes of islands caused by human actives such as urbanization and reclamations etc. Much more researchers have turned their focus to islands ecosystems monitoring using remote sensing images [2-4], and reasonable results have been achieved. The Pingtan city is located on the Haitan Island, which is the 5th largest island of China. As the starting point of the Maritime Silk Road and also an important strategic propeller to Taiwan, the state input on the economic development of the Pingtan has increased, and the infrastructure of the city has greatly improved. The extent of the city and the density of man-made features have increased obviously, which result in great pressure to ecosystem of the island and also the surrounding open sea as the increase of reclamation and the port constructions etc.. It is quite essential to monitor the spatio-temporal urban changes over the whole city using multi-temporal and wide range remote sensing images.

Numerous applications and methods have been used for detecting urban LULC types and changes through time [5-6]. The post-classification change detection was used in this study. The object-based classification method was performed to two high resolution Worldview-II images, while a post-classification change detection method was carried out in the following. Accuracy assessment was carried out and the performance of the results was discussed.

2. TEST SITE AND DATASETS

The study area is the city of Pingtan, Fujian province, China, and is located on the middle of Haitan Island which covers an area of more than 300km^2. The island is spread from 119°32′ E to 120° 10′ in longitude and 25° 15′ N to 25° 45′ N in latitude. It is a tourist place as the specific landscape and the long shorelines with white beach all around the whole island. The main LULC types in and around the city are man-made object as buildings, roads and natural featuresas farmland, forest, unutilized land and water.

A pair of high resolution SPOT images was used in the study, and the images were shown in Fig1. The SPOT5 image was acquired on December 26th 2006, while the SPOT6 image was acquired March 7th 2013.

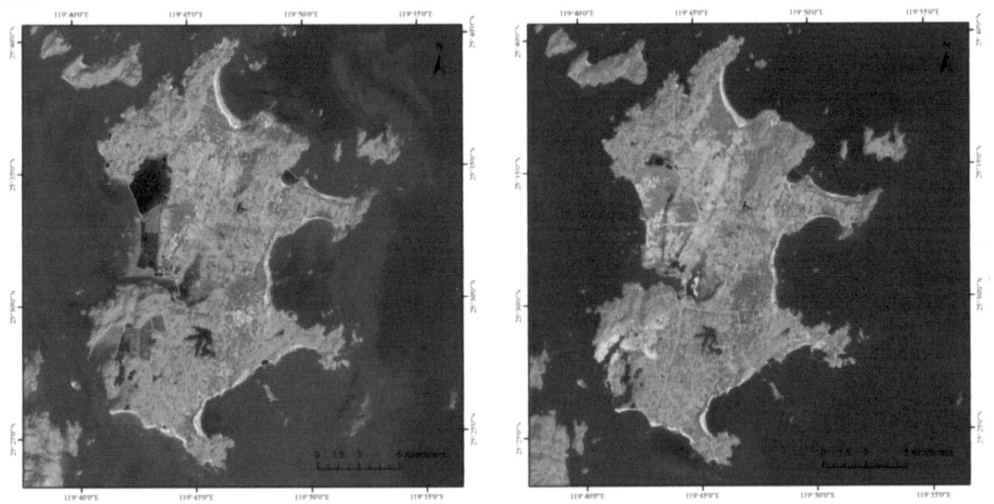

(a) (b)

Figure.1 Color images of the SPOT5-6 images, (a)-SPOT5 acquired on December 26th2006, and (b)-SPOT6 acquired on March 7th 2013.

3. METHODOLOGY

The object-based classification method is utilized to the high resolution SPOT images using PCI and ArcGIS software. The PCI software was used for the image pre-processing and object-based classification, and the ArcGIS software was used to extract the change information of the LULC types. Both images were segmented into five classes as man-made object, forest, farmland, unutilized land and water. An auxiliary ASTER GDEM was used to improve the classification accuracy of features along the coastal areas. Post-classification comparison change detection was carried out to the classified images. And the change information about the LULC types was analysed finally.

The procedure can be described as follows: (1) Image preprocessing was done to two SPOT images including radiometric calibration, data fusion, and geometric registration. (2) Object-based classification using PCI images processing software was done the processed images. (3) Change detection using ArcGIS spatial analysis was performed. (4) Result analysis and discussion was carried out finally. The flowchart is shown in Fig.2.

IOP Conf. Series: Earth and Environmental Science **57** (2017) 012056 doi:10.1088/1755-1315/57/1/012056

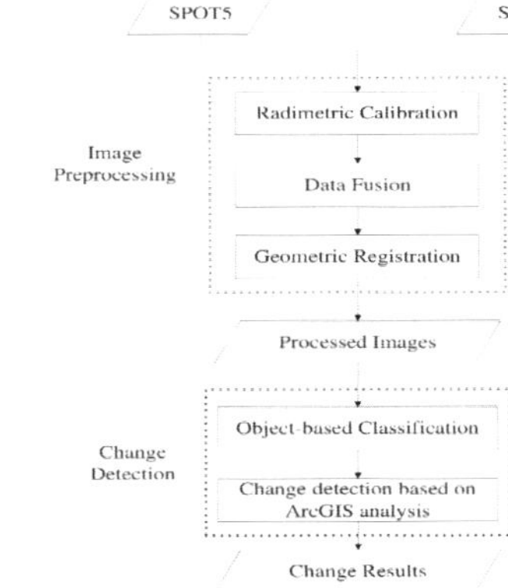

Figure.2 Flowchart of the change detection.

4. RESULT AND DISCUSSION

4.1 Object-based classification

Image pre-processing was done firstly to the bi-temporal SPOT data, and then the object-based classification using PCI software was conducted in the following. Five classes were segmented as man-made object, water, forest, farmland and unutilized land from both the two SPOT images. Accuracy assessment of the two classified images was evaluated with the random selected 100 testing samples for each image. The overall accuracy is 86.66% with the Kappa coefficient 0.83 for the SPOT5 image, while the overall accuracy is 80.00% with the Kappa coefficient 0.69 for the SPOT6 image. The results were achieved and shown in Fig.3.

The results show that the spatial distribution of the five classes is quite different which indicate significant changes have been occurred to the whole island. Urban or man-made object has greatly extended, large area of open water has changed to man-made objects, and the road network has been formed across the island preliminarily. The area of agriculture land has decreased obviously. Forest has also increased especially in the south part of the island. The changes of unutilized land are also quite significant with the area around the city center and the marine reclamation areas changed into man-made objects.

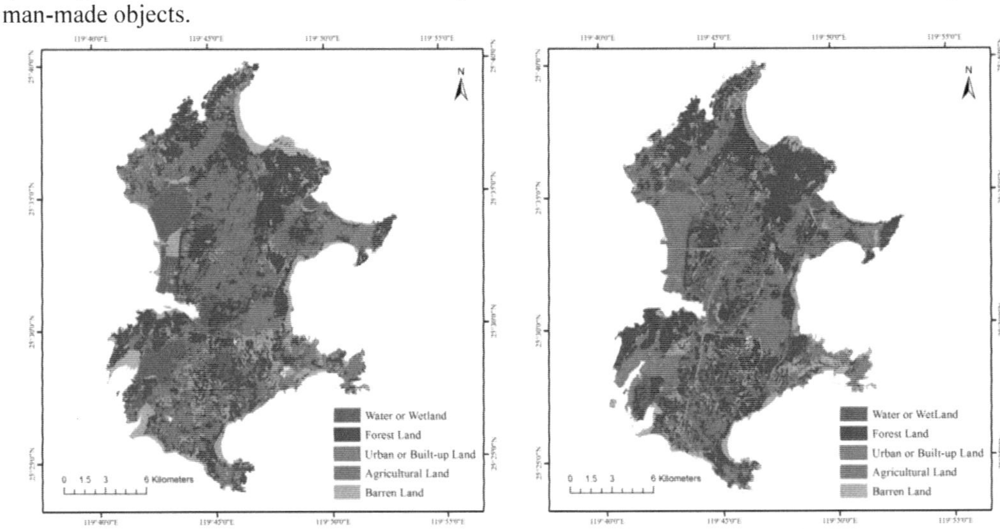

(a) 2006-12-26 (a) 2013-3-7

Figure.3 Result of the classification

4.2 Change detection

In order to get more precise change information and the possible reasons during this period, a post-classification change detection based on the ArcGIS spatial analysis was conducted. The result was shown in Fig.4.

Fig.4 shows the changes during the periods of the two images. The pink areas represent the LULC feature that stays the same, while the green ones indicates the features that changed from 2006 to 2013. Further analysis shows that great changes were occurred all over the whole island, and the most dramatic changes were detected on the west part. Accuracy assessment was conducted, and an overall accuracy of 70.00% with the Kappa coefficient 0.67 was achieved.

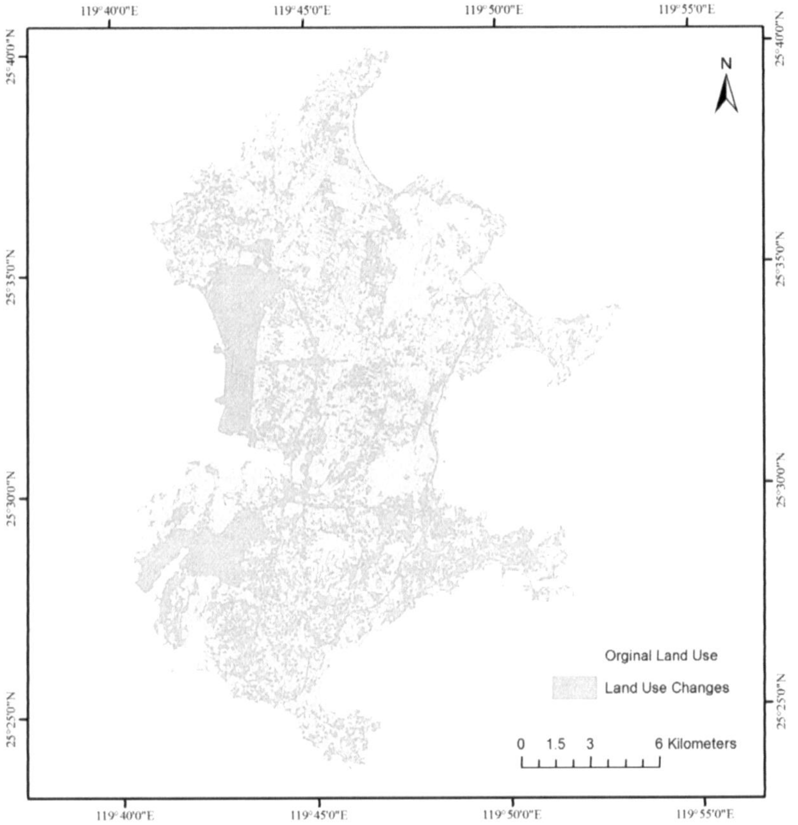

Figure.4 Results from the post-classification change detection

More analysis was conducted to identify the change types from the previous change image, and four different change images were got in Fig.5. The pink areas represent ground feature that are not changed. The blue, green, gold and grey color indicates the ground feature in 2006 is waterbody, forest, farmland and unutilized land respectively.

The first image in Fig.5 described changes from other ground features to man-made objects. It shows that a large amount of water areas has changed to man-made object, especially on the west side of the island. As the propose of " Pingtan comprehensive experimental zone" and the "Free trade zone" strategies, the development of the city infrastructure has become quite intense, and the pressure of limited land in the island pushed the development of reclamation industry, and forced the human beings to live near shore. The development of road network resulted in the changes from agriculture land and forest into man-made objects. The constructive activities also caused the changes from unutilized land into built-up areas.

The second image in Fig.5 shows the changes from the ground types to farmland. Further analysis revealed that most of the changes were from unutilized land to farmland, which maybe results from the growth of crops.

The third image in Fig.5 shows the changes to forest. We learn that most changes were from farmland and the unutilized land. This may be caused by the reforestations as the development of Pingtan International Tourism Island.

The last image in Fig.5 shows the changes to unutilized land. We find out that most changes were from water area, forest and farmland. Further analysis indicates that changes from water are also caused by human reclamation, and the changes from forest are probably deforestation caused by the strong and salty wind of the island in the east part of the island. Most changes from farmland were around the road construction areas.

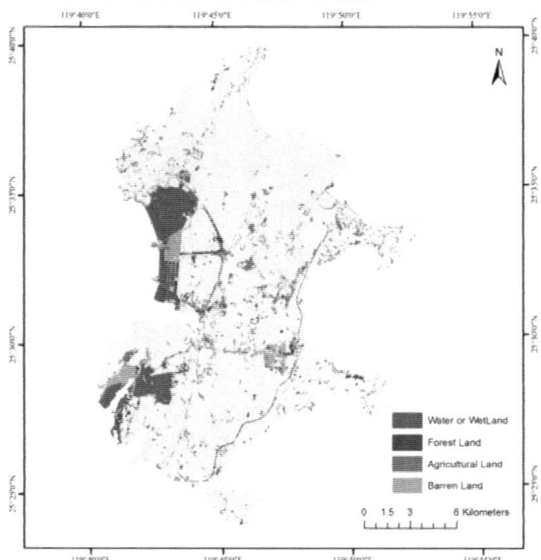

(a) Features To man-made object

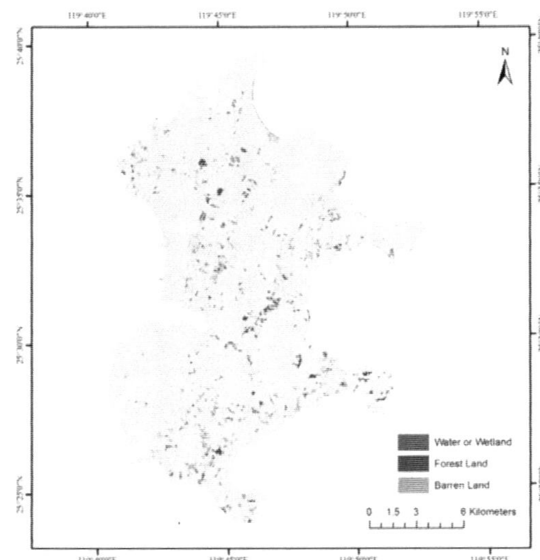

(b) Features To Agriculture Land

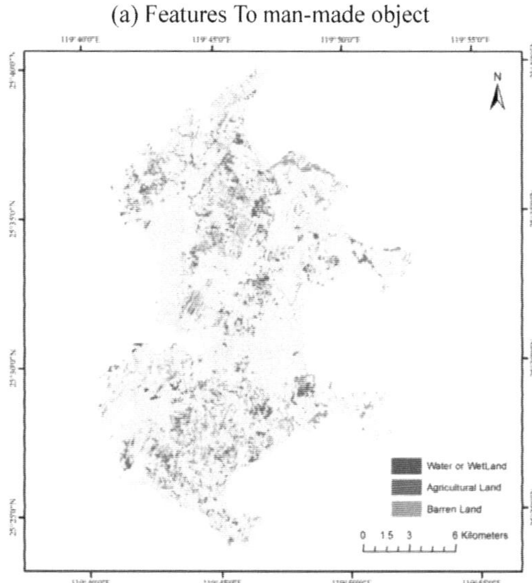

(c) Features To Forest

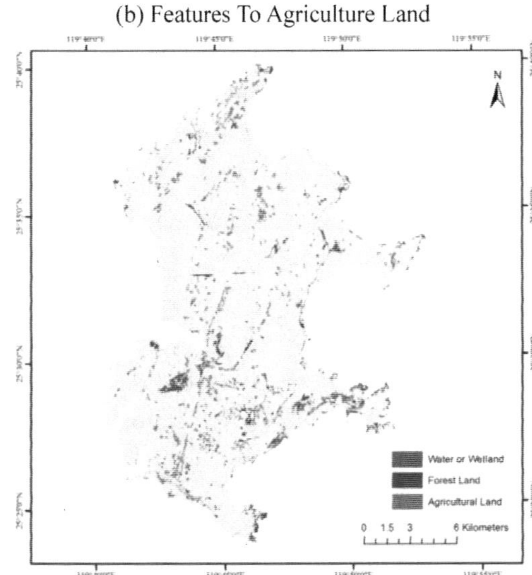

(d) Features To Unutilized Land

Figure.5 Change types of the ground features

5. CONCLUSIONS

In order to investigate the urban changes of the Pingtan Island, a pair of high resolution SPOT images was used. An object-based classification method using PCI software was performed firstly, and the post-classification comparison change detection method was carried out using the ArcGIS spatial analysis. Most of the changes occurred from 2006 to 2013 was detected, and the mainly change LULC type was the one changed into man-made object because of reclamation for port and CBDs constructions, and the type changed into forest came next. The main reason is possibly the propose of a series strategies to exploit the island with high state inputs and supports, which results in rapid development of the city area from the original inland to the reclamation land near the shore beach and the development of urban afforestation.

ACKNOWLEDGEMENTS

This work was financially supported by the National Marine Public Welfare Project (201505012).

References

[1] Thakur R. One Belt, One Road: China's New Strategic and Trade Policy [J]. 2015.

[2] Gevana D, Camacho L, Carandang A, et al. Land use characterization and change detection of a small mangrove area in Banacon Island, Bohol, Philippines using a maximum likelihood classification method[J]. Forest Science and Technology, 2015, 11(4): 197-205.

[3] Pourebrahim S, Hadipour M, Mokhtar M B. Impact assessment of rapid development on land use changes in coastal areas; case of Kuala Langat district, Malaysia [J]. Environment, Development and Sustainability, 2015, 17(5): 1003-1016.

[4] Ghosh M K, Kumar L, Roy C. Monitoring the coastline change of Hatiya island in Bangladesh using remote sensing techniques [J]. ISPRS Journal of Photogrammetry and Remote Sensing, 2015, 101: 137-144.

[5] Xu J.Y, Zhang L, Liao M.S, Wang H, Joint use of alos palsar and landsat TM images for urban change detection[C], Proceedings of Dragon 2 Final Results and Dragon 3 Kick-Off Symposium, 2013,704 SP.

[6] Xu J.Y, Zhang L, Wang Y, Wang H, Liao M.S, Urban change detection with polarimetric Advanced Land Observing Satellite phased array type L-band synthetic aperture radar data: a case study of Tai'an, China[J], Journal of Applied Remote Sensing, 2013, 7(1):073481.

Urbanization Process Monitoring in Northwest China based on DMSP/OLS Nighttime Light Data

K Wang[1], L Y Bai[1], J Z Feng[2]
[1]Institute of Remote Sensing and Digital Earth, Chinese Academy of Sciences, Beijing 100094, China;
[2]Key Labof Agri-information Service Technology, Ministry of Agriculture of P. R. China, Agricultural Information Institute, Chinese Academy of Agricultural Sciences, Beijing100081, China.

Abstract. In recent years, the DMSP/OLS nighttime light data have been widely applied to various fields such as monitoring and evaluation of urbanization, estimation of social economy, economical environment and health effects, hazards analysis, and fisheries research. The general urbanized level in China has rapidly developed since the 1990s, and the cities in northwest China, which were important population centres of the ancient silk road, have also been developed in a high speed thanks to China's national strategy of Western Development. Given the Xinjiang autonomous region as a core area of One Belt and One Road, it is very necessary to study the urbanization processes and changes of its urban system and the whole northwest region of China. In this paper, we extracted built-up areas of the cities in northwest China in 1992,1997,2002,2007, and 2012, evaluated urban expansion and spatial pattern through appropriate indexes, and also quantitatively analyzed the urbanized level of each city.The results showed that the cities in northwest China generally presented high strong and rapid expansion, but there were some large differences among cities. Urban expansion forms alternate with exterior expansion and interior filling, in general, the cities externally expandedafter 2002 and internally filledbefore 2002, meanwhile, there were a high positive correlation between urban built-up areas and population growth in Xinjiang autonomous.

1.Introduction

In order to promote globalized socio-economic development deeply, China presented a new international cooperation initiative, i.e., One Belt and One Road (OBOR), which promotes the trade relations between different countries and their urbanization development along the OBOR, and nowadays, it is widely responded and supported by the countries of the OBOR[1,2].

City serves as a complex and integrated object and its urbanization reflected the mankind activities and effects synthetically also contains an abundant meaning, including various factors as land, economy, population and so on. The monitoring of urbanization process is the premise to regulate every aspect in a city, and it is also the realistic basis for urban planning and making sustainable development strategy. Since the industrial revolution taking place in the 18th century, global urbanization rate has maintained a rapid growth. Today, there are more than 50% population living in cities around the world[3], and the increasing number of cities has contributed to the regional economic growth as well. At the same time, China' s urbanization process also has entered a stage of rapid development since the 1990s, and its rural population is gradually turning to the city population[4]. Along with the implementation of the OBOR, urbanization along the OBOR will have a big jump. Faced with the rapid development of urbanization, it is an urgent and important problem to monitor the urbanization process in a quick and effective method among a wide range, and to study the urban expansion modes, the spatial pattern and economic culture.

DMSP/OLS nighttime light data recording the night light levels in worldwide is acquired at night by Operational Linescan System (OLS) sensor in American defense meteorological satellite program (DMSP). Nighttime light signal can be enhanced thanks to its unique photomultiplier tube (PMT), so it was designed to observe the changes of night clouds at beginning, and then it is also used to monitor the fire points of cities, fish boats, gas combustions, etc. At present, the DMSP/OLS nighttime light data have been widely applied in various fields, such as economic evaluation[5, 6, 7], population estimation[8, 9], and research of urbanization expansion[10, 11, 12]. And their results show that the nighttime light data is a good embodiment for human activities and the comprehensive effects, that is to say, it can better character the urban socio-economic conditions, population density, population growth rate, urban expansion, etc. In the studies of urbanization, there are researches on extraction of urban scope, for example the research of Mi et al.[14], by consulting MODIS land cover product, they set a float threshold and extracted the central urban area; there also has some researches on spatial patterns, for instance, Wu et al.[16] extracted Zhengzhou's built-up area and analyzed its expansion intensity and direction features based on DMSP/OLS nighttime light data, and Zhang et al.[15] also used this data to study the spatial expansion in new first-tier cities.

In this paper, based on the DMSP/OLS nighttime light data in 1992, 1997, 2002, 2007 and 2012, we selected the expansion speed, expansion intensity, shape index and compactness index as our evaluation indexes referencing to Mi et al.'s urban extraction method, and then analyzed the urbanization process of northwest China in a long time series. Acting as important centers of the ancient silk road, the cities' urbanization levels in northwest China had been improved significantly. Given Xinjiang as the core area in the OBOR, it was very necessary to study the urbanization process and urban system evolution in this region as well as in the whole northwest China.

2.Dataandstudyaera

2.1 Data

DMSP/OLS nighttime light data was provided by National Geophysical Data Center under National Oceanic and Atmospheric Administration, and acquired in local 20:30 to 21:30[13]. The OLS sensor with total image breadth in 3000km and sampling distance in 0.55km, includes2 channels, that is visible light (VNIR, 0.4~1.0um) and thermal infrared (TIR, 10-13UM). This study selected the DMSP/OLS nighttime stable light data in 1992(F-10), 1997(F-12/F-14), 2002(F-12/F-14),

2007(F-15/F-16) and 2012(F-18) as a basis to research. This data with 0.008333 ° spatial resolution and 0~63 gray level belongs to visible light channel, and it mainly in cities, towns, and other places with stable light, which can eliminate fires and other transient events caused by flammable gas[14]. Considering the data in 1997, 2002, 2007 were all made up by two satellites, so we adopted average pixel value of two images as the subsequent analysis basis.

2.2 Study area

In this paper, we mainly studied urbanization processes of large and medium cities in northwest China (including Shanxi, Gansu, Ningxia, Qinghai and Xinjiang) (fig.1) in recent 20 years.

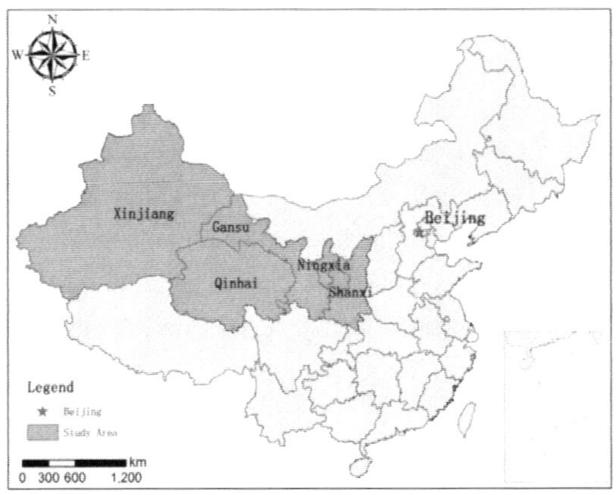

Fig.1 Study area

Cites with rapid development and in a large scale in recent years were selected as the major study regions, including Urumchi, Shihezi, Kuitun, Kramay, Hami, Xi'ning, Jiayuguan, Jinchang, Lanzhou, Yinchuan, Shizuishan, Xi'an and Hanzhong. Most of these cities are important centers in the ancient silk road, and once had prosperous histories, even in recent times, they also played important roles in the OBOR, and are the support cities in international trade.

3. Study method

Urbanization process is the procedure of economic and social structure evolution in a city during a period time, which has diversely measured indexes and generally can be embodied as a change in spatial and temporal scale. In order to study urbanization processes and analyze the changes of population and economy in northwest China comprehensively and exactly, we firstly adopted Mi et al.'s method to extract the urban scope, and then calculated the northwest cities' expansion speed, expansion intensity, shape index and compactness index. Based on these results, we finally analyzed each city's spatial expansion law and evolution process quantitatively, and revealed the changes mechanisms in its economy and social structure, that is to say, analyzed the northwest cities' urbanization tendency with a long time series data.

3.1 Analysis of urban expansion

1) Expansion Speed

Expansion speed shows the expansion speed of urban during the study period, and can reveal the expansion tendency and whole scale of urban[15]. Bigger expansion speed presents the bigger annual growth rate of urban area. Expansion speed can be calculated as follows:

$$AGA = \frac{U_{A(n+i)} - U_{Ai}}{n} \tag{1}$$

Where AGA =expansion speed, $U_{A(n+i)}$ =the area in the $(n+i)_{th}$ year, U_{Ai} =the area in the i_{th} year,and n = the time interval.

2) Expansion Intensity

Expansion intensity shows the expansion area in urban occupied percentage of total area during certain time interval, and it reveals growth rate of built-up area[16], which also can analyze the expansion intensity, speed and tendency of built-up area at different times[17]. It can be calculated as follows:

$$I = \frac{U_{A(n+i)} - U_{Ai}}{U_{Ai}} \times \frac{1}{T} \times 100\% \tag{2}$$

Where I =expansion intensity, $U_{A(n+i)}$ =the area in the $(n+i)_{th}$ year, U_{Ai} =the area in the i_{th} year,and T = the time interval.

3.2 Analysis of urban spatial pattern

1) Shape Index

Urban shape index is defined as the ratio between the area and square perimeter of built-up region, and it shows the shape feature of urban spatial exterior outline. With a bigger index, it shows that the shape of built-up area tends to roundness, conversely, it will tend to strip. It can be calculated as follows:

$$S = 4A / P^2 \tag{3}$$

Where S = shape index, A = built-up area,and P =the square along its exterior outline.

2) Compactness Index

Compactness index is a significant index to reflect urban spatial form, and it shows the aggregation extent of urban space. This index can quantificationallyanalyze urban space structure, and the bigger value shows the more compact structure, otherwise, the more loose structure. It can be calculated as follows:

$$BCI = 2\sqrt{\pi A} / P \tag{4}$$

Where BCI = shape index, A = built-up area,and P =the square along its exterior outline.

3.3 Correlation analysis of population and urban built-up area

Urban population is a significantly measured index to urbanization, which has a great influence on the development of economy and society and the utilization of resources. Positive correlation in urban population and built-up area is bigger showing that it has a more close relationship between them. On the contrary, negative correlation is bigger revealing that the urban population reduces along with the urban expansion, that it is to say, the population is falling, and then we can infer that the city has some changes, for example the change of its function. Certainly, urban population stops growing along with urban expansion when the correlation index is 0. It can be calculated as follows:

$$r = \frac{\sum_{i=1}^{n}(P_i - \overline{P})(A_i - \overline{A})}{\sqrt{\sum_{i=1}^{n}(P_i - \overline{P})^2}\sqrt{\sum_{i=1}^{n}(A_i - \overline{A})^2}} \quad (5)$$

Where P_i = urban population in the i_{th} year, $\overline{P}$ = the average urban population, A_i =built-up area in the i_{th} year,and $\overline{A}$ =the average built-up area.

4.Results and discussion

4.1 Analysis of urban expansion

We extracted the built-up areas scope of 13 significant cities in 1992, 1997, 2002, 2007 and 2012 based on DMSP/OLS nighttime light data (fig.2), and calculated the built-up areas (table 1).Through analyzing the extraction results and counting the geometric information, we calculated the expansion speed (fig. 3(a)) and expansion intensity(fig. 3(b)).

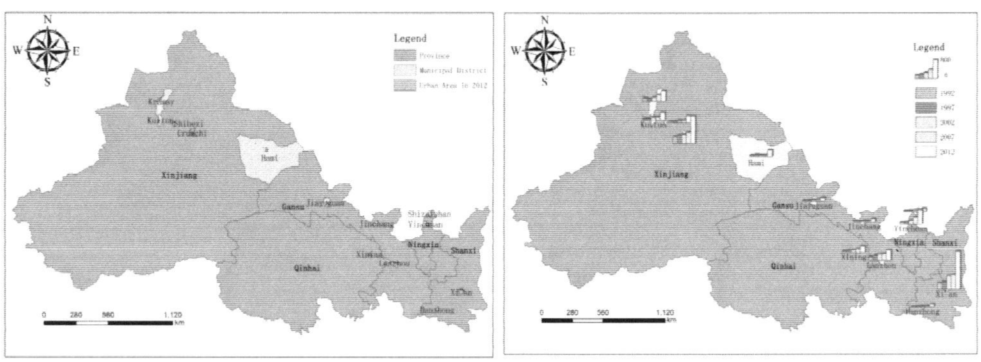

a. The built-up areas scope in 2012 b. The built-up areas of cities from 1992 to 2012

Fig.2 The built-up areas scope of each city

(1) According to fig.2 and table 1, cities had the largest built-up areas from 2007 to 2012. In these years, the built-up areas, theywere even beyond 1000km^2 in Xi'an and Urumchi, and between 400-600km^2 in Lanzhou, Yinchuan and Kramay, but below 300km2 in the rest cities. We also can get a conclusion from the fig.2 that the built-up areas in provincial capital and autonomous capital was the largest, especially in Xian and Urumchi, where area growth was high on the list of the whole cities.

(2) According to fig. 3(a), the total expansion speed in all cities reached 650km2/a from 2007 to 2012, and Xi'an and Urumchi had the highest values. In view of expansion speed from 1992 to 2012, Xi'an

had the most prominent character among all cities, especially from 1997 to 2002 and from 2007 to 2012, and Urumchi took the second place.

Table 1. Built-up area of each city from 1992 to 2012(km^2)

year	1992	1997	2002	2007	2012
Urumchi	287.91	359.21	374.57	470.99	1152.64
Shihezi	55.13	56.33	94.67	107.9	283.05
Kuitun	34.8	62.38	115.18	122.38	312
Kramay	134.94	101.77	166.82	271.05	433.87
Hami	58.72	91.34	66.54	64.67	280.56
Xi'ning	53.38	80.07	114.22	144.12	275.46
Jiayuguan	15.62	23.43	51.32	65.82	133.86
Jinchang	19.71	16.43	21.9	35.05	120.47
Lanzhou	154.82	215.23	271.44	292.65	459.11
Yinchuan	67.87	77.71	147.77	276.97	569.33
Shizuishan	40.8	48.51	60.64	83.79	135.62
Xi'an	265.37	309.93	572.22	620.91	1600.41
Hanzhong	11.24	16.35	26.58	36.8	104.26

Expansion speed, which is related to the urban scale in early time, is an absolute figure. Generally, city with a certain scale is easier to expand when it gets an advantage orientation. Oppositely, thanks to the limited scale of population and economy in small cities, it is also limited for their absolute expansion scale in a short time. That is to say, expansion intensity can give good explanations to these phenomena.

(3) We get a further analysis on urban expansion intensity. As shown in fig. 3(b), cities with the highest expansion intensity were Kuitun, Jiayuguan, Yinchuan and Hami in 1992-1997, 1997-2002, 2002-2007 and 2007-2012 respectively. Although these cities had limited urban scale in early time, their built-up areas expansion were more positive and has higher intensity, especially in Hami, its expansion intensity is more than 70% from 2007 to 2012. The top four cities in expansion intensity over the whole 20 years were Hanzhong, Kuitun, Jiayuguan and Yinchuan, whose expansion intensity with little difference in each other were approach to 40%, and their expansion intensity were much bigger than the other cities'.

(4) Taking a comprehensive analysis on expansion speed and intensity, we can draw a conclusion that Xi'an, Urumchi and Yinchuan, whose expansion speed and intensity were higher than other cities', had the fastest expansion. Meanwhile, these cities with faster expansion had a common feature because of the rapid urbanization process, there was a mergence between municipal district and its surrounding counties' centers, which caused the increase of built-up area to some degree.

The northwest region located in the underdeveloped area of China presents an inconsistent development in economy and society, which leaded to the absonant urbanization level. Additionally, influenced by the landform, location, policy and urban planning of every city in this region, the built-up areas showed significant differences in each city. However, along with the implement of the West Development strategy, the city in northwest region presented a good developmental tendency, and Xinjiang took the most remarkable progress, so its developmental level was relatively higher than the other cities'. On the contrary, some cities' expansive level were inferior to the others' because of the poor location with many mountains, which leaded to the cities must distribute a shape of strip and restrict their development caused by the limited expansion space, for example Lanzhou.

On the whole, the cities in northwest China maintained high expansion speed and have a good development from 1992 to 2012.

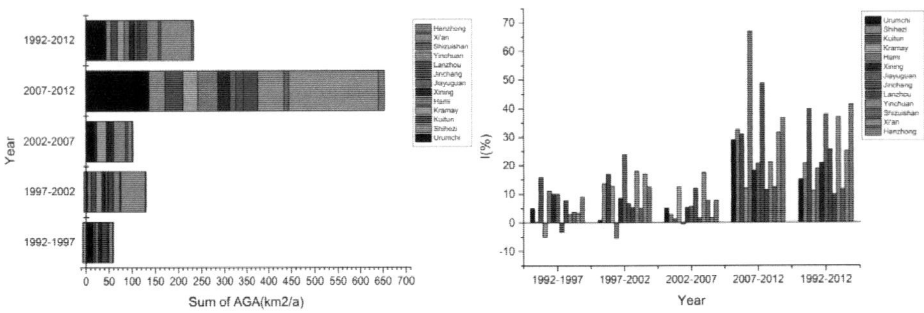

a. Expansion speedb. Expansion intensity

Fig. 3 Chart of expansion speed and expansion intensity

4.2 Analysis of spatial patterns of cities

Based on the extraction result of built-up areas, we selected the shape index and compactness index to evaluate the spatial pattern evolutions of cities, and the results are shown in fig.5.

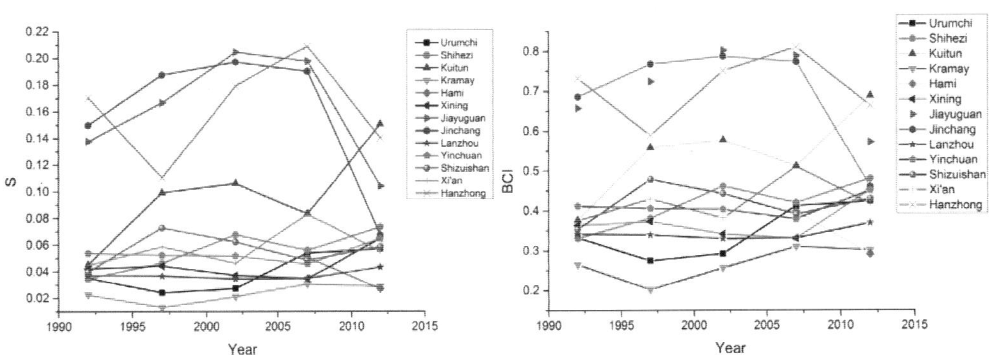

a. Shape Indexb. Compactness Index

Fig.4 Chart of analysis of urban spatial patterns

(1)According to fig 4(a), the shape index of each city presents a constant change along with the time. Some of the shape indexes in partial cities increases overall and their exterior outlines tended to be rounded, which means that their shapes of built-up regions turned strip into round, such as Kuitun and Shihezi; inversely, some of other cities' shape indexes decrease overall land their exterior outline tended to strip, which meant that their shapes of built-up regions became long and narrow, such as Lanzhou, Jiayuguan and Hanzhong; the rest cities' shape indexes remain unchanged and these cities maintained their original shape.

(2) In view of the compactness index as shown in fig. 4 (b), except in a few cities, they increased in

general before 2002, but it presents a fluctuation after 2002 and finally becomes smaller.

The reasons which caused this situation may be as follows. First, the urban scale has a big promotion than before, i.e., the larger urban scale and the faster expansion speed. In consequence, the cities with rapid expansion are mainly characterized by the externally sharp inflation and the suburbanization, which result in smaller compactness of exterior outline of cities. Second, because of the larger urban scale, the city is easier to be limited by the natural topography[18].

There are two forms of urban expansion, which are exterior expansion and interior filling. The expansion process of each city presents various characters, such as changing from large to small, varying from small to large or even revealing an alternant variation between large and small, and it showed that there are two expansion forms which appeared an alternative phenomenon in the process of urban expansion. In addition, the urban changes were mainly characterized by internal filling before 2002 while external expansion after 2002. What's more, the fluctuations of compactness indexes in Lanzhou and Xi'ning remain stable and were smaller than the other provincial capitals', while they kept fluctuant in Xi'an, Urumchi and Yinchuan, and the possible reason which resulted in this difference was the evenly urban expansion of Lanzhou and Xi'ning between exterior expansion and interior filling, inversely, the other cities were only outstanding on one aspect.

(3) Given the above factors, we can discover that the shape index and compactness index of each city in Xinjiang show more ideal results, which imply a more balanced development and higher spatial compactness in these cities. Meanwhile, these cities will have better opportunities for progress along with the development of Xinjiang acting as the core area in the OBOR. And the others, severing as the important cities in this line, will benefit from its leading effects, complete their industrial transformation and the establishment of emerging industries, and finally win a faster and better development.

4.3 Correlation analysis between population and built-up area

Severing as the core area in the OBOR and also playing an important role in the development of northwest China, we chose Xinjiang as the crucial region in our study and then selected its five key cities as the research objects for analyzing correlation between population and built-up areas, that are Urumchi, Shihezi, Kuitun, Kramay and Hami.

Based on the formula (5), we calculated the correlation coefficient between population and built-up areas in 1992, 1997, 2002, 2007 and 2012, and the results of Urumchi, Shihezi, Kuitun, Kramay and Hami are 0.833, 0.561, 0.710, 0.544 and 0.668 respectively (fig.5). According to the results, we can conclude that the population and built-up areas in Urumchi have a strongest correlation and the other cities' also have a good one.

The results showed that the population growth in every city relies on its urban expansion, especially in Urumchi; but the correlation indexes of Shihezi and Kramay were relatively low, which revealed that their urban expansion speed were faster than the population growth. Although the lower-density population will form a well-regulated city, it will restrict urban development as well.

In conclusion, severing as the core area in the OBOR, it was a crucial issue for Xinjiang to coordinate the relationship between urban population and expansion.

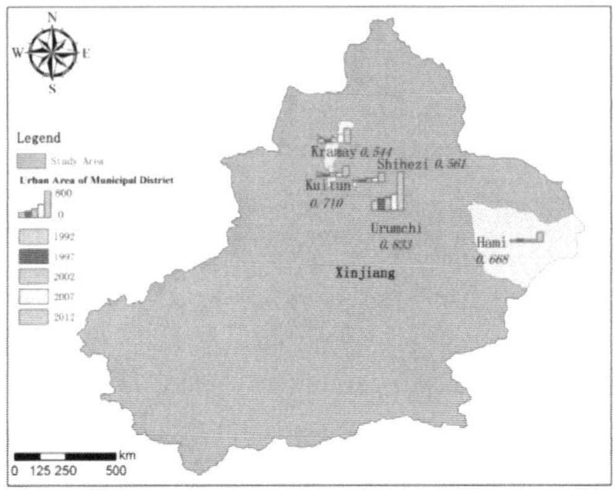

Fig.5correlation between population and built-up area in Xinjiang

5.Conclusion

Based on the DMSP/OLS nighttime data, this study monitored the urbanization process of cities in northwest China, and measured their urbanization level through some evaluation indexes. In addition, we mainly analyzed the correlation between population and built-up areasfor cities in Xinjiang.

1) According to the expansion speed and expansion intensity of each city in 1992-1997, 1997-2002, 2002-2007 and 2007-2012, we found that the cities in northwest China generally maintained a high expansion speed, but there were also great differences among them. Taking a comprehensive analysis on expansion speed and intensity, we can draw a conclusion that Xi'an, Urumchi and Yinchuan had the fastest expansion. However, because of restrict development caused by the poor location with many mountains, Lanzhou presented a lower expansion than other cities.

2) By using the shape index and compactness index to analyze the cities' spatial patterns in 1992, 1997, 2002, 2007 and 2012, we found that the exterior outlines of Kuitun and Shihezi tend to be round while exterior outlines of Lanzhou, Jiayuguan and Hanzhong tended to be strip , and the other cities' remained unchanged. There was a periodic change between exterior expansion and interior filling from 1992 to 2012, i.e., the cities externally expandedafter 2002 and internally filledbefore 2002. On the whole, Xinjiang with balanced progress in every aspect showed an ideally developmental tendency. At the same time, the other cities also have good development in the future.

3) Electing five key cities in Xinjiang to analyze the correlation between its population and built-up areas, we drew a conclusion that the population and built-up areas had a strong correlation in every city. The highest correlation index of Urumchi was 0.833 while it was lowest in Kramay with the value of 0.544.

The following questions should be given a further discussion: ① we only adopted an uniform analysis for all cities, however, the urbanization processes in different scale and different region cities present diverse characteristics and driving mechanism, which needs a following research; ② we used Mi et al.'s urban extraction method in this paper to finish our research, and the result accuracy by employing this method was 80%, which was proved to be feasible; all indexes used in this study were calculated based on the vectorial data, which would cause a negative urbanization to some degree, because there were some loss of information when conversing the raster data to vector data. In order to improve the accuracy of results, we will employ the pixel statistics method in the future research;

③The results showed that one reason, which leaded to the slower development of Lanzhou than the other provincial capitals, was the narrow topography, so it is important for our further works to discuss the topography factors when analyzing the urbanization process.

Acknowledgements
The work was supported by the Natural Science Foundation of China (NSFC) under Grant Nos. 41171340 and 41101390. This study was also supported by the Major International Cooperation and Exchange Project of National NSFC (Grant No. 41120114001), the Western Light Talent Culture Project (YBXM-2014-05), and the Science and Technology Innovation Program (for Agricultural InformationInstitute) of Chinese Academy of Agricultural Sciences.

References

[1] Zou J L, Liu C L, Yin, G Q and Tang Z P 2015 *Progress in Geography* **34** 598-605
[2] Liu X, Hu G, Ai B, Li X and Shi Q 2015 *Remote Sensing* **7** 17168-89
[3] Hao R F, Yu D F, Liu Y P and Sun Y 2014 *Journal of Beijing Normal University (Natural Science)* **50** 407-13
[4] Wang C P, Wang H W, Li C M and Dong R C 2012 *Act Ecologica Sinica* **32** 942-54
[5] Henderson J V, Storeygard A and Weil D N 2012 *American Economic Review* **102** 997-1028
[6] Elvidge C D, Sutton P S, Baugh K E, Tuttle B T, Howard A T, Erwin E H, Bhaduri B and Bright E, T 2009 *Computers & Geosciences* **35** 1652-60
[7] Keola S, Andersson M and Hall O 2015 *World Development* **66** 322-34
[8] Amaral S, Monteiro A M V, Camara G and Quintanilha J A 2006 *International Journal of Remote Sensing* **27** 855-70
[9] Gao Y, Wang H, Wang P T, Sun X Y and Lv T T 2013 *Resources Science* **35** 2517-23
[10] Wei Y, Liu H, Song W, Yu B L and Xiu C L 2014 *Landscape & Urban Planning* **128** 1-13
[11] Liao B, Wei K X and Song W W 2012 *Resource& Environment in the Yangtze Basin* **21** 1295-300
[12] Xu H, Yang H, Li X, Jin H and Li D 2015 *Sustainability* **7** 13469-99
[13] Su Y, Chen X, Wang C, Zhang H, Liao J, Ye Y and Wang C 2015 *GIScience & Remote Sensing* **52** 218-38
[14] Mi X N, Bai L Y, Tan X H, Lu N, Peng G P and Shao Q 2013 *Journal of Geo-Information Science* **15** 255-61
[15] Zhang Q Y 2015 *Journal of Geomatics* **40** 72-75
[16] Wu J S, He S B, Peng J, Huang X L and Zhang L Q 2014 *Geography and Geo-Information Science* **30** 20-25
[17] Zhuo L, Li Q, Shi P J, Chen J, Zheng J, Li X and Wang F 2007 *Acta Scientiarum Naturalium Universitatis Sunyatseni* **46** 98-102
[18] Liu Q P, Yang Y C, Fu D X, Li H Y and Tian H Z 2014 *Scinetia GeoGraphica Sinica* **34** 129-136

Surface Disturbance Analysis in Rare Earth Mining

H K. Li[1], L. Yang[2] and Z W. Liu[3]

College of Architecture and Surveying Engineering, Jiangxi University of Science and Technology, Ganzhou 341000, China

[1]giskai@126.com, [2]gisliusn@126.com, [3]Liuzw@jxust.edu.cn

Abstract. Mining ion-type rare-earth ore made the landscape and ecological environment degraded in mining area, and the tailing produced by rare-earth mining also led large areas land desertification, which resulted in surface temperature variations and significant differences in other types of mining disturbances. In order to analyse surface disturbance of rare-earth mining area, this paper applied the methods based on Normalized Difference Vegetation Index (NDVI) and Temperature different Coefficient (TDC) as the ecological disturbance indicator, compared and validated their applicability in Lingbei rare-earth mining area of Southern China. The results illustrated that, compared to NDVI, the TDC which reflected the characteristic of rare-earth mining technology has better discrimination of disturbance, especially for in-situ leach mining area. The places of tailing and the in-situ leach mining plants were the most dramatic mining disturbance. They had the biggest TDC value, followed by orchards and farmlands, reclamation plants, they had relatively small disturbance. And the last was the plant with the smallest TDC value. TDC in rare-earth mining could better correspond to the level of surface ecological disturbance. Therefore, TDC as the indicator of ecological disturbance factor had better performance than NDVI in rare-earth mining area.

1. Introduction

Surface mining activities led to enormous change of mine stripping topsoil, destruction of vegetation and soil physical and chemical properties in rare-earth mining area, and caused great disturbance of the ecosystems in mining areas, which directly brought about the degradation and destruction of mining landscape and ecological environment. For a long time, vegetation parameter was taken as a vital indicator of mining ecological disturbance, which was widely used (Peng and He et al., 2013). On the other hand, surface disturbance caused by surface mining activities could cause land surface temperature differentiation in rare-earth mining region. Hu's study shown that heat island effect in coal mining area was significant, thermal field variance index and vegetation coverage were good negative correlation, it could be an important factor which could assess the regional ecological environment (Hu and Zhao et al., 2010); In Pingshuo surface mining area, Zhang's study indicated that the perturbation intensity and range of disturbance affected the equilibrium of surface temperature spatial distribution, and existed the fierce game between low-temperature region and high-temperature region (Zhang, 2014). These studies indicate that the temperature variation of surface mining area could also be used as the indicator of mining ecological disturbance. Specific mining technology of ionic rare-earth directly caused the damage of the soil surface, vegetation and other environmental factors, which resulted in a large area vegetation of destruction and degradation and large-scale desertification in mining region. Due to the specific heat capacity of sand and vegetation are fairly different, temperature difference maybe better reflects disturbance characteristics of rare-earth mining than vegetation parameter. Selecting Lingbei rare-earth mining area in Southern China as a research area in this paper, we firstly calculated the land surface inversion temperature in the mining area. Furthermore, we got the TDC and analyzed the same location in the high-resolution aerial images. Lastly, we explored the relationship between TDC of different land types and the levels of surface ecological disturbance in mining area.

2. Data acquisition and pre-processing

Lingbei rare-earth mining is located in Dingnan County Jiangxi Province with approximately 200 square kilometers. It is shown as Figure 1. This mining has been mined for more than twenty years. In this study, we firstly pre-processed the Landsat-8 OLI imagery of October 8, 2014 and took it as the test data. Meanwhile, we selected a resolution of 0.5 meters high resolution mining aerial remote sensing data as the validation data collected in the December 10, 2014. To ensure the accuracy of verification results, the aerial imagery and the corrected Landsat OLI imagery should be registered. Subsequently, in order to make sample point's distribution uniformly, the error must be controlled within a cell.

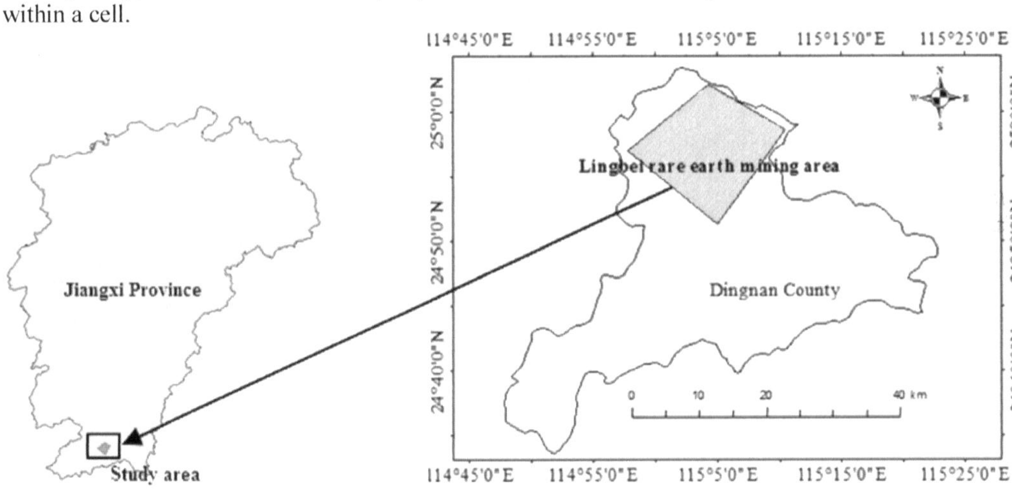

Figure 1. Study Area

3. Construction method of surface disturbance analysis

3.1. Surface temperature retrieval

Single-window Algorithm is a surface temperature inversion algorithm which was deduced by using the band 6 of Landsat TM based on the heat radiating surface conduction equations (Qin and Zhang et al., 2001). Due to this algorithm involved in the influence factors of the atmosphere and the surface in the calculation formula, it is easy to be calculated with high precision and widely used. The algorithm is calculated as the following equation (1), (2), (3).

$$T_s = [a(1 - C - D) + (b(1 - C - D) + C + D)T_b - DT_a] / C \tag{1}$$

$$C = \varepsilon\tau \tag{2}$$

$$D = (1 - \tau)[1 + (1 - \varepsilon)\tau] \tag{3}$$

Where T_s stands real surface temperature (K); a and b are constant; C and D are intermediate variable; ε is surface emissivity; τ is atmospheric transmittance; T_b is pixel brightness temperature that the sensor detected on the satellite height (K); T_a is the average atmospheric temperature.

There is a linear relationship between T_a and air temperature near the ground (t) (usually at 2m), and the t could be collected data from local weather stations. In accordance with the study area is located in the middle latitudes, the aver-age atmospheric temperature could be calculated by using the formula (4).

$$T_a = 16.0110 + 0.92621(t + 273.15) \quad (4)$$

From the above, the Single-window Algorithm's key is the parameter of T_b, ε and τ. The atmospheric transmittance could be estimated by atmospheric moisture content ω (g/cm^2) (Zhang and Chen et al., 2013). Water vapor in the atmosphere is mainly distributed in the troposphere, and the total amount of moisture in the troposphere air column has the relation as the formula (5).

$$\omega = 0.0981 \times (6.1078 \times 10^{\frac{7.5t}{t+273.3}}) \times RH + 0.1697 \quad (5)$$

Where t stands the air temperature near the ground (usually at 2m) (°C); RH is relative humidity; the air temperature and relative humidity data were collected from local weather stations.

The calculation of surface emissivity is used by the NDVI threshold method as formula (6).

$$\varepsilon = 0.004P_V + 0.986 \quad (6)$$

Where P_v is vegetation coverage which is calculated by dimidiate pixel method as formula (7).

$$P_V = [(NDVI - NDVI_{Soil}) / (NDVI_{Veg} - NDVI_{Soil})] \quad (7)$$

Where $NDVI$ stands normalized difference vegetation index. $NDVI_{soil}$ is the NDVI values of completely bare soil or some region without vegetation; $NDVI_{veg}$ is the NDVI values of the pixels completely covered by vegetation. We calculated the $NDVI_{soil}$ and $NDVI_{veg}$ by Li's literature (Li and Liu et al., 2013). Firstly, we extracted the corresponding pure pixel of the bare soil and vegetation by using pure pixel extraction function. Secondarily, the NDVI value of corresponding pure pixel could be extracted. Lastly, we could count average pure pixel NDVI value of bare soil and vegetation, and the vegetation coverage could be obtained by formula (7). Which it could overcome time and space uncertainty of $NDVI_{soil}$ and $NDVI_{veg}$ values, and enhance the P_v accuracy. Thanks to the geographical characteristics of the mining area that surface cover types such as water and towns in the study area are few, we calculated surface emissivity without consideration of water and towns, as a result, the natural surface emissivity could be calculated by formula (6). Brightness temperature could be calculated with the formula (8) and (9).

$$T_b = k_2 / \ln(1 + k_1 / L_{(\lambda)}) \quad (8)$$

$$L_{(\lambda)} = L_{\min(\lambda)} + (L_{\max(\lambda)} - L_{\min(\lambda)}) Q_{dn} / Q_{\max} \quad (9)$$

Where K_1 and K_2 are the preset value before satellite launch, K_1 is 774.89 $W/(m^2 * \mu m * sr)$ and K_2 is 1321.08 K ; $L_{(\lambda)}$ stands pixel spectral radiance at the sensor; Q_{max} stands the maximum value of the DN with the value of 255; Q_{dn} is the band pixel gray value; $L_{max(\lambda)}$ and $L_{min(\lambda)}$ are the maximum and minimum radiation intensity received by the sensor.

3.2. Calculation and classification of temperature difference degree

In order to better reflect the temperature difference among different types of ground features, we applied the thermal field variance index method of urban heat island research (Yang and Li et al., 2014). In Accordance with the similarity criterion, the degree of surface temperature in the mining area is characterized by the temperature difference coefficient. It is defined as the formula (10).

$$TVC(T) = (T - T_{mean}) / T_{mean} \quad (10)$$

Where T (°C) stands surface temperature inversion of a certain point in the mining area; T_{mean} (°C) stands the average surface temperature inversion in study area.

Due to the ionic rare-earth mining area are in southern hilly region, natural vegetation is lush and ecological situation is generally good, rare-earth mining is the most strong natural ecological disturbance in the region, and the ecological environment in the region brought very bad influence. The temperature difference coefficient of the mining area could assess the human disturbance level and ecological function of the natural surface in a certain extent. To clear strength of mine region temperature difference coefficients, we classified the similarity criterion based on the Sun's study on the intensity of Urban Heat Island (Sun and Lu, 2002). Different temperature divided into different grades and ecological disturbance level, as show in Table 1.

Table 1. The partition table of temperature differentiation and ecological evaluation threshold

Temperature difference coefficient	Temperature difference degree	Disturbance level	Ecological Assessment
<0.005	weaker	no	excellent
0.005-0.05	weak	weak	good
0.05-0.1	Mild	medium	general
0.1-0.15	Moderate	Stronger	Poor
0.15-0.2	Heavier	Strong	difference
>0.2	Severe	Very strong	Very poor

3.3. Method validation

Taking the Landsat OLI Imagery as the test data, we could calculate the surface temperature and the TVC in the study area with above method. We could classify the TVC in mining area as the Table 1. In order to study the relations between TDC and NDVI, we localized with the geographical links pixel-level ways in the study area of temperature differences degree images and NDVI images. A total of 111 samples were randomly collected based on various temperature different degrees, the NDVI and TVC of corresponding samples were recorded.

In order to reflect the relationship between NDVI and Temperature Difference Coefficient (TDC), we drew a line chart based on 111 samples of NDVI and TDC data in rare-earth mining area as shown in Figure 2. As we could see from Figure 2, on the whole NDVI and TDC are not related, especially in the middle part of the samples have no relationship. Which is different with the research of coal mining areas (Qiu and Hou, 2013.), therefore, the effect of temperature difference in rare-earth mining area has its own characteristics, not only related to vegetation degradation caused by rare-earth mining, and it is not only related to the vegetation degradation caused by rare-earth mining.

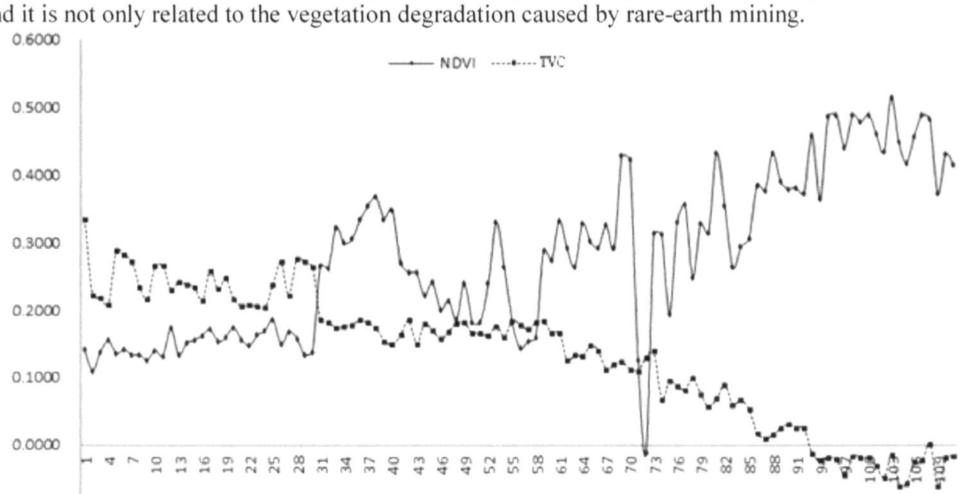

Figure 2. NDVI and TVC line chart of sampling points

From Figure 2, we could know that there is no obvious linear relationship between NDVI and TVC, and some samples appear that NDVI is high, but TDC also is high, there may be some kind of special relationship because of the rare-earth mining technology. To find the general rule of temperature difference in rare-earth mine region, we cropped aerial imagery in 10 December 2014, furthermore, found the 111 corresponding samples of different types of ground features and recorded the temperature difference degree through geographical links ways.

Then we could draw the figure of temperature difference degree in study area as shown in Figure 3. Analysed from the 111 corresponding samples of different types of ground features in Landsat-8 imagery and Figure 2, we could know that the region of the highest temperature different degrees is the tailing ground with less vegetation coverage. This region with a large area in rare-earth mining region is also very strong mining disturbance and poor ecological environment; the region of the higher temperature different degrees is the tailing ground and in-situ leaching area with more vegetation coverage. Due to the in-situ leaching technology that workers excavated the injection hole and poured ammonium sulfate liquid on the ground, which had made the vegetation degradation and the mountain water and soil pollution, and finally, it resulted in a larger ecological disturbance. If only just taking NDVI as the indicator, it is more difficult to distinguish the disturbance of different types of ground features, but TDC could better distinguish them; Orchards, farmlands, plant reclamation have a strong human disturbance with the result of human action on the natural terrain. And their NDVI is almost close to natural vegetation, but their TDC is at different levels, and they have a good ability to distinguish; tailing and in-situ leaching could be distinguished by TDC, and the TDC also has ability to distinguish the orchard, farmlands, plant reclamation in mining area and natural vegetation. Therefore, the TDC could be chose as an indicator to assess the level of ecological disturbance in the rare-earth mining area; TDC could better assess in-situ leaching disturbance such as the sample of 31, 37, 38 in Figure 2, but NDVI could not identify it, because the NDVI values of in-situ leaching vegetation and natural vegetation are in the same range.

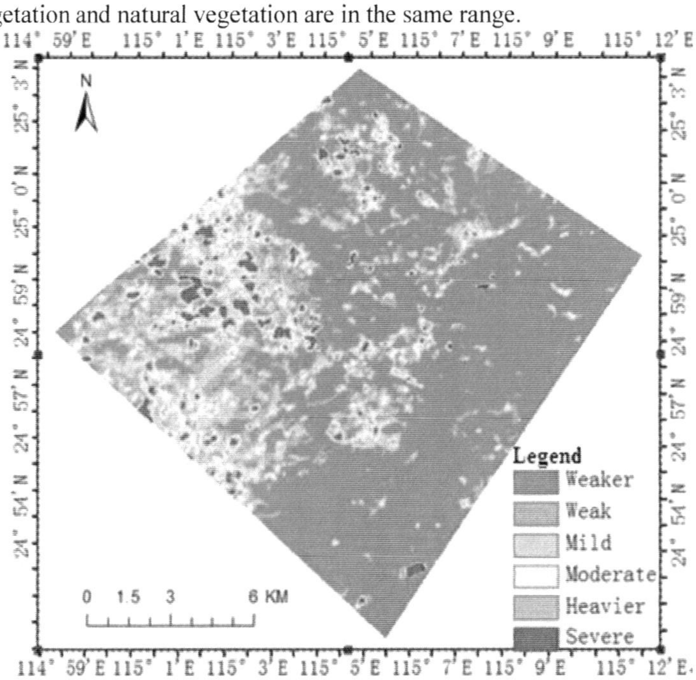

Figure 3. The temperature difference degree in study area

In summary, NDVI has obvious defects in the analysis of surface disturbance in rare-earth mine region, and TVC has obvious advantages. Especially most of rare-earth mining was using in-situ leaching process in Ganzhou City, the TVC can reflect the in-situ leaching characteristics. Thus, it can be used to distinguish the surface disturbance in mining area by the temperature difference degree method, which can provide a rapid quantitative method for the ecological evaluation in rare-earth mining area.

4. Conclusion

Taking Lingbei rare-earth mining area in Southern China as a study region in this paper, we calculated the surface inversion temperature and explored the relationship between TDC of different land types and the levels of surface ecological disturbance in mining area. We could conclude that compared to NDVI, TDC proposed in this paper could analyse rare-earth mining region disturbance recognition and better reflect the characteristics of the rare-earth mining area. Vegetation destruction of in-situ leaching area is relatively small, hence, in-situ leaching vegetation and general plants are difficult to distinguish with NDVI, but they could be distinguished by TDC because of in-situ leaching vegetation's the heavier lever of temperature difference degree. In-situ leaching vegetation area and tailing are the most severe disturbance area in rare-earth mining region, they also have the highest value of TDC. Followed by orchards, farmlands, plant reclamation, their ecological disturbance is relatively small with TDC value of successive reduction. Finally, the general vegetation has a minimum TDC value. TDC could preferably correspond with surface disturbance level in rare-earth mine region, so as an indicator of ecological disturbance in the mining area, TDC has better performance than NDVI.

Acknowledgments

This paper is support by the National Natural Science Foundation of China (No: 41261093 and 51464014).

References

Hu W L, Zhao P and Dong Z Y. Thermal environmental effect of coal mining area and its ecological significance based on TM data[J]. JOURNAL OF HEFEI UNIVERSITY OF TECHNOLOGY, 2010, 33(5): 741-744. (In Chinese)

Li H K, Liu X S, Li B and Li F S. Vegetation Coverage Variations and Correlation with Geomorphologic Factors in Red Soil Region: A Case in South Jiangxi Province[J]. SCIENTIA GEOGRAPHICA SINICA, 2013, 34(1): 103-109. (In Chinese)

Peng Y, He G J and Jiang W. Eco-environment Quality Evaluation of Rare Earth Ore Mining Area Based on Remote Sensing Techniques[J].Geo-Informatics in Resource Management and Sustainable Ecosystem. Springer Berlin Heidelberg, 2013:246-257.

Qiu W W and Hou H P. Study on Surface Temperature Variation Caused by Ecological Disturbance in the Mining Area Based on RS[J]. MINING R & D, 2013,33(2): 02:68-71+83.(In Chinese)

Qin Z H, Zhang M H, Karnieli A and Berliner P. Mono-window algorithm for retrieving land surface temperature from Landsat TM6 data[J]. ACTA GEOGRAPHICA SINICA-CHINESE EDITION, 2001, 56(4): 466-474. (In Chinese)

Sun S M, Lu C Y. Study on Monitoring Intensity of Urban Heat Island and Taking it as an Indicator for Urban Ecosystem by Remote Sensing[J]. Journal of Xiamen University (Natural Science), 2002, 41(1): 66-70.(In Chinese)

Yang W R, Li F and He Y. Characteristic change and analysis of urban heat island in Beijing, China, Summer, 2003 — 2011[J]. ACTA ECOLOGICA SINICA, 2014, 34(15): 4390-4399.(In Chinese)

Zhang C C, Chen D H and Dong H T. Land Surface Temperature Retrieval based on Landsat-5 TM Data in Baisha Irrigation of Henan province[J]. REMOTE SENSING TECHNOLOGY AND APPLICATION,2013,06:964-968.(In Chinese)

Zhang Y L. Open Cast Mine Monitoring and Ecological Effect Assessment through Remote Sensing — A Case Study from Pingshuo[D]. China University of Geosciences (Beijing), 2014. (In Chinese)

Institute of Physics
Dirac House, Temple Back
Bristol BS1 6BE UK

ISSN: 1755-1307
ISBN 978-1-5108-3743-0

9 781510 837430

CURRAN ASSOCIATES INC.
proceedings.com

International Scientific-Technical Conference on Innovative Engineering Technologies, Equipment and Materials 2016 (ISTC-IETEM 2016)

IOP Conference Series: Materials Science and Engineering
Volume 240

Kazan, Russia
7 – 9 December 2016

IOP
Institute of Physics